Theoretical foundation for large-scale computations for nonlinear material behavior

MECHANICS OF ELASTIC AND INELASTIC SOLIDS
Editors: S. Nemat-Nasser and G.Æ. Oravas

G.M.L. Gladwell, Contact problems in the classical theory of elasticity. 1980. ISBN 90-286-0440-4.

G. Wempner, Mechanics of solids with applications to thin bodies. 1982. ISBN 90-286-0880-X.

T. Mura, Micromechanics of defects in solids. 1982. ISBN 90-247-2560-7.

R.G. Payton, Elastic wave propagation in transversely isotropic media. 1983. ISBN 90-247-2843-6.

S. Nemat-Nasser, H. Abé, S. Hirakawa, Hydrawlic fracturing and geothermal energy. 1983. ISBN 90-247-2855-X.

S. Nemat-Nasser, R.J. Asaro, G.A. Hegemier (eds.), Theoretical foundation for large-scale computations for nonlinear material behavior. 1984. ISBN 90-247-3092-9.

Theoretical foundation for large-scale computations of nonlinear material behavior

Proceedings of the Workshop on the Theoretical Foundation for Large-Scale Computations of Nonlinear Material Behavior
Evanston, Illinois, October 24, 25, and 26, 1983

Editor-in-Chief
Siavouche Nemat-Nasser
Northwestern University, Evanston, Illinois

Editors
Robert J. Asaro
Brown University, Providence, Rhode Island
Gilbert A. Hegemier
University of California, San Diego, La Jolla, California

Sponsored by
NATIONAL SCIENCE FOUNDATION
DEFENSE ADVANCE RESEARCH PROJECTS AGENCY

1984 **MARTINUS NIJHOFF PUBLISHERS**
a member of the **KLUWER** ACADEMIC PUBLISHERS GROUP
DORDRECHT / BOSTON / LANCASTER

Distributors

for the United States and Canada: Kluwer Academic Publishers, 190 Old Derby Street, Hingham, MA 02043, USA
for the UK and Ireland: Kluwer Academic Publishers, MTP Press Limited, Falcon House, Queen Square, Lancaster LA1 1RN, England
for all other countries: Kluwer Academic Publishers Group, Distribution Center, P.O. Box 322, 3300 AH Dordrecht, The Netherlands

Library of Congress Cataloging in Publication Data

Main entry under title:

Theoretical foundation for large-scale computations
    for nonlinear material behavior.

    (Mechanics of elastic and inelastic solids ; 6)
    1. Materials--Congresses. 2. Solids--Congresses.
3. Numerical calculations--Congresses. I. Nemat-
Nasser, S. II. Asaro, Robert J. III. Hegemier,
Gilbert A. IV. National Science Foundation (U.S.)
V. United States. Defense Advanced Research Projects
Agency. VI. Series.
TA401.3.T48 1983   620.1'12   84-20596

ISBN-13:978-94-009-6215-6 e-ISBN-13:978-94-009-6213-2
DOI: 10.1007/978-94-009-6213-2

Copyright

PREFACE

This book contains the proceedings of a workshop on the Theoretical Foundation for Large-Scale Computations of Nonlinear Material Behavior, held under the auspices of the National Science Foundation (NSF) and the Defense Advance Research Projects Agency (DARPA), at Northwestern University, October 24-26, 1983. The main objective of this workshop was to provide a forum for the exchange of information and views on major issues relating to the fundamentals of characterizing the inelastic constitutive material behavior. Comments on the Aims of the Workshop, by Drs. William Snowden and Thomas Bache, pp. 1-5, outline reasons for holding this workshop, and provide further background.

The format of the workshop was designed to optimize the interaction between researchers whose primary interest is material characterization and numerical analysts whose primary interest is the development and practical use of large computer codes. The program of the workshop and a list of the workshop participants are found at the end of these proceedings.

The format of the workshop was as follows: There were three general lectures (Chapters I, II, III) which provided a broad outline of the theoretical foundations, the numerical implementation, and the large-scale computations of nonlinear material response. There were eight thirty-minute general talks, each followed by a fifteen-minute general discussion. Then followed a fifteen-minute prepared formal discussion by another participant, who had received a summary and materials related to the general talk several weeks ahead of the meeting and subsequently had prepared a discussion stressing significant aspects and bringing to focus complementary viewpoints. These formal discussions were each then followed by another fifteen-minute general discussion. There were two sessions for short presentations and discussions (Chapters V and XIII). The workshop concluded with a session (Chapter XIV) in which each session chairman presented a summary of some of the major points and each member of the scientific committee made a few final comments, with Dr. William Snowden presenting the closing remarks.

A preparatory book containing copies of selected articles by each speaker, as well as a short summary of each talk and other relevant information, was issued to the participants at the beginning of the workshop. This preparatory book served as a resource and a vehicle for familiarizing the attendees with the terms, scope, and concepts to be covered by the speakers.

All discussions and comments were recorded by a professional stenographer and also recorded on a tape. The transcripts were edited by each discusser. These proceedings include major portions of all discussions. A serious attempt was made to adhere to the original wording, and to preserve the spirit of what was actually said. Nevertheless, in order to provide continuity and avoid repetitive statements, some modifications had to be introduced. Any errors, omissions, or deficiencies, therefore, should be blamed on the chief editor, who edited the final version of the text.

Preface

The workshop was organized by a committee consisting of:

R. J. Asaro, Brown University

T. C. Bache, Science Applications, Inc.

G. A. Hegemier, University of California, San Diego

T. J. R. Hughes, Stanford University

S. Nemat-Nasser, Northwestern University (coordinator)

W. E. Snowden, DARPA

M. C. Stickley, The BDM Corporation

G. B. Taggart, The BDM Corporation

K. Thirumalai, NSF

On behalf of the committee, I wish to thank Mrs. Erika Ivansons, who skill-fully and graciously helped this writer to coordinate all local arrangements and assisted with the editing tasks. Sincere thanks are also due Miss Vera Fisher, who typed the final manuscript, and Mrs. Suzan van Damme, who helped with various editorial efforts, as well as to Mrs. Barbara Fabian and Dr. Hideyuki Horii, who — together with Mrs. Ivansons and Mrs. van Damme — cheerfully and competently took care of many essential functions and made the workshop a success.

S. Nemat-Nasser
Wilmette, Illinois
October 1983

CONTENTS

Contents

COMMENTS ON THE AIMS OF THE WORKSHOP

William E. Snowden and Thomas C. Bache*

DARPA/Defense Sciences Office, Arlington, Virginia

The Defense Advanced Research Projects Agency (DARPA), in collaboration with the National Science Foundation (NSF), is pleased to sponsor this important workshop on the "Theoretical Foundation for Large-Scale Computations of Non-Linear Material Behavior." The research-support activities of both DARPA and NSF have long represented a significant commitment to the advance of science and technology. DARPA supports basic and applied research and development work in a wide range of technical areas for eventual application to problems related to the national defense. The NSF has an even broader charter in overseeing much fundamental research in many diverse fields of science and engineering. It seems most appropriate, therefore, that an academic setting has been selected for this forum, and we wish to thank our hosts and the supporting staff at Northwestern University for their hospitality and organizational efforts that have made this meeting possible.

One of the many areas of common interest to both DARPA and NSF, that of computer modeling of complex physical phenomena, includes the more specific subject of this workshop. This area has witnessed remarkable development over the past few decades, and both agencies are proud to have played important roles in marking that progress. We maintain a keen interest in supporting further advances: for DARPA, to improve and expand the capabilities of the cognizant research community to carry out detailed computations on problems of concern to the Department of Defense; for NSF, particularly from the perspective of its Division of Civil and Environmental Engineering, to develop an improved fundamental understanding of the response of soils and structures to dynamic loading. The DARPA and NSF interests are clearly complementary.

Problems addressed using large-scale computations that are of interest to DARPA and NSF are many and varied. Much of the early impetus for development of computer modeling as a research tool derived from the need to solve complex problems associated with the development of advanced weapons systems in the national defense laboratories. The list of defense-related applications has grown considerably over the years. Notable examples of such applications include:

(1) analyses of structural impacts at very high velocities (e.g., meteorite impacts on satellites and other space structures, and impacts involving aircraft or interceptor rockets);

(2) analyses involving blast loading and effects on soils or structures (e.g., as might be encountered in qualification tests for advanced rocket motors);

---

*Presently with SAI and a visiting research fellow at the UK Ministry of Defence, Blacknest.

(3)   studies related to the design and evaluation of numerous conventional ordnance systems (e.g., chemical energy munitions, including shaped charges and explosively formed penetrators);

(4)   analyses of penetrator/target interactions and advanced armor systems;

(5)   examination of technical missile basing issues and site hardening for defense systems;

(6)   combustion and explosion phenomena (e.g., deflagration-to-detonation transitions, and explosion initiation and growth);

(7)   explosive welding, and, most recently, the dynamic compaction of materials.

Other applications for large-scale computations include calculations of strong ground motions resulting from earthquakes, and analyses of structures designed to withstand the resulting forces. For most of these problems, material properties and deformation/fracture behavior are of considerable importance, and successful modeling of a given problem is critically dependent on both our understanding of those properties and our ability to devise a reasonable physical model and the corresponding mathematical algorithms necessary for implementation in computer simulation programs.

Modeling of non-linear material behavior associated with stress waves in solids is the focal point for this workshop. A general assessment of the current state of this critical technical area was recently made by a panel of solid mechanics specialists, who worked with members of the scientific staff of the BDM Corporation, McLean, VA under DARPA sponsorship. Members of this panel include Professors Sia Nemat-Nasser of Northwestern University; G. A. Hegemier of the University of California, San Diego; and Robert J. Asaro of Brown University. These members were recently joined by Professor Thomas Hughes of Stanford University, an expert in the area of numerical implementation of constitutive models. All of these individuals are participants in this meeting. The panel was asked to review the general area of elasto-plastic and visco-plastic material descriptions and their implementation in large computer simulation programs. In addition, it was requested that the current extent of communication and collaboration between the solid mechanics and numerical analysis communities be considered and recommendations made for enhancing our overall capabilities in large-scale computational solid mechanics, particularly for defense-related problems.

The initial work of the panel is described in a recent report issued by the BDM Corporation ("Report and Recommendations for Improving Constitutive Relations Used in Computer Codes," S. Nemat-Nasser, G. A. Hegemier, and R. J. Asaro, Report BDM/W-83-44-TR, The BDM Corporation, McLean, VA, July 15, 1983). One specific recommendation contained in that report is that, "as a first step toward cultivating productive exchange and collaboration between concerned groups," a workshop be held in the fall of 1983 at which prominent solid mechanics researchers and developers and users of computer simulation programs are brought together to discuss critical material modeling issues and exchange views on perceived problems of immediate concern. This meeting represents the action taken in response to that recommendation. The work of the panel is expected to continue, however, and another recommendation suggests broader and more intensive future activities. A concerted effort will be made to assemble relevant known results involving constitutive relations for metals, certain engineered materials, and geo-materials, and to present these results in a unified format that includes discussion of their range of applicability — but also their limitations — in order to properly identify areas in need of further research. It is hoped that discussions forthcoming in this workshop will provide important input representing a first step toward that goal.

We reiterate the prime motivation for holding this workshop: the recognition that we must continue to improve our capability for carrying out realistic large-scale computations involving non-linear behavior of materials subjected to dynamic loading. Such calculations play an important role in modern engineering science by guiding the design and interpretation of experiments, helping to both identify and quantify the important material and geometric parameters and the effects of varying these parameters on some overall system behavior. This approach is the traditional one for mathematical models of physical phenomena in general. However, large-scale computations have also come to be used extensively to describe material behavior in situations where model-verifying experiments are extremely difficult or impossible. This role is philosophically quite different from traditional modeling, and perhaps one that classical natural scientists might find somewhat unsettling. In these instances computational results may tend to take on a life of their own and in some sense become reality. This problem is probably more common in the geophysical sciences, in part because the earth is so resistant to direct, controlled experiments, but also because geoscientists are often interested in problems where the loading is hypothetical and cannot be directly simulated. An example is the problem of designing civil structures to withstand strong ground motions associated with earthquakes. The tendency to accept computational results as reality occurs when dealing with engineering materials as well, perhaps because of the expense and/or difficulty of carrying out direct experiments. We recognize the need for —and the potential value of —large-scale computations of this nature. But we further realize that the best fundamental scientific knowledge available or attainable must be incorporated into the computer simulation programs used to address these problems.

As a starting point, the various elements of the constitutive models must be based on sound continuum mechanics theories. Where possible, model verification by appropriate experimental data must be established. The implications of observed micromechanical deformation processes that occur in real materials but which have only recently begun to be incorporated into continuum models must be carefully considered. Then, the material models must be implemented within the computer simulation programs using mathematical techniques that minimize physical compromise. This is, of course, nothing more than an application of the scientific method. The approach is important when the problem at hand involves interpolation, i.e., where the physical bounds fall within the range of previous experimentation. But it is critical when the problem requires extrapolation into untested realms. Computations in areas where reliability can only be deduced or presumed become more credible.

Concern as to the credibility of large-scale computations of non-linear material behavior is not limited to the technical community, a point which deserves some emphasis. Technical input based largely on theoretical considerations often forms the basis for crucial political and economic decisions. Prime examples involve site selection and design of nuclear power plants, and basing of missile defense systems. Large-scale computations can play a key role in deciding questions related to these issues; this is true whether the conclusions made based on the analyses are correct, or if they are only perceived to be correct. Thus, credibility of the computations is a crucial requirement for making important policy decisions. That credibility will be recognized and maintained only if the computational methodology develops not as a separate discipline of the engineering sciences accessible only to a relatively small number of experts, but rather as an interdisciplinary product that integrates the most advanced results from relevant theoretical solid mechanics, experimental materials, and mathematical/numerical research.

The participants in this workshop are collectively capable of providing that critical expertise and research input. Four readily identifiable groups are represented. One of these groups is concerned with the development of constitutive models from a primarily theoretical perspective. A second group is comprised of solid mechanics/materials researchers whose work involves experimental/empirical investigation of dynamic material behavior. The third group specializes in numerical implementation of constitutive models. Finally, the fourth group includes the developers and users of complex computer programs who are together responsible for producing some of the best large-scale simulations of non-linear dynamic material behavior reported to date. Most of the people here today can be associated with more than one of these groups. It is unlikely that any single individual could claim considerable expertise in all of the noted areas.

An important objective of the workshop, then, is to foster further communication among the represented groups for the express purpose of accelerating advances in material modeling for large-scale computations. There are frequent meetings and a number of technical journals with this same noble goal. Why might this meeting be different? It is our hope that this small and select assembly can work together for common benefit in a forum specifically organized for frank and open discussion of a few critical technical areas. We all recognize the many positive aspects of our present capabilities. We all are aware of certain limitations. Both sides should be heard, and from assorted individual and group perspectives.

Consider the situation of those individuals responsible for carrying out large-scale computations. They deal with very complex problems that frequently challenge the frontiers of knowledge of dynamic materials response. But an answer is required in a timely fashion. For many problems, perhaps, it is only necessary that the results, however approximate, be better than no answer at all. (Of course, one seldom hears calculated results presented in those terms.) In reviewing the literature for help, the problem solver is likely to find that most of the available constitutive models require assumptions that do not apply to the problem at hand. Thus, construction of the required constitutive models proceeds in an *ad hoc* fashion, made consistent with the most basic physical laws and compatible with whatever limited experimental data are available. Over a period of several years, constitutive models developed in this manner can become unwieldy, their justification fragmented and incomplete. But the models provide answers, so their application persists. What is the alternative to this approach?

For help, we must turn to the theoretical solid mechanics community. Here are the researchers who engage in developing rigorous and consistent theories of dynamic material behavior, and we should make use of them. Unfortunately, these theoreticians may have difficulties in understanding the organization/implementation methodologies contained in the large computer simulation programs, and, if they do comprehend, in correctly assessing them within the framework of established continuum mechanics and then suggesting improvements. The theoretical and computational communities may thus be subjected to forces that inevitably tend to separate them in language and emphasis. All we can do is to continually work to bridge that gap. It should also be noted that between these two groups are the materials experimentalists, whose results may resist precise interpretation by either side —another problem of communication.

But better communication — a goal of this workshop —is not enough. It is really direct and frequent collaboration that allows a free flowing of ideas among the various disciplines. Some collaborative efforts already exist in this area: the generally recognized complexity of the material modeling problem has demanded that this be so. However, one satisfying outcome of this workshop would be the creation of new collaborative efforts attributed to communication stimulated here.

To conclude these introductory remarks, a brief statement of what the sponsoring agencies and the organizing committee hopes will be achieved at this workshop is in order. First, we hope to see a clear description of the theoretical framework constraining the development of constitutive models to be used in large-scale computations of non-linear dynamic material behavior. Presumably, we should be able to be quite confident of computed results based on constitutive models developed within this framework. It is important that assumptions of the models be clearly stated, and that we understand and discuss the consequences of relaxing these assumptions. In addition, from the calculators we hope to gain a further appreciation of the range and complexity of the practical problems currently being addressed. What *are* the constitutive models actually being used for specific computations, and what is the theoretical or experimental basis for their selection? Finally, from those researchers conducting experiments to determine constitutive behavior, we hope to learn what kinds of measurements can now be made to validate theoretical predictions.

In closing, we would like to offer a quotation, attributed to Albert Einstein, as a restrained appeal for free and open discussion of both the positive and negative aspects of our current modeling capabilities: "The search for truth implies also a duty. One must not conceal any part of what one has recognized to be true."

CHAPTER I

THEORETICAL FOUNDATIONS OF PLASTICITY

S. Nemat-Nasser

Northwestern University, Evanston, Illinois

ABSTRACT

The fundamental continuum aspects of the rate-independent elastic-plastic behavior of materials are discussed. First the general structure of the corresponding rate constitutive relations at finite strains and rotations is examined, bringing into focus the effect of plastically-induced material spin. The choice of stress and strain measures is then reviewed, and it is emphasized that all objective stress-rates equal the Jaumann rate plus terms linear in stress and deformation rate, when the current state is used as the reference state. The induced anisotropy due to stressing is shown to preclude an isotropic relation between the Jaumann rate of Cauchy (or Kirchhoff) stress and the deformation rate. The equations of the classical plasticity theory at finite strains and rotations are presented next, and questions of associative and non-associative flow rules, dilatancy and pressure-sensitivity, plastic anisotropy, Drucker's postulate and its relation to normality and convexity, and finally the non-coaxiality of the plastic strain rate tensor and the stress tensor are discussed. Then some thermodynamic bases of both rate-dependent and rate-independent inelasticity are considered, emphasizing conditions under which flow potentials for inelastic strain-rates exist. Finally, the problem of strain-softening is briefly discussed, and it is pointed out that both for metals and geo-materials, such softening may often be the result of highly localized damage within the test specimen which no longer remains macroscopically homogeneous, and, hence, should not be regarded as a "representative sample."

1. INTRODUCTION

One of the most significant problems of current concern is an adequate and physically sound description of the thermomechanical constitutive behavior of technological materials and geo-materials at high pressures and temperatures. A great deal of progress has already been made in this and related areas, especially in relation to the basic structure of continuum constitutive relations. A thorough overview has been given by Nemat-Nasser (1983); see also related review articles by Havner (1982), Hill (1978), Nemat-Nasser (1975,1982), and Rice (1975). Although some aspects of rate-dependent constitutive relations have been briefly examined in some of these review articles, these articles are basically concerned with deformation regimes, where rate effects are not dominant. It is well known that metals as well as geo-materials exhibit rate effects, even at moderate temperatures. Therefore, the assumption of rate-independency is more a matter of convenience than of physical reality. On the other hand, for certain materials and at suitably low temperatures, a great deal can be learned by the assumption of a rate-independent elasto-plastic response.

The purpose of this article is to briefly examine some of the fundamental continuum aspects of both the rate-independent and the rate-dependent elasto-plastic behavior of materials.

Beginning with the classical concepts in plasticity, we examine in the manner proposed by Hill (1958,1959,1967) rate-independent plasticity in both the stress and the strain-rate space. We then supplement these with some recent results on the non-associative flow rule, particularly relevant to frictional, plastically dilatant materials (such as pressure-sensitive metals and geo-materials) and examine the physical implications of pressure effects. Addition-ally, we review the question of the non-coaxiality of the plastic strain-rate and the stress tensors, contrasting it with the non-associative flow rule. Both the isotropic and the kinematic hardening rules are briefly discussed.

Then, we deal with some thermodynamic bases of both rate-independent and rate-dependent inelasticity, emphasizing conditions under which flow potentials exist.

Finally, we briefly discuss the problem of strain-softening, and point out that both for metals and geo-materials, such softening may not always be the result of a continuum material softening, but rather it often stems from highly localized damage within the test specimen which no longer remains macroscopical-ly homogeneous, and therefore, should not be considered at such a stage to be a "representative sample."

2.   RATE-INDEPENDENT PLASTICITY

In the context of rate-independent plasticity, inelastic deformation is often viewed to stem from rate-independent slip on crystallographic planes, twinning, and void formation and growth in crystalline solids, relative slip and rolling of individual grains in granular materials, and frictional sliding of pre-existing cracks, as well as the growth of these cracks in rocks. Such in-elastic deformation is invariably accompanied by elastic deformation which serves to produce compatible overall deformation under a prescribed set of kine-matical and dynamical boundary data. In this section we examine the correspond-ing continuum constitutive relations for fully non-linear finite strains and rotations. The results, however, can immediately be specialized and interpreted in the context of the usual infinitesimal deformations.

2.1 Notation

Both index and direct notation are used. For simplicity, a fixed rectangu-lar Cartesian background coordinate system is employed. The coordinate axes are denoted by x_i or X_A, $i, A = 1, 2, 3$. Repeated indices are summed (unless stated otherwise), and when direct notation is used, double contraction is denoted by a double dot, and tensor products are denoted by a circle with a cross inside. For example, with $\mathcal{L}$ a fourth-order tensor, and D and τ second-order tensors, the components of $\mathcal{L}:D$, $D \otimes \tau$, and $D\tau$, respectively, are: $\mathcal{L}_{ijkl}D_{kl}$, $D_{ij}\tau_{kl}$, and $D_{ik}\tau_{kj}$. Various quantities are defined wherever they first occur.

2.2 Basic Constitutive Equations

When all micro-defects and micro-events which currently are instrumental in inelastic deformation are "locked" (a thought experiment), the incremental response would be elastic and totally recoverable. The difference between the actual incremental deformation and the corresponding elastic one is the plastic incremental deformation which is produced because in the actual process, micro-defects are not locked and micro-events are fully operative.

Let $v_i = v_i(x,t)$ be the velocity field, and $v_{i,j} \equiv \partial v_i / \partial x_j$ be the velocity gradient. The deformation rate tensor, $\underset{\sim}{D}$, and the spin tensor, $\underset{\sim}{W}$, have the following components:

$$D_{ij} = \tfrac{1}{2}(v_{i,j} + v_{j,i}), \quad W_{ij} = \tfrac{1}{2}(v_{i,j} - v_{j,i}). \tag{2.1}$$

Since rate-independent plasticity is involved, the time paramater t can be any suitable monotone "load parameter," and the incremental quantities can be defined in the usual manner for a given increment of t. Therefore, we deal directly with the rate quantities in the sequel.

From the comments made at the beginning of this subsection, it follows that

$$\underset{\sim}{D} = \underset{\sim}{D}^e + \underset{\sim}{D}^p, \quad \underset{\sim}{W} = \underset{\sim}{W}^e + \underset{\sim}{W}^p, \tag{2.2}$$

where superscripts e and p, respectively, denote the elastic and the plastic contribution to the total rate quantity.

Let $\underset{\sim}{\tau}$ and $\underset{\sim}{\sigma}$ denote the Kirchhoff and Cauchy stresses, respectively, $\underset{\sim}{\tau} = (\rho_0/\rho)\underset{\sim}{\sigma}$, where ρ_0 and ρ are mass densities, respectively, in some reference ground state (not necessarily undeformed or unstressed) and in the current one. Denote by superimposed $\circ$ the Jaumann rate co-rotational with the material; e.g.,

$$\overset{\circ}{\underset{\sim}{\tau}} = \overset{\cdot}{\underset{\sim}{\tau}} - \underset{\sim}{W}\underset{\sim}{\tau} + \underset{\sim}{\tau}\underset{\sim}{W} \tag{2.3}$$

is the objective Jaumann rate of the Kirchhoff stress. Moreover, denote by $\overset{\circ}{\underset{\sim}{\tau}}{}^*$ the Jaumann rate of the Kirchhoff stress co-rotational with the *elastic* distortion, i.e.,

$$\overset{\circ}{\underset{\sim}{\tau}}{}^* = \overset{\cdot}{\underset{\sim}{\tau}} - \underset{\sim}{W}^e\underset{\sim}{\tau} + \underset{\sim}{\tau}\underset{\sim}{W}^e. \tag{2.4}$$

In (2.3) and (2.4), superposed dot denotes the material derivative.

The change in stress stems from elastic lattice distortion in crystalline solids. If $\underset{\sim}{\mathcal{L}}$ is the current elastic modulus tensor of the crystalline solid, we have

$$\overset{\circ}{\underset{\sim}{\tau}}{}^* = \underset{\sim}{\mathcal{L}}:\underset{\sim}{D}^e. \tag{2.5}$$

Then, by direct substitution from (2.1) to (2.5), we obtain the following general rate constitutive relations (Nemat-Nasser, 1982):

$$\overset{\circ}{\underset{\sim}{\tau}} = \underset{\sim}{\mathcal{L}}:\underset{\sim}{D} - \{\underset{\sim}{\mathcal{L}}:\underset{\sim}{D}^p + \underset{\sim}{W}^p\underset{\sim}{\tau} - \underset{\sim}{\tau}\underset{\sim}{W}^p - \underset{\sim}{\tau}D_{kk}^p\}. \tag{2.6}$$

For most metals, the components of the elastic modulus tensor, $\underset{\sim}{\mathcal{L}}$, are several orders of magnitude larger than the stress that the solid can support. Therefore, the terms in the braces in Eq. (2.6) which are proportional to the stress can be ignored, in comparison with the term which involves the elastic modulus tensor. For geo-materials, on the other hand, this is not the case, and indeed, the stress-dependent terms should not be ignored without further specific consideration. In this context, therefore, in general one must construct a constitutive relation for the plastic deformation rate tensor, D^P, as well as for the plastically induced spin, W^P, as has been observed by Mandel (1971), and recently discussed by Dafalias (1983) and Loret (1983).

2.3 Choice of Stress and Strain Measures

For finite deformation problems, the choice of proper stress and strain measures, to a certain extent, is a matter of convenience and taste, although care is required in order to avoid pitfalls. In this connection, Hill's contributions (1968,1972) do provide guidance and should be consulted; see also Havner (1982) and Nemat-Nasser (1982,1983).

Denote particle positions in the reference and current states by X and x, respectively. (The reference state may be a deformed, stressed state.) The deformation is given by the mapping $x = x(X,t)$, where t is time. The deformation gradient is $F = \partial x/\partial X$, and the velocity gradient is $L = \dot{F}F^{-1}$, where $L_{ij} \equiv D_{ij} + W_{ij} \equiv \partial v_i/\partial x_j$. By polar decomposition, $F = RU = VR$, $U^2 = F^T F$, $V^2 = FF^T$, and $R^{-1} = R^T$ with det $R = +1$.

If the principal values (stretches) and directions of the tensor U are $\lambda_{(a)}$ and $N^{(a)}$, $a = 1, 2, 3$, respectively, then we have

$$U = \sum_{a=1}^{3} \lambda_{(a)} N^{(a)} \otimes N^{(a)}, \quad F = \sum_{a=1}^{3} \lambda_{(a)} n^{(a)} \otimes N^{(a)},$$

$$R = \sum_{a=1}^{3} n^{(a)} \otimes N^{(a)}, \quad V = \sum_{a=1}^{3} \lambda_{(a)} n^{(a)} \otimes n^{(a)}, \tag{2.7}$$

where $n^{(a)}$, $a = 1, 2, 3$, are the principal directions of V. $N^{(a)}$ and $n^{(a)}$, respectively, define the Lagrangian and the Eulerian triads. R rotates $N^{(a)}$ into $n^{(a)}$.

A broad class of strain measures, which includes many commonly used ones, is defined by (Hill, 1968)

$$E = \sum_{a=1}^{3} f(\lambda_{(a)}) N^{(a)} \otimes N^{(a)}, \tag{2.8}$$

$$f(1) = 0, \quad f'(1) = 1, \quad \text{and} \quad f(\lambda) \text{ monotone.}$$

To each E from this class, there corresponds a conjugate stress measure, say, S, such that $S:\dot{E}$ is the rate of stress-work per unit reference volume. Stress-rates, $\dot{S}$, associated with this class of stress measures are all objective. We have the following important relation between *any* such $\dot{S}$ and the Jaumann rate of the Kirchhoff stress $\overset{\circ}{\tau}$:

$$\dot{S} = \overset{\circ}{\tau} - \tfrac{1}{2}[f''(1) + 1][\sigma D + D\sigma] \tag{2.9}$$

when the current state is used as the reference state, and, hence, all stress measures equal σ and all strain-rate measures reduce to D.

Equation (2.9) shows that the Jaumann rate of the Kirchhoff stress, $\overset{\circ}{\tau}$, is identical with $\dot{S}$ when $E = \ln U$ and, hence, $f''(1) = -1$.

Consider the stress measure,

$$\hat{\tau} = R^T \underset{\sim}{\tau} R,$$ (2.10)

and its flux $\overset{\curlywedge}{\underset{\sim}{\tau}}$, written as

$$\overset{\curlywedge}{\underset{\sim}{\tau}} \equiv R\overset{\cdot}{\hat{\tau}}R^T = R(R^T \underset{\sim}{\tau} R)\overset{\cdot}{} R^T = \overset{\circ}{\underset{\sim}{\tau}} + \underset{\sim}{\tau}\underset{\sim}{\epsilon} - \underset{\sim}{\epsilon}\underset{\sim}{\tau},$$ (2.11)

where (Nemat-Nasser, 1983)

$$\underset{\sim}{\epsilon} = \sum_{a,b=1}^{3} \frac{\lambda_{(b)} - \lambda_{(a)}}{\lambda_{(b)} + \lambda_{(a)}} D_{(ab)} \underset{\sim}{n}^{(a)} \otimes \underset{\sim}{n}^{(b)};$$ (2.12)

$D_{(ab)}$ are the components of $\underset{\sim}{D}$ in the Eulerian triad.

The objective flux $\overset{\curlywedge}{\underset{\sim}{\tau}}$ is the rate of change of stress as observed in the Lagrangian frame. If the material has certain inherent special symmetry properties which are altered in the course of deformation, it may be more effective to use $\overset{\curlywedge}{\underset{\sim}{\tau}}$ to define the corresponding constitutive relations. Suppose, as an example, that the material is initially isotropic, and we wish to use a simple linear relation between a stress flux and the corresponding strain-rate. The expression

$$\overset{\curlywedge}{\underset{\sim}{\tau}} = 2\mu\hat{\underset{\sim}{D}} + \lambda D_{kk}\underset{\sim}{\delta}, \quad \hat{\underset{\sim}{D}} = R^T \underset{\sim}{D}R,$$ (2.13)

with μ and λ constant (or dependent on scalar measures), is a good relation; $\underset{\sim}{\delta}$ stands for the Kronecker delta with components δ_{ij}. This gives

$$\overset{\circ}{\underset{\sim}{\tau}} = 2\mu\underset{\sim}{D} + \lambda D_{kk}\underset{\sim}{\delta} + \underset{\sim}{\epsilon}\underset{\sim}{\tau} - \underset{\sim}{\tau}\underset{\sim}{\epsilon}$$ (2.14)

which is no longer an isotropic relation, unless $\underset{\sim}{\tau}$ is hydrostatic; the anisotropy in (2.14) is stress-induced. For application to geo-materials, or when special deformation regimes are involved, the last two terms in (2.14) may become important.

2.4 Classical Plasticity at Finite Strains (Hill, 1958,1959,1967)

Classical plasticity, as generalized by Hill (1958,1959), assumes the existence of a yield surface in either the stress or the strain-rate space, within which the material state is regarded to be elastic, and on which the material is elastic-plastic. For workhardening (or worksoftening) materials, the stress-rate is

$$\overset{\cdot}{\underset{\sim}{S}} = \underset{\sim}{\mathcal{L}}^0:[\overset{\cdot}{\underset{\sim}{E}} - \overset{\cdot}{\underset{\sim}{E}}^P],$$ (2.15)

where $\underset{\sim}{S}$ and $\underset{\sim}{E}$ are any conjugate stress and strain tensors, $\underset{\sim}{\mathcal{L}}^0 = \partial\phi/\partial\underset{\sim}{E}\partial\underset{\sim}{E}$ is the elasticity tensor (the second gradient of the free energy, ϕ, with respect to the strain with micro-defects all "locked," Nemat-Nasser, 1982), and $\overset{\cdot}{\underset{\sim}{E}}^P$ is the plastic part of the strain-rate.

Let $\underset{\sim}{\ell}$ be normal to the yield surface and $\underset{\sim}{\lambda}$ be in the $\overset{\cdot}{\underset{\sim}{E}}^P$-direction in the strain-rate space. We express $\overset{\cdot}{\underset{\sim}{E}}^P$ in terms of these quantities as $\overset{\cdot}{\underset{\sim}{E}}^P = \underset{\sim}{\lambda}(\underset{\sim}{\ell}:\overset{\cdot}{\underset{\sim}{E}})$, and write

$$\overset{\cdot}{\underset{\sim}{S}} = \begin{cases} \underset{\sim}{\mathcal{L}}^0:\overset{\cdot}{\underset{\sim}{E}} & \text{when } \underset{\sim}{\ell}:\overset{\cdot}{\underset{\sim}{E}} \leq 0 \\ \underset{\sim}{\mathcal{L}}^0:[\underset{\sim}{I} - \underset{\sim}{\lambda} \otimes \underset{\sim}{\ell}]:\overset{\cdot}{\underset{\sim}{E}} & \text{when } \underset{\sim}{\ell}:\overset{\cdot}{\underset{\sim}{E}} \geq 0, \end{cases}$$ (2.16)

where $\underset{\sim}{I}$ is the identity tensor, $\underset{\sim}{\ell}:\overset{\cdot}{\underset{\sim}{E}} \leq 0$ corresponds to elastic unloading, and

$\underset{\sim}{\ell}:\dot{\underset{\sim}{E}} \geq 0$ corresponds to plastic loading.

It is convenient to introduce the instantaneous elastic-plastic modulus tensor, $\underset{\sim}{L}$, by

$$\underset{\sim}{L} = \underset{\sim}{\mathcal{L}}^0:[\underset{\sim}{I} - \lambda \otimes \underset{\sim}{\ell}], \qquad (2.17a)$$

and its inverse, the instantaneous elastic-plastic compliance $\underset{\sim}{M}$, by

$$\underset{\sim}{M} = \underset{\sim}{L}^{-1} = \underset{\sim}{\mathbb{m}}^0:[\underset{\sim}{I} + \mu \otimes \underset{\sim}{m}], \qquad (2.17b)$$

where $\underset{\sim}{\mathbb{m}}^0$, the elastic compliance, is the inverse of $\underset{\sim}{\mathcal{L}}^0$. Direct calculation shows that

$$\mu = \underset{\sim}{\mathcal{L}}^0:\lambda/(1 - \underset{\sim}{\ell}:\lambda), \quad \underset{\sim}{m} = \underset{\sim}{\ell}:\underset{\sim}{\mathbb{m}}^0, \qquad (2.18)$$

where $\underset{\sim}{\ell}:\lambda \neq 1$ is assumed; $\underset{\sim}{\ell}:\lambda = 1$ corresponds to the elastic-perfectly-plastic case.

Relations (2.16) may now be expressed as

$$\dot{\underset{\sim}{E}} = \begin{cases} \underset{\sim}{\mathbb{m}}^0:\dot{\underset{\sim}{S}} & \text{when } \underset{\sim}{m}:\dot{\underset{\sim}{S}} \leq 0 \\ \underset{\sim}{\mathbb{m}}^0:[\underset{\sim}{I} + \mu \otimes \underset{\sim}{m}]:\dot{\underset{\sim}{S}} & \text{when } \underset{\sim}{m}:\dot{\underset{\sim}{S}} \geq 0 \end{cases} \qquad (2.19)$$

which is the more familiar form of the elastic-plastic rate constitutive relation. When the direction $\lambda \sim \underset{\sim}{\mathbb{m}}^0:\mu$ coincides with the direction $\underset{\sim}{m}$ which is normal to the yield surface in the stress space, we have the associated flow rule. In general, if $f(\underset{\sim}{S},...) = 0$ and $g(\underset{\sim}{S},...) = 0$ define the yield surface and the flow potential, respectively, then we have

$$\underset{\sim}{m} \sim \partial f/\partial \underset{\sim}{S} \quad \text{and} \quad \lambda \sim \partial g/\partial \underset{\sim}{S}. \qquad (2.20)$$

For the associative flow rule, we normalize $\underset{\sim}{m}$ such that $\underset{\sim}{m}:\underset{\sim}{m} = 1$, and set $\lambda = \underset{\sim}{m}/\bar{H}$ and $\mu = \underset{\sim}{\ell}/H$, where H is the usual hardening parameter and $\bar{H} = H + \underset{\sim}{\ell}:\underset{\sim}{m}$. Then, in terms of the Jaumann rate of the Kirchhoff stress, $\overset{\circ}{\underset{\sim}{\tau}}$, and the deformation rate, $\underset{\sim}{D}$, we obtain

$$\overset{\circ}{\underset{\sim}{\tau}} = \{\underset{\sim}{\mathcal{L}} - \frac{\alpha}{H} \underset{\sim}{\ell} \otimes \underset{\sim}{\ell}\}:\underset{\sim}{D},$$

$$\underset{\sim}{D} = \{\underset{\sim}{\mathbb{m}} + \frac{\alpha}{H} \underset{\sim}{m} \otimes \underset{\sim}{m}\}:\overset{\circ}{\underset{\sim}{\tau}}, \qquad (2.21)$$

where $\alpha = 0$ for elastic loading or unloading, and $\alpha = 1$ otherwise. In (2.21), $\underset{\sim}{\mathcal{L}}$ and $\underset{\sim}{\mathbb{m}} = \underset{\sim}{\mathcal{L}}^{-1}$ are the elasticity and the compliance tensors with the *current* state as the reference state.

2.5 Non-Associative Flow Rule; Dilatancy and Pressure-Sensitivity

For frictional materials, classical plasticity with plastic strain-rate, D^P, normal to the yield surface, leads to results which are not in accord with experimental observations; see Nemat-Nasser (1980,1983) and Nemat-Nasser and Shokooh (1980). Simple micro-mechanical modeling of pressure-dependent, slip-induced plasticity (Nemat-Nasser, Mehrabadi, and Iwakuma, 1980) shows that if the plastic strain-rate is derivable from a potential function (say, $D^P \sim \partial g/\partial \tau$), then, in general, this potential function, g, would be distinct from the yield surface, f.

A <u>Modified</u> J_2-<u>Plasticity</u> <u>Theory</u> (Nemat-Nasser and Shokooh, 1980):

Let f and g be defined by*

$$f \equiv \tau - F(I, \Delta, \gamma) : \text{ yield surface,}$$

$$g \equiv \tau + G(I, \Delta, \gamma) : \text{ flow potential,}$$

$$\tau = (\tfrac{1}{2} \tau'_{ij} \tau'_{ij})^{\frac{1}{2}} \equiv \sqrt{J_2}, \quad I = \tau_{kk},$$

$$\Delta = \int_0^t \frac{\rho_0}{\rho} D^P_{kk} \, dt, \text{ and } \gamma = \int_0^t (2 D^{P'}_{ij} D^{P'}_{ij})^{\frac{1}{2}} \, dt; \qquad (2.22)$$

here Δ is the total plastic volumetric strain (measured relative to a reference state of mass-density ρ_0), γ is the usual effective strain, and prime denotes the deviatoric part. Since the material is assumed to be rate-independent, any monotone parameter may be used for t. For continued plastic flow we use γ and from (2.22) and the consistency condition $\dot{f} = 0$ we obtain

$$\frac{d\tau}{d\gamma} - \frac{\partial F}{\partial I} \frac{dI}{d\gamma} = H, \qquad (2.23)$$

$$H = 3 \frac{\rho_0}{\rho} \frac{\partial G}{\partial I} \frac{\partial F}{\partial \Delta} + \frac{\partial F}{\partial \gamma} ; \qquad (2.24)$$

H is the hardening parameter. It then follows from $\underset{\sim}{D}^P = \Lambda \, \partial g / \partial \underset{\sim}{\tau}$ that

$$\underset{\sim}{D}^P = \frac{1}{H} \{ \frac{\underset{\sim}{\mu}}{\sqrt{2}} + \frac{\partial G}{\partial I} \underset{\sim}{\delta} \} \{ \frac{\underset{\sim}{\mu}}{\sqrt{2}} - \frac{\partial F}{\partial I} \underset{\sim}{\delta} \} : \overset{\circ}{\underset{\sim}{\tau}}, \quad \underset{\sim}{\mu} = \frac{\sqrt{2}}{2} \frac{\tau'}{\tau} . \qquad (2.25)$$

The quantity $-\partial F/\partial I$ is the pressure-sensitivity parameter. For geo-materials, for example, it represents the overall friction coefficient and is strictly positive.

The "dilatancy parameter" $3 \, \partial G/\partial I = D^P_{kk}/\dot{\gamma}$, on the other hand, is the measure of the volumetric strain-rate, and may be positive, negative, or zero, depending on the loading regime. Hence, for geo-materials, the use of an associated flow rule involves a basic contradiction, because then one will have $-\partial F/\partial I = \partial G/\partial I$, unless, of course, other assumptions are imposed.

By equating the rate of energy dissipation and the total rate of plastic work, one may obtain a relation between $\partial G/\partial I$ and $-\partial F/\partial I$; Dorris and Nemat-Nasser (1982) and Nemat-Nasser and Shokooh (1980). For example, in *simple shearing*, one obtains

$$\frac{1}{3} \frac{D^P_{kk}}{\dot{\gamma}} = \frac{\partial G}{\partial I} = \frac{\partial F}{\partial I} - \frac{\tau}{I} , \qquad (2.26)$$

when tension and expansion are regarded positive. Under compression, $I < 0$, shearing produces initial compaction, as commonly observed; note that $\partial F/\partial I < 0$. As τ is increased, the term $-\tau/I > 0$ eventually dominates the first term on the right-hand side of (2.26), leading to plastic volumetric expansion.

From (2.24), it is seen that the hardening parameter H consists of two parts: a density hardening given by

---

*Note that g and f have similar projections onto the deviatoric stress space, so that normality in the deviatoric space is preserved.

$$h_1 = 3 \frac{\rho_0}{\rho} \frac{\partial G}{\partial I} \frac{\partial F}{\partial \Delta} \qquad (2.27)$$

which may be positive, negative, or zero, depending on the sign of the dilatancy parameter, $3 \, \partial G/\partial I = D_{kk}^p/\dot{\gamma}$; and a distortional hardening,

$$h = \frac{\partial F}{\partial \gamma} \geq 0. \qquad (2.28)$$

Since $\partial F/\partial \Delta < 0$ when volume expansion is regarded positive*, $h_1 > 0$ when $D_{kk}^p < 0$, and $h_1 < 0$ when $D_{kk}^p > 0$.

As a specific example, consider simple shearing at constant confining pressure, $p = -I/3$. Take pressure and compaction positive and, ignoring the elastic strains, note that the volumetric strain-rate $\dot{v}/v$ is given by $\dot{v}/v = \dot{\Delta}/(1 - \Delta)$, so that $\partial G/\partial p \equiv -3 \, \partial G/\partial I = (1/v)dv/d\gamma = \partial F/\partial p - \tau/p$. Define $\bar{\tau} = \tau/p$ and from (2.23), (2.24), and (2.26) obtain

$$\frac{d\bar{\tau}}{d\gamma} = \frac{1 - \Delta}{p} \frac{\partial F}{\partial \Delta} \left[\frac{\partial F}{\partial p} - \bar{\tau} \right] + \frac{1}{p} \frac{\partial F}{\partial \gamma} ,$$

or

$$\frac{d\bar{\tau}}{d\gamma} = a[M - \bar{\tau}] + \hat{h}, \qquad (2.29)$$

where $a \equiv (1 - \Delta)(\partial F/\partial \Delta)/p > 0$, $\hat{h} = (\partial F/\partial \gamma)/p > 0$, and $M = \partial F/\partial p > 0$; M is the coefficient of overall friction. For frictional granules, the distortional hardening is a decreasing function of plastic strain γ. Let $M^* = \lim(M + \hat{h}/a)$ as γ becomes large, and note that $M^* \geq M$. Figure 1 is a schematic representation of (2.29): $\bar{\tau}$ increases with increasing γ, attains a peak value, and then decreases to a limiting value defined by $M^* \geq M$. From 0 to A, $\dot{v}/v = M - \bar{\tau}$ is positive and the material is compacting. After A, $\dot{v}/v$ is negative, and we have expansion. The maximum rate of expansion occurs at the peak stress, and then this rate decreases monotonically with decreasing stress.

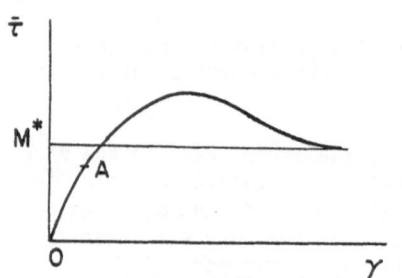

Figure 1. Normalized shear stress vs. shear strain for frictional granules.

Going back to the general case, we note that the inelastic response of most pressure-sensitive materials is affected by the magnitude of the intermediate stress; Dorris and Nemat-Nasser (1982), Mogi (1966), and Nemat-Nasser (1980). To account for this, the third stress invariant must also occur in the yield function and the flow potential. Dorris and Nemat-Nasser give a theory of this kind, identify all the corresponding parameters, discuss their physical meaning, and show how these parameters can be fixed in terms of simple conventional triaxial tests.

2.6 Plastic Anisotropy

The evolution of micro-defects in the course of a given load-history, in general, leads to plastic anisotropy. The defects serve as local stress and strain concentrators, producing local yielding under rather small, macroscopic

---

*Note that Nemat-Nasser and Shokooh (1980) regard volume expansion negative, in which case $\partial F/\partial \Delta > 0$, i.e., the material's resistance to flow increases as it densifies.

stress levels. Upon unloading, and reverse loading in particular, the presence
of residual stresses and strains may produce plastic flows at macroscopic
stresses quite different from those required to initiate yielding of a virgin
sample. To (approximately) simulate this fact, kinematic hardening has been
introduced.

For a combined kinematic-isotropic hardening, the yield surface is assumed
to move as well as to expand during the course of plastic loading. The motion
of the yield surface is defined by the location of its center in, say, the
stress-space. Let β be a deviatoric stress defining this center, $\beta_{kk} = 0$. Then
a J_2-plasticity theory with kinematic-isotropic hardening, for example, is defined by the following set of equations:

$$f \equiv \hat{\tau} - F(I,\Delta,\gamma) : \text{ yield surface,}$$

$$g \equiv \hat{\tau} + G(I,\Delta,\gamma) : \text{ flow potential,}$$

$$\hat{\tau} = [\tfrac{1}{2}(\underset{\sim}{\tau}' - \underset{\sim}{\beta}):(\underset{\sim}{\tau}' - \underset{\sim}{\beta})]^{\frac{1}{2}}, \quad \hat{\underset{\sim}{\mu}} = \frac{\sqrt{2}}{2}\frac{(\underset{\sim}{\tau}' - \underset{\sim}{\beta})}{\hat{\tau}}, \tag{2.30}$$

$$\underset{\sim}{D}^P = \frac{1}{H}\{\frac{\hat{\underset{\sim}{\mu}}}{\sqrt{2}} + \frac{\partial G}{\partial I}\underset{\sim}{\delta}\}\{\dot{\hat{\tau}} - \frac{\partial F}{\partial I}\dot{I}\},$$

$$\frac{d\hat{\tau}}{d\gamma} - \frac{\partial F}{\partial I}\frac{dI}{d\gamma} = H \equiv 3\frac{\rho_0}{\rho}\frac{\partial G}{\partial I}\frac{\partial F}{\partial \Delta} + \frac{\partial F}{\partial \gamma}.$$

The last equation characterizes the evolution of the size of the yield surface.
Observe that the last two equations may be replaced by

$$\underset{\sim}{D}^P = \frac{1}{H}\{\frac{\hat{\underset{\sim}{\mu}}}{\sqrt{2}} + \frac{\partial G}{\partial I}\underset{\sim}{\delta}\}\{\frac{\hat{\underset{\sim}{\mu}}}{\sqrt{2}} - \frac{\partial F}{\partial I}\underset{\sim}{\delta}\}: \overset{\circ}{\underset{\sim}{\tau}},$$

$$\tag{2.31}$$

$$\frac{\hat{\underset{\sim}{\mu}}:\overset{\circ}{\underset{\sim}{\tau}}'}{\sqrt{2}} - \frac{\partial F}{\partial I}\dot{I} = \dot{\gamma}\,\hat{H} \equiv \dot{\gamma}\left[\frac{\partial F}{\partial \Delta}\frac{\dot{\Delta}}{\dot{\gamma}} + \frac{\partial F}{\partial \gamma} + \frac{\hat{\underset{\sim}{\mu}}:\overset{\circ}{\underset{\sim}{\beta}}}{\sqrt{2}\dot{\gamma}}\right],$$

which follow from (2.30) by algebraic manipulations. When F is constant (no
isotropic hardening), this form should be used.

To complete the formulation, we give an evolutionary equation for β. The
simplest (rate-independent) equation of this kind is a linear relation between
$\underset{\sim}{\beta}$ and $\underset{\sim}{D}^P$. An isotropic relation, though the simplest, would *not* be appropriate
because the existence of micro-heterogeneities introduces residual stresses and
strains which render the current state plastically anisotropic. On the other
hand, the state at which $\beta = 0$ is, in fact, (macroscopically) plastically isotropic, and thus analogous to (2.14), we may write

$$\overset{\circ}{\underset{\sim}{\beta}} = A\,\underset{\sim}{D}^{P'} + \varepsilon\underset{\sim}{\beta} - \beta\underset{\sim}{\varepsilon}, \tag{2.32}$$

where the material parameter A, in general, depends on the effective strain γ
and possibly on stress.

Note that, with $\hat{\tau}$ given by (2.30)$_3$, the size of the yield surface is calculated by integrating (possibly numerically)

$$\frac{d\hat{\tau}}{d\gamma} = H + \frac{\partial F}{\partial I}\frac{dI}{d\gamma}, \tag{2.33}$$

and the location of its center is obtained by integrating (2.32). The material
parameters $\partial F/\partial\Delta$, $\partial F/\partial\gamma$, $\partial F/\partial I$, and A, in general, depend on variables Δ, γ,

and I, as well as on the temperature T; note that $\partial G/\partial I$ is not an independent function. These parameters may be fixed by comparison with experimental results. Examples are given by Dorris and Nemat-Nasser (1982), Nemat-Nasser and Shokooh (1980), and Rowshandel and Nemat-Nasser (1983).

The preceding discussion and examples involved normality of the plastic strain-rate to the flow potential function g which is distinct from the yield surface, f = 0. We observed that for pressure-sensitive materials, this is a basic requirement; otherwise one obtains contradictions. It should, however, be noted that the projections of the yield surface f = 0 and the flow potential g = const. onto the deviatoric stress space are, indeed, *similar* in all the examples we have considered, so that $\underset{\sim}{D}^{p'}$ is in fact normal to the *yield surface* in the $\underset{\sim}{\tau}'$-space, i.e.,

$$\underset{\sim}{D}^{p'} = \Lambda \frac{\partial g}{\partial \underset{\sim}{\tau}'} \sim \frac{\partial f}{\partial \underset{\sim}{\tau}'} . \tag{2.34}$$

Hence, for this class of materials, the non-association is in relation to the pressure and dilatancy only, i.e.,

$$D^p_{kk} = 3 \overset{\cdot}{\gamma} \frac{\partial G}{\partial I} \neq -3 \overset{\cdot}{\gamma} \frac{\partial F}{\partial I} . \tag{2.35}$$

2.7 Drucker's Postulate; Normality and Convexity

Drucker considers a stressed, elasto-plastically deformed solid or structure and postulates that the net work done by an external agency in applying and removing an incremental set of loads or stresses to this solid should be non-negative. As pointed out by Drucker (1950) and others, e.g. Martin (1975), the postulate is *not* a thermodynamic requirement. It does not necessarily apply to all materials whose response may be classified as "plastic." However, once Drucker's postulate is accepted for a given class of materials (at small strains) then it immediately leads to the normality of the plastic strain-rate on the yield surface at smooth points and to the convexity of the yield surface, and it confines the direction of plastic strain-rates within the outward normals at vertices and corners of the yield surface. Drucker's postulate precludes non-convex yield surfaces when the elastic properties remain the same in the course of plastic flow (which is essentially the case for most metals up to moderate strains). When the elastic properties change because of plastic flow (which is often the case for most geo-materials and also concrete), Drucker's postulate does not necessarily imply convexity, but does yield normality. When strains and rotations are large, Drucker's postulate is not free from ambiguity, inasmuch as it can be stated in terms of objective rates of different stress measures conjugate to different strain measures; see Hill (1968), Nemat-Nasser (1983), and Subsection 2.3. *Here we confine attention to small strains* and denote the linearized strain tensor by$\overset{*}{\underset{\sim}{\varepsilon}}$ and the stress tensor by $\underset{\sim}{\sigma}$.

Consider an admissible stress state, $\underset{\sim}{\sigma}^a$, and a cycle of loading to $\underset{\sim}{\sigma}^b$ and unloading back to $\underset{\sim}{\sigma}^a$. The net work over this *stress* cycle, according to Drucker's postulate, must be non-negative,

$$\oint(\sigma^b_{ij} - \sigma^a_{ij})d\varepsilon_{ij} \geq 0. \tag{2.36}$$

---

*This should *not* be confused with $\underset{\sim}{\varepsilon}$ defined by (2.12).

Let this stress cycle involve an increment of plastic strain, $d\varepsilon^P$, and assume that the elastic properties remain unchanged. Then (2.36) yields

$$(\sigma_{ij}^Y - \sigma_{ij}^a)d\varepsilon_{ij}^P \geq 0 \qquad (2.37)$$

which must hold for any state $\underset{\sim}{\sigma}^a$ in or on the yield surface, and any $\underset{\sim}{\sigma}^Y$ on the yield surface. Hence $d\varepsilon^P$ must be normal to the yield surface at smooth points, and, at corners, it must be within the outward normals to the surface; see Figs. 2a,b. Moreover, (2.37) excludes concavity at any point on the yield surface, $f = 0$; see Figs. 3a,b,c,d.

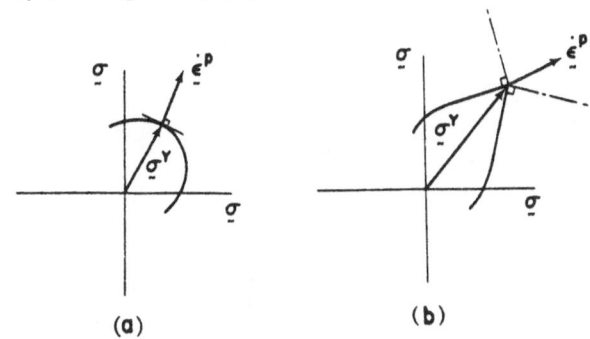

We thus have

$$\dot{\varepsilon}^P = \Lambda \frac{\partial f}{\partial \underset{\sim}{\sigma}} \qquad (2.38)$$

from Drucker's postulate, with f = 0 a convex surface.

(a)   (b)

Figure 2. (a) Normality rule, and the convexity of a smooth yield surface.
(b) Convexity at a vertex.

It can be shown (Martin, 1975) that the normality rule and the convexity of the yield surface lead to the uniqueness of (well-posed) incremental boundary-value problems when the material work-hardens. For an elastic-perfectly-plastic material, the stress increment is unique, but not necessarily the strain increment.

<u>Comments on Physical Basis of Normality</u>: A simple model for elastic-plastic deformation of single crystals results if we assume that plasticity stems from slip over active crystallographic slip planes, and elasticity from lattice distortion which produces a compatible overall deformation. Let there be N active slip systems. The αth slip system is defined by the unit normal $\underset{\sim}{n}^\alpha$ of the slip plane, and the unit vector $\underset{\sim}{s}^\alpha$ in the direction of the slip. The plastic strain rate then is (Bishop and Hill, 1951, Taylor and Elam, 1926, and others*)

$$\dot{\varepsilon}_{ij}^P = \sum_{\alpha=1}^{N} \tfrac{1}{2}(n_i^\alpha s_j^\alpha + n_j^\alpha s_i^\alpha)\dot{\gamma}^\alpha, \qquad (2.39)$$

where $\dot{\gamma}^\alpha$ is the slip rate. Note from (2.39) that $\dot{\varepsilon}_{ii}^P = 0$. According to this model, the slip induced plastic flow is incompressible.

The resolved shear stress, τ^α, acting in the slip direction is

$$\tau^\alpha = \sigma_{ij} n_i^\alpha s_j^\alpha = \sigma_{ij} \tfrac{1}{2}(n_i^\alpha s_j^\alpha + n_j^\alpha s_i^\alpha) \qquad (\alpha \text{ not summed}).$$

According to Schmid's law, for an active slip, $\tau^\alpha = \tau_Y^\alpha$, where τ_Y^α is the current

---

*See, e.g., Asaro (1979, 1983), Havner and Shalaby (1977), Nemat-Nasser (1983), Nemat-Nasser, Mehrabadi, and Iwakuma (1980), and Rice (1971, 1975) for rather comprehensive accounts.

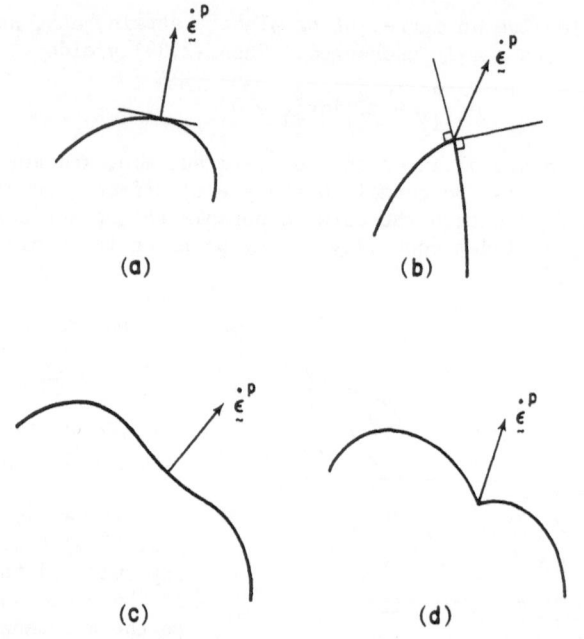

Figure 3.   (a) and (b) are yield surfaces which *do not*
violate Drucker's postulate;
(c) and (d) are yield surfaces which *do*
violate Drucker's postulate.

local yield stress in shear. The slip is inactive if $\tau^\alpha < \tau_Y^\alpha$. The rate of
plastic dissipation is

$$\sigma_{ij}\dot{\varepsilon}_{ij}^P = \sigma_{ij} \sum_{\alpha=1}^{N} \tfrac{1}{2}(n_i^\alpha s_j^\alpha + n_j^\alpha s_i^\alpha)\dot{\gamma}^\alpha = \sum_{\alpha=1}^{N} \tau^\alpha \dot{\gamma}^\alpha = \sum_{\alpha=1}^{N} \tau_Y^\alpha \dot{\gamma}^\alpha. \qquad (2.40)$$

Now, consider a stress state σ_{ij}^* inside the current yield surface. The corresponding resolved shear stress, $\tau^{\alpha*}$, for the αth slip system then satisfies
$\tau^{\alpha*} \le \tau_Y^\alpha$. Thus

$$(\sigma_{ij} - \sigma_{ij}^*)\dot{\varepsilon}_{ij}^P = \sum_{\alpha=1}^{N} (\tau_Y^\alpha - \tau_Y^{\alpha*})\dot{\gamma}^\alpha \ge 0 \qquad (2.41)$$

which reduces to (2.37) if we note that σ_{ij} is on the yield surface ($\sigma_{ij} = \sigma_{ij}^Y$)
and σ_{ij}^* is inside or on the yield surface ($\sigma_{ij}^* \equiv \sigma_{ij}^a$).

The above model may be modified to include non-Schmid effects, plastic
volumetric strain, and friction. This will immediately show that the plastic
strain-rate can no longer be normal to the yield surface (which, for frictional
materials, will depend on pressure). In this case, one may still have a plastic
potential, g, such that

$$\dot{\varepsilon}^P = \Lambda \frac{\partial g}{\partial \sigma}, \qquad (2.42)$$

but this plastic potential, as discussed before, in general will not coincide
with the yield surface.

2.8 Non-Coaxiality of Plastic Strain-Rate and Stress Tensors

When the plastic potential function g depends on stress through stress invariants only, then it follows from (2.42), or more generally, from $\underset{\sim}{D}^p \sim \partial g/\partial \underset{\sim}{\tau}$, that the plastic strain-rate is coaxial with the stress tensor, i.e., the two tensors have the same principal directions. It is well known that deviation from this coaxiality may have profound effects on a predicted material response; Iwakuma and Nemat-Nasser (1982), Nemat-Nasser, Mehrabadi, and Iwakuma (1980), Rudnicki and Rice (1975), and Stören and Rice (1975). Indeed, analytic calculations of shear band formation in a necked bar by Iwakuma and Nemat-Nasser (1982) show that the introduction of a small plastic strain-rate component tangential to the J_2-yield surface can reduce the predicted critical axial load by an order of magnitude. More specifically, consider the pressure-insensitive J_2-yield function, $f \equiv \tau - F(\gamma)$, and let the plastic strain-rate be defined by

$$\underset{\sim}{D}^{p'} = \frac{1}{H} \frac{\underset{\sim}{\mu}}{\sqrt{2}} \frac{\underset{\sim}{\mu}:\underset{\sim}{\tau}}{\sqrt{2}} + A\overset{\circ}{\underset{\sim}{\mu}}, \quad \underset{\sim}{\mu} = \frac{\sqrt{2}}{2} \frac{\underset{\sim}{\tau}'}{\tau}, \quad \tau^2 = \tfrac{1}{2} \underset{\sim}{\tau}':\underset{\sim}{\tau}', \tag{2.43}$$

where H = F'. Since μ is a unit vector in the $\underset{\sim}{\tau}'$-direction, $\overset{\circ}{\mu}$ is normal to μ and hence to τ'. Moreover, since μ is normal to the yield surface, $\overset{\circ}{\mu}$ is tangent to the yield surface. Iwakuma and Nemat-Nasser (1982) show that, in a uniaxial tension, the presence of the non-coaxiality term $A\overset{\circ}{\mu}$ can change the critical value of the axial load at localization by an order of magnitude. Thus, introduction of terms of this kind must be for good physical reasons, and must be guided by sound reasoning. We note that micromechanical modeling does naturally lead to non-coaxiality of a very special kind; see Christoffersen, Mehrabadi, and Nemat-Nasser (1981), Mandel (1947,1965), Nemat-Nasser (1983), Nemat-Nasser, Mehrabadi, and Iwakuma (1980), and Spencer (1964,1982).

3. THERMODYNAMIC BASIS OF NORMALITY

As pointed out in Section 2, Drucker's postulate which leads to normality is not a thermodynamic principle or requirement. There are, however, certain thermodynamic assumptions at the micro-level which lead to the existence of a flow potential for both rate-dependent and rate-independent inelasticity. An overview with references is given by Nemat-Nasser (1983). Here we give a brief account.

Since the inelastic response of solids stems from micro-structural changes, one reasonable way to account for this is to introduce a set of *internal* state variables, collectively denoted by $\underset{\sim}{\xi}$, in such a manner that the response is *elastic* when ξ's remain unchanged, i.e., when no micro-structural changes occur. While this may not be the most general approach, it serves to illustrate the kind of assumptions which are required in order to obtain flow potentials for inelastic strain-rates. (These assumptions are in addition to the basic requirement of non-negative dissipation.) For simplicity, we assume that ξ's are n independent scalar parameters, ξ_α, $\alpha = 1,2,\ldots,n$.

Since for $\underset{\sim}{\xi}$ = constant the response is elastic, the Helmholtz free energy $\phi = \phi(\underset{\sim}{E},\theta;\underset{\sim}{\xi})$ and the Gibbs function $\psi = \psi(\underset{\sim}{S},\theta;\underset{\sim}{\xi})$ exist with the following properties*:

---

*The differentiation denoted by the operator "$\partial/\partial\ldots$" implies derivative with respect to the indicated variable with *all* other variables held fixed.

$$\underset{\sim}{S} = \frac{\partial \phi}{\partial \underset{\sim}{E}}, \quad \eta = -\frac{\partial \phi}{\partial \theta}, \quad \underset{\sim}{\Lambda} = -\frac{\partial \phi}{\partial \underset{\sim}{\xi}}, \tag{3.1}$$

$$\underset{\sim}{E} = \frac{\partial \psi}{\partial \underset{\sim}{S}}, \quad \eta = \frac{\partial \psi}{\partial \theta}, \quad \underset{\sim}{\Lambda} = \frac{\partial \psi}{\partial \underset{\sim}{\xi}}, \tag{3.2}$$

where θ and η are the temperature and entropy-density, respectively; Λ's are the thermodynamic *forces* conjugate to ξ's; and ϕ and ψ are related by a Legendre transformation as follows:

$$\phi + \psi = S_{ij} E_{ij}; \tag{3.3}$$

$\underset{\sim}{S}$ and $\underset{\sim}{E}$ are the conjugate stress and the material strain, respectively.

By definition, the inelastic strain-rate is caused by the changes in internal variables, ξ's, only. Hence, we have

$$\dot{\underset{\sim}{E}}^P = \frac{\partial \underset{\sim}{E}}{\partial \xi_\alpha} \dot{\xi}_\alpha, \quad \alpha = 1, 2, \ldots, n \quad (\alpha \text{ summed}), \tag{3.4}$$

and, in view of (3.2),

$$\dot{\underset{\sim}{E}}^P = \frac{\partial^2 \psi}{\partial \xi_\alpha \partial \underset{\sim}{S}} \dot{\xi}_\alpha = \frac{\partial \Lambda_\alpha}{\partial \underset{\sim}{S}} \dot{\xi}_\alpha \quad (\alpha \text{ summed}). \tag{3.5}$$

The evolution of the internal variables must be prescribed on the basis of the involved physical processes. This must be guided by experimental observations. The only basic requirement is that the dissipation rate be non-negative, i.e.,

$$\Lambda_\alpha \dot{\xi}_\alpha \geq 0 \quad (\alpha \text{ summed}). \tag{3.6}$$

Depending on the type of evolutionary equations used, the inelastic strain-rate, $\dot{\underset{\sim}{E}}^P$, may or may not admit a potential.

In some applications it may be reasonable to regard each flux, $\dot{\xi}_\alpha$, dependent *explicitly* only on its own conjugate force Λ_α, as well as (in general) on the temperature θ, but not on other forces nor explicitly on the overall strain, $\underset{\sim}{E}$; note, however, the implicit dependence on $\underset{\sim}{E}$ through Λ_α. The Schmid law is of this kind. Other physical examples are discussed by Rice (1971,1975). In cases of this kind, or even in more general settings, one may express $\dot{\xi}$'s as

$$\dot{\underset{\sim}{\xi}} = \Lambda \frac{\partial \Omega}{\partial \underset{\sim}{\Lambda}}, \tag{3.7}$$

where $\Omega = \Omega(\underset{\sim}{\Lambda}, \theta)$. From (3.7) and (3.5) it then follows that

$$\dot{\underset{\sim}{E}}^P = \Lambda \frac{\partial \Omega}{\partial \Lambda_\alpha} \frac{\partial \Lambda_\alpha}{\partial \underset{\sim}{S}} = \Lambda \frac{\partial g}{\partial \underset{\sim}{S}}, \tag{3.8}$$

$$g(\underset{\sim}{S}, \theta; \underset{\sim}{\xi}) \equiv \Omega(\underset{\sim}{\Lambda}(\underset{\sim}{S}, \theta; \underset{\sim}{\xi}), \theta). \tag{3.9}$$

Note that (3.7), and hence (3.8), are always valid if the fluxes, $\dot{\underset{\sim}{\xi}}$, are related linearly to the forces, $\underset{\sim}{\Lambda}$, and the Onsager reciprocal relations hold (the classical irreversible thermodynamics). In this case, the flow potential g is proportional to the (isothermal) rate of entropy production associated with the plastic flow. In general, however, (3.7), and hence (3.8), are additional constitutive assumptions, beyond any basic thermodynamic requirements. The normality rule (3.8) does not stem from the second law of thermodynamics.

4. STRAIN-SOFTENING

The strain-softening observed in laboratory experiments on certain metals
(e.g., mild steel) and geo-materials (e.g., rocks and concrete) essentially
stems from a dramatic change in the sample's deformation pattern, and often
involves a transition from a more or less (macroscopically) homogeneous deformation state into a highly localized heterogeneous state. Therefore, the descending portion of most stress - strain curves does not pertain to a continuum
constitutive response, but rather to the response of a highly heterogeneous
"structure." Since this matter will be discussed at length by others within
these proceedings, here I will cite an example in rock mechanics by way of
illustration. Figure 4, taken from Hallbauer, Wagner, and Cook (1973) shows
the change in the distribution of the microcracks in rock samples, tested in a

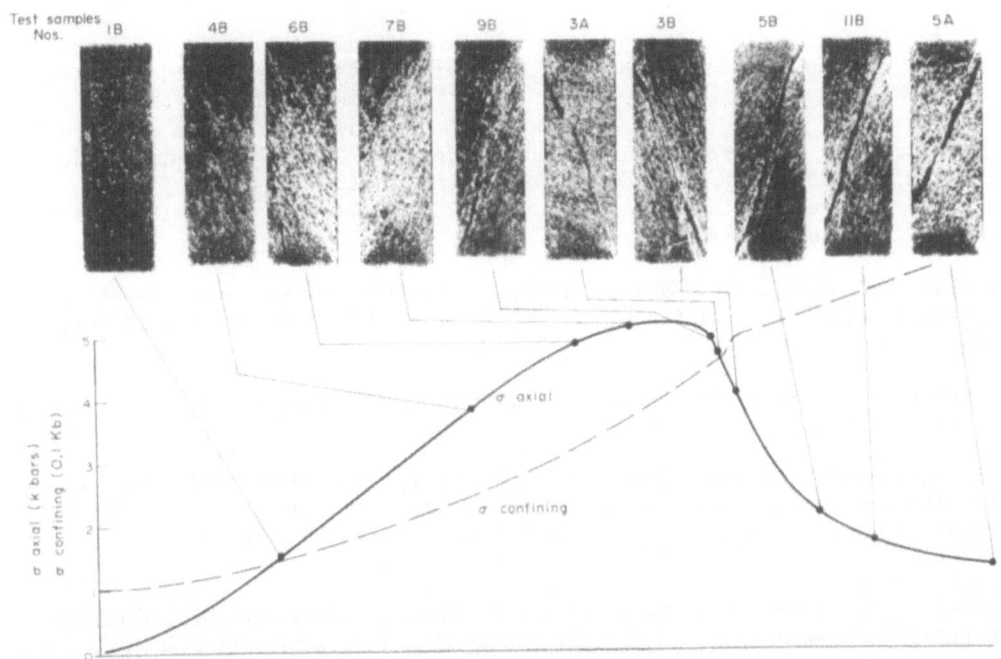

Figure 4. Stress-strain curve and photograph of test specimens at different
stress levels (from Hallbauer, Wagner, and Cook, 1973).

very stiff triaxial machine. As is clearly seen and thoroughly discussed by the
authors, over the ascending portion of the stress-strain curve, the distribution
of the microcracks is more or less homogeneous throughout the sample, but the
cracks begin to cluster more and more within a narrow band (which eventually
forms the faulting plane), as the peak stress is approached. Close to the peak
stress, a strong band of microcracks has already emerged, and the declining portion of the stress-strain curve then essentially corresponds to the sliding
resistance of the two blocks to which the sample has been reduced. Recently,
this phenomenon has been modeled analytically, and the model has been verified
experimentally by Nemat-Nasser and Horii (1982) and Horii and Nemat-Nasser (1983).

ACKNOWLEDGMENT

     This work was completed as part of consulting services the author provided
to The BDM Corporation.

REFERENCES

Asaro, R.J. (1979), "Geometrical Effects in the Inhomogeneous Deformation of
Ductile Single Crystals," *Acta Met.*, *27*, 445-453.

Asaro, R.J. (1983), "Crystal Plasticity," *J. Appl. Mech.* (50th Anniv. Issue), *50*,
921-934.

Bishop, J.F.W. and R. Hill (1951), "A Theory of the Plastic Distortion of a
Polycrystalline Aggregate under Combined Stresses," *Phil. Mag.*, [7] *42*, 414-427.

Christoffersen, J., M.M. Mehrabadi and S. Nemat-Nasser (1981), "A Micromechanical Description of Granular Material Behavior," *J. Appl. Mech.*, *48*, 339-344.

Dafalias, Y.F. (1983), "On the Evolution of Structure Variables in Anisotropic
Yield Criteria at Large Plastic Transformations," Colloque International du
C.N.R.S. No. 351 in *Critères de Rupture des Matériaux a Structure Interne
Orientée*, Villard-de-Lans, France, June 1983, J.P. Boehler (ed.), Editions du
C.N.R.S., Paris, in press.

Dorris, J.F. and S. Nemat-Nasser (1982), "A Plasticity Model for Flow of Granular
Materials under Triaxial Stress States," *Int. J. Solids Structures*, *18* (6), 497-
531.

Drucker, D.C. (1950), "Some Implications of Work Hardening and Ideal Plasticity,"
Q. Appl. Math., *7* (4), 411-418.

Hallbauer, D.K., H. Wagner and G.W. Cook (1973), "Some Observations Concerning
the Microscopic and Mechanical Bahaviour of Quartzite Specimens in Stiff, Triaxial Compression ests," *Int. J. Rock Mech. Min. Sci. & Geomech. Abstr.*, *10*,
713-726.

Havner, K.S. (1982), "The Theory of Finite Plastic Deformation of Crystalline
Solids," in *Mechanics of Solids: The Rodney Hill 60th Anniversary Volume*, H.G.
Hopkins and M.J. Sewell (eds.), Pergamon Press, Oxford, 265-302.

Havner, K.S. and A.H. Shalaby (1977), "A Simple Mathematical Theory of Finite
Distortional Latent Hardening in Single Crystals," *Proc. Roy. Soc. London*, *A358*,
47-70.

Hill, R. (1958), "A General Theory of Uniqueness and Stability in Elastic-
Plastic Solids," *J. Mech. Phys. Solids*, *6*, 236-249.

Hill, R. (1959), "Some Basic Principles in the Mechanics of Solids without a
Natural Time," *J. Mech. Phys. Solids*, *7*, 209-225.

Hill, R. (1967), "The Essential Structure of Constitutive Laws for Metal Composites and Polycrystals," *J. Mech. Phys. Solids*, *15*, 79-95.

Hill, R. (1968), "On Constitutive Inequalities for Simple Materials. I," *J. Mech.*

Phys. Solids, 16, 229-242; "II," *J. Mech. Phys. Solids, 16,* 315-322.

Hill, R. (1972), "On Constitutive Macro-Variables for Heterogeneous Solids at Finite Strain," *Proc. Roy. Soc. London, A326,* 131-147.

Hill, R. (1978), "Aspects of Invariance in Solid Mechanics," in *Advances in Applied Mechanics,* C.-S. Yih (ed.), Academic Press, New York, *18,* 1-75.

Horii, H. and S. Nemat-Nasser (1983), "Compression-Induced Micro-Crack Growth in Brittle Solids: Axial Splitting and Shear Failure," submitted for publication.

Iwakuma, T. and S. Nemat-Nasser (1982), "An Analytical Estimate of Shear Band Initiation in a Necked Bar," *Int. J. Solids Structures, 18* (1), 69-83.

Loret, B. (1983), "On the Effects of Plastic Rotation in the Finite Deformation of Anisotropic Elastoplastic Materials," *Mechanics of Materials, 2,* 287-304.

Mandel, J. (1947), "Sur les Lignes de Glissement et le Calcul des Déplacements dans la Déformation Plastique," *Comptes rendus de l'Academie des Sciences, 225,* 1272-1273.

Mandel, J. (1965), "Généralisation de la Théorie de Plasticité de W.T. Koiter," *Int. J. Solids Structures, 1* (3), 273-295.

Mandel, J. (1971), *Plasticité Classique et Viscoplasticité,* Courses and Lectures, No. 97, CISM, Udine, Springer, New York.

Martin, J.B. (1975), *Plasticity: Fundamentals and General Results,* M.I.T. Press, Cambridge, MA.

Mogi, K. (1966), "Pressure Dependence of Rock Strength and Transition from Brittle Fracture to Ductile Flow," *Bull. Earthquake Res. Inst., 44,* 215-232.

Nemat-Nasser, S. (1975), "Continuum Bases for Consistent Numerical Formulations of Finite Strains in Elastic and Inelastic Structures," in *Finite Element Analysis of Transient Nonlinear Structural Behavior,* T. Belytschko, J.R. Osias and P.V. Marcal (eds.), AMD-*14,* ASME, New York, 85-98.

Nemat-Nasser, S. (1980), "On Constitutive Behavior of Fault Materials," in *Solid Earth Geophysics and Geotechnology,* S. Nemat-Nasser (ed.), AMD-*42,* ASME, New York, 31-37.

Nemat-Nasser, S. (1982), "On Finite Deformation Elasto-Plasticity," *Int. J. Solids Structures, 18* (10), 857-872.

Nemat-Nasser, S. (1983), "On Finite Plastic Flow of Crystalline Solids and Geomaterials," *J. Appl. Mech.* (50th Anniv. Issue),*50,* 1114-1126.

Nemat-Nasser, S. and H. Horii (1982), "Compression Induced Nonplanar Crack Extension with Application to Splitting, Exfoliation and Rockburst," *J. Geophys. Res., 87* (B8), 6805-6821.

Nemat-Nasser, S., M.M. Mehrabadi and T. Iwakuma (1980), "On Certain Macroscopic and Microscopic Aspects of Plastic Flow of Ductile Materials," in *Three-Dimensional Constitutive Relations and Ductile Fracture,* Proc. IUTAM Symp., Dourdan, France, S. Nemat-Nasser (ed.), North-Holland Publ. Co., 157-172.

Nemat-Nasser, S. and A. Shokooh (1980), "On Finite Plastic Flows of Compressible Materials with Internal Friction," *Int. J. Solids Structures, 16* (6), 495-514.

Rice, J.R. (1971), "Inelastic Constitutive Relations for Solids: An Internal-Variable Theory and Its Application to Metal Plasticity," *J. Mech. Phys. Solids, 19,* 433-455.

Rice, J.R. (1975), "Continuum Mechanics and Thermodynamics of Plasticity in Relation to Microscale Deformation Mechanisms," Chapter 2 in *Constitutive Equations in Plasticity,* A.S. Argon (ed.), M.I.T. Press, Cambridge, MA, 23-79.

Rowshandel, B. and S. Nemat-Nasser (1983), "Finite Strain Rock Plasticity: Stress Triaxiality, Pressure, and Temperature Effects," Earthquake Research and Engineering Laboratory Technical Report No. 83-5-53, Department of Civil Engineering, Northwestern University, Evanston, IL.

Rudnicki, J.W. and J.R. Rice (1975), "Conditions for the Localization of Deformation in Pressure-Sensitive Dilatant Materials," *J. Mech. Phys. Solids, 23,* 371-394.

Spencer, A.J.M. (1964), "A Theory of the Kinematics of Ideal Soils under Plane Strain Conditions," *J. Mech. Phys. Solids, 12,* 337-351.

Spencer ,A.J.M. (1982), "Deformation of Ideal Granular Materials," *Mechanics of Solids, The Rodney Hill 60th Anniversary Volume,* H.G. Hopkins and M.J. Sewell (eds.), Pergamon Press, Oxford, 607-652.

Stören, S. and J.R. Rice (1975), "Localized Necking in Thin Sheets," *J. Mech. Phys. Solids, 23,* 421-441.

Taylor, G.I. and C.F. Elam (1926), "The Distortion of Iron Crystals," *Proc. Roy. Soc. London, A112,* 337-361.

GENERAL DISCUSSION

CURRAN: Sia, I would like that you would reschedule the agenda here and give an equally comprehensive lecture on the micromechanical side, on which I know you are now working.

NEMAT-NASSER: I guess I don't need to respond to that. I'll just say thank you.

WILKINS: In regard to the tension test where you shaved off the outer boundaries with the result that the elongation to failure increased, this relates to the plastic size effect. This effect can be introduced into computer simulation programs that can model fracture. During the tension test, when necking starts, the hydrostatic tension in the central region increases with respect to positions near the free surfaces. Using the concept that fracture is the result of cumulative damage from the initiation and coalescence of voids, the process is enhanced as the hydrostatic tension is increased. By shaving off the boundaries and starting over again, the pressure must build up again to a high negative value before the same damage conditions are reached. Before fracture can occur, the damage must span some finite dimension. As the lateral dimensions of the specimen are decreased, the extent of the hydrostatic tension is decreased.

The fracture mode tends to shift from high tension, low elongation to low tension, high elongation, i.e. brittle to ductile. The characteristic dimension of the material where this occurs has been determined by G. Irwin to be the square of the fracture toughness over the yield stress. The larger the number is, the larger will be the specimen size where increased elongation can be achieved by removing material from the lateral boundaries. With a ceramic, the characteristic dimension is very small. The effect of super-elongation will not occur at any reasonable dimension test size.

NEMAT-NASSER: Thank you very much. I certainly appreciate your comment. I did not intend that example as an illustration of the process of failure but as an illustration that the test may not relate to a homogeneously deformed solid.

There are various explanations, including the effect of hydrostatic tension, as you have pointed out, of failure phenomenon. There have been a number of calculations which produce in fact not only the shape of the neck but also the shape of the final failure. I did not intend to discuss those, and I'm happy that you pointed out a few very important aspects of the failure mechanics.

BALADI: You said that the softening behavior is the result of development of cracking bands and shear bands. We know from the lab data that you can apply more than twenty percent strain on a sample containing very loose sand and never develop softening. On the other hand, the stress-strain curves for very dense sand show a well-defined peak stress at relatively low strains. With increasing strain the sand develops softening behavior while the sample is still in the homogeneous state (uniform stress). So softening is real, and I don't think it is the result of the formation of the shear bands alone.

NEMAT-NASSER: Yes. Softening is perhaps real. Also, it has been observed in mild steel. If you take a sample of mild steel, encase it in another metal with positive hardening, extend it and then calculate the stress, you can observe a very nice, smooth softening and then again hardening. So it is a fact of life that some materials, because of certain micromechanical instabilities which begin with plastic flow, do soften. However, not all. The examples that I presented are very clear illustrations of nonhomogeneous deformation, i.e., structural instability rather than softening.

A VOICE: Selective examples.

NEMAT-NASSER: Not necessarily so because rock mechanics is infested with this kind of results.

SANDLER: You indicated that the introduction of a non-associated component in the flow rule could produce certain sensitivity in any calculations or analyses that you might perform. That is, I think, certainly a well-accepted fact.

You also pointed out that in the modeling of dilatancy, you're driven to the adoption of a non-associated flow rule. But that, it seems, is a function of the yield condition that you assume for the frictional behavior of the material. There are ways of assuming different yield conditions where the alternative of compaction at low shear stresses and dilatancy at higher shear stresses can be introduced, still within the context of an associated flow rule.

This is a very important point because while it's true that the introduction of non-associated flow components produces sensitivity, actual sensitivities may not arise with respect to the compaction or dilatancy of material as it's being sheared. Therefore, it may be important to distinguish between introducing non-associated behavior in order to model instabilities as opposed to introducing non-associated behavior to model frictional behavior. Now, we don't know the yield function a priori. It's not something we can measure; it's something we can only infer from a set of experimental results. It is possible, by appropriate choices, to assume yield functions in such a way that you get the appropriate sequence of compaction and dilatant behavior without giving up the requirement of normality.

NEMAT-NASSER: I fully agree that it is possible to have a yield function which matches any experimental result. But it is at the expense of a large number of parameters or functions. What I intended to do here as illustration is to pick a yield function which has been extensively used over the past twenty, twenty-five years —and that is the J_2 plasticity —and make a very small modification to it. And at the same time, the modification is done in such a manner that every parameter that emerges can be given a physical interpretation, either macroscopically or at least by some microscopic consideration; for example, the pressure sensitivity of the flow potential is in fact the dilatancy parameter, which can be measured, and the pressure sensitivity of the yield surface is in fact the overall coefficient of friction. This is quite simple. I have seen plasticity for soils with several yield surfaces moving all over the place in the stress-space with a hundred eighty or hundred ninety different parameters at each nodal point. Of course, you can fit anything you want with that. I'm not objecting to those models. They may have their own merits.

A VOICE: I object to them.

NEMAT-NASSER: As I pointed out, my model involves associative flow rule in the deviatoric space. The model is not sensitive because of non-association! The sensitivity comes from adding a term to the plastic strain-rate, which is tangent to the yield surface in the deviatoric space.

ASARO: No, I agree with what Dr. Sandler said. I think there is another way of saying it, though, more fundamentally. Maybe you didn't understand his question. I think what you're referring to is, if all kinetic laws for plastic dilatancy as well as deviatoric effect of plastic strain only involve work conjugate stress measures or forces, that is if those kinetic laws only involve precisely work conjugate stress measures, then that's right. Then you'll have pressure sensitivity and normality.

NEMAT-NASSER: Yes, but then the friction and dilatancy are mixed up!

HERRMANN: I was very glad that you brought up this question of strain localization either in necks or in faulting in a triaxial specimen. It seems to me that, as you pointed out, this is really an instability of the machine specimen configuration. You can easily show that, for example, in the triaxial test, by changing the aspect ratio of the specimen. That does more things than introduce end effects. For example, when you make a short specimen, it inhibits the ability for a shear band to go across that specimen. That instability is now inhibited and you can easily move the peak of the stress-strain curve to the right. Now, in view of that, I'm interested in your comment as to how one might introduce a yield surface because the whole idea of placing a yield surface at a global instability of the system is totally inconsistent. You certainly can't place a yield surface at the place where this curve gets to its peak. Certainly, you need the yield surface somewhere else.

The second thing that happens, of course, is shown in the pictures of

Hallbauer et al. There are cracks that formed early in the process, and those cracks localize as you approach the peak. But in fact, the cracks alter the elastic behavior of the material. They're a permanent effect which introduces, among other things, frictional effects in the "elastic range."

Where do you suppose you should place the yield surface, and in which way should you identify how it moves from experiment?

NEMAT-NASSER: For what kind of material?

HERRMANN: For that kind of brittle material.

NEMAT-NASSER: I hope we will discuss this at this workshop, so I'm not going to venture comments at this time.

VARDOULAKIS: I would like to comment on this strain-softening. I would like to present some experimental results from triaxial tests on sand samples. Do you think it is possible?

BUDIANSKY: Yes. Well, I think we might look forward to seeing them tonight, because there are many more people who would like to comment now.

PISTER: Sia, I know you like controversy, so let me ask you, did you consider the use of the term stress-induced anisotropy a necessary or a convenient condition?

NEMAT-NASSER: A convenient condition.

PISTER: It seems to me it is likely to obscure the main issue. Every material is anisotropic, with few exceptions, except for trivial cases, away from its natural state. So it seems to me that the use of the term just mixes the whole thing up.

NEMAT-NASSER: I agree. I want to make one comment. In relation to isotropic tensors one has to be a little careful: An isotropic relationship between two tensors may represent a non-isotropic material response.

KRAJCINOVIC: Apparently, from your talk, one thing was obvious — that a judicious selection of parameters can give you just about anything, hardening, softening, everything else. However, I think that leaves unloading as being probably the best criterion for deciding whether you can use the theory of plasticity in a particular case. I wondered why you didn't comment on that because some of those materials, geomaterials, in unloading do not conform to our traditional criteria for plasticity, according to which the unloading path is parallel to the initial loading path.

NEMAT-NASSER: Well, there is elastic degradation or the change of the elasticity due to either the formation of micro-cracks, for example, as was mentioned here, or other microscopic defects that evolve in the course of loading. If you take account of those judiciously by including elastic degradation, then I would imagine that these materials do satisfy the usual definition of unloading.

KRAJCINOVIC: But then it's a combination of plasticity and something else.

NEMAT-NASSER: There is a great body of literature on estimating the instantaneous moduli of solids with various defects. As you know, Budiansky has done a lot of work on that; also Hutchinson. We have also done some work. There is recognition that the evolution of micro-defects does affect elasticity. If you take that into account, you can define unloading.

VALANIS: I was hoping that you might discuss some questions of uniqueness of the initial value problems, specifically since the lecture was addressed to the identification and modeling of materials. One obviously can consider an array of all kinds of models. And recently, I think there has been some kind of hesitancy to discuss those models in conjunction with the problem of uniqueness of the initial value problem of the material.

NEMAT-NASSER: I thought that I did at least mention that some of the classical assumptions or postulates were motivated for this very purpose, and the deviation from these assumptions can really create problems. I hope that during the workshop there will be some discussion of the uniqueness —

VALANIS: Are you aware of any uniqueness proofs in plasticity with the exception of Hill's uniqueness proof, in the case of non-associated flow rules?

NEMAT-NASSER: You take Drucker's postulate and you state a well-posed boundary value problem for rates, then you immediately show the uniqueness under certain conditions, but Drucker's postulate also yields normality.

BUDIANSKY: He asked anything else.

NEMAT-NASSER: I do not know anything beyond Hadamard's results and the results as generalized by Hill and by some other people.

BUDIANSKY: I know a theorem for a theory you probably don't approve. In small strains there is uniqueness for J_2 deformation theory with a power hardening law.

VALANIS: Just the case of the associated flow rule.

BUDIANSKY: No, this is deformation theory. Completely violates an associated flow rule.

VALANIS: Oh, the deformation theory.

BUDIANSKY: Yes. I said power hardening to make sure there is no unloading.

CHAPTER II

NUMERICAL IMPLEMENTATION OF CONSTITUTIVE MODELS: RATE-INDEPENDENT DEVIATORIC PLASTICITY

Thomas J.R. Hughes

Stanford University, Stanford, California

ABSTRACT

Basic concepts concerning the numerical implementation of rate-independent deviatoric plasticity theories are presented. Both small and finite deformations are dealt with. Emphasis is placed on the central problem of computational plasticity, that is, given a deformation history, find the corresponding stress history by integrating the constitutive equations. Mathematical well-posedness requires non-classical specification of elastic and plastic processes. This problem is discussed for Mises, Tresca and non-convex "starlike" yield surfaces, associative and non-associative flow rules, and strain hardening and softening. The radial-return concept is emphasized in the algorithmic descriptions. A unified treatment of a class of finite-deformation theories is presented. Accurate and efficient procedures for calculating kinematical quantities necessary in finite deformation analysis are described.

1. INTRODUCTION

Presently, large-scale inelastic computations, involving small and finite deformations, are routinely performed by finite difference and finite element procedures. Most widely-used programs are continually upgraded as better methodology appears and although much progress has been made there are still numerous questions of basic theory and numerical adaptation which need to be answered. Despite this fact, inelastic calculations are being used to design many critical engineering systems. Consequently, there is considerable need for scrutiny of this methodology and improvement is called for wherever theoretical or numerical deficiencies are found to remain.

The purposes of this paper are to present an overview of the most commonly used methods and to introduce some new ideas which bear on this topic. Attention is confined herein to the integration of elastic-plastic constitutive equations. That is, a deformation history is assumed *given*, and the problem is to accurately and efficiently compute the corresponding stress history. This may be described as *the central problem of computational plasticity* as it corresponds to the main role played by the constitutive equations in actual computations: The discretized momentum equations generate incremental motions which are, in turn, used by the constitutive routines to calculate new stresses; lack

29

of satisfaction of equilibrium creates new motions, and so forth. There are, of course, many other important computational ingredients in the overall procedure. However, these are particular to the type of solution algorithm and involve the constitutive theory in, at most, a limited way. For example, in *implicit* methods the constitutive theory may be used to define "tangent matrices." In typical procedures, in which it is required that the residual forces converge to zero at each step, tangent arrays are part of the *strategy* used to develop the solution, but in the end have no effect on the accuracy of it. On the other hand, the integration of the constitutive equation is a fundamental ingredient in *all* solution algorithms. In *explicit* procedures, which are widely used in large-scale inelastic analysis, there are no tangent matrices and the integration of the constitutive equations is the only function to which the constitutive theory is put.

Thus, mathematically, the fundamental problem is to integrate coupled systems of ordinary differential equations at each point where stresses are required. Due to the constant "switching" between elastic and plastic behavior, there is an inherent lack of smoothness which necessitates non-classical integration procedures.

The specific topics dealt with in the remainder of the paper are outlined as follows: Section 2 is concerned with problems which may be described within the context of small-deformation theory. We begin in Section 2.1 with the elastic-perfectly-plastic case with Mises yield surface and associative flow rule. Particular attention is paid to formulating the theory in a mathematically well-posed fashion. To do this we must abandon classical loading and unloading conditions and replace them by meaningful definitions of elastic and plastic processes. (This is a repeated theme throughout this paper.) These are the theoretical counterparts of what have been used numerically out of necessity for approximately twenty years. Next, we describe perhaps the most widely used and effective algorithm for integrating constitutive equations of this type: the *radial-return method*. The basic procedure is applicable to the three-dimensional case as well as axisymmetry and plane strain. The theory is generalized in Section 2.2 to include strain hardening and softening. This is again facilitated by non-classical definitions of elastic and plastic processes. The radial-return implementation of the theory is then described. We follow in Section 2.3 with a description of a somewhat overlooked phenomenon which accompanies the use of yield surfaces with corners, such as Tresca's, with an *associative* flow rule. Implementational aspects of associative and non-associative rules are presented next. A nonconvex yield surface is briefly discussed in Section 2.4.

Section 3 deals with topics from finite-deformation theory. After dispensing with preliminaries, in Section 3.1 we show how theories involving various objective rates of "rotational" type can be cast into a uniform format, form-identical to small-deformation theory. Once the relevant kinematical quantities are calculated, numerical integration of the constitutive equations can be performed by the algorithms of Section 2. Thus the only additional contributions to the numerical error are the approximations made in calculating strain increments and rotational transformations. Accurate and efficient procedures for calculating these quantities are presented in Section 3.2.

Length limitations have precluded inclusion of a number of other important topics such as: consistent definition of tangent arrays; rate-dependent effects; pressure-sensitive theories; multiple yield-surface theories; finite-deformation inelastic beam and shell analysis (for a few preliminary ideas see Hughes and Liu, 1981); and some fundamental issues in the development of finite-deformation constitutive theories. We hope to discuss these in future works.

2. TOPICS IN SMALL-DEFORMATION THEORY

2.1 The Elastic-Perfectly-Plastic Case

2.1.1 Theory; definition of elastic and plastic processes

The first step in the development of an algorithmic interpretation of a given theory is a precise mathematical statement of the problem to be solved. This appears to be obvious, but leads to non-classical forms of loading and unloading conditions for even the simplest cases. To illustrate this, consider an elastic-perfectly-plastic solid with Mises yield surface. The *constitutive equation* is

$$\dot{\sigma}_{ij} = c_{ijk\ell}(\dot{\epsilon}_{k\ell} - \dot{\epsilon}^{p\ell}_{k\ell}) \tag{1}$$

where

$$c_{ijk\ell} = \mu(\delta_{ik}\delta_{j\ell} + \delta_{i\ell}\delta_{jk}) + \lambda\,\delta_{ij}\delta_{k\ell} \tag{2}$$

in which σ_{ij} = Cauchy stress; $\epsilon_{k\ell}$ = strain; $\epsilon^{p\ell}_{k\ell}$ = plastic strain; δ_{ij} = Kronecker delta; λ,μ = Lamé parameters; and a superposed dot denotes time differentiation. Throughout the summation convention is employed. We also often use "direct notation." For example, (1) is written as

$$\dot{\underset{\sim}{\sigma}} = \underset{\sim}{c} \cdot (\dot{\underset{\sim}{\epsilon}} - \dot{\underset{\sim}{\epsilon}}^{p\ell}), \tag{3}$$

where the dot following $\underset{\sim}{c}$ refers to the dot product between second-rank tensors.

The *Mises yield function* is defined by

$$f(\underset{\sim}{\sigma}) = k^2 \tag{4}$$

where

$$f(\underset{\sim}{\sigma}) = J_2(\sigma') = \tfrac{1}{2}\|\sigma'\|^2 = \tfrac{1}{2}\,\sigma'_{ij}\sigma'_{ij} \tag{5}$$

in which σ'_{ij} is the deviatoric stress and k is a parameter which defines the "size" of the yield surface (i.e., $R = \sqrt{2}\,k$ is the radius of the Mises cylinder). The admissible states of stress are those interior to, or on, the surface.

We shall further employ an *associative flow rule*, namely

$$\dot{\underset{\sim}{\epsilon}}^{p\ell} = \begin{cases} \Lambda\,\underset{\sim}{N} & \text{"plastic process"} \\ \underset{\sim}{0} & \text{"elastic process"} \end{cases} \tag{6}$$

where $\underset{\sim}{N}$ is the unit outward normal to the yield surface, that is,

$$\underset{\sim}{N} = (\partial f/\partial\underset{\sim}{\sigma})/\|\partial f/\partial\underset{\sim}{\sigma}\| \tag{7}$$

and Λ is defined by the consistency condition. For the moment, we are intentionally vague about what is meant by "elastic and plastic processes."

To proceed further it is worthwhile to review the role of the constitutive theory. Basically it is an input-output device which generates stress histories from given strain histories. Equation (3) serves as a definition of $\dot{\underset{\sim}{\sigma}}$. Consequently, the stress can be advanced in time once the right-hand side of (3) is known. The right-hand side is determined by the history of (total) strain and specification of $\underset{\sim}{\epsilon}^{p\ell}$, which in turn depends on the state of stress during a "plastic process," but is zero otherwise.

Once the definitions of elastic and plastic processes are delineated, the evolutionary structure of the constitutive theory is completely specified. The following assertions state the obvious:

(i) A process is said to be (instantaneously) *elastic* if the current stress state is strictly interior to the yield surface.

(ii) If the current stress is on the yield surface, and if the *trial rate-of-stress*, $\dot{\sigma}^{tr} = c \cdot \dot{\varepsilon}$, points towards the interior side of the tangent plane to the yield surface, then the process is also *elastic*. Otherwise the process is said to be *plastic*.

Condition (ii) invokes the elastic constitutive equation whenever it makes sense to do so, that is, whenever the elastically-induced stress is not attempting to get outside of the yield surface. Otherwise, a plastic process is invoked and the consistency condition is used to define Λ so that σ remains on the yield surface. This can all be put in analytical terms and leads to the following evolution, or initial-value, problem: *Given an initial value of σ and a time-history of ε, find the time-history of σ such that the equations in Box 1 are satisfied.*

Box 1. Summary of elastic-perfectly-plastic constitutive theory with Mises yield surface and associative flow rule.

Constitutive equation:
$$\dot{\sigma} = c \cdot (\dot{\varepsilon} - \dot{\varepsilon}^{p\ell}) \tag{8}$$

Flow rule:
$$\dot{\varepsilon}^{p\ell} = \begin{cases} 0 & \text{if (E)} \\ \Lambda N & \text{if (P)} \end{cases} \tag{9}$$

Consistency condition:
$$\Lambda = N \cdot \dot{\sigma}^{tr}/(N \cdot c \cdot N) = N \cdot \dot{\varepsilon} \tag{10}$$

Trial rate-of-stress:
$$\dot{\sigma}^{tr} = c \cdot \dot{\varepsilon} \tag{11}$$

Definition of an elastic process (E):
$$f(\sigma) < k^2, \quad \text{or} \quad f(\sigma) = k^2 \text{ and } N \cdot \dot{\sigma}^{tr} \leq 0 \tag{12}$$

Definition of a plastic process (P):
$$f(\sigma) = k^2 \text{ and } N \cdot \dot{\sigma}^{tr} > 0 \tag{13}$$

Unit normal:
$$N = \sigma'/R \tag{14}$$

Remark 1. The equations of Box 1 may be succinctly written as an *abstract initial-value problem*:
$$\dot{\sigma}(t) = f(\sigma(t), t), \quad \sigma(0) = \text{given}. \tag{15}$$

The material point is fixed so it does not appear as an argument. The explicit t-dependence emanates from $\dot{\varepsilon}$ which is considered a given function of t. The classical theory of ordinary differential equations provides existence and uniqueness theorems for an f which is Lipschitz continuous in σ and C^0 in t. In the present case, f is potentially discontinuous in both arguments and thus lies outside the realm of applicability of classical o.d.e. theory. Problems of this type arise in optimal control theory and issues of existence and uniqueness are discussed, for example, in Young (1969). It is somewhat surprising that sufficiently general theorems to accomodate plasticity do not seem to be known at this time.

Remark 2. The fundamental quantity which distinguishes elastic and plastic processes is the sign of the normal component of the trial rate-of-stress, $\underset{\sim}{N} \cdot \dot{\underset{\sim}{\sigma}}^{tr}$. By virtue of $\underset{\sim}{c}$ being isotropic and $\underset{\sim}{N}$ being deviatoric, $\underset{\sim}{N} \cdot \dot{\underset{\sim}{\sigma}}^{tr} = 2\mu \underset{\sim}{N} \cdot \dot{\underset{\sim}{\epsilon}}$.

Remark 3. Observe that the classical "loading and unloading conditions" (see, e.g., Naghdi, 1960) play no role whatsoever in the theory. One is often left with the impression in classical treatises that these conditions are the crux of the evolutionary character of plasticity, but this is not the case as the equations of Box 1 illustrate. We shall come back to this point again when we consider hardening and softening in Section 2.2. Correct definitions of elastic and plastic processes have appeared previously in Hill (1967) although they seem to have attracted little attention.

Remark 4. The following geometric interpretation of (8) is useful in developing numerical algorithms: *In a plastic process, $\dot{\underset{\sim}{\sigma}}$ is the orthogonal projection of $\dot{\underset{\sim}{\sigma}}^{tr}$ onto the tangent plane of the yield surface.* (This follows from the normality of $\dot{\underset{\sim}{\epsilon}}^{pl}$ and isotropy of the elastic coefficients.) Use of a Mises yield surface allows us to substitute the word "radial" for "orthogonal" in this interpretation; see Fig. 1. For further discussion on the use of projection operators in plasticity see Pinsky et al. (1983).

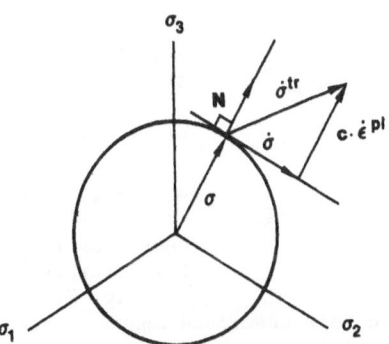

Figure 1. π-plane illustration of $\dot{\underset{\sim}{\sigma}}$ for the elastic-perfectly-plastic case with Mises yield surface and associative flow rule.

2.1.2 A numerical algorithm based on the "radial-return" concept

Many numerical algorithms can be developed for the present theory. The attributes that one strives for are accuracy, reliability, efficiency and ease of computer implementation. The particular algorithm to be described satisfies all of these attributes. It was presented by Wilkins in 1964 (Wilkins, 1964) and apparently is the first algorithm to be developed for computational plasticity. Subsequently developed algorithms, which are invariably more complicated, have also been shown to fall short of Wilkins' method with respect to accuracy (Krieg and Krieg, 1977). Consequently, the algorithm has become a "standard" in finite difference and finite element programs employed for large-scale inelastic analysis. HEMP (Wilkins, 1964), HONDO (Key, Beisinger and Krieg, 1978), and the DYNA and NIKE codes (Hallquist, 1979, 1981a, 1981b, 1982) are but a few of the widely-used programs which may be mentioned in this regard.

The problem to be solved is as follows: *Given the strain history in the form of a series of increments, calculate the stress history in the form of a series of stress states at discrete times.* Only the latest stress is stored for each material point. Thus the algorithm must update the stress given the old stress and a strain increment.

The first step in the algorithm is to calculate the elastically-induced trial stress:

$$\sigma_{n+1}^{tr} = \sigma_n + c \cdot \Delta\varepsilon. \tag{16}$$

In (16) the subscript refers to the time level. That is, σ_n is the approximation of the stress at time t_n. The strain increment corresponds to the change in strain during the time interval $[t_n, t_{n+1}]$. The trial stress is tested to see if it is inside or outside the yield surface. If it falls within, or on, the yield surface the process was clearly an elastic one and the updated stress is given simply by the trial stress. In this case we have

$$\sigma_{n+1} = \sigma_{n+1}^{tr}. \tag{17}$$

If, on the other hand, the trial stress is outside the yield surface, some procedure is necessary to generate a corresponding σ_{n+1} on the yield surface. In the radial-return process, σ_{n+1} is defined as the intersection of the line connecting the center of the yield surface to the trial stress with the yield surface; see Fig. 2. This is a simple discrete analog of the geometric idea described above in Remark 4. Note that this process takes place in the deviatoric plane and thus the dilatational components are unaffected. The algorithm is summarized in Table 1.

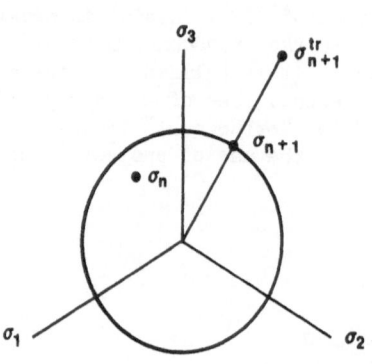

Figure 2.   Illustration of the radial-return process.

Table 1.   Wilkins' radial-return algorithm for the elastic-perfectly-plastic case with Mises yield surface and associative flow rule.

Step 1.   Calculate the trial stress: $\sigma_{n+1}^{tr} = \sigma_n + c \cdot \Delta\varepsilon$

Step 2.   Calculate the mean stress: $\sigma = \frac{1}{3} \text{ trace } \sigma_{n+1}^{tr}$

Step 3.   Calculate the deviatoric trial stress:

$$\eta = \sigma_{n+1}^{tr} - \sigma I$$

Step 4. $A = \eta \cdot \eta$

Step 5.   If $A \leq R^2$, then $\sigma_{n+1} = \sigma_{n+1}^{tr}$.   Otherwise,

$$\sigma_{n+1} = \left(\frac{R}{A^{\frac{1}{2}}}\right)\eta + \sigma I.$$

Remark 1. Clearly, very few numerical operations are necessary to advance the stress. The algorithm is trivially exact for all elastic processes. In general it is convergent in the small (i.e., the integration error vanishes as $\| \Delta\varepsilon \| \to 0$), a minimal requirement, and it is also asymptotically exact in the large (i.e., the integration error also vanishes in the limit $\| \Delta\varepsilon \| \to \infty$). Exact results are also obtained for "radial processes" in which there exists a constant x such that $x(\sigma_{n+1}^{tr})' = \sigma_n'$. Furthermore, the maximum error is significantly smaller than other proposed schemes. See Krieg and Krieg (1977) for a complete error analysis.

Remark 2. The algorithm of Table 1 is applicable to the three-dimensional case, and the two-dimensional cases of plane strain and axisymmetry.

2.2 Strain Hardening and Softening

2.2.1 Theory

In this section we consider generalizations of the previous ideas to the case of strain hardening and softening. We shall first, as before, carefully articulate the equations that we wish to solve. The correct definitions of elastic and plastic processes will enable us to accomodate softening in contra-distinction to the prevalent notion that softening lies outside the realm of "stress-space" plasticity (see, e.g., Naghdi and Trapp, 1975, Caulk and Naghdi, 1978, and Casey and Naghdi, 1981). The theory considered retains the constitu-tive equation previously introduced, the Mises yield function and the normality condition. However, a linear combination of isotropic and kinematic hardening is assumed. Thus k is allowed to vary and the center of the yield surface in deviatoric space, or "back stress," α, needs also to be introduced as a vari-able. For simplicity, we limit the discussion to linear hardening and softening. The evolution problem to be solved is stated as follows: *Given initial values of σ, k and α, and a time history of ε, find the time histories of σ, k and α such that the equations in Box 2 are satisfied.*

Box 2. Summary of elastic-plastic constitutive theory with Mises yield surface, associative flow rule, and linear combination of isotropic and kinematic hardening or softening.

Constitutive equation:
$$\dot{\sigma} = c \cdot (\dot{\varepsilon} - \dot{\varepsilon}^{pl}) \tag{18}$$

Evolution equation for the radius of the yield surface:
$$\dot{R} = \frac{\sqrt{2}}{\sqrt{3}} \beta H' \dot{\varepsilon}^{pl} \tag{19}$$

Evolution equation for the back stress:
$$\dot{\alpha} = \frac{2}{3} (1 - \beta) H' \dot{\varepsilon}^{pl} \tag{20}$$

Flow rule:
$$\dot{\varepsilon}^{pl} = \begin{cases} Q & \text{if (E)} \\ \Lambda N & \text{if (P)} \end{cases} \tag{21}$$

continued

Consistency condition:

$$\Lambda = \frac{1}{2\mu} \frac{1}{\left(1 + \frac{H'}{3\mu}\right)} \, \underset{\sim}{N} \cdot \dot{\underset{\sim}{\sigma}}^{tr} = \frac{1}{\left(1 + \frac{H'}{3\mu}\right)} \, \underset{\sim}{N} \cdot \dot{\underset{\sim}{\varepsilon}} \qquad (22)$$

Trial rate-of-stress:

$$\dot{\underset{\sim}{\sigma}}^{tr} = \underset{\sim}{c} \cdot \dot{\underset{\sim}{\varepsilon}} \qquad (23)$$

Definition of an elastic process (E):

$$f(\underset{\sim}{\xi}) < k^2, \quad \text{or} \quad f(\underset{\sim}{\xi}) = k^2 \quad \text{and} \quad \underset{\sim}{N} \cdot \dot{\underset{\sim}{\sigma}}^{tr} \le 0 \qquad (24)$$

Definition of a plastic process (P):

$$f(\underset{\sim}{\xi}) = k^2 \quad \text{and} \quad \underset{\sim}{N} \cdot \dot{\underset{\sim}{\sigma}}^{tr} > 0 \qquad (25)$$

Unit normal:

$$\underset{\sim}{N} = \underset{\sim}{\xi}'/R \quad \text{where} \quad \underset{\sim}{\xi} = \underset{\sim}{\sigma} - \underset{\sim}{\alpha} \qquad (26)$$

Remark 1. H' and β are material properties. H' is the plastic modulus and corresponds to the (constant) slope of the effective stress vs. effective plastic strain curve under radial loading conditions.* Thus we have:

$$H' > 0 \iff \text{strain hardening}$$
$$H' = 0 \iff \text{perfect plasticity}$$
$$H' < 0 \iff \text{strain softening}$$

If $H' = 0$ we reduce to the elastic-perfectly-plastic theory considered previously. The parameter β determines the proportion of isotropic and kinematic hardening or softening. Note

$$\beta = 0 \iff \text{kinematic}; \quad \beta = 1 \iff \text{isotropic}.$$

Intermediate values involve both mechanisms. β may be determined from the reverse-loading yield point in a simple cyclic experiment (see, e.g., Key, Stone and Krieg, 1981). If $H' < 0$ and $\beta > 0$, a stopping criterion needs to be introduced to ensure that the radius of the yield surface does not become negative. This criterion must be appended to the equations of Box 2 to render the theory meaningful in this case.

Remark 2. As was previously emphasized, the present theory is applicable to softening. The reason for this is that correct definitions of elastic and plastic processes are employed instead of the classical "loading and unloading criteria" in terms of $\dot{\underset{\sim}{\sigma}}$. Naghdi and colleagues have noticed this deficiency of the classical theory and have chosen to abandon the entire "stress space" formalism in favor of a "strain space" theory. This seems unwarranted as it is only the classical loading and unloading criteria which need to be replaced. Nevertheless, there may be some other advantages to strain space theories as pointed out in the works of Yoder and colleagues (see Yoder and Iwan, 1981; Iwan and Yoder, 1983; and Yoder and Whirley, 1983). Theories of the type described

---

*In radial loading from a zero state, $\dot{\bar{\sigma}} = H' \dot{\bar{\varepsilon}}^{p\ell}$, where $\bar{\sigma} = \sqrt{3}(J_2(\underset{\sim}{\sigma}'))^{\frac{1}{2}}$ and $\dot{\bar{\varepsilon}}^{p\ell} = 2(J_2(\dot{\underset{\sim}{\varepsilon}}^{p\ell})/3)^{\frac{1}{2}}$.

in this paper, in which correct definitions of elastic and plastic processes are employed, have been used by numericists since the inception of computational plasticity out of necessity because no other criteria make any sense. It should be manifestly apparent that defining the loading and unloading criteria in the classical way in terms of $\dot{\sigma}$ is potentially tautological because it is the purpose of these criteria to define the right-hand-side of (3) so that $\dot{\sigma}$ may in turn be defined. In examining the strain space formalism, Moss (1983) has made similar observations.

Remark 3. Viscoplastic phenomena lie outside the scope of the present paper. However, we note that Sandler and Wright (1983) have pointed out serious shortcomings in inviscid strain-softening theories. These observations have important consequences in the modelling of geological materials, in particular, concrete.

2.2.2 The radial-return algorithm

Krieg and Key (1976) generalized the radial-return concept to the present case. The procedure is now widely used in large-scale inelastic analysis. The basic idea is to approximate the normal vector $\underset{\sim}{N}$ by $(\xi_{n+1}^{tr})' / \| (\xi_{n+1}^{tr})' \|$ where $(\xi_{n+1}^{tr})' = (\sigma_{n+1}^{tr} - \underset{\sim}{\alpha}_n)'$. The consistency condition is then applied in the large to the time-discrete counterparts of (18)-(20) (resp.):

$$\sigma_{n+1} = \sigma_{n+1}^{tr} - 2\mu \tilde{\Lambda} \underset{\sim}{N} \tag{27}$$

$$R_{n+1} = R_n + \frac{2}{3} \beta H' \tilde{\Lambda} \tag{28}$$

$$\alpha_{n+1} = \alpha_n + \frac{2}{3}(1 - \beta)H' \tilde{\Lambda} \underset{\sim}{N} \tag{29}$$

This results in

$$\tilde{\Lambda} = \frac{1}{2\mu} \frac{1}{\left(1 + \frac{H'}{3\mu}\right)} \left(\| (\xi_{n+1}^{tr})' \| - R_n \right). \tag{30}$$

The algorithm is summarized in Table 2.

Table 2. Krieg and Key's radial-return algorithm for the elastic-plastic case with Mises yield surface, associative flow rule and linear combination of isotropic and kinematic hardening or softening.

Step 1. Calculate the trial stresses:

$$\sigma_{n+1}^{tr} = \sigma_n + \underset{\sim}{c} \cdot \Delta\varepsilon; \quad \xi_{n+1}^{tr} = \sigma_{n+1}^{tr} - \underset{\sim}{\alpha}_n$$

Step 2. Calculate the mean component of ξ_{n+1}^{tr}:

$$\xi = \frac{1}{3} \text{ trace } \xi_{n+1}^{tr}$$

Step 3. Calculate the deviatoric part of ξ_{n+1}^{tr}:

$$\underset{\sim}{\eta} = \xi_{n+1}^{tr} - \xi \underset{\sim}{I}$$

Step 4. $A = \underset{\sim}{\eta} \cdot \underset{\sim}{\eta}$.

continued

Step 5. If $A \le R_n^2$ (i.e., elastic process), then $\sigma_{n+1} = \sigma_{n+1}^{tr}$,
         $R_{n+1} = R_n$, and $\alpha_{n+1} = \alpha_n$.
         Otherwise (i.e., plastic process), continue.

Step 6. Calculate the approximation to the normal:
$$N = \eta / \| \eta \|$$

Step 7. Calculate $\tilde{\Lambda}$:

$$\tilde{\Lambda} = \frac{1}{2\mu} \; \frac{1}{\left(1 + \dfrac{H'}{3\mu}\right)} \; \left(\| \eta \| - R_n \right)$$

Step 8. Update:

$$\sigma_{n+1} = \sigma_{n+1}^{tr} - 2\mu \, \tilde{\Lambda} \, N$$

$$R_{n+1} = R_n + \tfrac{2}{3} \, \beta \, H' \tilde{\Lambda}$$

$$\alpha_{n+1} = \alpha_n + \tfrac{2}{3} (1 - \beta) H' \tilde{\Lambda} \, N$$

Remark 1. The plastic strain tensor and/or the effective plastic strain may be computed from the following formulae if desired (resp.):

$$\varepsilon_{n+1}^{p\ell} = \varepsilon_n^{p\ell} + \tilde{\Lambda} \, N \tag{31}$$

$$\bar{\varepsilon}_{n+1}^{p\ell} = \bar{\varepsilon}_n^{p\ell} + \sqrt{\tfrac{2}{3}} \, \tilde{\Lambda} \tag{32}$$

Remark 2. Yoder and Whirley (1983) have analyzed the accuracy of the radial-return method in the presence of hardening and compared it with several other algorithms. The superiority of radial return was even more dramatically apparent in this case than for the elastic-perfectly plastic case (cf. Krieg and Krieg, 1977).

Remark 3. The present theory is not so complicated that analytical solutions are impossible. The idea is to solve the evolution equations for σ_{n+1}, R_{n+1} and α_{n+1}, given σ_n, R_n and α_n, under the assumption that $\dot{\varepsilon} = \Delta\varepsilon/\Delta t = $ constant over the time step. The exact solution represents the "ultimate algorithm." Krieg and Krieg (1977) obtained an exact solution for the elastic-perfectly-plastic case. Yoder and Whirley (1983) obtained an exact solution for the case of kinematic hardening, and asymptotic solutions for the case of isotropic hardening. It turns out that evaluation of these analytical solutions is about twice as costly computationally as radial return (Yoder and Whirley, 1983). Consequently, their use in most practical computing would not warrant their additional cost. However, they are clearly superior to radial-return subiterative strategies and thus, if high precision is required, they may be recommended.

2.3 Yield Surfaces with Corners

As an example of a yield surface with corners, or vertices, consider the Tresca yield surface which is defined by*

---

*We return to the elastic-perfectly-plastic case to simplify the exposition.

$$g(\underset{\sim}{\sigma}) = k_T \tag{33}$$

where

$$g(\underset{\sim}{\sigma}) = \tfrac{1}{2} \max_{i,j} |\sigma_i - \sigma_j| \tag{34}$$

in which k_T is a parameter which characterizes the size of the yield surface and σ_i, $i = 1,2,3$, denote the eigenvalues of $\underset{\sim}{\sigma}$. The Tresca yield surface is a regular hexagonal cylinder with pressure axis as generator (see Figure 3). The admissible stress states are those interior to, or on, the yield surface. Note that, as for the Mises condition, the admissible stresses for the Tresca condition constitute a *convex set* (i.e., if $\underset{\sim}{\sigma}_0$ and $\underset{\sim}{\sigma}_1$ are admissible states of stress, then so are all points along the straight line segment joining $\underset{\sim}{\sigma}_0$ and $\underset{\sim}{\sigma}_1$, namely, all $\underset{\sim}{\sigma}_\theta = \theta\,\underset{\sim}{\sigma}_1 + (1 - \theta)\underset{\sim}{\sigma}_0$, where $\theta \in [0, 1]$). The Tresca condition predates the Mises condition and at one time was thought to represent a more fundamentally correct description of plastic yielding in metals. However, the combined tension-torsion tests of Taylor and Quinney in 1931 demonstrated that the Mises surface fit experimental data for copper, aluminum and mild steel better then the Tresca surface (see Hill, 1950, for further elaboration). Nevertheless, the Tresca surface retained some

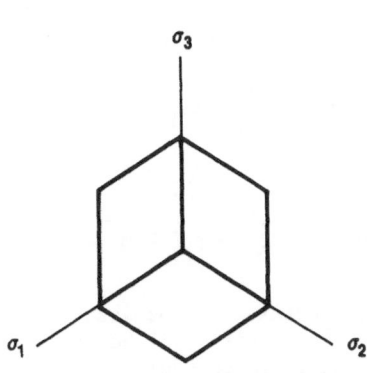

Figure 3. Tresca yield surface.

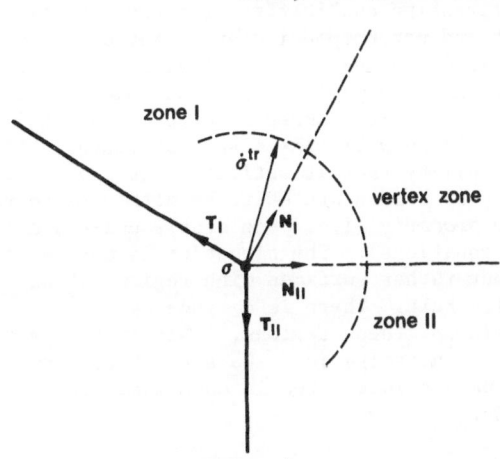

Figure 4.

popularity because it facilitated the hand calculation of some simple problems. It does not seem to be widely used in computing. However, the Mohr-Coulomb condition is widely used in modelling geological materials and it shares certain properties with the Tresca condition, namely it is composed of flat intersecting surfaces. The Mohr-Coulomb condition, which accounts for pressure dependence, is outside the scope of the present paper, but some of the ideas to be presented pertain to it as well.

We wish to consider the flow rule for the Tresca surface. Let us assume that normality is invoked wherever the normal is unambiguously defined (i.e., everywhere except the vertices). Something special needs to be done at the vertices. We assume that σ is at a vertex, as illustrated in Figure 4, and consider the various possibilities. As for the Mises yield surface, we call upon the elastic constitutive equation whenever possible. Thus an *elastic process at a vertex* may be defined by:

$$\dot{\sigma}^{tr} \cdot N_I \leq 0 \quad \text{and} \quad \dot{\sigma}^{tr} \cdot N_{II} \leq 0 \tag{35}$$

Whenever the trial rate-of-stress violates (35), the process is said to be *plastic*. In this case there are three possibilities corresponding to the three zones in Figure 4 into which $\dot{\sigma}^{tr}$ may be directed. The situation is summarized as follows:

$$
\dot{\varepsilon}^{pl} =
\begin{cases}
\Lambda \, N_I & \text{if } \dot{\sigma}^{tr} \cdot T_I \nparallel 0 \text{ and } \dot{\sigma}^{tr} \cdot T_{II} < 0 \\[2ex]
\Lambda \, N_{II} & \text{if } \dot{\sigma}^{tr} \cdot T_I < 0 \text{ and } \dot{\sigma}^{tr} \cdot T_{II} \nparallel 0 \\[2ex]
\Lambda_I N_I + \Lambda_{II} N_{II} & \text{if } \dot{\sigma}^{tr} \cdot T_I < 0 \text{ and } \dot{\sigma}^{tr} \cdot T_{II} < 0
\end{cases}
\tag{36}
$$

In the first two cases, σ will move away from the vertex. In the last case, σ will remain at the vertex; thus, necessarily, $c \cdot \dot{\varepsilon}^{pl} = 2\mu \dot{\varepsilon}^{pl}$ will be parallel to $(\dot{\sigma}^{tr})'$. As usual, consistency conditions may be employed to evaluate the Λ's. Observe that (36) provides a *continuous* representation of $\dot{\varepsilon}^{pl}$ as $\dot{\sigma}^{tr}$ passes through the boundaries of the vertex zone (see Fig. 4).

The Tresca yield surface and normality, in the form described above, combine to produce an interesting qualitative property. Consider a strain history in which $\dot{\varepsilon}$ is constant and not perpendicular to any of the six flat surfaces composing the Tresca yield surface. Assume that $\dot{\varepsilon}$ has a non-vanishing deviatoric component so that σ reaches the yield surface at some instant. Then σ will move along the yield surface until it reaches a vertex at which $\dot{\sigma}^{tr}$ points into the vertex zone. Subsequently, σ will remain at that vertex. If we randomly assign a constant $\dot{\varepsilon}$, the probability that it satisfies the above defining conditions is one. Thus the tendency of stress states to be *attracted* to vertices may be described as a *generic property* (i.e., one stable under small, random perturbations of the data and equations). The situation is not particular to yield surfaces with vertices, but rather surfaces with regions of high curvature combined with an associative flow rule. There is a tendency of stress states to be attracted to, and remain in, these regions. This is the physical response of the type of theory under investigation. We are not concerned here whether or not it is "real," but we are interested in numerical algorithms which correctly reproduce this response.

Prior to this section we have reported upon the compelling success of the radial-return concept used in conjunction with a Mises yield surface. Can this concept be used here? Of course it can, but it will not converge to solutions of the above theory! This can be seen by consideration of constant $\dot{\varepsilon}$ as described previously. The stress state will approach a point on the yield surface at which $(c \cdot \Delta\varepsilon)' = 2\mu\Delta\varepsilon'$ and σ_n' are parallel. The vertices are not exceptional with respect to this property (neither is any other point on the yield surface for that matter). Consequently, the attractor property of the vertices is not possessed by the radial-return algorithm and thus convergence is impossible.

The radial-return algorithm in this case is converging to the solution of another theory, namely, one in which the flow rule takes on the *non-associative* form:

$$\dot{\varepsilon}^{p\ell} = \begin{cases} 0 & \text{if (E)} \\ \Lambda\ \sigma' & \text{if (P)} \end{cases} \tag{37}$$

where an *elastic process* (E) is defined as

$$g(\sigma) < k_T, \quad \text{or} \tag{38}$$

$$g(\sigma) = k_T \quad \text{and} \tag{39}$$

$$\begin{cases} \dot{\sigma}^{tr} \cdot N \leq 0 \quad \text{(at a regular point)} \tag{40} \\ \dot{\sigma}^{tr} \cdot N_I \leq 0 \quad \text{and} \quad \dot{\sigma}^{tr} \cdot N_{II} \leq 0 \quad \text{(at a vertex)} \tag{41} \end{cases}$$

and a *plastic process* (P) is simply one that is not elastic. From (37) it is clear that the Mises yield function plays the role of a *plastic potential* in this case.

A simple, convergent algorithm for the original theory may be easily conceived. If σ^{tr}_{n+1} falls within a regular zone, σ_{n+1} is defined to be its orthogonal projection onto the yield surface; if σ^{tr}_{n+1} falls within a vertex zone, σ_{n+1} is located at the vertex (see Fig. 5). Implementation of this theory is clearly more complicated than radial return. Thus for numerical reasons, it is tempting

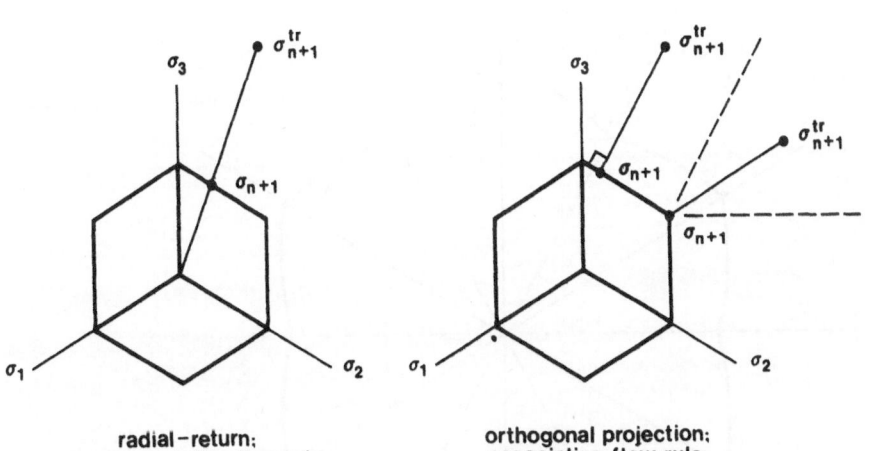

radial-return;
non-associative flow rule

orthogonal projection;
associative flow rule

Figure 5.   Algorithmic counterparts of associative and non-associative flow rules for a Tresca yield surface.

to apply radial return. However, it is known that significantly different response may occur when non-associative flow rules are employed (e.g., localization of deformation may occur in the presence of strain hardening, which is impossible in the associative case). Thus caution must be exercised. It would seem a worthwhile study to precisely characterize and contrast the behaviors of the two theories described in this section. The results would provide valuable guidelines for numericists interested in yield surfaces other than Mises.

2.4   Non-convex Yield Surfaces

Recently, Wilkins et al. (1980) have introduced a non-convex yield surface. This yield surface is constructed so that independently measured uniaxial and shear yield stresses are exactly matched. Between these points a simple inter-polation scheme is employed to define the regular regions of the yield surface (see Fig. 6). The admissible stress states constitute a *starlike set* with respect to the center, that is, if $\underset{\sim}{\sigma}$ is admissible then so are all points along the straight line segment joining $\underset{\sim}{\sigma}$ and the center, namely all $\underset{\sim}{\sigma}_\theta = \theta \underset{\sim}{\sigma}$, where $\theta \in [0, 1]$. Furthermore, if $\underset{\sim}{\sigma}$ is a stress outside, or on, the yield surface then the straight line segment $\underset{\sim}{\sigma}_\theta = \theta \underset{\sim}{\sigma}$, $\theta \in [0, 1]$, intersects the yield sur-face at exactly one point. The upshot is that a radial return algorithm is, ostensibly, easily implemented. Clearly, this algorithm is allied with the non-associative flow rule introduced previously (see (37)-(41)).

This theory is so unusual from the classical perspective that many funda-mental questions have been raised. The analytical consequences of employing a theory of this type do not seem to be fully appreciated at this time and thus caution is advisable. [It should be mentioned that Wilkins et al. (1980) report very good correlation with experiments.] This, however, is not our main concern here. What we wish to point out to numericists is that starlike yield surfaces

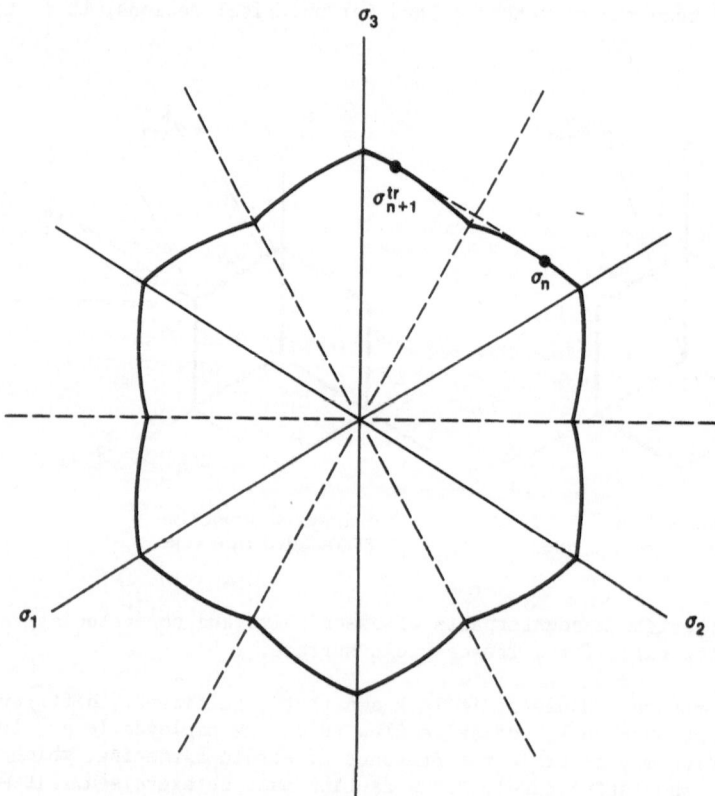

Figure 6.   The starlike yield surface of Wilkins et al. (1980).

pose some implementational problems not possessed by convex yield surfaces. Consider, for example, the stress states depicted in Fig. 6. The elastically-induced trial stress, σ_{n+1}^{tr}, is on, or just slightly inside, the yield surface and so, of course, is σ_n. However, the line segment connecting them is not. Thus, this process may not be considered elastic. (In the case of a convex yield surface, σ_{n+1}^{tr} inside, or on, the yield surface guarantees that the process was elastic.) Obviously, this is a complicating condition from the standpoint of implementation. One would like to omit this possibility, but this appears to be a violation of the theory. It would be helpful if an analysis was performed to assess the effect of elastically "jumping over" the interior vertices.

3. TOPICS IN FINITE-DEFORMATION THEORY

3.1 Theory

In this section we wish to consider a class of finite-deformation constitutive theories which include the models most commonly used in computational plasticity. For the most part we employ the notations of Marsden and Hughes (1983). The set-up is as follows: x will denote the present location of a material particle located at X at time $t = 0$. We write $x = \phi(X, t) = \phi_t(X)$, where ϕ is called the *motion* and ϕ_t is the *configuration* at time t. Spatial coordinates are denoted by x_i and material coordinates X_A. For simplicity, we assume both coordinate systems are Cartesian. Indices of tensors with "legs" in the spatial domain are denoted by lower-case letters (e.g., i, j, k, ℓ) whereas indices of tensors with "legs" in the material domain are denoted by upper-case letters (e.g., A, B, C, D). The material time derivative is denoted by a superposed dot. For example, $v = \dot{\phi}$ is the *velocity*. The *spatial gradient of velocity*, ℓ, decomposes into the *rate-of-deformation*, d, and *spin*, or *vorticity*, ω, as follows:

$$\ell_{ij} = v_{i,j} = \partial v_i / \partial x_j \tag{42}$$

$$\ell = d + \omega \tag{43}$$

$$d = d^T \tag{44}$$

$$\omega = -\omega^T . \tag{45}$$

The spin measures the angular velocity of the principal axes of the rate-of-deformation. The *deformation gradient* is defined by

$$F_{iA} = \phi_{i,A} = \partial\phi_i / \partial X_A . \tag{46}$$

The deformation gradient and velocity gradient are related by

$$\dot{F} = \ell \, F . \tag{47}$$

We recall the ("right") polar decomposition

$$F = R \, U \tag{48}$$

where R is proper-orthogonal and U is symmetric and positive-definite; R is a measure of the material rotation and U measures deformation. Let

$$w = \dot{R} \, R^{-1} = \dot{R} \, R^T = -w^T . \tag{49}$$

The preceding relations may be combined:

$$\underset{\sim}{\ell} = \underset{\sim}{d} + \underset{\sim}{\omega} = \underset{\sim}{\delta} + \underset{\sim}{w} \tag{50}$$

where

$$\underset{\sim}{\delta} = \underset{\sim}{R} \, \dot{\underset{\sim}{U}} \, \underset{\sim}{U}^{-1} \, \underset{\sim}{R}^T. \tag{51}$$

In general, $\underset{\sim}{\delta}$ is unsymmetric. In a rigid motion $\underset{\sim}{d}$ and $\underset{\sim}{\delta}$ vanish; thus, $\underset{\sim}{\omega} = \underset{\sim}{w}$ for this case. When the motion is non-rigid, $\underset{\sim}{\omega}$ and $\underset{\sim}{w}$ can be quite different (see e.g. Dienes, 1979).

In finite-deformation plasticity it is common to introduce $\underset{\sim}{d}$ as a measure of strain rate. A strain tensor, $\underset{\sim}{\varepsilon}$, may then be *defined* by:

$$\dot{\underset{\sim}{\varepsilon}}(\underset{\sim}{X}, \, t) = \underset{\sim}{d}(\underset{\sim}{X}, \, t) \tag{52}$$

where $\underset{\sim}{\varepsilon}(\underset{\sim}{X}, \, 0)$ is given. Equivalently,

$$\underset{\sim}{\varepsilon}(\underset{\sim}{X}, \, t) = \underset{\sim}{\varepsilon}(\underset{\sim}{X}, \, 0) + \int_0^t \underset{\sim}{d}(\underset{\sim}{X}, \, \tau)d\tau. \tag{53}$$

(*Warning:* Here $\underset{\sim}{\varepsilon}$ is definitely *not* the same as the small-deformation strain tensor used throughout Section 2. We retain this notation for reasons which will be made clear later on.) For the case of uniaxial extension, or compression, in which

$$\underset{\sim}{F}(\underset{\sim}{X}, \, t) = \begin{bmatrix} \lambda(t) & 0 & 0 \\ 0 & 1 & 0 \\ 0 & 0 & 1 \end{bmatrix} \tag{54}$$

where $\lambda(0) = 1$ and $\lambda(t) > 0$, $\underset{\sim}{\varepsilon}$ becomes "logarithmic strain," viz.

$$\underset{\sim}{\varepsilon}(\underset{\sim}{X}, \, t) = \begin{bmatrix} \ln\lambda(t) & 0 & 0 \\ 0 & 0 & 0 \\ 0 & 0 & 0 \end{bmatrix} \tag{55}$$

Given a skew-symmetric tensor, $\underset{\sim}{\Omega} = -\underset{\sim}{\Omega}^T$, we may generate a group of rotations $\underset{\sim}{R}$, by solving

$$\dot{\underset{\sim}{R}} = \underset{\sim}{\Omega} \, \underset{\sim}{R} \tag{56}$$

$$\underset{\sim}{R}\big|_{t=0} = \underset{\sim}{I}. \tag{57}$$

As an example, we note that $\underset{\sim}{w}$ may be thought of as the generator of $\underset{\sim}{R}$ (see (49)). Likewise, $\underset{\sim}{\omega}$ generates a rotation group, say $\underset{\sim}{P}$:

$$\dot{\underset{\sim}{P}} = \underset{\sim}{\omega} \, \underset{\sim}{P} \tag{58}$$

$$\underset{\sim}{P}\big|_{t=0} = \underset{\sim}{I}. \tag{59}$$

Objectivity

Consider a *superposed rigid-body motion*:

$$\bar{x}_i = c_i(t) + Q_{ij}(t)x_j \tag{60}$$

where Q is proper-orthogonal. (This may be equivalently thought of as a time-dependent change of spatial coordinates.) Let $A_{ij\ldots k \, AB\ldots C}$ be a general tensor of mixed type; $A_{ij\ldots k \, AB\ldots C}$ is said to be *objective* if, under the mapping $\underset{\sim}{x} \mapsto \bar{\underset{\sim}{x}}$, it transforms according to the following rule:

$$\bar{A}_{ij...k\ AB...C} = Q_{i\ell}\ Q_{jm}\ \cdots\ Q_{kn}\ A_{\ell m...n\ AB...C} \tag{61}$$

Note that material indices are unaffected. □

A class of constitutive equations for finite-deformation plasticity, frequently used in computations, may be written in the following form reminiscent of (3):

$$\sigma^* = c \cdot (d - d^{p\ell}) \tag{62}$$

where $d^{p\ell}$ is the *plastic rate-of-deformation* and σ^* is an objective rate of Cauchy (i.e. true) stress. Both d and c are objective, that is,

$$\bar{d}_{ij} = Q_{ik}\ Q_{j\ell}\ d_{k\ell} \iff \bar{d} = Q d Q^T \tag{63}$$

$$\bar{c}_{ijk\ell} = Q_{ia}\ Q_{jb}\ Q_{kc}\ Q_{\ell d}\ c_{abcd}. \tag{64}$$

(In fact, c being isotropic implies the stronger condition $\bar{c} = c$.) We also assume $d^{p\ell}$ will be defined by a flow rule which renders it objective. Objectivity requires σ^* transform as follows:

$$\bar{\sigma}^* = Q \sigma^* Q^T. \tag{65}$$

This is a minimal requirement; there are infinitely many objective rates.

An example of an objective rate is the *Jaumann rate*:

$$\sigma^* = \dot{\sigma} + \sigma \omega - \omega \sigma. \tag{66}$$

This rate is also associated with the names of Zaremba and Noll. It has also been referred to as the "co-rotational" stress rate, but this name seems inappropriate. There are a number of objections that one can raise about the use of the Jaumann rate. In particular, its use in conjunction with kinematic hardening leads to oscillatory stress response in simple shear (Nagtegaal and de Jong, 1981).

Dienes (1979) and Johnson and Bammann (1982) have advocated the use of an objective rate akin to (66) in which ω is replaced by w:

$$\sigma^* = \dot{\sigma} + \sigma w - w \sigma. \tag{67}$$

The earliest uses of this rate in constitutive theory apparently are those of Green and Naghdi (1965) and Green and McInnis (1967). Consequently, we shall refer to (67) as the *Green-Naghdi rate*. The Green-Naghdi rate has replaced the Jaumann rate in the HONDO, NIKE and DYNA codes due to its superior performance in simple shear.

The Jaumann and Green-Neghdi rates fall into a category of rates which are generated by a skew-symmetric matrix $\dot{\Omega}$:

$$\sigma^* = \dot{\sigma} + \sigma \dot{\Omega} - \dot{\Omega} \sigma. \tag{68}$$

Objectivity requires that $d = 0$ implies $\dot{\Omega} = \omega$. By virtue of the fact that $\dot{\Omega}$ generates a group of rotations [R, see (56)], (68) takes on a particularly simple form when transformed by R. Let

$$\sigma_R = R^T \sigma R, \tag{69}$$

then it is easily shown that

$$\dot{\sigma}_R = R^T \sigma^* R \tag{70}$$

where σ^* is given by (68). Thus a somewhat complicated expression [i.e. (68)]

becomes a rather simple time derivative under the appropriate change of coordinates. This suggests that the entire theory and implementation will take on canonically simple forms if transformed to the $\hat{R}$-system.* This idea is summarized by the contents of Box 3 in which a finite deformation analog of the small-deformation theory of Box 2 is presented.

Box 3. Summary of a finite-deformation elastic-plastic constitutive theory with Mises yield surface, associative flow rule, and linear combination of isotropic and kinematic hardening or softening.

Constitutive equation:

$$\dot{\sigma}_{\hat{R}} = c_{\hat{R}} \cdot (d_{\hat{R}} - d_{\hat{R}}^{p\ell}) \tag{71}$$

Evolution equation for the radius of the yield surface:

$$\dot{R} = \frac{\sqrt{2}}{\sqrt{3}} \beta H' \, \bar{d}^{p\ell} \tag{72}$$

Evolution equation for the back stress:

$$\dot{\alpha}_{\hat{R}} = \frac{2}{3}(1 - \beta)H' \, d_{\hat{R}}^{p\ell} \tag{73}$$

Flow rule:

$$d_{\hat{R}}^{p\ell} = \begin{cases} 0 & \text{if (E)} \\[2mm] \Lambda N_{\hat{R}} & \text{if (P)} \end{cases} \tag{74}$$

Consistency condition:

$$\Lambda = \frac{N_{\hat{R}} \cdot \dot{\sigma}_{\hat{R}}^{tr}}{\left(N_{\hat{R}} \cdot c_{\hat{R}} \cdot N_{\hat{R}} + \frac{2}{3} H'\right)} = \frac{1}{2\mu} \frac{1}{\left(1 + \frac{H'}{3\mu}\right)} N_{\hat{R}} \cdot \dot{\sigma}_{\hat{R}}^{tr} \tag{75}$$

$$= \frac{1}{\left(1 + \frac{H'}{3\mu}\right)} N_{\hat{R}} \cdot d_{\hat{R}} \qquad \text{(isotropy of } c_{\hat{R}}) \tag{76}$$

Trial rate-of-stress:

$$\dot{\sigma}_{\hat{R}}^{tr} = c_{\hat{R}} \cdot d_{\hat{R}} \tag{77}$$

Definition of an elastic process (E):

$$f(\xi_{\hat{R}})^{\dagger} < k^2, \quad \text{or}$$

$$f(\xi_{\hat{R}}) = k^2 \quad \text{and} \quad N_{\hat{R}} \cdot \dot{\sigma}_{\hat{R}}^{tr} \leq 0 \tag{78}$$

Definition of a plastic process (P):

$$f(\xi_{\hat{R}}) = k^2 \quad \text{and} \quad N_{\hat{R}} \cdot \dot{\sigma}_{\hat{R}}^{tr} > 0 \tag{79}$$

continued

---

*This fact seems to be well known to numericists. Hallquist has used this idea in his numerical codes and Nagtegaal and Veldpaus (1983) have employed it in their study of theories based on the Jaumann rate.

†Note $f(\xi_{\hat{R}}) = f(\xi)$.

Unit normal:

$$\underset{\sim}{N}_{R} = \underset{\sim}{\xi}'_{R}/R \quad \text{where} \quad \underset{\sim}{\xi}_{R} = \underset{\sim}{\sigma}_{R} - \underset{\sim}{\alpha}_{R} \tag{80}$$

Strains:

$$\dot{\underset{\sim}{\varepsilon}}_{R} = \underset{\sim}{d}_{R} \tag{81}$$

$$\dot{\underset{\sim}{\varepsilon}}_{R}^{p\ell} = \underset{\sim}{d}_{R}^{p\ell} \tag{82}$$

$$\dot{\bar{\varepsilon}}^{p\ell} = \bar{d}^{p\ell} \tag{83}$$

Remark 1. By virtue of the form of the kinematic-hardening evolution equation, it is clear that it, like the constitutive equation, is posed in terms of the same objective rate; that is, (73) may be written equivalently as

$$\underset{\sim}{\alpha}^{*} = \dot{\underset{\sim}{\alpha}} + \underset{\sim}{\alpha}\,\underset{\sim}{\Omega} - \underset{\sim}{\Omega}\,\underset{\sim}{\alpha} = \tfrac{2}{3}(1 - \beta)H'\,\underset{\sim}{d}^{p\ell}. \tag{84}$$

Remark 2. Note that if $\underset{\sim}{d} = \underset{\sim}{0}$, then $\dot{\underset{\sim}{\sigma}}_{R} = \underset{\sim}{0}$. That is, the stress is constant in the rotating system and, in particular, the invariants of the Cauchy stress are constant. Thus Prager's condition (Prager, 1961) is satisfied for all rates of the form (68). Likewise, $\underset{\sim}{\alpha}_{R}$ is constant as are the invariants of $\underset{\sim}{\alpha}$.

Remark 3. Nagtegaal (1983) has proposed a family of rates of the form (68) which generalize the results of Lee, Mallett and Wertheimer (1983). A member of this family,

$$\underset{\sim}{\Omega} = \underset{\sim}{\omega} + \begin{cases} \underset{\sim}{0} & \text{if } T = 0 \\ T^{-1}(\underset{\sim}{d}\,\underset{\sim}{\alpha} - \underset{\sim}{\alpha}\,\underset{\sim}{d}) & \text{if } T \neq 0 \end{cases} \tag{85}$$

where

$$T = \max_{i,j} |\alpha_i - \alpha_j| \qquad (\text{"Tresca intensity of } \underset{\sim}{\alpha}\text{"}) \tag{86}$$

is successful in handling the simple-shear kinematic-hardening problem in the sense that the stress response increases monotonically with strain whereas the slopes of the stress-strain curves decrease monotonically. This is in accord with the response of most ductile metals. However, (85) does not seem suitable for general applications as the following scenario illustrates: Consider the case of pure kinematic hardening (i.e. H' > 0 and β = 0). Assume initially that $\underset{\sim}{\sigma}$ and $\underset{\sim}{\alpha}$ are $\underset{\sim}{0}$. Assume further that the stress is brought to the yield surface and slight yielding occurs so that $\underset{\sim}{\alpha}$ is moved to a nonzero location. Immediately elastically unload. During subsequent elastic processes the response is altered by the $O(\|\underset{\sim}{\alpha}\|)$ perturbation to $\underset{\sim}{\omega}$ in (85). That is, the effect of an $O(\|\underset{\sim}{\alpha}\|) \ll 1$ perturbation is $O(1)$. This discontinuous behavior needs to be corrected if rates of this type are to be employed in large-scale inelastic codes.

Remark 4. Objective rates are manifestations of the Lie derivative [see Marsden and Hughes (1983), Section 1.6, and references therein, and Simo and Marsden (1983)] and thus they are ordinary (material) time derivatives of some trans-formed quantity. This is the fundamental reason behind (70). However, the class of objective rates is larger than that described by (68). In general, the transformation will involve deformation as well as rotation, consequently, Prager's condition may be violated.

Remark 5. When restricted to an elastic process, the constitutive equation (71) does not correspond to a *hyperelastic material*. Thus it is objectionable on fundamental grounds. Nevertheless (71) is typical of equations currently used in practice.

Remark 6. As this paper was being completed, Yannis Dafalias informed me of a new proposal of his for $\dot{\Omega}$ [see e.g. Dafalias (1983)]. Unfortunately, sufficient time was not available prior to the deadline for manuscripts to present an evaluation of this new rate. It is anticipated that further attempts will be made to develop improvements to the predictive capability of this class of theories by introducing different definitions of $\dot{\Omega}$.

3.2 Implementation

The beauty of the theory in Box 3 is that it is form-identical to the small-deformation theory presented in Box 2. Thus implementation may proceed in identical fashion to before [see Table 2] if all quantities are appropriately transformed. The accuracy of the overall procedure is thus affected by the accuracy of the computation of $\underset{\sim}{R}$. In addition

$$\Delta\underset{\sim}{\varepsilon}(X) = \int_{t_n}^{t_{n+1}} \underset{\sim}{d}(X, \tau)d\tau \tag{87}$$

will require approximate evaluation and this will also affect the overall accuracy. In order to formally achieve second-order accuracy, we shall provide the radial-return algorithm [i.e. Table 2] with the strain increment rotated into the midpoint orientation:

$$(\Delta\underset{\sim}{\varepsilon}_{\underset{\sim}{R}})_{n+\frac{1}{2}} = \underset{\sim}{R}^T_{n+\frac{1}{2}} \Delta\underset{\sim}{\varepsilon}\underset{\sim}{R}_{n+\frac{1}{2}} . \tag{88}$$

That is, $(\Delta\underset{\sim}{\varepsilon}_{\underset{\sim}{R}})_{n+\frac{1}{2}}$ is to be used in place of $\Delta\underset{\sim}{\varepsilon}$ in Step 1. Likewise, $\underset{\sim}{\sigma}_n$, $\underset{\sim}{\alpha}_n$, $\underset{\sim}{\sigma}_{n+1}$ and $\underset{\sim}{\alpha}_{n+1}$ in Table 2 are to be viewed as the following transformed quantities (resp.):

$$(\underset{\sim}{\sigma}_{\underset{\sim}{R}})_n = \underset{\sim}{R}^T_n \underset{\sim}{\sigma}_n \underset{\sim}{R}_n \tag{89}$$

$$(\underset{\sim}{\alpha}_{\underset{\sim}{R}})_n = \underset{\sim}{R}^T_n \underset{\sim}{\alpha}_n \underset{\sim}{R}_n \tag{90}$$

$$(\underset{\sim}{\sigma}_{\underset{\sim}{R}})_{n+1} = \underset{\sim}{R}^T_{n+1} \underset{\sim}{\sigma}_{n+1} \underset{\sim}{R}_{n+1} \tag{91}$$

$$(\underset{\sim}{\alpha}_{\underset{\sim}{R}})_{n+1} = \underset{\sim}{R}^T_{n+1} \underset{\sim}{\alpha}_{n+1} \underset{\sim}{R}_{n+1} . \tag{92}$$

In summary we shall require computation of $\Delta\underset{\sim}{\varepsilon}$, $\underset{\sim}{R}_n$, $\underset{\sim}{R}_{n+\frac{1}{2}}$ and $\underset{\sim}{R}_{n+1}$ in the finite-deformation procedure. The remaining calculations are identical to those of the small-deformation case, Table 2.

The kinematical quantities must be computed from the configurations at t_n and t_{n+1} as these are the only available data. Let $\underset{\sim}{x}_{n+\alpha}$ denote the one-parameter family of configurations which linearly interpolate $\underset{\sim}{x}_n = \underset{\sim}{\phi}(X, t_n)$ and $\underset{\sim}{x}_{n+1} = \underset{\sim}{\phi}(X, t_{n+1})$:

$$\underset{\sim}{x}_{n+\alpha} = (1 - \alpha)\underset{\sim}{x}_n + \alpha\underset{\sim}{x}_{n+1}, \quad \alpha \in [0, 1]. \tag{93}$$

The *displacement increment* over the step is written

$$\underset{\sim}{u} = \underset{\sim}{x}_{n+1} - \underset{\sim}{x}_n . \tag{94}$$

The gradient of $\underset{\sim}{u}$ may be used to calculate desired kinematical quantities:

$$\underset{\sim}{G}_\alpha = \partial \underset{\sim}{u}/\partial \underset{\sim}{x}_{n+\alpha} . \tag{95}$$

Strain increment

The approximate formula for the strain increment is taken to be

$$\Delta \underset{\sim}{\varepsilon} = \tfrac{1}{2}(\underset{\sim}{G}_\alpha + \underset{\sim}{G}_\alpha^T). \tag{96}$$

The quality of the approximation is clearly affected by the choice of α. It turns out that the midpoint configuration (i.e. $\alpha = \tfrac{1}{2}$) is the most accurate choice. In this case (96) is a second-order approximation to the exact formula (87). To illustrate this point, consider the following:

Example

Take the case of uniaxial extension, or compression, in which the deformation gradient is given by (54). The exact strain increment, (87) is defined by

$$\Delta \varepsilon_{11}^{exact} = \ln(1 + \gamma) \tag{97}$$

where

$$\gamma = \frac{\lambda(t_{n+1})}{\lambda(t_n)} - 1. \tag{98}$$

The approximate version, (96), is

$$\Delta \varepsilon_{11}^{approx.} = \gamma/(1 + \alpha \, \gamma). \tag{99}$$

Equations (97) and (99) can be compared by expanding in γ:

$$\Delta \varepsilon_{11}^{exact} = \gamma - \frac{\gamma^2}{2} + \frac{\gamma^3}{3} - O(\gamma^4) \tag{100}$$

$$\Delta \varepsilon_{11}^{approx.} = \gamma - \alpha \, \gamma^2 + \alpha^2 \, \gamma^3 - O(\gamma^4). \tag{101}$$

Note that (101) is a first-order approximation to (100) for all α. If $\alpha = 1/2$, the approximation is second-order. This is representative of the general case. The high accuracy of the approximation is illustrated in Fig. 7.   □

Given the importance of shearing deformations in plasticity, one might be curious as to the accuracy of (96) in the case of simple shear for which

$$\underset{\sim}{F}(\underset{\sim}{X}, t) = \begin{bmatrix} 1 & \kappa(t) & 0 \\ 0 & 1 & 0 \\ 0 & 0 & 1 \end{bmatrix} \tag{102}$$

It is somewhat surprising that in this case (96) is exact for all α!

Further insight into the nature of the approximate incremental strain tensor may be gained from the following analysis. Define the *incremental finite strain tensor*, $\Delta \underset{\sim}{e}_\alpha$, by way of:

T. J. R. Hughes

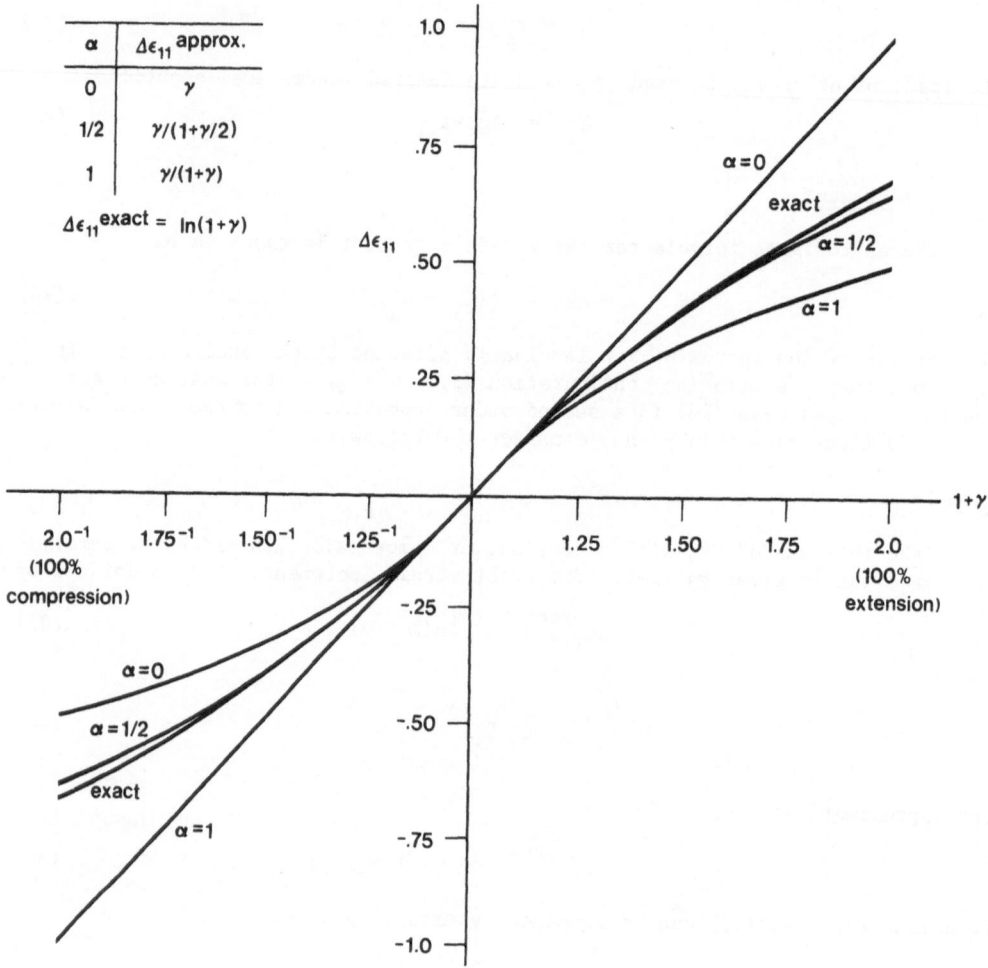

Figure 7. Accuracy of the midpoint strain increment ($\alpha = \frac{1}{2}$) in the uniaxial case.

$$2dx_{n+\alpha}^{T} \; \Delta e_{\alpha} dx_{n+\alpha} \;\; = \;\; dx_{n+1}^{T} \; dx_{n+1} - dx_{n}^{T} \; dx_{n}. \qquad (103)$$

This leads to the explicit formula

$$\Delta e_{\alpha} \;\; = \;\; \frac{1}{2}(G_{\alpha} + G_{\alpha}^{T} + (1 - 2\alpha)G_{\alpha}^{T} \; G_{\alpha}). \qquad (104)$$

The most important cases are summarized below:

α	$2\Delta e_{\alpha}$	
0	$G_0 + G_0^{T} + G_0^{T}G_0$	(incremental Lagrangian strain)
1	$G_1 + G_1^{T} + G_1^{T}G_1$	(incremental Eulerian strain)
$\frac{1}{2}$	$G_{\frac{1}{2}} + G_{\frac{1}{2}}^{T}$	(midpoint strain)

Note that each strain tensor possesses precisely the same information. The choice $\alpha = \frac{1}{2}$ leads to the simplest representation of this information. This was the tensor advocated previously. Keep in mind, however, that the only relevant criterion for the present purposes is how well the midpoint strain approximates the integral in (96), which was discussed previously.

Rotations

There are two cases corresponding to whether or not a discrete approximation of $\dot{\Omega}$ is needed to calculate $\dot{R}$. In the case of the Green-Naghdi rate $\dot{R} = \dot{R}$ and thus the polar decomposition can be directly used to calculate $\dot{R}$. Otherwise, $\dot{R} = \dot{\Omega} R$ needs to be integrated. We consider the two cases separately:

(1) $\dot{R} = \dot{R}$.

The deformation gradient at time t_n is $F_n(X) = F(X, t_n)$, and likewise at t_{n+1}. The polar decomposition may then be written as

$$F_n = R_n U_n. \tag{105}$$

In most texts on continuum mechanics, cumbersome calculations are described for explicitly carrying out the polar decomposition. These would be too inefficient for large-scale computations. In Marsden and Hughes (1983) the Cayley-Hamilton theorem is used to derive an explicit formula for U in terms of $C = F^T F$ for *two dimensions*:

$$U_n = \frac{1}{\sqrt{\text{trace } C_n + 2\sqrt{\det C_n}}} (C_n + \sqrt{\det C_n}\, I) \tag{106}$$

where $\det C_n$ is the determinant of C_n. An analogous formula for three dimensions has been developed by Stephenson (1983a). With the aid of (106), R_n may be explicitly calculated:

$$R_n = F_n U_n^{-1}. \tag{107}$$

There are several possibilities for computing $R_{n+\frac{1}{2}}$. The one we shall describe here is based on the polar decomposition. One needs to calculate the deformation gradient of the midpoint configuration:

$$F_{iA}(X, t_{n+\frac{1}{2}}) = \frac{\partial \phi_i}{\partial X_A}(X, t_{n+\frac{1}{2}}) \tag{108}$$

where

$$\phi(X, t_{n+\frac{1}{2}}) = \frac{1}{2}\big(\phi(X, t_n) + \phi(X, t_{n+1})\big). \tag{109}$$

In terms of $F_{n+\frac{1}{2}} = F(X, t_{n+\frac{1}{2}})$, the calculations are identical to (106) and (107). This procedure has been implemented in the NIKE codes.

(2) $\dot{R} \neq \dot{R}$.

In this case we must integrate $\dot{R} = \dot{\Omega} R$ over the time interval $[t_n, t_{n+1}]$. We assume R_n is known and our first objective is to obtain an expression for R_{n+1}. Furthermore, we assume that an incremental approximation to the integral of $\dot{\Omega}$ over $[t_n, t_{n+1}]$ is available. For the Jaumann and Nagtegaal rates, the incremental approximations are given as follows (resp.):

$$\Delta \underset{\sim}{\Omega} = \Delta \underset{\sim}{\psi} \overset{def.}{=} \frac{1}{2}(\underset{\sim}{G}_\alpha - \underset{\sim}{G}_\alpha^T) \tag{110}$$

$$\Delta \underset{\sim}{\Omega} = \Delta \underset{\sim}{\psi} + \begin{cases} \underset{\sim}{0} & \text{if } T_n = 0 \\ T_n^{-1}(\Delta \underset{\sim}{\varepsilon}\, \underset{\sim}{\alpha}_n - \underset{\sim}{\alpha}_n\, \Delta \underset{\sim}{\varepsilon}) & \text{if } T_n \neq 0 \end{cases} \tag{111}$$

As in the case of the incremental strain calculation, use of the midpoint, or average, configuration (i.e. $\alpha = \frac{1}{2}$) is crucial to the accuracy of the overall procedure. Hughes and Winget (1980) have proposed the following formula for evaluating $\underset{\sim}{R}_{n+1}$:

$$\underset{\sim}{R}_{n+1} = \underset{\sim}{q}\, \underset{\sim}{R}_n \tag{112}$$

where

$$\boxed{\underset{\sim}{q} = (\underset{\sim}{I} + (\underset{\sim}{I} - \alpha\, \Delta\underset{\sim}{\Omega})^{-1}\, \Delta\underset{\sim}{\Omega}).} \tag{113}$$

This formula is designed to satisfy the criterion of *incremental objectivity* set forth by Hughes and Winget (1980) as a discrete analog of the objectivity of the governing constitutive theory. Incremental objectivity requires that if the motion over $[t_n, t_{n+1}]$ is rigid, that is, if

$$\underset{\sim}{x}_{n+1} = \underset{\sim}{c} + \underset{\sim}{Q}\, \underset{\sim}{x}_n, \tag{114}$$

where $\underset{\sim}{c}$ and $\underset{\sim}{Q}$ are constant, and $\underset{\sim}{Q}$ is proper-orthogonal, then

$$\underset{\sim}{\sigma}_{n+1} = \underset{\sim}{Q}\, \underset{\sim}{\sigma}_n\, \underset{\sim}{Q}^T, \tag{115}$$

that is, σ_{n+1} is exactly updated. Incremental objectivity precludes the generation of spurious stresses in rigid motions. It may seem like an obvious criterion, but a number of algorithms which have been developed for large-deformation plasticity do not satisfy it [see e.g. McMeeking and Rice (1975)].

Following Hughes and Winget (1980), one can easily prove the following theorem for the procedures described in this paper.

THEOREM. *Assume that:*

   (a) *the motion over $[t_n, t_{n+1}]$ is rigid,*

   (b) $\alpha = \frac{1}{2}$; *and*

   (c) $\underset{\sim}{Q} + \underset{\sim}{I}$ *is nonsingular;*

then   (i) $\Delta\underset{\sim}{\varepsilon} = \underset{\sim}{0}$;

    (ii) $\underset{\sim}{q} = \underset{\sim}{Q}$; *and*

   (iii) $\underset{\sim}{\sigma}_{n+1} = \underset{\sim}{Q}\, \underset{\sim}{\sigma}_n\, \underset{\sim}{Q}^T$, *(i.e. the algorithms are incrementally objective).*

Remark 1. Note that (iii) is equivalent to $(\underset{\sim}{\sigma}_R)_{n+1} = (\underset{\sim}{\sigma}_R)_n$, that is, $\underset{\sim}{R}_{n+1}^T\, \underset{\sim}{\sigma}_{n+1}\, \underset{\sim}{R}_{n+1} = \underset{\sim}{R}_n^T\, \underset{\sim}{\sigma}_n\, \underset{\sim}{R}_n$.

Remark 2. The algorithms presented herein go beyond Hughes and Winget (1980). There, only the concept of incremental objectivity was emphasized.

Remark 3. The q-formula can be expressed more simply without the need for inverting a matrix. Let $\alpha\Delta\underset{\sim}{\Omega}$ be represented as follows:

$$\alpha \Delta \underset{\sim}{\Omega} = \underset{\sim}{\theta} = \begin{cases} \begin{bmatrix} 0 & \theta \\ -\theta & 0 \end{bmatrix} & \text{(two dimensions)} \\ \\ \begin{bmatrix} 0 & \theta_3 & -\theta_2 \\ -\theta_3 & 0 & \theta_1 \\ \theta_2 & -\theta_1 & 0 \end{bmatrix} & \text{(three dimensions)} \end{cases} \tag{116}$$

In the three-dimensional case, define

$$\theta^2 = \theta_1^2 + \theta_2^2 + \theta_3^2. \tag{117}$$

Then, the Cayley-Hamilton theorem can be used to derive the following results:

$$
\begin{aligned}
\underset{\sim}{q} &= \underset{\sim}{I} + \frac{1}{\alpha(1 + \theta^2)} (\underset{\sim}{\theta} + \underset{\sim}{\theta}^2) && \text{(three dimensions)} \\
\underset{\sim}{q} &= \underset{\sim}{I} + \frac{1}{\alpha(1 + \theta)} (\underset{\sim}{\theta} - \theta^2 \underset{\sim}{I}) && \text{(two dimensions)}
\end{aligned}
\tag{118}
$$

These equations are useful in analytical dynamics [see e.g. Hamel (1949) and Rosenberg (1980)].

Remark 4.   Hypothesis (c) is a "black-hole" condition which precludes the collapse of the midpoint configuration when Q corresponds to a rotation of 180°. See Fig. 8 and Hughes and Winget (1980) for further discussion.   □

Now we turn our attention to the calculation of $\underset{\sim}{R}_{n+\frac{1}{2}}$ given $\underset{\sim}{q}$. Define a proper-orthogonal matrix $\underset{\sim}{p}$ such that

$$\underset{\sim}{R}_{n+1} = \underset{\sim}{p} \, \underset{\sim}{R}_{n+\frac{1}{2}} \tag{119}$$

and

$$\underset{\sim}{R}_{n+\frac{1}{2}} = \underset{\sim}{p} \, \underset{\sim}{R}_n. \tag{120}$$

Thus

$$\underset{\sim}{R}_{n+1} = \underset{\sim}{p}^2 \, \underset{\sim}{R}_n, \tag{121}$$

that is, $\underset{\sim}{p}^2 = \underset{\sim}{q}$. Consequently, we wish to calculate the proper-orthogonal square-root of a proper-orthogonal matrix. The Cayley-Hamilton theorem may be used for this purpose. In *two-dimensions* we have

$$\underset{\sim}{p} = \frac{1}{\sqrt{2 + \text{trace } \underset{\sim}{q}}} (\underset{\sim}{q} + \underset{\sim}{I}) \tag{122}$$

which is actually a special case of (106). The three-dimensional case is described by Stephenson (1983b).

Another possibility in two dimensions is to use a "half-angle formula." Represent $\underset{\sim}{q}$ by

$$\underset{\sim}{q} = \begin{bmatrix} \cos \theta & \sin \theta \\ -\sin \theta & \cos \theta \end{bmatrix}. \tag{123}$$

Calculate $\theta/2$ by

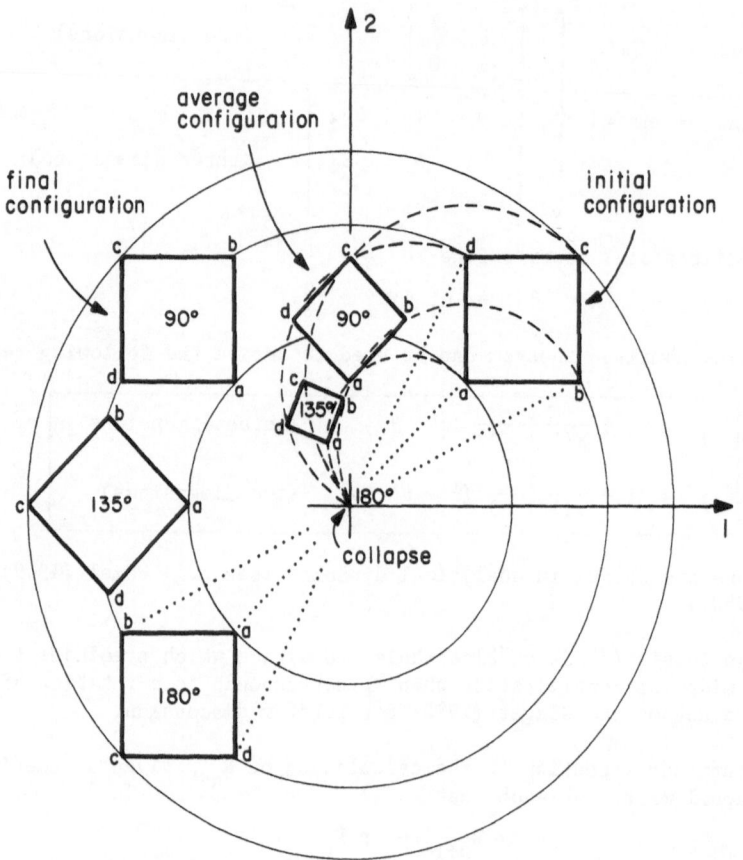

Figure 8. Behavior of the midpoint, or average, configuration
($\alpha = \frac{1}{2}$) as a function of rotation.

$$\frac{\theta}{2} = \text{arc cos} \sqrt{\frac{1 + \cos \theta}{2}} \,. \tag{124}$$

Then

$$\underset{\sim}{p} = \begin{bmatrix} \cos \frac{\theta}{2} & \sin \frac{\theta}{2} \\ -\sin \frac{\theta}{2} & \cos \frac{\theta}{2} \end{bmatrix} \,. \tag{125}$$

This procedure has been used by Key in the HONDO code [Key, Stone and Krieg
(1981)].

REFERENCES

Casey, J., and P.M. Naghdi (1981), "On the Characterization of Strain Hardening
in Plasticity," *J. Appl. Mech.*, *48*, 285-296.

Caulk, D.A., and P.M. Naghdi (1978), "On the Hardening Response in Small Deformation of Metals," *J. Appl. Mech.*, *45*, 755-764.

Dafalias, Y.F. (1983), "Corotational Rates for Kinematic Hardening at Large Plastic Deformations," *J. Appl. Mech.*, *50*, 561-565.

Dienes, J.K. (1979), "On the Analysis of Rotation and Stress Rate in Deforming Bodies," *Acta Mech.*, *32*, 217-232.

Green, A.E., and B.C. McInnis (1967), "Generalized Hypo-Elasticity," *Proc. Roy. Soc. Edinburgh*, *A57*, 220-230.

Green, A.E., and P.M. Naghdi (1965), "A General Theory of an Elastic-Plastic Continuum," *Archive Rational Mech. & Anal.*, *18*, 251-281.

Hallquist, J.O. (1979), "NIKE2D: An Implicit, Finite-deformation, Finite Element Code for Analyzing the Static and Dynamic Response of Two-dimensional Solids," University of California, Lawrence Livermore National Laboratory, Rept. UCRL-52678.

Hallquist, J.O. (1981a), "User's Manual for DYNA3D and DYNAP — Nonlinear Dynamic Analysis of Solids in Three Dimensions," University of California, Lawrence Livermore National Laboratory, Rept. UCID-19156.

Hallquist, J.O. (1981b), "NIKE3D: An Implicit, Finite-deformation, Finite Element Code for Analyzing the Static and Dynamic Response of Three-dimensional Solids," University of California, Lawrence Livermore National Laboratory, Rept. UCID-18822.

Hallquist, J.O. (1982), "User's Manual for DYNA2D — An Explicit Two-dimensional Hydrodynamic Finite Element Code with Interactive Rezoning," University of California, Lawrence Livermore National Laboratory, Rept. UCID-18756, Rev. 1.

Hamel, G. (1949), *Theoretische Mechanik*, Springer, Berlin.

Hill, R. (1950), *Plasticity*, Clarendon Press, Oxford.

Hill, R. (1967), "On the Classical Constitutive Relations for Elastic/Plastic Solids," *Recent Progress in Applied Mechanics* (The Folke-Odqvist Volume) Wiley, New York.

Hughes, T.J.R., and W.K. Liu (1981), "Nonlinear Finite Element Analysis of Shells: Part I. Three-dimensional Shells," *Comp. Meths. Appl. Mech. Eng.*, *26*, 331-362.

Hughes, T.J.R., and J. Winget (1980), "Finite Rotation Effects in Numerical Integration of Rate Constitutive Equations Arising in Large-deformation Analysis," *Int. J. Numer. Meths. Eng.*, *15*, No. 12, 1862-1867.

Iwan, W.D., and J. Yoder (1983), "Computational Aspects of Strain-Space Plasticity," *J. Eng. Mech.*, *109*, No. 1, 231-243.

Jaumann, G. (1911), "Geschlossenes System physikalischer und chemischer Differentialgesetze," *Sitz. der Akad. Wiss. Wien*, (IIa) *120*, 385-550.

Johnson, G.C., and D.J. Bammann (1982), "A Discussion of Stress Rates in Finite Deformation Problems," SAND-82-8821, Sandia National Laboratories.

Key, S.W., Z.E. Beisinger and R.D. Krieg (1978), "HONDO II — A Finite Element

Computer Program for the Large Deformation Dynamic Response of Axisymmetric Solids," Sandia National Laboratories, Albuquerque, NM, Report No. 78-0422.

Key, S.W., C.M. Stone and R.D. Krieg (1981), "Dynamic Relaxation Applied to the Quasi-static, Large Deformation, Inelastic Response of Axisymmetric Solids," pp. 585-620 in *Nonlinear Finite Element Analysis in Structural Mechanics*, W. Wunderlich et al. (eds.), Springer-Verlag, Berlin.

Krieg, R.D., and S.W. Key (1976), "Implementation of a Time-Independent Plasticity Theory into Structural Computer Programs," pp. 125-137 in *Constitutive Equations in Viscoplasticity: Computational and Engineering Aspects*, J.A. Stricklin and K.J. Saczalski (eds.), AMD-*20*, ASME, New York.

Krieg, R.D., and D.B. Krieg (1977), "Accuracies of Numerical Solution Methods for the Elastic-Perfectly Plastic Model," *J. of Pressure Vessel Tech.*, ASME, *99*, 510-515.

Lee, E.H., R.L. Mallett and T.B. Wertheimer (1983), "Stress Analysis for Kinematic Hardening in Finite Deformation Plasticity," *J. Appl. Mech.*, *50*, 554-560.

Marsden, J.E., and T.J.R. Hughes (1983), *Mathematical Foundations of Elasticity*, Prentice-Hall, Englewood Cliffs, New Jersey.

McMeeking, R.M., and J.R. Rice (1975), "Finite-element Formulations for Problems of Large Elastic-plastic Deformation," *Int. J. Solids & Struct.*, *11*, 601-616.

Moss, W.C. (1983), "On the Computational Significance of the Strain Space Formulation of Plasticity Theory," UCRL-8892, Lawrence Livermore National Laboratory.

Naghdi, P.M. (1960), "Stress-Strain Relations in Plasticity and Thermoplasticity", pp. 121-169 in *Proceedings of the 2nd Symposium on Naval Structural Mechanics*, Pergamon Press.

Naghdi, P.M., and J.A. Trapp (1975), "The Significance of Formulating Plasticity Theory with Reference to Loading Surfaces in Strain Space," *Int. J. Eng. Sci.*, *13*, 785-787.

Nagtegaal, J.C. (1983), "A Note on the Construction of Spin Tensors," preprint.

Nagtegaal, J.C., and J.E. de Jong (1981), "Some Aspects of Nonisotropic Work-Hardening in Finite Deformation Plasticity," *Proc. of the Workshop on Plasticity of Metals at Finite Strain: Theory, Computation and Experiment*, E.H. Lee and R.L. Mallett (eds.), Division of Applied Mechanics, Stanford University, Stanford, California.

Nagtegaal, J.C., and F.E. Veldpaus (1983), "On the Implementation of Finite Strain Plasticity Equations in a Numerical Model," preprint.

Pinsky, P.M., M. Ortiz and K.S. Pister (1983), "Rate Constitutive Equations in Finite Deformation Analysis: Theoretical Aspects and Numerical Integration," *Comp. Meths. Appl. Mech. & Eng.*, to appear.

Prager, W. (1961), "An Elementary Discussion of Definitions of Stress Rate," *Q. Appl. Mech.*, *18*, 403-407.

Rosenberg, R.M. (1980), *Analytical Dynamics of Discrete Systems*, Plenum, New York, 2nd ed.

Sandler, I., and J. Wright (1983), private communication.

Simo, J.C., and J.E. Marsden (1983), "On the Rotated Stress Tensor and the Material Version of the Doyle-Ericksen Formula," PAM-163/DEF, Center for Pure and Applied Mathematics, University of California, Berkeley, California (to appear in the *Archive for Rational Mechanics and Analysis*).

Stephenson, R. (1983a), "Polar Decomposition," preprint.

Stephenson, R. (1983b), private communication.

Wilkins, M.L. (1964), "Calculation of Elastic-plastic Flow," *Methods of Computational Physics*, *3*, B. Alder et al. (eds.), Academic Press, New York.

Wilkins, M.L., R.D. Streit and J.E. Reaugh (1980), "Cumulative-Strain Damage Model of Ductile Fracture: Simulation and Prediction of Engineering Fracture Tests," University of California, Lawrence Livermore Laboratory, Rept. UCLR-53058.

Yoder, P.J., and W.D. Iwan (1981), "On the Formulation of Stress-Space Plasticity with Multiple Loading Surfaces," *J. Appl. Mech.*, *48*, 773-778.

Yoder, P.J., and R.G. Whirley (1983), "On the Numerical Implementation of Elasto-plastic Models," preprint.

Young, L.C. (1969), *Lectures on the Calculus of Variations and Optimal Control Theory*, W.B. Saunders Co., Philadelphia-London-Toronto.

GENERAL DISCUSSION

PHILLIPS: I would like to ask a question about the star-like non-convex yield surface. Did anybody ever find experimentally a star-like non-convex yield surface?

HUGHES: Well, I probably should pass the floor to Mark Wilkins. But I heard him speak about this awhile back, and that's exactly what he found, and that's why he generated that yield surface; so perhaps he could amplify.

WILKINS: We look to experiments to generate constitutive relations, and then by judgment we develop constitutive models that explain the experiment. The experimental results suggested the shape of the yield surface. There really isn't that much of a departure from usual yield surfaces. There is just a slight deviation from the von Mises and the Tresca yield surfaces. The experiment gave the results; it wasn't a theory.

BUDIANSKY: Could you take a minute or two to comment on how you would assess the utility of calculational procedures based on convected coordinates and the details of covariant and contravariant stress and strain components? They seem to have dropped out of popularity, and I wonder if you can give an assessment.

HUGHES: Well, there are certain fundamental problems that permeate this subject and different coordinate representations are just different ways of manifesting these objects. Whatever ways you choose, there are ways of doing it right. You can do it according to your choice. It's just like representing a tensor in different coordinate systems. It doesn't affect accuracy or anything else, so it seems not to be an issue anymore.

BUDIANSKY: Was there or was there not an implication that in all of the exposition that you gave, one would be working, at any instant of time, with ordinary rectangular coordinates at each point, for the representation of, say, the stresses? And in fact, is that the way it goes into the code?

HUGHES: Well, yes, what I was doing here is essentially not thinking about situations where it might be advantageous to use preferred systems, other than the one that I specifically emphasized for "rotational" theories. There is a clear preferred system in this case, namely, a time dependent rotating system.

BUDIANSKY: Am I wrong? It does seem to me that if you work with imbedded convected coordinates, issues of rotation don't seem to enter explicitly.

HUGHES: The type of rotation effects that I introduced here, showed not to be thought of as purely geometrical artifacts of finite deformations. These are constitutive hypotheses. They affect material response. That's different than just some choice of coordinates for a given theory. I'm speaking about theories. Different spins give you different theories.

SANDLER: I just wanted to comment on that question. Everything done here is local in the sense that it's done on an element level so that you can always operate within a Cartesian system of tensors which applies locally for the element. You need not get into the complication of arbitrary non-Cartesian coordinate systems which might represent the global situation.

HUGHES: You never have to do that. Even the thrust of modern differential geometry acknowledges this point. Then one constructs local Cartesian frames and studies their orientation as they are moved about the geometrical object. You adopt a Cartesian point of view. So I really think this is the "frill" in the subject and not the essential.

NEMAT-NASSER: You didn't make reference to them, but some of the slides that I put on the screen in the last part of my talk, were to emphasize that all objective stress rates are Jaumann rate plus other terms which are linear in stress and deformation rate. So in that sense you're changing your constitutive relation when you use different stress rate with the same functional form. For example, the rate that you attributed to Green-Naghdi is precisely the one that I had put on the board.

HUGHES: Sure.

NEMAT-NASSER: And of course, using the Eulerian and Lagrangian triads, you can define that stress rate explicitly in terms of the Jaumann rate and the components of the Cauchy stress and the off-diagonal components of the deformation rate in the Eulerian triad (see Eq. (2.11) of Chapter I).

HUGHES: That's true. That part is true. But I differ with one thing you said. There are objective rates that are outside of that realm that you described. You could have, instead of, say, rate of deformation, a plastic rate of deformation and still construct an objective rate. We used to say there are infinitely many objective rates. I think there are at least two or three times that many now. *(Laughter)*

TING: One of the problems in writing a correct incremental constitutive law is the choice of rotation frame, as is mentioned, which insures the objectivity. There are so many you can choose. Would you care to comment which one is the best one in terms of physical consideration?

HUGHES: There will be a couple more talks I think which will address this topic. Let me just give you a few historical remarks from the computational viewpoint. In all the codes, the Jaumann rate was used for a long time. When the results for simple shear in hypoelasticity and kinematic hardening became known, a number of people switched to the Green-Naghdi-Dienes rate.

There are a number of people who do not believe that this rate is correct and others have been proposed; for example, the rate given by Lee and his colleagues. Their proposition, though, is very much tuned to a particular problem, namely simple shear.

Nagtegaal has attempted to generalize the Lee rate, but I do not think the extension is viable. However, it is still being worked on. Recently Yannis Defalias has proposed a new version which involves a plastic deformation. There seems to be an effort to link this with micromechanical ideas, so I assume that a few more of this type will be generated by way of micromechanical inspiration. It's a very important class, though. It's the class that has been used computationally.

I would have liked to talk about rates outside of this class, but I didn't even allow myself to mention the Kirchhoff stress. I didn't want to perturb the main class that has been used computationally.

PISTER: I know there are a few more other questions, but can I ask, is there no one here that's going to rise to the defense of the classical formulation of plasticity?

HUGHES: Yes, what about my elastic and plastic processes? I'm claiming that the classical loading conditions are tautological.

CLIFTON: I'll address that topic in the form of a question. Is it clear that with this definition of elastic processes and plastic processes you can actually march a solution forward in time, unambiguously? I mean, isn't it true that your trial stress would say that you should use, say, the plastic algorithm and then you would complete the calculations of that step and learn that you should have used the elastic one?

HUGHES: No. The evolutionary structure is exactly right with these recipes. If you sit down and play with these ideas for a little while, you'll see that this is really the correct way to render the initial value problem at a stress point well-posed. There is absolutely no doubt about it. I see what your concern is, but I don't see exactly what the difficulty is.

NEEDLEMAN: Actually, I had two comments or questions. One concerns the class of constitutive laws at finite strains. If one were to consider isotropic hardening, then isn't it a question of how one is representing the elasticity? Because if one were to look at the rigid plastic isotropic hardening case, the question of stress rate never enters at all. There is just no issue of stress rate. You're only differentiating scalars in the rigid plastic isotropic hardening case. So it's purely a question of how you represent the elasticity.

HUGHES: Yes. I mean, the stress rate is not significant for pure isotropic hardening. I agree with that. I'm taking the general case here, kinematic hardening included.

NEEDLEMAN: But then it's a question of how you represent the rotation of the center of the yield surface more than a question of how you represent the stress rate.

HUGHES: I think these things are intimately linked.

PISTER: That's right.

NEEDLEMAN: No.

HUGHES: If you don't have the stress and center rotating together, I think you're going to concoct some extremely strange theories.

NEEDLEMAN: You can do that even if you have the picture going together. My second comment concerns what Prof. Budiansky said earlier about convected coordinates. You could look at a class of convected coordinate formulations, and there one is dealing with the deformation gradient rather than the rotation.

HUGHES: Are you thinking of a polar-decomposition spin or the deformation gradient itself?

NEEDLEMAN: The deformation gradient itself. Now, that has the advantage, I think, of avoiding the polar decomposition theorem entirely. All those calculations drop out. The disadvantage is that you deal with the non-Cartesian coordinate systems, so it's a trade-off.

HUGHES: That's not such a big deal.

NEEDLEMAN: No, I think it's not a big deal.

HUGHES: There are certain funny things that accompany those types of theories. I did exclude that class of theories where deformation plays a role in the structure. But you get outside the realm of Prager's condition.

NEEDLEMAN: That's not true. You can enforce that condition. So it's a question, I think, that's a computational one, of whether you want to go through the polar decomposition theorem.

HUGHES: If that is what you meant, then you're talking about a different constitutive equation.

BUDIANSKY: No, no, no, no, no. Same constitutive equation.

HUGHES: Well, then I come back to what I originally said. Coordinates play no role. Coordinates are just different ways of looking at the same object.

BUDIANSKY: We're just talking about the relative merits of the different numerical implementations which would flow. Alan put his finger on one — work with convected coordinates and you never need calculate a rotation matrix.

HUGHES: In the type of theory I described, the relevant system of convected coordinates is the "R-system." That's exactly what we're working with.

BUDIANSKY: That's right, but you come at it from a completely different view-point.

HUGHES: Well, I'm taking the viewpoint that there is a class of models that have been proposed in terms of rotational-like convected rates. Here we have a box full of them. Now we want to develop a formulation which turns out to be a convected formulation in terms of the right system for that class of models. And that's the system that I wrote down. And, believe me, the implementation is really clean if you do it the way I indicated. However, I think now I appreciate your remarks more.

ODEN: A comment relevant to the algorithms that you were talking about involves the notion of linear programming and non-linear programming, which mostly grew up in the fifties. In 1963, Fichera derived the first variational inequality showing that continuum problems with convex constraints could be formulated as an inequality in a variational inequality setting. It took five or six years before people realized that classical plasticity could also be set forth in that framework because in general, you have something like the classical optimization problem which arose from minimizing a function with a convex constraint. If you have a convex yield function, you can regard that as a constraint on the stress space.

Now, the interesting point of this observation is that once you admit that this class of problems can be formulated within this mathematical framework, it brings to bear two bodies of knowledge that have particular relevance to this symposium. The first is that virtually all the linear programming algorithms are directly applicable to the plasticity problems. As a matter of fact, your radial correction term in the linear programming literature is known as standard — successive over-relaxation with radial projections. So there is a family of algorithms that are designed in particular to handle linear programming problems that would fall into this general category.

What is equally interesting is that in the seventies, a broader class of plasticity problems was also put in this framework, primarily by Necas and his co-workers in Czechoslovakia. These have not necessarily lent themselves to the development of new algorithms, but they have presented a framework whereby it was possible to make certain theoretical conclusions that were not heretofore possible.

In particular, Kirk mentioned this morning the questions of uniqueness and existence. Many questions of existence and uniqueness for plasticity problems can be answered easily within this framework using the mathematical apparatus of variational inequalities.

As you expand this class of problems to include larger and larger classes of materials in finite deformations, then the structure of these inequalities and the associated algorithms become more complicated. In particular, if you look at kinematic hardening problems, you encounter so-called quasi-variational inequalities. These frequently do not have unique solutions, and the algorithms for this class of problems are still under development. If you also consider possibilities of non-differentiability of the yield surface and also losses of convexity, you encounter a completely new class of variational inequalities for which not much is known, called hemivariational inequalities. It could be that this point of view might also lead to new algorithms for this class of problems. I think it certainly will lead to a better understanding of the mathematical foundations of plasticity.

MOSS: I just wanted to make a brief comment that the loading and unloading criteria you showed are identical to the strain space loading and unloading criteria proposed by Naghdi. There is no difference.

HUGHES: Really, nothing is going on in strain space. It's only the loading and unloading criteria which need to be properly posed. The stress space picture stays intact. And, in fact, as I pointed out, the fundamental quantity is the normal component of the trial rate-of-stress. The fact that it reduces to strain rate follows from the assumption of isotropic elastic moduli. But in general, if the moduli were anisotropic, for example, then I think it's clear that you would still want to work in stress space.

In the anisotropic case the strain space picture would look very complicated whereas the stress space picture would still be simple.

MOSS: But you're still looking at a trial elastic stress that is outside the yield surface; that's what Naghdi is pointing out in his papers.

HUGHES: Well, it seems like he's doing more to me. I don't want to impugn strain space plasticity. I think the work of Yoder and Iwan is extremely interesting in that they've found some new algorithmic possibilities. It's also led them to some exact solutions of the stress space problem. So there are things that are to be derived from this approach. But to claim that the whole classical edifice has come crumbling down in favor of something fundamentally new is not true. It's just that the classical loading and unloading conditions are incorrect.

KEY: With reference to Prof. Clifton's remarks, I think he identified an unfortunate collision in terminology. In solving boundary value problems and iterating for an equilibrium solution, the temporary solutions that one deals with are frequently referred to as trial solutions. And that's a completely uncoupled notion to the trial elastic states that you're referring to here. Do I need to elaborate more on that?

CLIFTON: No, that was not the confusion to me. I see now what the situation is: you know the strain rates at one time, and so from those, and those only at one element, you compute the stress rate. Then when you later march in time, you only use strain rates at that time in determining the stress rates at the later time. So one never asks the question of whether at the previous time the stress state had been inside the yield surface and the elastic law had been used.

HUGHES: That's right. All of that memory is forgotten. And it's clear from the mathematical structure of the theory that that is irrelevant. It's a consequence of the plasticity theory.

PREVOST: I just wanted to comment on your remark on the unloading function to be the same in stress space and strain space formulations. If you write the loading function in stress space as one over H prime times N sigma dot, and if you back substitute the stress rate which you get out of the constitutive equation, you find exactly what you have as the loading function except it's divided by H prime plus the inner product of N times the elastic moduli. So it's exactly the same loading function.

HUGHES: I always believe you implicitly, Jean, but I can't follow verbal equations.

PREVOST: Let me try again. The loading and unloading in classical plasticity is defined in terms of the product of the normal to the yield surface and the stress rate.

HUGHES: But it's in terms of the stress rate sigma dot.

PREVOST: That's right. I call the normal N, then it is in terms of N times sigma dot.

   Now, if you define the loading function as N times sigma dot but as one over H prime (the plastic modulus), and back-substitute sigma dot which you get out of the plasticity equation, you get exactly your loading criterion.

TRULIO: I'd like to ask you about a possible computational problem that has to do with non-ideal plasticity, hardening or softening.

   Generally you have a parameter or parameters to govern the motion of the yield surface, and a natural way to define such a parameter is in terms of the plastic strain that's taken place. Often that's done conceptually, anyway — by watching the plastic strain (or the deviatoric part of it) trace out a path in strain space and then getting the length of that path. When you trace that path from a complete calculation, the discretization error is largest for the highest frequency modes of the mesh. Cycle to cycle variations dominate that error. So you replace what should be a very smooth curve by a jagged line in plastic strain space. The length of that line will now not converge to the smooth curve value as fast as the rest of the system of equations. In fact, such a result, for instance from the use of a linear artificial viscosity, might not converge on the correct length at all. Have you run into such a thing? What do you do about it?

HUGHES: I've never really experienced that. Maybe Jerry Goudreau or Sam might have had more experience on that.

TRULIO: Have I made the point clear?

HUGHES: Yes, I think you have. Have you actually seen this in calculations?

TRULIO: Yes, just out of curiosity, having thought of this potential problem, I took a case where there was shear failure (but not a case of hardening) and just looked at the plastic strain increment.

HUGHES: What kind of plasticity elements are you using?

TRULIO: It was a simple calculation of uniaxial strain.

HUGHES: Explicit calculation?

TRULIO: Explicit, right.

HUGHES: Radial return --

TRULIO: Right.

HUGHES: And you calculate the plastic strain in the right way?

TRULIO: Yes, and there is a discretization error.

HUGHES: I think there are a couple of ways it can be calculated.

TRULIO: Well, yes, you can imagine smoothing over a number of cycles. I don't think it's done that way.

HUGHES: Yes. Well, I guess I'd like to see the details and explore the issue further.

CHAPTER III

LARGE SCALE COMPUTATIONS

G. L. Goudreau*

University of California, Lawrence Livermore National Laboratory

ABSTRACT

The central ingredients of general codes for large scale nonlinear material computation are reviewed as a focus and complement to other presentations on the theory and implementation of constitutive models. Spatial discretization by the simplest finite elements is discussed relative to integral differences and higher order finite element alternatives. Explicit and implicit transient integration strategies are evaluated. Contact sliding algorithms, rezoning, and vectorization are included as important aspects of large scale computing. A few examples are included.

1. INTRODUCTION

The spatial discretization and time integration of the governing field equations of continuum mechanics are presented and reviewed as applied to the construction of general purpose computer programs for large scale computation. At this NSF/DARPA workshop, this is the third general lecture, the first two of which address the theoretical foundation of constitutive theory, and the numerical implementation of material models. Thus this lecture will only highlight material models, but will address general code interface requirements for constitutive subroutines.

Many high pressure and hypervelocity impact problems, such as explosive forming of metallic jets and armor penetration, severely tax the ability of Lagrangian codes to follow the physical process, even with extensive rezoning, and can best be simulated by Eulerian and mixed Euler-Lagrange codes. However, that added algorithmic complexity is not treated here. The issue of material constitutive modeling is most clearly treated in a Lagrangian formulation where the computational grid tracks the material point.

Time integration of dynamic and quasi-static response is discussed in both explicit and implicit treatment of the same formulation of nonlinear continua. Although explicit and implicit computation is here treated as two parallel or alternative choices, algorithms do exist, though not widely used, to admit simultaneous treatment of implicit and explicit regions of the same problem.

---

*Work performed under the auspices of the U.S. Department of Energy by the Lawrence Livermore National Laboratory under contract number W-7405-Eng-48.

With a bias toward "large scale" computation, the development to follow
will skip much in the finite element and general numerical literature of computational mechanics. The long history of methods for linear analysis will only
be referenced for their insights and contribution to the nonlinear problem.
Many small deformation developments have not been generalizable to large deformation. Finally, this is not a comprehensive survey of nonlinear continuum computer codes. The requirement of computational feasibility and efficiency for
the large scale nonlinear problem leads us to bypass the higher order element
literature, and focus on the simple elements.

This treatment will tend to emphasize the codes developed at the national
laboratories, not that these are the only centers for large scale Lagrangian
computation. A fairly unique combination of efficient computing with portable
public domain software allows a more open critique of the methodology. This
facilitates academic researchers in formulating theoretical models and numerical
algorithms contributing to the domain of realistic computation.

A survey of the Livermore development of two and three dimensional explicit
finite element hydrocodes was presented by Goudreau at FENOMECH (Goudreau and
Hallquist, 1981, and published in 1982). This presentation builds upon that,
integrating the status of implicit continuum computation.

2. EVOLUTION OF LARGE DEFORMATION LAGRANGIAN CONTINUUM CODES

The "hydrocodes" were born in the fifties at the U.S. National laboratories
with the numerical computation of high pressure shock wave propagation and
nuclear weapons applications. Insight into the early two dimensional work can
be found in the 1964 Journal of Computational Physics, where the Lagrangian
codes of Wilkins, Maenchen and Sack, and Noh are first published. Many hydrocodes have been written over the years, at government laboratories and private
computational companies. HEMP at Livermore (Wilkins, 1964) and its counterpart
TOODY at Sandia (Bertholf, 1968) served as prototypes of the integral finite
difference "hydrocode."

Independently, finite element methodology was being developed to allow the
static and vibration analysis of structural systems. Implicit codes with their
element and global stiffness matrices ensued. Linear equation solvers and unconditionally stable time integration schemes were important. The technology
which developed worked well for linear systems, but only with the advent of the
even larger computers of the last decade have large nonlinear problems been
attempted. No doubt the SAP series of linear structural codes (Wilson, 1970,
and Bathe, 1974) served as code organization tutorials as much as production
computing tools, and influenced a generation of finite element code developers.

Fewer nonlinear implicit codes exist which realistically and economically
treat finite deformation of solids. NONSAP (Bathe, 1974) served as an early
guide for implicit treatment of nonlinear finite element analysis. This led to
the more complete ADINA (Bathe, 1978) and its Lagrangian and Updated Lagrangian
treatment, although the material models admitted only small strains and moderate
rotations. MARC (Marcal, Hibbitt, et al., 1967) quickly became a dominant com-
mercial code in the nonlinear analysis area, but geared toward small strain and
quasi-static problems, and reactor industry applications. Our own NIKE codes
(Hallquist, 1979, 1981) also represent such technology, and were born out of
weapons structural design requirements. An extensive literature exists on the

nonlinear solution of plate and shell structures, and is not considered here.

Although implicit finite element codes often have an explicit option, these are generally very inefficient and no competition for the optimally coded explicit hydrocode in treating the high frequency regime.

An important contribution to explicit finite element technology developed at Sandia was HONDO (Key, 1974). His constant stress quadrilateral element exploited earlier coding by Wilson and Farhoomand (1970) which was obscured by an implicit formulation without hourglass control. Key's use of this software in an explicit context with his own hourglass control, automatic time stepping, and robust material library provided the first efficient explicit finite element large deformation code for solid continua.

Little attempt appears to have been made in applying higher order elements to stress wave propagation. The simplest quads, triangles, hexahedra, and tetrahedra have been preferred. As will be seen, however, the problem of one point vs selective multipoint integration remains to be resolved.

Three dimensional explicit codes have not been historically successful, because of the tremendous costs imposed by size and restrictive timesteps. HEMP3D (Wilkins, 1975) was an early effort, generalizing his two dimensional integral difference work. A small strain finite element code (TRANAL, 1979) has been heavily used by Weidlinger for ground motion and soil/structure analysis.

Engineering code development began at Livermore in 1976. Several years of experience using the finite difference hydrocodes on problems of large deformation impact of complex solid bodies revealed the limits of logically regular zoning and one-way sliding. Three-dimensional impact requirements were driving engineering development even more strongly. Our first effort in three dimensions found control of the hourglass problem of the one point integrated eight node hexahedron both expensive and inadequate.

The early two and three dimensional explicit DYNA codes (Hallquist, 1976, 1978) served as test vehicles for comparison of variable node quadratic elements with 2 × 2 quadrature against one point integrated elements with myriad hourglass treatments, as well as an element assembly of the HEMP finite difference equations (Wilkins, 1964, Hallquist, 1978). By 1980, DYNA2D was streamlined to the one point quadrilateral with an area weighted treatment of axisymmetry. DYNA3D provided a comparably optimized three-dimensional version (Hallquist, 1982).

The use of triangular and tetrahedral elements in an explicit finite element treatment was exploited in the EPIC codes (Johnson, 1976, 1981), and will be discussed. A constant pressure four node quadrilateral later became the standard for use in our implicit NIKE codes (Hallquist, 1979, 1981). All avoid hourglassing, but are expensive, especially in an explicit context.

3. LAGRANGIAN FORMULATION

Although some choose to set the physical formulation in a global or integral statement, the simplest context in which to contrast finite elements with finite differences is to start with the differential balance laws. This in no way prevents a global method of approximate solution. The linear momentum and energy balance laws in local form are:

$$\underset{\sim}{\nabla} \cdot \underset{\sim}{\sigma} + \underset{\sim}{b} = \rho \; \underset{\sim}{a}$$

$$\underset{\sim}{\sigma} \cdot \underset{\sim}{\nabla}\underset{\sim}{v} + \underset{\sim}{\nabla} \cdot \underset{\sim}{q} + r = \dot{E} \tag{1}$$

where $\underset{\sim}{v}$ is the particle velocity $\dot{\underset{\sim}{x}}$, $\underset{\sim}{a}$ the particle acceleration $\dot{\underset{\sim}{v}}$, $\underset{\sim}{\sigma}$ is the Cauchy stress tensor, q the heat flux vector per current area, ρ the current density, E the internal energy and ∇ is the divergence operator per current configuration $\underset{\sim}{x}$. $\nabla\underset{\sim}{v}$ is the velocity gradient, $\underset{\sim}{b}$ the body force, and r is the internal heat generation rate apart from stress power. The time derivatives are material, and thus simple partials for a Lagrangian frame where coordinates are assigned to material points. It is important to note that for Lagrangian codes, the continuity equation is identically satisfied whereas in an Eulerian frame it becomes the critical equation to be solved. Lagrangian codes must deal with the convective terms of the continuity equation when their configuration is rezoned.

On the time scale of impact, stress waves, and high frequency response, heat conduction is normally not important. When the gradient of the heat flux is dropped from Eq. (1b), the resulting adiabatic energy balance equation becomes a point equation, that is, involves no direct spatial coupling. Internal energy can then be regarded as an internal variable of the constitutive model and the energy balance equation as its rate equation. For energy dependent mechanical material models the energy balance and stress constitutive equation must be solved simultaneously. This contrasts with the usual uncoupled thermomechanics where the temperature is assumed known and mechanical property dependence on temperature can be computed directly.

4. SPACE/TIME DISCRETIZATION

Both finite element and finite difference Lagrangian codes assign a finite mesh on a reference configuration (usually the initial configuration) and track particles through time. That is, the current configuration is expressed as a function of the reference configuration and time

$$\underset{\sim}{x}(\underset{\sim}{X},t) = \underset{\sim}{x}[\underset{\sim}{X}(\underset{\sim}{s},n),t] = \sum_1^n \phi_i(\underset{\sim}{s}) \; \bar{\underset{\sim}{x}}_i(t) \tag{2}$$

and a displacement field may be defined

$$\underset{\sim}{u} \equiv \underset{\sim}{x} - \underset{\sim}{X}. \tag{3}$$

For the finite element method spatial discretization is achieved by a global orthogonalization of the error in the differential equation with respect to the assumed basis

$$\int \phi^T (\underset{\sim}{\nabla} \cdot \underset{\sim}{\sigma} + \underset{\sim}{b} - \rho \; \underset{\sim}{a}) dV = 0 \tag{4}$$

which leads to the matrix form

$$\underset{\sim}{M} \; \underset{\sim}{a} + \underset{\sim}{F} = \underset{\sim}{P} \tag{5}$$

where

$$\underset{\sim}{F} = \sum_e \underset{\sim}{L}^t \int_{V_e} \underset{\sim}{B}^t \; \underset{\sim}{\sigma} dV_e. \tag{6}$$

$B = \nabla\phi$ is the linearized strain displacement matrix with respect to the current configuration $\underset{\sim}{x}$, and $\underset{\sim}{L}$ is the Boolean operation of the arbitrary finite element assembly. The selection of basis functions of minimum support distinguishes finite elements from global Galerkin forms, in that the integrals can be computed by a sum of simple element integrals.

The above is valid regardless of choice of reference configuration or time integration scheme. We now consider the choice of integration scheme, and put off to the next section the choice of finite element.

4.1 Explicit Time Integration

The time centered integration rule

$$a_n = M^{-1}[P_n - F_n]$$
$$v_{n+\frac{1}{2}} = v_{n-\frac{1}{2}} + \Delta t \, a_n \qquad\qquad (7)$$
$$x_{n+1} = x_n + \Delta t \, v_{n+\frac{1}{2}}$$

advances the motion explicitly from the current acceleration, where the right-hand side $P - F$ contains all the effort, especially F from Eq. (6) where localization of the motion through a strain measure allows application of a point constitutive law to that locally homogeneous strain history for the determination of stress and then discrete forces.

The conditional stability limit is governed by the Courant condition of a finite difference grid (expressed as a characteristic length divided by the sound speed). For more general finite element formulations, this is the highest eigenvalue of the generalized tangent matrix which can be estimated from the maximum of each element. Analytic studies of the primitive quadrilateral and hexahedron are an important step in better understanding the general case, but so far appear too conservative (Flanagan and Belytschko, 1983).

4.2 Implicit Time Integration

The application of an implicit algorithm to a nonlinear dynamic or quasi-static process requires an iterative solution of the end point or current geometry for a small or large increment depending on dynamics, path dependence of constitutive law, and significant features in the loading history. Accuracy considerations usually limit the load increment or time step size since an inaccurate solution will ultimately not converge. Three iterative schemes are in general use:

- BFGS quasi-Newton method,
- Broyden's quasi-Newton method,
- modified Newton-Raphson.

A line search is recommended with each of these schemes along with automatic stiffness reformations as needed to avoid slow or nonconvergence. Generally, the quasi-Newton methods require fewer iterations than the modified Newton-Raphson method since unlike the latter method, they have superlinear local convergence. Following NIKE2D (Hallquist, 1983), but generally indicating a preferred approach, we first consider the quasi-static case, which introduces the key difference of an implicit treatment, and then show that implicit dynamic algorithms fit simply into this structure.

To obtain the solution at load increment $n+1$ given the solution at load increment n, linearized equations of the form

$$K_t(x^n)\Delta u_0 = P(x^n)^{n+1} - F(x^n) \qquad\qquad (8)$$

are assembled where

K_t = symmetric positive-definite tangent stiffness matrix based on the geometry at n,

$\underset{\sim}{P}$ = external loads vector based on the applied loading at step $n+1$ but the geometry at n,

$\underset{\sim}{F}$ = stress divergence vector based on the displaced state and stresses at load step n,

$\underset{\sim}{x}^n$ = coordinate vector at time n,

$\Delta\underset{\sim}{u}_0$ = desired increment in displacements.

Solution of Eq. (1) yields $\Delta\underset{\sim}{u}_0$. The coordinate vector is updated

$$\underset{\sim}{x}^{n+1} = \underset{\sim}{x}^n + s_0\Delta\underset{\sim}{u}_0, \tag{9}$$

where s_0 is a parameter between 0 and 1 found from the line search discussed below. Iteration for equilibrium now begins using

$$\underset{\sim}{K}_{t_j}(\underset{\sim}{x}_j^{n+1})\Delta\underset{\sim}{u}_i = \underset{\sim}{P}(\underset{\sim}{x}_i^{n+1})^{n+1} - \underset{\sim}{F}(\underset{\sim}{x}_i^{n+1}) = \underset{\sim}{Q}_i^{n+1}, \quad (j \le i) \tag{10}$$

where the subscript i denotes the iteration number. After each iteration, convergence is checked and assumed if

$$\| \Delta\underset{\sim}{u}_i \| /u_{max} \le \varepsilon_d \quad \text{and} \quad |\Delta\underset{\sim}{u}_i^t \underset{\sim}{Q}_i / |\Delta\underset{\sim}{u}_0^t \underset{\sim}{Q}_0| < \varepsilon_e \tag{11}$$

where the vertical bars denote the Euclidean norm, u_{max} is the maximum Euclidean norm of the displacement vector obtained over the n-completed load increments including the present iterate, and ε_d and ε_e are input parameters that are typically 10^{-2} to 10^{-3} or smaller. The tolerance on the energy norm, ε_e, is discussed by Bathe (1982). If convergence is not attained in the current iteration and the solution is not divergent, the coordinate vector is updated:

$$\underset{\sim}{x}_{i+1}^{n+1} = \underset{\sim}{x}_i^{n+1} + s_i \Delta\underset{\sim}{u}_i. \tag{12}$$

Sometimes convergence fails to occur within the allowable number of iterations or the solution diverges. Divergence is usually defined as

$$\| \underset{\sim}{Q}_0^{n+1}\| < \| \underset{\sim}{Q}_{i+1}^{n+1}\| . \tag{13}$$

Lack of convergence or divergence causes NIKE2D to reform $\underset{\sim}{K}_t$ using the current estimate of the geometry at $n+1$ before continuing with the equilibrium iterations.

The line search to determine s_i is a very important but expensive ingredient of the quasi-Newton methods. In NIKE2D, the line search is also applied to the modified Newton method. We search for a value of s_i such that the inner product

$$\Delta\underset{\sim}{u}_i^t \underset{\sim}{Q}_i \to 0. \tag{14}$$

The line search is used for stiffening systems ($s_i < 1$) where stability is often a problem. For stable softening systems ($s_i > 1$) convergence, although sometimes slow until the stiffness matrix is reformed, is usually obtained, and the line search is not needed. The numerical implementation of the line search is straightforward (Matthies and Strang, 1979). We allow up to 10 right-hand side evaluations to find s_i.

During the equilibrium iterations the stiffness matrix inverse is implicitly updated by recursively applying

$$K_{t_i}^{-1} = (I + w_i \, v_i^t) K_{t_{i-1}}^{-1} \, (I + v_i \, w_i^t) \qquad \text{(BFGS)} \qquad (15)$$

$$K_{t_i}^{-1} = (I + w_i' \, v_i') K_{t_{i-1}}^{-1} \qquad \text{(Broyden)} \qquad (16)$$

as the iterations progress. Here, I is the identity matrix, v_i and w_i are update vectors defined by Matthies and Strang, and v_i' and w_i are obvious in Walker's report. We note that unlike the BFGS update, the Broyden update is nonsymmetric.

For the dynamic problem, the linearized equilibrium equations may be written in the form

$$M\ddot{u}^{n+1} + D\dot{u}^{n+1} + K_t(x^n)\Delta u_0 = P(x^n)^{n+1} - F(x^n), \qquad (17)$$

where M = lumped mass matrix,

D = diagonal damping matrix,

$u^{n+1} = x^{n+1} - X$ = nodal displacement vector,

$\dot{u}^{n+1}$ = nodal point velocities at time $n+1$,

$\ddot{u}^{n+1}$ = nodal point accelerations at time $n+1$.

Equation (17) is solved by unconditionally stable, one-step, time integration schemes of the form of Newmark (1959) or Hilber and Hughes (1977). Dissipation and overshoot are important aspects of these algorithms (Goudreau, 1973, Hilber and Hughes, 1978).

Substitution into Eq. (17) leads to

$$K^*\Delta u_0 = P(x^n)^{n+1} - F^*(x^n), \qquad (18)$$

where

$$K^* = K_t + c_0 M + c_1 D, \qquad (19a)$$

$$F^* = F - M[c_2\dot{u}^n + c_3\ddot{u}^n] - D[c_4\dot{u}^n + c_5\ddot{u}^n]. \qquad (19b)$$

The solution of Eq. (18) yields Δu_0, the displacement, velocity, and acceleration vectors are updated

$$\ddot{u}_1^{n+1} = c_0\Delta u_0 - c_2\dot{u}^n - c_3\ddot{u}^n, \qquad (20a)$$

$$\dot{u}^{n+1} = \dot{u}^n + c_6\ddot{u}^n + c_7\ddot{u}^{n+1}, \qquad (20b)$$

$$x^{n+1} = x^n + \Delta u_0, \qquad (20c)$$

where $c_0 - c_7$ are given by the particular transient integration algorithm. Equilibrium iterations begin with

$$K^*\Delta u_i = P(x_i^{n+1})^{n+1} - F^*(x_i^{n+1}), \qquad (21)$$

where

$$F^* = F(x_i^{n+1}) + M\ddot{u}_i^{n+1} + D\dot{u}_i^{n+1}. \qquad (22)$$

At this point the method is essentially the same as the static algorithm. Although the Newmark method is still used at Livermore, and generally elsewhere as well, the Hilber-Hughes "α" damping is recommended.

5. ELEMENT CHOICE

Lagrangian finite difference codes invariably use the integral difference method. Derivatives in the differential equations of momentum and the strain equations are represented by contour integrals, and algebraic terms by an arithmetic nodal averaging scheme.

5.1 Primitive Element in 2D/Axisymmetry

The comparison between primitive finite elements and the integral difference method will be illustrated in axisymmetric geometry. The 2D Cartesian equations are an obvious special case. We start with the momentum equations in the absence of body forces

$$\rho a_r = \frac{\partial \sigma_{rr}}{\partial r} + \frac{\partial \tau_{rz}}{\partial z} + \frac{\sigma_{rr} - \sigma_{\theta\theta}}{r}$$

$$\rho a_z = \frac{\partial \tau_{rz}}{\partial r} + \frac{\partial \sigma_{zz}}{\partial z} + \frac{\tau_{rz}}{r} \tag{23}$$

The integral difference methods represent both element centered strain derivatives and node centered stress derivatives by contour integrals (Wilkins, 1964). Alternatively, a primitive constant stress finite element is easily considered in terms of element forces

$$\{F\}_e = \left\{ \int_{V_e} B^T \, dV \right\} \underset{\sim}{\sigma}_e. \tag{24}$$

A detailed comparison of the one point quadrilateral finite element with the integral difference method has been presented (Goudreau and Hallquist, 1982) and will not be repeated here. The key result is that they yield the same result in the plane case, and finite element provides the added insight into discretizing the nongradient (arithmetic) terms of axisymmetry. Integral difference stencils can be assembled on a topologically arbitrary grid, but appears not to have been tried, except by Hallquist. Wilson and Key used 2 × 2 Gauss quadrature (still constant stress). One point integration is equally accurate and much faster, and recommended for a volumetric formulation. Also, area integrals are not usually volume weighted, leading to a time dependent "mass vector" which is actually constant mass per time dependent centroidal radius. See Goudreau and Hallquist (1982) for detailed assessment of the axisymmetric problem and the "area Galerkin" solution.

$$\{F\}_e = \left\{ \int_A B^T r \, dA \right\} \underset{\sim}{\sigma}_e = 4r_0 \, \underset{\sim}{B}_0^T \, \underset{\sim}{\sigma}_e. \tag{25}$$

Pre-vectorized DYNA2D on the CDC7600 was two to three times faster than HONDO, in large part due to this simplification.

The biggest disadvantage to one point integration is the need to control the zero energy modes which arise, called hourglassing modes. This is a subject in itself, and is deferred to the next section and the 3D case.

5.2 Three-Dimensional Solid Element

The development of DYNA3D in 1976 was our first large deformation code effort. Because early hourglass formulations, such as that in HEMP3D and our extension of the HONDO algorithm to three dimensions, dominated the cost of the primitive one point hexahedra, we explored the twenty node serendipity brick, and variable node specializations. Multipoint integration precluded the hour-glass modes. This first version had constant stress 4-8 node linear solid ele-ments, and 16 and 20 node quadratic solid elements with $2 \times 2 \times 2$ Gaussian quad-rature. Though reliable, this method required a special mass lumping scheme and also proved very expensive. Higher order elements seemed to be impractical for shock wave propagation because of numerical noise resulting from the ad hoc way mass is lumped to generate a diagonal mass matrix. Subsequent work opti-mized the coding of the one point element, and a cheap effective hourglass control was developed. The new version was released in 1979. On the CRAY the vectorized speed was 50 times faster, .67 minutes per million mesh cycles.

Modern 3D hexahedrons are a straightforward application of the linear iso-parametric hexahedron element using one point Gaussian quadrature (Zienkiewicz, 1977). What is important is recognition of the compact coding possible, and the added gain from straightforward vectorization of that coding. Vectorization will be discussed separately in the context of "chunking" a group of elements. Here we focus on the operation count of a compact implementation of one point quadrature for a single element. Then more complex formulations can be judged on the basis of their cost effectiveness.

The central cost item of the element, as in 2D, is the construction and use of the $\underset{\sim}{B}$ matrix for force projection and velocity strains.

$$\underset{\sim}{F}_e = \int_{V_e} \underset{\sim}{B}^T \underset{\sim}{\sigma} \, dV = \left[\int_{V_e} \underset{\sim}{B}^T \, dV\right]\underset{\sim}{\sigma} = \underset{\sim}{B}_0^T V \underset{\sim}{\sigma} \qquad (26)$$

with

$$\underset{\sim}{B}_0 = (\nabla\underset{\sim}{\phi})_0, \qquad \text{(Hallquist, 1983)} \qquad (27)$$

and the symmetric stress tensor represented as a vector

$$\{\underset{\sim}{\sigma}\} = <\sigma_{11}\sigma_{22}\sigma_{33}\sigma_{12}\sigma_{13}\sigma_{23}>^T. \qquad (28)$$

Perhaps the biggest advantage to one point integration is its extreme efficiency. An antisymmetry property in the terms of the strain displacement matrix at $\xi = \eta = \zeta = 0$ (see Fig. 1)

$$\frac{\partial\phi_1}{\partial\underset{\sim}{x}} = -\frac{\partial\phi_7}{\partial\underset{\sim}{x}} \qquad \frac{\partial\phi_3}{\partial\underset{\sim}{x}} = -\frac{\partial\phi_5}{\partial\underset{\sim}{x}}$$

$$\frac{\partial\phi_2}{\partial\underset{\sim}{x}} = -\frac{\partial\phi_8}{\partial\underset{\sim}{x}} \qquad \frac{\partial\phi_4}{\partial\underset{\sim}{x}} = -\frac{\partial\phi_6}{\partial\underset{\sim}{x}} \qquad (29)$$

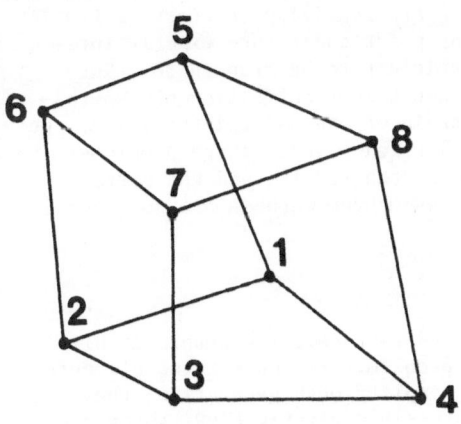

FIG. 1

reduces the amount of effort required to compute this matrix by more than 25 times over 8 point integration. This cost savings extends to strain and element nodal force calculations where the num-ber of multiplies is reduced by a factor of 16. Because only one constitutive evaluation is needed, the time spent determining stresses is reduced by 8.

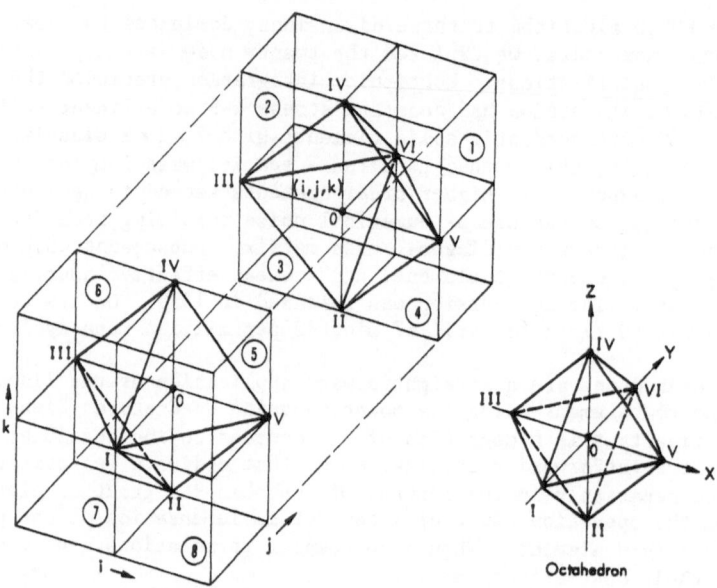

FIG. 2

Generalization from 2D to 3D is not unique, as was seen just going from the 2D Cartesian case to axisymmetry. The HEMP3D integral difference algorithm was published (Wilkins, 1975), but has never reached a production code status, and lacks slide plane capability. In our effort to obtain direct comparisons with the finite element alternatives, the HEMP3D equations were programmed directly as an alternate solution overlay in DYNA3D, assembled element by element, eliminating the need for special node treatment at boundaries. The analog to the nodal contour integral in 2D now becomes an octahedral surface integral, seen in Fig. 2. This leads to a loss of the antisymmetry property of the element forces seen in 2D, and retained by the constant stress finite element in 3D. Likewise, the calculation of the zone centered strain rates is a hexahedral surface integral over the six faces of the zone, thus leading Wilkins to considerably more work than required for the one point finite element.

The twenty-four modes of the hexahedron's vector displacement response contain twelve zero energy hourglass modes, which are receiving increased attention (Flanagan and Belytschko, 1981). Although one point quadrature exactly integrates the 2D quadrilateral, Flanagan shows this not to be true in 3D. They propose an exact analytic integration which then allows definition of generalized strains different from the centroidal strain of one point integration. We have implemented their algorithm in DYNA3D and tried several large problems. Our experience finds no noticeable difference in the result, and the extra effort not justified. Operation counts for these three methods are reported in Goudreau and Hallquist (1982), Table 1.

5.3 Triangles and Tetrahedra

Johnson advocates the use of triangular and tetrahedral elements in his EPIC codes as a solution to the hourglassing problem. Although these elements are sometimes used as transition elements to simplify mesh generation, they generally are not recommended. During incompressible plastic flow, these elements lock-up in the constant volume bending modes. These modes generate unrealistic tensile and compressive volumetric strains in the component triangular and tetrahedral elements that define the quadrilateral and hexahedral elements, respectively, in the mesh. This is illustrated by the bar impact example of the NIKE2D manual which shows the locking of a triangular element mesh compared with one point DYNA2D and constant pressure NIKE2D. Quadrilaterals defined by four crossed triangles are better behaved but are still relatively stiff.

Constant pressure hexahedrons based on the average volume change of the component tetrahedrons (Johnson) avoid locking but require multiple constitutive evaluations per element and may not be cost effective. Two point integration with constant pressure in DYNA2D and DYNA3D successfully avoided the hourglass problem at low pressures. At high pressures, strength is not a factor and 2 point integration simply increases the cost. A similar experience might be expected with a constant pressure hexahedron based on tetrahedra.

5.4 Constant Pressure Quadrilateral

Whereas the explicit codes started with the constant stress element of the hydrocodes, implicit codes start with the linear isoparametric and move up to higher order and down to one point.

The original development of NIKE2D is based to some extent on work by Bathe, Wilson, and Iding (1974), Key (1974), Wilkins (1964), and Nagtegaal, Parks, and Rice (1974). Late in 1979 we implemented BFGS and Broyden's quasi-Newton methods (Strang, 1979). These methods substantially improved the convergence behavior.

The 1983 version of NIKE2D is almost fully vectorized for machines like the CRAY-1 and is programmed to run fully core contained except for the largest problems. A stiffness matrix, suggested by Hughes (1980), consistent with the constant pressure element, has improved convergence in many problems.

Spatial discretization is accomplished by using three or four node elements that are integrated with a 1 or 2 point Gauss quadrature rule. At this time, one point integration is discouraged. Hourglass formulations are being considered to stabilize the elements when one point quadrature is used.

Strain increments at a point in a constant pressure element are defined for axisymmetry problems by

$$\Delta\varepsilon_{rr} = \partial\Delta v/\partial r^{n+\frac{1}{2}} + \phi, \tag{30a}$$

$$\Delta\varepsilon_{zz} = \partial\Delta w/\partial z^{n+\frac{1}{2}} + \phi, \tag{30b}$$

$$\Delta\varepsilon_{\theta\theta} = \Delta v/r^{n+\frac{1}{2}} + \phi, \tag{30c}$$

$$\Delta\varepsilon_{rz} = (\partial\Delta v/\partial z^{n+\frac{1}{2}} + \partial\Delta w/\partial r^{n+\frac{1}{2}})/2, \tag{30d}$$

$$\phi = \Delta\varepsilon_v^{n+\frac{1}{2}} - (\partial\Delta v/\partial r^{n+\frac{1}{2}} + \partial\Delta w/\partial z^{n+\frac{1}{2}} + \Delta v/r^{n+\frac{1}{2}})/3, \tag{30e}$$

where $\Delta\varepsilon_v$ is the average volumetric strain increment

$$\Delta\varepsilon^{n+\frac{1}{2}}_v = \frac{\frac{1}{3}\int_{V^{n+\frac{1}{2}}} (\partial\Delta v/\partial r^{n+\frac{1}{2}} + \partial\Delta w/\partial z^{n+\frac{1}{2}} + \Delta v/r^{n+\frac{1}{2}})dV^{n+\frac{1}{2}}}{\int_{V^{n+\frac{1}{2}}} dV^{n+\frac{1}{2}}}, \tag{31}$$

and

$$V^{n+\frac{1}{2}} = \text{volume at the midstep.}$$

Δv and Δw are displacement increments in the r and z directions, respectively, and

$$r^{n+\frac{1}{2}} = (r^n + r^{n+1})/2, \tag{32a}$$

$$z^{n+\frac{1}{2}} = (z^n + z^{n+1})/2. \tag{32b}$$

To satisfy the condition that rigid body rotations cause zero straining we use the geometry at the midstep. This was discussed in the 1979 manual as well.

Since the bulk modulus is constant in the plastic and viscoelastic materials models, constant pressure elements result. In the thermo-elastic-plastic material, a constant temperature is assumed over the element. In the soil and crushable foam material, an average relative volume is computed for the element at load increment $n+1$, and the pressure and bulk modulus associated with this relative volume is used at each integration point.

The foregoing procedure requires that the strain-displacement matrix, $\underset{\sim}{B}$, be evaluated twice for each integration point. First, $\underset{\sim}{B}$ is evaluated at $n+\frac{1}{2}$ and is used to compute the strain increments in Eqs. (30). Then, $\underset{\sim}{B}$ is reevaluated at $n+1$ and is used to compute the stress divergence, and if necessary, the stiffness matrix. For a single element, these latter integrals are defined by

$$\underset{\sim}{F} = \int_{V^{n+1}} \bar{\underset{\sim}{B}}^{n+1^t} \underset{\sim}{\sigma}^{n+1} dV^{n+1}, \tag{33}$$

$$\underset{\sim}{K} = \int_{V^{n+1}} \bar{\underset{\sim}{B}}^{n+1^t} \underset{\sim}{E}^{n+\frac{1}{2}} \bar{\underset{\sim}{B}}^{n+1} dV^{n+1} \tag{34}$$

where $\underset{\sim}{E}$ is the symmetric part of the Cauchy stress tangent modulus tensor. For plasticity we now use the rotated Cauchy stress with symmetrized modulus tensor $\underset{\sim}{C}$. Thus,

$$\underset{\sim}{E}^{n+\frac{1}{2}} = \underset{\sim}{T}^{n+\frac{1}{2}} \underset{\sim}{C}^{n+\frac{1}{2}} \underset{\sim}{T}^{n+\frac{1}{2}^t} \tag{35}$$

is the transformation that brings $\underset{\sim}{C}$ into the rotated configuration, and $\bar{\underset{\sim}{B}}$ is the mean dilatation strain-displacement matrix formed from the terms of $\underset{\sim}{B}$ as clearly explained by Hughes (1980).

6. SLIDE LINES WITH CONTACT-IMPACT

Finite element codes with contact-impact capability are applicable to a class of structural problems where adjacent components may independently slide, separate, and impact along interfaces.

Four algorithms have been used in the literature:

- nodal constraint,
- penalty,
- distributed parameter,
- Lagrangian multiplier

methods. The first three of these have been used in the Livermore DYNA and NIKE codes by Hallquist (1978-1983). Since the proper choice of penalty can produce as accurate a result as Lagrangian multiplier and the latter requires an extra variable and a more complicated equation solver, we will consider only the first three.

Interfaces in two dimensions are defined by specifying the string of nodes along each side of the interface. In three dimensions all triangular and quadrilateral segments that comprise each side are listed in arbitrary order. One side of the interface is designated as the slave side, and the other is designated as the master side. Due to the symmetry of the penalty method, this distinction is irrelevant. In the distributed parameter and nodal constraint methods the slave nodes are constrained to slide on the master surface after impact and to remain on the master surface until a tensile interface force develops.

6.1 Nodal Constraint Method

Constraints are imposed into the global equations by a transformation of the nodal displacement components of the slave nodes along the contact interface. This transformation has the effect of eliminating the normal degree of freedom of the slave node and distributing its normal force component to adjacent master nodes. To preserve the efficiency of the explicit time integration, the mass is lumped to the extent that only the global degrees of freedom of each master node are coupled. Impact and release conditions have been imposed to insure momentum conservation (Hughes et al. 1976).

Problems can arise with this method when master surface zoning is finer than slave surface zoning. Here certain master nodes can penetrate through the slave surface without resistance and create a kink in the slide line.

6.2 Penalty Method

The penalty method is used in the explicit programs, HONDO II, DYNA2D, and DYNA3D, and the implicit programs, NIKE2D and NIKE3D. The method consists of placing normal interface springs between all pentrating nodes and the contact surface. With the exception of the spring stiffness matrix which must be assembled into the global stiffness matrix, the implicit and explicit treatments are similar. HONDO II requires the user to choose a restoring force modulus for each side of the interface whereas the NIKE programs and DYNA3D compute a unique modulus for each slave and master segment based on the thickness and bulk modulus of the element in which it resides.

In contrast to the nodal constraint method, the penalty method approach was found to incite little if any mesh hourglassing. This lack of noise is undoubtably attributable to the symmetry of the approach. Momentum is exactly conserved without the necessity of impact and release conditions. Furthermore, no special treatment of intersecting interfaces is required, greatly simplifying the implementation.

As discussed later, the interface stiffness is chosen to be less than the stiffness of the interface elements normal to the interface. As a result, the explicit stability condition is not affected. However, if interface pressures become large, unacceptable penetration may occur. By scaling up the stiffness and scaling down the time step size, such problems may still be solved using the penalty approach.

6.3 Distributed Parameter

This method is optional in DYNA2D, and a specialization of it is the sliding only option in DYNA3D. Motivation for this approach came from the TENSOR (1964) and HEMP (1964) programs which displayed fewer mesh instabilities than DYNA2D with the nodal constraint algorithm. The first DYNA2D implementation of this last algorithm is described in detail by Hallquist (1976) and more recently (1983).

In the distributed parameter formulation one-half the slave element mass of each element in contact is distributed to the covered master surface area. Also, the internal stress in each element determines a pressure distribution for the master surface area that received the mass. After completing this distribution of mass and pressure, we can update the acceleration of the master surface. Constraints are then imposed on slave node accelerations and velocities to insure their movement along the master surface. Unlike the finite difference hydro programs, slave nodes are not allowed to penetrate; therefore, we do not need "put back on" logic. In another simplification our calculation of the slave element relative volume ignores any intrusion of the master surfaces. HEMP and TENSOR consider the master surface in this calculation.

6.4 Slave Search in Three Dimensions

The slave search is common to all interface algorithms implemented in DYNA3D and NIKE3D. This search finds for each slave node its closest point on the master surface. Lines drawn from a slave node to its closest point will be perpendicular to the master surface, unless the point lies along the intersection of two master segments.

Consider a slave node, n_s, sliding on a piecewise smooth master surface and assume that a search of the master surface has located the master node, m_s, lying closest to n_s. Figure 3 depicts a portion of a master surface with nodes m_s and n_s labeled.

Assume that a master segment has been located for slave node, n_s, and that n_s is not identified as lying on the intersection of two master segments. Then the identification of the "contact point," defined as the point on the master segment which is closest to n_s, becomes non-trivial. Each master surface segment, s_i, is given the usual isoparametric representation,

$$\underset{\sim}{r} = f_1(\xi,\eta)\underset{\sim}{i}_1 + f_2(\xi,\eta)\underset{\sim}{i}_2 + f_3(\xi,\eta)\underset{\sim}{i}_3 \tag{36}$$

where

$$f_i(\xi,\eta) = \sum_{j=1}^{4} \phi_j\, x_i^j, \qquad \phi_j = \tfrac{1}{4}(1 + \xi\xi_j)(1 + \eta\eta_j). \tag{37}$$

ξ_j, η_j take on their nodal values at (± 1, ± 1), and x_i is the nodal coordinate of the jth node in the ith direction; see Fig. 3.

The position vector $\underset{\sim}{t}$ drawn to slave node n_s defines a contact point with

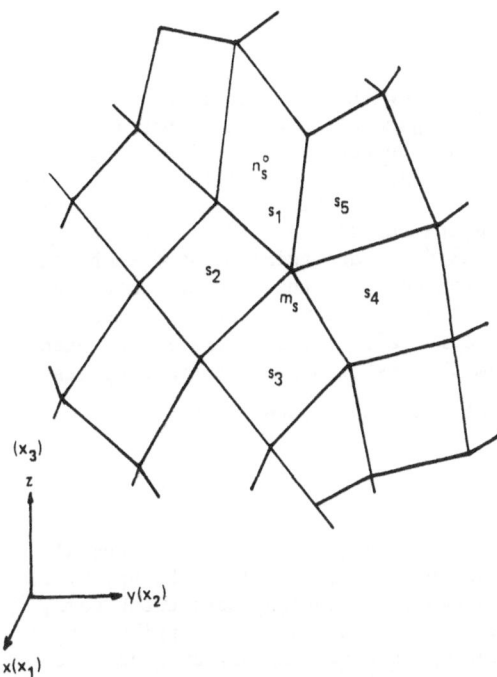

FIG. 3. In this figure, five master segments can harbor slave node n_s given that m_s is the closest master node.

coordinates (ξ_c, η_c) on s_i that must satisfy

$$\frac{\partial \underset{\sim}{r}}{\partial \xi}(\xi_c, \eta_c) \cdot [\underset{\sim}{t} - \underset{\sim}{r}(\xi_c, \eta_c)] = 0 \quad (38a)$$

$$\frac{\partial \underset{\sim}{r}}{\partial \eta}(\xi_c, \eta_c) \cdot [\underset{\sim}{t} - \underset{\sim}{r}(\xi_c, \eta_c)] = 0. \quad (38b)$$

Equations (38a) and (38b) are readily solved for ξ_c and η_c.

6.5 Sliding with Closure and Separation

Because this is perhaps the most general and most used interface algorithm, we choose to discuss it first. In applying this penalty method each slave node is checked for penetration through the master surface. If the slave node does not penetrate, nothing is done. If it does penetrate, an interface force is applied between the slave node and its contact point. The magnitude of this force is proportional to the amount of penetration. This may be thought of as the addition of an interface spring (Hallquist, 1983).

The stiffness factor k_i for master segment s_i is given in terms of the bulk modulus, K_i, the volume V_i, and the face area, A_i, of the element that contains s_i as

$$k_i = \frac{f_{SI} K_i A_i^2}{V_i} \quad (39)$$

where f_{SI} is a scale factor for the interface stiffness and is normally defaulted to .10. Larger values may cause instabilities unless the time step size is scaled back in the time step calculation.

6.6 Implicit Treatment

The approach in the two foregoing sections also applies to the implicit programs. A 15 × 15 stiffness matrix is added to the global stiffness matrix for each interface slave node in contact with the master surface. A slave node can couple with up to four master nodes of its harboring master segment. As with the explicit treatment the slave and master surfaces are reversed to obtain a symmetric treatment.

Since no restrictions exist on the amount of relative interface movement, the coupling between slave nodes and master segments changes continuously. Consequently, we must recompute the column height vector for our active column solver and redefine the block structure of the stiffness matrix each time we reform. Normally a reformation is performed at the beginning of each time step or load increment. Reformations also are done during equilibrium iterations whenever convergence is slow or divergence is detected.

6.7 Tied Interfaces

Sudden transitions in zoning are permitted with tied interfaces. This can often reduce the amount of effort required to generate meshes. Since the constraints are imposed only on the slave nodes, the more coarsely meshed side of the interface is recommended as the master surface. Implementation of the tied interface constraints is very straightforward. The interpolated contact point for each slave node is computed once since its relative position never changes. Every time step for each interface the nodal force and nodal mass of each slave node is distributed to the master nodes which comprise the segment containing the contact point. After the summation over all slave nodes is completed, the acceleration of the master surface is computed. The acceleration of each slave node, a_{i_s}, is then interpolated from the master segment containing its contact points. Velocities and coordinates are then updated normally.

7. REZONING

Lagrangian codes generally function well for problems when the material flow is smooth. When flow becomes turbulent or when material breaks up, i.e., simply-connected regions become multi-connected, Lagrangian codes break down, and Eulerian codes are a necessity. Between these two extremes, applications exist for which either code type may be appropriate but Lagrangian codes are usually preferred because of speed and accuracy considerations. Lagrangian codes may experience numerical difficulties if thinning of zones causes the stability controlled time step size to collapse. Often the time step is controlled in regions of the mesh where the solution is of little interest and the controlling elements may simply be deleted from the mesh. If, however, the time step is controlled in regions where the solution is important to the global problem, rezoning may be used to obtain a solution that would otherwise require an Eulerian code.

The basic approach to rezoning consists of three steps:

1. Generate nodal values on the old mesh for all variables to be remapped.

2. Rezone one or more materials either interactively or automatically.

3. Initialize remeshed regions by interpolating from nodal point values of the old mesh.

For each variable to be remapped, we seek to approximate it globally by

$$g(r,z) = \sum_{i=1}^{n} \phi_i(r,z)g_i \qquad (40)$$

where n is the number of nodes, ϕ_i is a piecewise continuous global basis function and g_i is a nodal point value. Given a variable to be remapped, $h(r,z)$, a least squares best fit is found by minimizing the functional

$$\Pi = \int (g - h)^2 \, dA \qquad (41)$$

with respect to the nodal values

$$\frac{\partial \Pi}{\partial g_i} = 0, \quad i = 1,2,\ldots,n \qquad (42)$$

leading to the set of matrix equations

$$\underset{\sim}{M} \underset{\sim}{g} = \underset{\sim}{f} \tag{43}$$

where

$$\underset{\sim}{M} = \sum_e \underset{\sim}{m}^e = \sum_e \int \underset{\sim}{\phi} \underset{\sim}{\phi}^t \, dA \tag{44}$$

$$\underset{\sim}{f} = \sum_e \underset{\sim}{f}^e = \sum_e \int h \, \underset{\sim}{\phi} \, dA. \tag{45}$$

One point quadrature is used in the area integrals and M is diagonalized. The foregoing procedure is repeated for all element variables including stresses, plastic strains, internal energy, and density.

In step 2 a new mesh is generated for a material to be rezoned by dekinking slidelines and material boundaries, adjusting the spacing of boundary nodes, and moving the interior nodes to achieve a smooth mesh.

In step 3, new element centered values in the rezoned region are found from the expression

$$h_{new} = \int g \, dA / \int \, dA \tag{46}$$

when the integrals are evaluated by Gaussian quadrature.

8. EXPLICIT/IMPLICIT TRADEOFF

The more expensive 2×2 constant pressure quad has not been considered seriously for explicit codes because the hourglass problem is under control.

The kernel of implicit costs has always been the direct factorization and solution of linear equations.

Detailed operation counts for DYNA3D were published by Goudreau (1982), (1983) and Hallquist (1983) and will not be repeated here. Optimized choice of implicit hourglass control is still under study, and will be reported by Hallquist soon, as will vectorization of NIKE3D.

Proper appreciation of the explicit/implicit tradeoff requires relating the right-hand side (or function evaluation) cost to the factorization and resolution costs. Profile linear equation solvers have been optimized for the CRAY-1, the CDC-7600, and VAX 11-780. Not only have operation counts been optimized, but advantage of the CRAY-1 FORTRAN vectorization in scalar product, scalar vector, and "element chunking" strategies are employed. The one to two million word memories on the CRAY now allow virtually all two-dimensional implicit problems to be run in-core. Three-dimensional problems are extremely expensive both in CPU and IO and await new solvers such as "element by element" (Hughes, 1983) before strongly nonlinear problems can be routinely solved.

9. VECTORIZATION PROCEDURE

A fourfold increase in speed through vectorization has been attained on CRAY-1 in the explicit DYNA codes. Such increases can be realized by recoding the solution phase to process vectors in place of scalars. It is necessary to process elements in groups rather than individually as had been done in the past on earlier computers.

Vector registers on CRAY-1 are 64 words long; consequently, vector lengths of 64 or some multiple of 64 are appropriate. In the DYNA codes, groups of 128 elements are utilized. Since the CRAY compiler, CFT (1978), vectorizes FORTRAN DO-loops, the CRAY sources may be programmed in FORTRAN, thus allowing the CRAY source to be implemented on both CRAYs and other machines.

Conceptually, vectorization is straightforward. Each scalar operation that is normally executed once for one element, is repeated for each element in the group. This means that each scalar is replaced by an array, and the operation is put into a DO-loop. For example, the nodal force calculation for the hexahedron element appears in a scalar version of DYNA3D as:

$$E11 = SGV1*PX1+SGV4*PY1+SGV6*PZ1$$

$$E21 = SGV2*PY1+SGV4*PX1+SGV5*PZ1$$

$$E31 = SGV3*PZ1+SGV6*PX1+SGV5*PY1$$

$$\vdots$$

and in the vectorized version as:

```
      DO 110 I = LFT, LLT
      E11(I) = SGV1(I)*PX1(I)+SGV4(I)*PY1(I)+SGV6(I)*PZ1(I)
      E21(I) = SGV2(I)*PY1(I)+SGV4(I)*PX1(I)+SGV5(I)*PZ1(I)
                        :
                        :
  110 E34(I) = SGV3(I)*PZ4(I)+SGV6(I)*PX4(I)+SGV5(I)*PY4(I)
```

where $1 \leq LFT \leq LLT \leq 128$. Elements LFT to LLT inclusive reference the same material model.

Gather-scatter operations have not been successfully vectorized. In the gather operation, variables needed for processing the element group are pulled from global arrays into local vectors. For example, the gather operation:

```
      DO 10 I = LFT, LLT
      X1(I)  = X(1,IX1(I))
                   :
                   :
   10 VZ8(I) = V(3,IX8(I))
```

initializes the nodal velocity and coordinate vectors for each element in the subgroup LFT to LLT. In the scatter operation, element nodal forces are added to the global force vector. No other scatter operations are needed. With the exception of gather-scatter, nearly every other operation can be programmed to vectorize with the CFT compiler.

It is this important added pay off for simple elements that has helped tip the balance away from higher order elements.

10. MATERIAL MODELS

Most codes are organized to accept a wide range of material models. However the ability for each having an arbitrary number of history variables and invoking its own strain measure is important. The common interface is based on the fact that the motion is advanced explicitly by global momentum balance, thus dictating the local point strains and rotations. The requirement of the constitutive routine is to provide the current multi-axial state of stress and other internal variables α, given the history and current value of the motion (strain)

$$f(\sigma_n, \alpha_n, \dot{\varepsilon}_{n+\frac{1}{2}}, \varepsilon_n, \Delta\varepsilon_n; \sigma_{n+1}, \alpha_{n+1}) = 0. \tag{47}$$

This general form allows both incremental and total strain forms of material models. It allows locally explicit stress relations, or implicit forms such as strain rate formulas. It also provides for coupling of the stress and internal variable equations, such as internal energy. The code should allocate storage for the required number of internal variables for each element of the given type. These same models can be used in the implicit as well as explicit codes; the only difference being the deformation variables (incremental strain and strain rate) are assumed to be trial values, and the output stresses (and tangent moduli) are also trial values to be used in a global iterate to improve the assumed motion. The explicit codes also require the tangent modulus, to be used in the computation of the stability constraint which determines the time step.

Material models will only be classed in terms of the usual division into bulk or deviatoric forms. The former contains the classical equation of state work where pressure is given as a function of relative volume and energy. References such as Woodruff's KOVEC (1976) and Hallquist's DYNA manuals describe code implementation and use.

The hydrodynamic shocks generated by hardening equations of state present severe problems of numerical stability. A bulk viscosity based on the work of Richtmyer and Morton is added to the pressure to smooth shocks over several zones (Wilkins, 1980). This can be regarded as an algorithmic supplement to the material model, and used only as needed.

The internal energy equation does not include heat conduction but its point form can be complicated by the release of chemical energy from the burn of high explosives. This explosive energy can be released in a programmed way, based on arrival time from specified lighting points or lines and transit at a prescribed detonation velocity. Burn fraction can then be activated over a few time steps. Or lighting can be induced mechanically, with volume compression from a traveling shock wave triggering the burn fraction in the energy equation. More recent developments of reactive high explosives solve a more complex internal variable equation for the burn fraction, coupled with the energy equation and pressure equation of state (Cochran, 1979). These detailed models are generally available in one-dimensional codes, but have proved difficult to implement fully and efficiently in two dimensions. DYNA2D has recently achieved an efficient implementation of the strongly coupled algorithm.

Hydrodynamic models usually have a simple plasticity model controlling the stress deviators — models such as Steinberg-Guinan (1978) where shear modulus and yield strength are pressure and energy dependent

$$G = G_0 \ f(p,E) \qquad \sigma_y = \sigma_0'(\varepsilon^{-p}) \ f(p,E) \tag{48}$$

where σ_0' is a strain hardening function, and f includes pressure hardening, linear thermal softening, and an exponential cut off at thermal melt. Bammann at Sandia (1983) has developed a powerful viscoplastic alternative.

Deviatoric models will serve as the principal part of this workshop, and will not be further discussed. Typical code libraries can be found in the DYNA

and NIKE manuals, the work of Key (1974), and Krieg and Key (1976). Geologic
models as well as metal plasticity are of paramount current interest.

11. EXAMPLES

11.1 Lateral Impact of a Canister on a Rail

Three-dimensional dynamic calculations are very useful in determining the
response of a complex design during environments. An example of this is the
calculations which were done to assess a canister's resistance to lateral rail
impacts. These calculations were done in cooperation with the canister
designers and the use of computer modeling to simulate tests streamlined the
test program by allowing testing at or near the design's limits.

In the modeled test, a canister was dropped onto a rail which ran perpendicular to the axis of the canister. The point of impact on the canister was
midway between the two stainless steel honeycomb rings which support the
shipping can within the canister. Impacting at this point minimizes the energy
dissipation of the mitigator system, thus requiring the plastic deformation of
the canister outer case to dissipate all the kinetic energy.

The model used for this analysis is shown in Fig. 4-1. We assumed vertical
planes of symmetry along the rail and along the canister's axis. Assuming a
plane of symmetry along the rail prohibits the rotation of the model. This is
a worst case assumption since rotation can dissipate significant amounts of
energy during impact. Included in the model are slide surfaces between 1) the
canister's outer case and the mitigator rings and 2) the mitigator rings and
the shipping can.

This model was calculationally impacted at various initial velocities into
the perfectly rigid rail shown in Fig. 4-1. The initial kinetic energy of the
canister assembly is absorbed by the plastic deformation of the outer case. As
the outer case deforms, the clearance between the outer case and the shipping
can reduces. If the plastic deformation does not dissipate all the kinetic
energy before the clearances are reduced to zero, then the shipping can will
impact the outer case. This would be of great concern.

The calculations tried to identify the initial velocity at which the plastic deformation would dissipate all the kinetic energy as the clearances were
reduced to zero. This calculational condition is shown in Fig. 4-2. After
reaching this point of maximum deformation, the canister rebounds off the rail.
As the model rebounds, the clearance between the outer case and the shipping can
grows. Figure 4-3 shows the final shape of the model after it has lifted off
the rail.

After allowing for energy dissipation due to rotation, the calculated
damage was within 1% of the measured damage. Thus, we were able to determine
the canister's resistance to such impacts with a single test. The total CRAY-1
computer time used in the analysis was approximately eleven hours. This includes model development, three calculational impacts and all graphics. The
ability to use computer modeling to simulate tests resulted in significant cost
and time reductions.

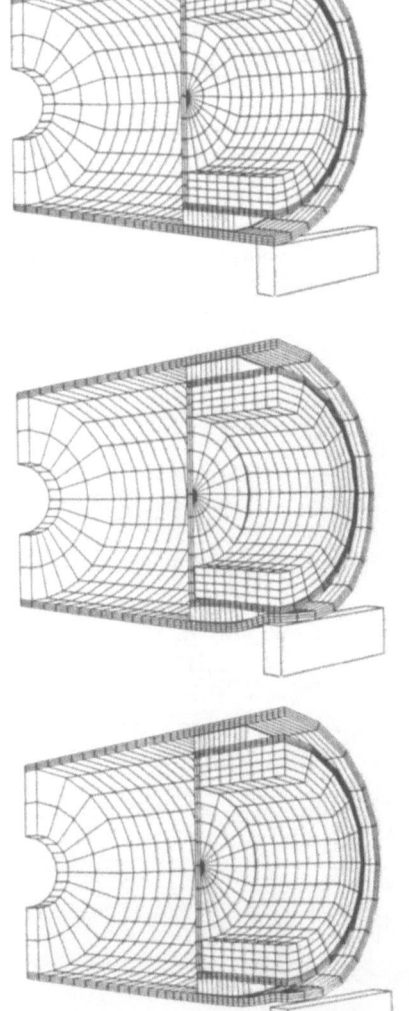

FIG. 4-1. Calculations were made to assess a canister's resistance to lateral rail impacts.

FIG. 4-2. For a lateral rail impact, the plastic deformation of the canister's outer case must dissipate all the kinetic energy.

FIG. 4-3. As the computational model rebounds off the rail, the clearance between the canister outer case and the shipping can grows.

FIGURE 4

11.2 Silo/Soil Island Response Calculation

We have been using DYNA3D to study the response of buried structures to the large scale ground motion that would result from detonation of a nuclear device. The procedure that we use involves the concept of a soil island, i.e., we assume that the structure does not affect the ground motion outside some local neighborhood surrounding the structure. The ground motion from the explosion is calculated using a two-dimensional Lagrangian hydrocode, TENSOR, (Burton, 1982) with sophisticated geologic material models. Output from this two-dimensional simulation then provides boundary conditions for the DYNA3D model in the form of nodal position time histories for each node on the soil island boundary (see

Fig. 5-1. The air blast overpressure at the surface of the soil island is calculated in both TENSOR and DYNA3D using Brode's formulas. We present here a
generic simulation.

Figure 5-2 shows contours of pressure in the soil island and silo structure
shortly after the shock front from the air blast has passed the far edge of the
soil island.

Further details are contained in Goudreau, Bailey, et al. (1983).

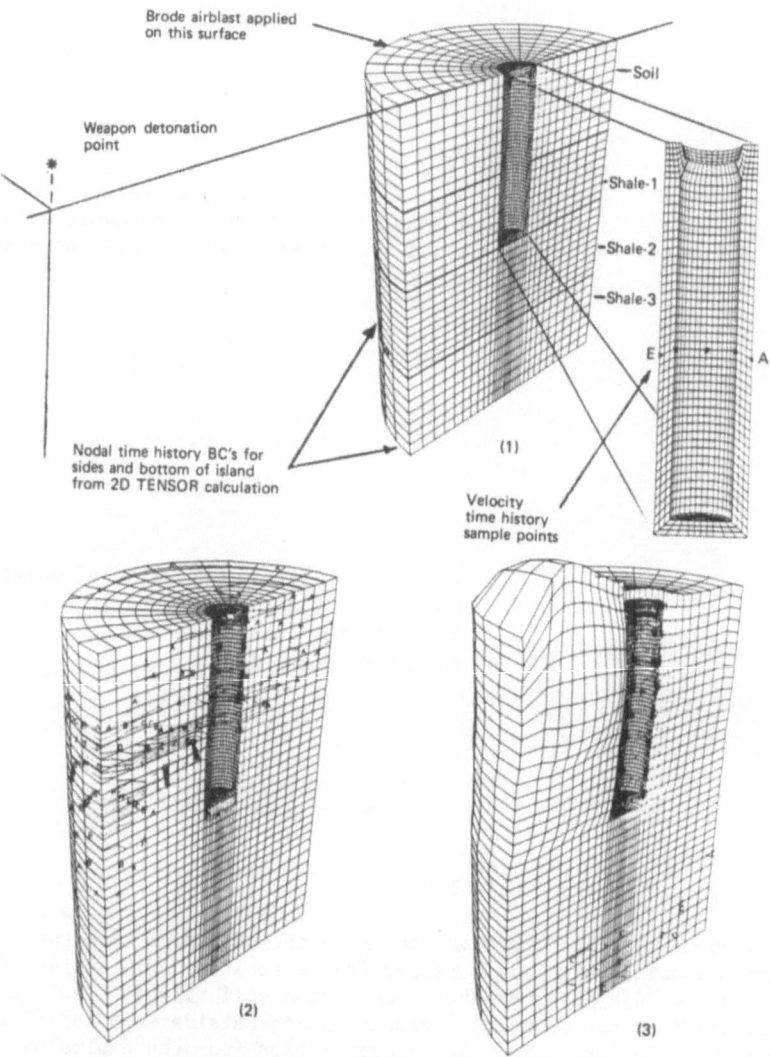

FIG. 5. Silo-soil island response calculation.
         (1) DYNA3D model
         (2) Pressure contours at time of air blast arrival
         (3) Pressure contours at time of peak ground motion

REFERENCES

Bammann, D.J. (1983), "An Internal Variable Model of Viscoplasticity," to appear in the Proceedings of the Workshop on Media with Microstructure and Wave Propagation, Michigan Technical University, January, 1983.

Bathe, K.J., E.L. Wilson, and F.E. Peterson (1974), "SAPIV —A Structural Analysis Program for Static and Dynamic Response of Linear Systems," EERC 73-11, University of California, Berkeley.

Bathe, K.J., E.L. Wilson, and R.H. Iding (1974), "NONSAP, A Structural Analysis Program for Static and Dynamic Responses of Nonlinear Systems," Rept. No. UCSESM 74-3, University of California, Berkeley.

Bathe, K.J. (1975), Revised (1978), "ADINA — A Finite Element Program for Automatic Dynamic Incremental Nonlinear Analysis," Rept. 82448-1. Acoustics and Vibration Laboratory, Mechanical Engineering Department, Massachusetts Institute of Technology, Cambridge, MA.

Bathe, K.J. (1982), *Finite Element Procedures in Engineering Analysis*, Prentice-Hall.

Bertholf, L.D. (1968), "TOODY II, A Computer Program for Two-Dimensional Wave Propagation," Sandia National Laboratories, Albuquerque, NM, Rept. SC-RR-68-41.

Burton, D.E., L.A. Lettis, Jr., J.B. Bryan, and N.R. Frary (1982), "Physics and Numerics of the TENSOR Code," University of California, Lawrence Livermore National Laboratory, Rept. UCID-19428.

Chiesa, M. and M. Callabresi (1981), "Nonlinear Analysis of a Mitigating Steel Nose Cone," *Comp. Struct.*, *13*, 295.

Chun, R.C. et al. (1981), "Uncertainty in Soil-Structure Interaction Analysis of a Nuclear Power Plant, a Comparison of Linear and Nonlinear Analysis Methods," SMiRT-6, University of California, Lawrence Livermore National Laboratory, Rept. UCRL-84199.

Cochran, S.G. and J. Chan (1979), "Shock Initiation and Detonation Models in One and Two Dimension," University of California, Lawrence Livermore National Laboratory, Rept. UCID-18024.

Cray-1 Computer System CFT Reference Manual (1978), Cray Research Incorporated, Bloomington, MN, Publication No. 2240009.

Farhoomand, I. and E.L. Wilson (1970), "A Nonlinear Finite Element Code for Analyzing the Blast Response of Underground Structures," U.S. Army Waterways Experiment Station, Contract Report N-70-1.

Flanagan, D.P. and T. Belytschko (1981), "A Uniform Strain Hexahedron and Quadrilateral with Orthogonal Hourglass Control," *Int. J. Numer. Meths. Eng.*, *17*, 679.

Flanagan, D.P. and T. Belytschko (1983), "Eigenvalues of the Uniform Strain Hexahedron and Quadrilateral," Sandia National Laboratories, Albuquerque, NM, Rept. SAND-83-0441J.

Goudreau, G.L. and R.L. Taylor (1973), "Evaluation of Numerical Methods in Elastodynamics," *J. Comp. Meths. Appl. Mech. Eng.*, *2*, 69.

Goudreau, G.L. and J.O. Hallquist (1979), "Synthesis of Hydrocode and Finite Element Technology for Large Deformation Lagrangian Computation," *Proc. CAFEM-5*, West Berlin, Germany.

Goudreau, G.L. and J.O. Hallquist (1982), "Recent Developments in Large Scale

Finite Element Hydrocode Technology," *J. Comp. Meths. Appl. Mechs, Eng., 30.* Presented at FENOMECH '81, Stuttgart, West Germany.

Goudreau, G.L. (1983), "Ispra Courses on Structural Dynamics —Lecture Notes — Element Choices for Explicit and Implicit Nonlinear Finite Element Computation," University of California, Lawrence Livermore National Laboratory, Rept. UCRL-89172, presented in Ispra, Italy, May 1983.

Goudreau, G.L., R.A. Bailey, J.O. Hallquist, R.C. Murray, and S.J. Sackett (1983), "Efficient Large-Scale Finite Element Computations in a CRAY Environment," University of California, Lawrence Livermore National Laboratory, Rept. UCRL-89385, prepared for submittal to the 1983 ASME Winter Meeting, Boston, MA., November 1983.

Hallquist, J.O. (1976), "A Procedure for the Solution of Finite-Deformation Contact-Impact Problems by the Finite Element Method," University of California, Lawrence Livermore National Laboratory, Rept. UCRL-52066.

Hallquist, J.O. (1976), "Preliminary User's Manuals for DYNA3D and DYNAP (Nonlinear Dynamic Analysis of Solids in Three Dimensions)," University of California, Lawrence Livermore National Laboratory, Rept. UCID-17268.

Hallquist, J.O. (1978), "DYNA2D —An Explicit Finite Element and Finite Difference Code for Axisymmetric and Plane Strain Calculations (User's Guide)," University of California, Lawrence Livermore National Laboratory, Rept. UCRL-52429.

Hallquist, J.O. (1978), "A Numerical Treatment of Sliding Interfaces and Impact (1978)," in *Computational Techniques for Interface Problems,* K.C. Park and D.K. Gartling (eds.), AMD Vol. 30, ASME, New York.

Hallquist, J.O. (1979), "NIKE2D: An Implicit, Finite-Deformation, Finite-Element Code for Analyzing the Static and Dynamic Response of Two-Dimensional Solids," University of California, Lawrence Livermore National Laboratory, Rept. UCRL-52678.

Hallquist, J.O. (1979), "Implicit Treatment of the Large Deformation Response of Inelastic Solids with Slide-Lines," presented SMiRT5, Vol. M, Berlin.

Hallquist, J.O. (1981), "NIKE3D: An Implicit, Finite-Deformation, Finite Element Code for Analyzing the Static and Dynamic Response of Three-Dimensional Solids," University of California, Lawrence Livermore National Laboratory, Rept. UCID-18822.

Hallquist, J.O. (1980), Rev. 1 (1982), "User's Manual for DYNA2D —An Explicit Two-Dimensional Hydrodynamic Finite Element Code with Interactive Rezoning," University of California, Lawrence Livermore National Laboratory, Rept. UCID-18756.

Hallquist, J.O. (1982), "DYNA3D User's Manual (Nonlinear Dynamic Analysis of Solids in Three Dimensions)," University of California, Lawrence Livermore National Laboratory, Rept. UCID-19592.

Hallquist, J.O. (1983), "Theoretical Manual for DYNA3D," University of California, Lawrence Livermore National Laboratory, Rept. UCID-19401.

Hallquist, J.O. (1983), "NIKE2D: A Vectorized, Implicit, Finite Deformation, Finite Element Code for Analyzing the Static and Dynamic Response of 2D Solids," University of California, Lawrence Livermore National Laboratory, Rept. UCID-19677.

Hallquist, J.O. (1983), "Lecture Notes for Bulk and Hourglass Viscosities in Wave Propagation Codes," University of California, Lawrence Livermore National Laboratory, Rept. UCRL-89156, presented in Ispra, Italy, May 1983.

Hallquist, J.O. (1983), "Lecture Notes for Contact-Impact Algorithms for Large Deformation Finite Element Analysis," University of California, Lawrence Livermore National Laboratory, Rept. UCRL-89157, presented in Ispra, Italy, May 1983.

Hilber, H.M., T.J.R. Hughes, and R.L. Taylor (1977), "Improved Numerical Dissipation for Time Integration Algorithms in Structural Dynamics," *Earthquake Engineering and Structural Dynamics, 5*, 283.

Hilber, H.M., T.J.R. Hughes, and R.L. Taylor (1978), "Collocation, Dissipation and 'Overshoot' for Time Integration Schemes in Structural Dynamics," *Earthquake Engineering and Structural Dynamics, 6*, 99.

Hughes, Thomas J.R., R.L. Taylor, J.L. Sackman, A.C. Curnier, and W. Kanoknukulchai (1976), "A Finite Element Method for a Class of Contact-Impact Problems," *J. Comp. Meths. Appl. Mechs. Eng., 8*, 249.

Hughes, Thomas J.R. (1979), "Implicit-Explicit Finite Element in Nonlinear Transient Analysis," *Comp. Meths. Appl. Mech. Eng., 17/18*, 159.

Hughes, Thomas J.R. (1980), "Generalization of Selective Integration Procedures to Anisotropic and Nonlinear Media," *Int. J. Numer. Meths. Eng., 15*, No. 9, 1413.

Hughes, Thomas J.R. (1982), "Stress-Point Algorithm for a Pressure-Sensitive Multiple-Yield-Surface Plasticity Theory," University of California, Lawrence Livermore National Laboratory, Rept. UCID-19339.

Hughes, Thomas J.R. (1983), "An Element-by-Element Solution Algorithm for Problems of Structural and Solid Mechanics," *Comp. Meths. Appl. Mech. Eng., 36*, No. 2, 241.

Johnson, G.R. (1976), "Analysis of Elastic-Plastic Impact Involving Severe Distortions," *J. Appl. Mechs., 43*, No. 3, 439.

Johnson, G.R. (1981), "Recent Developments and Analyses Associated with the EPIC-2 and EPIC-3 Codes," presented at Struct. Mat'ls. Conf. of Winter Meeting ASME, Washington, D.C.

Key, S.W. (1974), "HONDO — A Finite Element Computer Program for the Large Deformation Dynamic Response of Axisymmetric Solids," Sandia National Laboratories, Albuquerque, NM, Rept. 74-0039.

Key, S.W. (1974), "A Finite Element Procedure for Large Deformation Dynamic Response of Axisymmetric Solids," *Comp. Meths. Appl. Mech. Engr., 4*, 195.

Key, S.W., Z.E. Beisinger, and R.D. Krieg (1978), "HONDO II — A Finite Element Computer Program for the Large Deformation Dynamic Response of Axisymmetric Solids," Sandia National Laboratories, Albuquerque, NM., Rept. No. 78-0422.

Krieg, R.D. and S.W. Key (1976), "Implementation of a Time Dependent Plasticity Theory into Structural Computer Programs," *Constitutive Equations in Viscoplasticity: Computational and Engineering Aspects* (ASME), New York, NY, AMD-20, 125.

Lee, E.L. and C.M. Tarver (1979), "A Phenomenological Model of Shock Initiation in Heterogeneous Explosives," University of California, Lawrence Livermore National Laboratory, Rept. UCRL-83618.

Maenchen, G. and S. Sack (1964), "The TENSOR Code," *Meths. Comp. Phys., 3*, Academic Press, 181.

MARC Programmers and Users Manual, MARC Analysis Research Corp., Palo Alto, CA.

Marcal, P.V. and I.P. King (1967), "Elastic-Plastic Analysis of Two-Dimensional Stress Systems by the Finite Element Method," *Int. J. Mech. Sci., 9*, 143.

Matthies, H. and G. Strang (1979), "The Solution of Nonlinear Finite Element Equations," *Int. J. Numer. Meths. Eng., 14*, No. 11, 1613.

Nagtegaal, J.C., D.M. Parks, and J.R. Rice (1974), "On Numerically Accurate Finite Element Solutions in the Fully Plastic Range," *Comp. Meths. Appl. Mech. Eng., 4*, 153.

Nagtegaal, J.C. (1982), "On the Implementation of Inelastic Constitutive Equations with Special Reference to Large Deformation Problems," *Comp. Meths. Appl. Mech. Eng., 33*, 469.

Newmark, N.M. (1959), "A Method of Computation for Structural Dynamics," *Proc. ASCE, 85*, EM3.

Noh, W.F. (1964), "CEL: A Time-Dependent, Two-Space Dimensional, Coupled Eulerian-Lagrange Code," *Meths. Comp. Phys., 3*, Academic Press, 117.

Park, K.C. (1975), "Evaluating Time Integration Methods for Nonlinear Dynamic Analysis," in *Finite Element Analysis of Transfer Non-linear Behavior* (T. Belytschko, J.R. Osias, and P.V. Marcal, Eds.). Appl. Mech. Symposia Series, ASME, New York.

Steinberg, D.J. and M.W. Guinan (1978), "A High-Strain-Rate Constitutive Model for Metals," University of California, Lawrence Livermore National Laboratory, Rept. UCRL-80465.

Swegle, J.W. (1978), "TOODY IV — A Computer Program for Two-Dimensional Wave Propagation," Sandia National Laboratories, Rept. SAND-78-0552.

Taylor, R.L., E.W. Wilson, and S.J. Sackett (1981), "Direct Solution of Equations by Frontal and Variable Band, Active Column Methods," *Nonlin. Fin. El. Anal. Struct. Mech.*, Springer, Berlin.

TRANAL User's Guide, Part I (Small Strain, Small Displacement Version) (1979), Weidlinger Associates, New York, Rept. DNA 4960F.

Tuft, D.B. and C.S. Godfrey (1972), "Computer Analysis of a Three-Dimensional Mass Focus Projectile Device," University of California, Lawrence Livermore National Laboratory, Rept. UCRL-87678.

Walker, H.F. (1979), "Numerical Solution of Nonlinear Equations," University of California, Lawrence Livermore National Laboratory, Rept. UCID-18285.

Wilkins, M.L. (1964), "Calculation of Elastic Plastic Flow," *Meths. Comp. Phys., 3*, Academic Press, 211.

Wilkins, M.L. (1969), "Calculation of Elastic-Plastic Flow," University of California, Lawrence Livermore National Laboratory, Rept. UCRL-7322, Rev. 1.

Wilkins, M.L., R.E. Blum, et al. (1975), "A Method for Computer Simulation of Problems in Solid Mechanics and Gas Dynamics in Three Dimensions and Time," University of California, Lawrence Livermore National Laboratory, Rept. UCRL-51574, Rev. 1.

Wilkins, M.L. (1980), "Use of Artificial Viscosity in Multidimensional Fluid Dynamic Calculations," *J. Comp. Phys., 36*, 281.

Wilson, E.L. (1970), "SAP — A General Structural Analysis Program," SESM Report 70-20, Dept. of Civil Engineering, University of California, Berkeley.

Woodruff, J.P. (1976), "KOVEC User's Manual," University of California, Lawrence Livermore National Laboratory, Rept. UCID-17306.

Zienkiewicz, O.C. (1977), *The Finite Element Method*, McGraw-Hill.

GENERAL DISCUSSION

SANDLER: You spoke about explicit and implicit codes and dynamic and static situations. Is there ever a case of a strongly nonlinear material in which a three-dimensional dynamic situation is well handled by an implicit code?

GOUDREAU: What you want to ask is the computational cost of direct solving in 3D? Well, we've done problems of periodic symmetry. For example, axisymmetric problems that have either periodic bolts in a flange, or involving ring connections where there's some local periodic perturbation of an axisymmetric geometry, and one can model a pie-shaped section where a three-dimensional model is not constrained by band width. We've done problems on the order of an hour of CRAY time. We're in that regime right now.

I should use this opportunity to comment on my talk that in two dimensions, the availability of one million and two million word memory CRAYs have now allowed us to do ninety-five percent of our two-dimensional implicit problems in core without IO. That has really made this approach competitive. In 3D, you're right, equation solving costs make a majority of problems still very, very difficult to do in a nonlinear case where you may want to do hundreds of factorizations.

I should point out that iterative schemes are really being studied aggressively. Hughes' work in "element by element" is coming of age. I think we'll see within the next year or two significant iterative strategies that will make 3D implicit possible.

SCOTT: Where do you get your material information? Do you have a strong coupling with people that are actually testing the explosives mechanically under combined stresses?

GOUDREAU: Yes. I should have used a view-graph to indicate some sort of material library. But basically, what the general codes do is localize the motion, define a strain measure and just exactly in the sense in which Tom Hughes' presentation presented the strain history available to a constitutive point, go in and determine typically a bulk law. We tend to separate the bulk and deviatoric parts of the model.

So we go in with the volumetric strain, compute the pressure or the mean stress and then go into a deviatoric routine which takes the strain deviators, and computes the stress deviators, and then combine these results. The whole library of all the models that you'll be hearing about here, is available.

SCOTT: I'm not talking about essentially a library of models. I mean, there is some organization that you talk to that does mechanical tests and determines what the explosive charges actually do?

GOUDREAU: Oh, right, that's right. It's not vectorized. We have this problem of, to what extent do you want to automate the material library? You know, there are those that want to take big codes like this and push the button and run.

My philosophy is to make the assembly of material data into the code difficult so that the analyst has to take some responsibility for deciding what model and its domain of applicability because typically, these models and the way they're implemented in the codes do not have adequate bounds on the domain of applicability of the model.

So unless the analyst himself decides how to draw his stress-strain curve when he puts in his input or what temperature range he's going to quantify the data, he's liable to do calculations that are going to be way beyond the applicability of the model. We're not to the point where all materials and all regimes can have their material models so automated that the whole data base will come in and do everything for you. Some aspects of the hydrodynamics problem have been done that way. But I would say that in our case of strength and deviatoric plasticity, we're a long way from fully automating the modeling.

SCOTT: You're using words that I really don't identify.

GOUDREAU: We get the data from somebody who knows something about it.

PISTER: I think you're still missing Scott's question: it is that people do actually test and you simulate the test, don't you?

GOUDREAU: That's right.

PISTER: And your predictions correlate with the test results?

GOUDREAU: Yes. The philosophy, though, is to use the tests, full-scale or intermediate size tests, to check the models. We don't do a lot of that type of normalization, but that can be done. You really want to use lab tests as the basis for characterizing materials.

SCOTT: I'm still unsatisfied, but you may wish to cut me off.

PISTER: Why don't you nail him afterwards, Ron?

NEEDLEMAN: A quick question to get a feeling for the order of magnitude of computer time involved — Was it thirty-six milliseconds or microseconds per zone per cycle?

GOUDREAU: Right, a thousand zones, a thousand cycles.

NEEDLEMAN: Is a zone an element?

GOUDREAU: Yes.

NEEDLEMAN: Is a cycle an iteration or a time step?

GOUDREAU: That was an explicit cost, and so that would be a time step. There is no iteration.in that representation.

NEEDLEMAN: So it's per time step per element.

GOUDREAU: That's right.

NEEDLEMAN: And is it milliseconds or microseconds?

GOUDREAU: Those were microseconds.

ODEN: I'm always impressed, Jerry, with your color three-dimensional graphics. But nevertheless, I am also concerned that someone is not seriously worried about the quality of some of these large-scale computations. I mean, I appreciate the fact that to handle these large problems, that things such as under-integration are necessary and you can realize very large factors of savings in computations, but as you well know, this introduces very serious, spurious modes in the solution. One can, therefore, introduce hourglassing viscosity to try to overcome these. But, I dare say, that it is not known what this artificial viscosity does to the structure and quality of the solution.

GOUDREAU: Right.

ODEN: It's not even known for the simplest linear problems. I hope that in the interests of solving problems faster and faster you don't ignore also a thrust of your research into areas really designed to determine qualitatively what's going on with these kinds of approximations.

GOUDREAU: We're very interested in that.

ODEN: Hourglassing type instabilities are predominant in the hydromechanics literature. It's well known that, if you introduce mechanisms to dissipate them, you may also dissipate significant parts of the solution.

GOUDREAU: I think part of that is because the user is usually given a free parameter for the intensity of the hourglass control. As a result, when he sees that he is not containing it adequately, he jacks up that parameter until he effectively has damped out the oscillation. At that point, if he had calculated the global energy balance, he would have seen the large dissipative effect of that viscosity. There's still a lot of work that can be done on this: norms on the solution that help track that kind of unboundedness. That's why we've looked at Flanagan-Belytschko work and are very open to it.
    I don't think in explicit computation we really will consider multi-point integration. But in our implicit codes, primarily because we were not successful in the early days with implicit control of hourglassing, we did all our implicit work with multi-point, two by two by two quadrature with the under-integrated pressure. That has no hourglass problem, at least for problems of moderate material strength.
    You realize that in the hydrodynamic element, even in that element, the deviators can get washed out at high pressure and the hourglass modes can reappear.
    At least in this engineering computation we've had parallelism of implicit calculation with the multi-point integration and the explicit one with the hourglass problem. We've tried to watch that, perhaps with not as much rigor as we would like. Again that's one of the problems: large computing capability tends to be in the hands of the designer and the design analyst rather than in the hands of the algorithm developer or even the material modeler. I would like to see more use of large codes in a more academic setting, get you people access to the CRAYs of the world.

TRULIO: A comment. I'd just like to point out that there are some basic, one might say in principle, difficulties with internal consistency in discretization that would drive you toward simple local field structures. The one you assumed is a homogeneous strain as in most of the codes.

GOUDREAU: That's right.

TRULIO: That goes along with homogeneous stress. There is really no way to have this in axisymmetric fields. That's because the hoop strain can't possibly be homogeneous. Homogeneous strain requires a linear displacement field. A linear displacement field means a variable hoop strain throughout the element, which would mean a variable hoop stress and other stress components throughout the element. That degree of consistency is beyond you for geometric reasons.

GOUDREAU: So I don't follow your conclusion on that. You're saying that in axisymmetry, you have a unique problem that hasn't been licked yet? Is that what you're saying?

TRULIO: That problem arises in axial symmetry. It happens not to come up in plane motion, but the symmetry can prevent you from fully implementating a local field structure assumption. Going to higher order schemes, for instance, will not get rid of that problem.

GOUDREAU: Right. I have no comment to that. Is there anyone else that wants to comment?

ODEN: I don't think so; you do not assume a homogeneous strain field in an element necessarily. Your strain field varies throughout the element. It need not be linear, right?

TRULIO: To make it internally consistent will require that whatever field structure you assume —

ODEN: What do you mean by internally consistent?

TRULIO: Well, in the plane of motion, an azimuthal plane, let's say, you assume some simple field structure, much simpler than the actual one. You may take the field as spatially quadratic. You may go to two or three terms in a Fourier expansion. But the structure you assume has implications that are hard to live with, if carried through rigorously. It means continuous variation of both stress and strain in axisymmetric fields.

GOUDREAU: I think there is a certain redundancy in discretization as to whether you're approximating the stress/strain relation spatially or the strain/displacement relationship. And whether you think of it as a mean strain or point strain or single stress — an average stress over a variable strain field; I think you can say the same thing several ways.

KRIEG: I don't have a question, just a comment. I find it interesting to note that in one of the analyses that you did out there at Livermore, there were something like a quarter of a billion calls to the constitutive model for one calculation, for one of those problems. A quarter of a billion calls.

GOUDREAU: A billion? Is that ten to the ninth? That's seven thousand seconds [calculating at the blackboard]. I was just giving an estimate of a run time of thirty microseconds per mesh cycle. Now, you're taking the number of cycles times the number of zones, so it's the same thing, seven thousand seconds.

BALADI: The solution of the example you have presented is sensitive to the conditions along the impact area, i.e., the stress distribution on the impact area. Do you know these conditions? If you do not, what did you assume?

GOUDREAU: Are you asking me whether the discretization was biased towards the point of impact? Slightly in that problem, and I think normally it is. But this particular study did look at sort of random impacts. So they took the same model and did drop it at half a dozen different locations along the length of the impact bed. Sometimes you do have to localize your resolving power.

A lot of the analysts would like to have enough CRAYs that you didn't have to worry about that. You put enough zones everywhere so that whatever is going to happen, will happen.

BALADI: In most codes the treatment of the impact area is arbitrary. I do not know anybody who could describe accurately the condition between the two bodies at the moment of impact. Can you comment on this? The question is, do you know the condition at the point of impact?

GOUDREAU: I think when you have extreme — I didn't get into rezoning — but clearly when you're into a very high flow problem, Lagrangian approach presents a problem; Lagrangian code will solve that, but zones will diminish rapidly.

HUGHES: Jerry, since you're involved in unprecedented calculations of very critical engineering designs, I wonder if you could make a few remarks to this group as to what you think the crucial issues are in material modeling. The calculations are only as good as the material models that you're using. What, theoretically and experimentally, needs to be done to upgrade the models you're using, and what are the things you worry about in some of the models you're using?

GOUDREAU: Isn't that what this workshop is supposed to answer? I think that understanding the portability of material models between codes is very important.

I think having — what I call — a stress point driver in a university environment (that is a little routine and we have it — Sam and I keep talking about it; maybe he's got one now) is important. This is a little portable one point code that takes in the arbitrary multiaxial strain history and evaluates a constitutive routine.

Perhaps you need to sit at one of your new modern microcomputers with nice graphics, and joy-stick your material model through its multiaxial strain history. For this you don't need a CRAY! You're just doing one stress point. If you can get into that mode and talk to the people with the large codes, then we can port these subroutines around and use them.

Tom Hughes' question is about constitutive modeling independently of any codes, and you all could contribute in that respect. I'd say that with respect to this question, I think that better estimates of the tangent operator are useful. I think we've been very lucky that BFGS has compensated for the very approximate tangent operators that we've gotten out of our constitutive routine.

So understanding that sensitivity of the stress at time $N+1$ to the motion at $N+1$ in terms of the tangent operator, consistent linearization ideas, I think that's important. And I didn't mean to imply that everything is well understood.

I'm very interested in strain-softening. I have thirty more computer slides generated on the example that has a strain-softening implication, although maybe later in the week after we've heard from Ivan we can discuss that example.

I think there's a lot of outstanding issues in material modeling.

HUGHES: I want you to continue telling me about them. What about in the large deformation routines? Where do you feel you're on unsafe ground?

GOUDREAU: Yes, yes, and yes.

HUGHES: What about your large deformation models?

GOUDREAU: Most of the uncertainties so far in the finite deformation issue, say the finite deformation issues you raised on the choice of rate, I believe are more local and almost failure oriented things. I can show in the nose cone example that the overall deformation of the body is well characterized with Jaumann rate, okay? We can get eighty percent strains in that thing and not worry about whether Jaumann rate is good enough. But if you want to know what the stress is at a point, for example if you caught that nose cone down there, you've got grooves in that thing.

It's the inverse crack blunting problem. You start with a groove and you collapse it into a crack and you shear it simultaneously.

You've got extreme stress concentration at a point. We think it heats up enough that we get thermal softening locally at a point. And occasionally these things break and we don't know why.

So I mean, understanding material behavior well enough that you can accurately predict the point behavior incipient to failure, that's where a lot of research is going on. Mark Wilkins is doing a lot, and I'm sure a lot of others are.

In the engineering analysis community, we still cannot believe in general the fracture indication in a large complicated three-dimensional analysis.

To be able to do the large plastic flow itself, to understand how loads redistribute and get a general idea of collapse or failure of a structure, I think is only where we're at; and there is a lot more to be done.

BUDIANSKY: In connection with axisymmetrical calculations, have you ever looked to see whether the axisymmetric pattern wants to quit being axisymmetric and how sensitive this desire would be to your constitutive modeling?

GOUDREAU: Are you talking about an axisymmetric problem, say, in 3D, with, say, some axisymmetric load and look at sensitivity to the loading, let's say, or to anything, even material? No, I haven't done that.

BUDIANSKY: I would just suggest that there are issues here that I feel are not being addressed. People have been trying to get you to say, and I think with great fortitude you've not been saying, that there is an issue concerned with the choice of constitutive equation, which addresses whether you really can get in trouble or not by being just a little bit off, a little bit different in what you picked.

GOUDREAU: You're saying globally.

BUDIANSKY: Yes, globally. Now, Sia has touched on this and other people have touched on this, a question of corners, questions of non-coaxiality. Put in a little bit of change and suddenly what you think is axisymmetric might quit or you get a shear band or something like that.

GOUDREAU: Right.

BUDIANSKY: I think these are the issues that people have been trying to get into the discussion. Fracture is important. Yes, no doubt about it. But that's not really what is —

PISTER: If I can interject, aren't you really asking, what are the partial derivatives of the designed performance, however defined, with respect to problem parameters? We look at all sorts of other kind of continuity in our systems, but very often we forget the sensitivity of the solution of the problem with respect to the data of the problem.

BUDIANSKY: Yes.

GOUDREAU: Let me give you some examples I'm aware of, and I'll try to address this. For example, Ray Krieg was just telling me about their large field, geologic problems where they've tried to randomly vary material property spatially based on the uncertainty in the parameters. That's just uncertainty in the parameters of standard models.

Mark Wilkins, I believe, has done some work in varying the parameters of a plasticity model where you try to look at either instabilities or whatever would come from that sensitivity question.

But I think to some extent you're asking me about features of models; could one introduce corners in a yield surface, that one normally isn't doing, and see whether that qualitative change in the constitutive routine has radical global implications.

BUDIANSKY: Yes. Well, that's right. And it's not so new.

GOUDREAU: New to large computations.

BUDIANSKY: Well, perhaps.

GOUDREAU: That's all I'm saying.

BUDIANSKY: Because we know — I know there are people here with a large background of dramatic experience in the kind of phenomena I'm talking about.

I mean, Alan Needleman can tell you that if he puts a little non-coaxiality into a constitutive equation, he can get a sheet under large strains to develop shear bands; and when he doesn't, it won't develop shear bands.

Now, that's very important when you talk about large deformation of sheets. Now, in your business, I suppose it's rather important to know whether when you're squishing something down it will continue to squish smoothly or suddenly kick off sideways, things like that.

GOUDREAU: Absolutely, right.

BUDIANSKY: And this is a major partial derivative, if you'd like, with respect to constitutive modeling.

GOUDREAU: Either we put the codes in the hands of the researchers to be able to study these issues or you pose some problems for us to run.

CHAPTER IV

ON AN IMPLEMENTATION OF FINITE STRAIN PLASTICITY
IN TRANSIENT DYNAMIC LARGE-DEFORMATION CALCULATIONS

Samuel W. Key

RE/SPEC Inc., P.O. Box 14984
Albuquerque, New Mexico 87191

1. INTRODUCTION

In setting out to develop a computational procedure for the large deformation dynamic response of solids, there are four separate issues with which one
must deal. They are the underlying mechanics, the constitutive theories of interest, the spatial discretization, and the time integration scheme. In actual
fact, decisions in one area will affect the possibilities in another, but it is
nonetheless useful to examine them individually. Here, only the issue of a
finite strain plasticity constitutive model will be considered. The behavior of
two separate finite strain plasticity models will be considered in two separate
homogeneous deformations. The two plasticity models differ only in their use of
separate invariant time derivatives of the stress. While superficially these
would appear to be similar physical models, the use of separate invariant time
fluxes portends the possibility of different behavior. Indeed, in this case,
significantly different stress strain behavior results.

2. UNDERLYING MECHANICS

The treatment in this section is general for the sake of brevity and com-
pleteness. The treatment of continuum mechanics found in Truesdell and Toupin
(1960) is used. The nonpolar case is considered.

2.1 Equations of Motion

A body V is given which occupies a finite region of Euclidean space. Subjected to prescribed body forces and surface tractions, the body V undergoes
the motion $x^i = \chi^i(X^\alpha, t)$. The particles of the body are identified by the coordinates X^α. They are referred to as material coordinates, and the relations of
the particles to the coordinates X^α does not change in time. The places in space
which the particles occupy during the motion are identified by the coordinates
x^i. The function χ^i describes the motion of the particles X^α through space as a
function of time t. It is the motion χ^i which is sought.

The place occupied by the body at $t = 0$ is taken as the reference configuration. In this configuration, the body is assumed to be strain free, though

not necessarily stress free. Only material coordinates X^α which coincide with the spatial coordinates x^i in the reference configuration are considered. Thus, in the reference configuration, $\chi^i(X^\alpha,0) \equiv X^\alpha$.

The problem is stated in terms of the principle of virtual work. The differential form

$$\delta\pi = \int_V \rho\ddot{x}^k \delta x_k \, d\upsilon + \int_V t^{km} \delta x_{k,m} \, d\upsilon - \int_V \rho f^k \delta x_k \, d\upsilon - \oint_{S^1} s^k \delta x_k \, da \qquad (1)$$

is to vanish at all points along the path of motion for all variations δx_k satisfying the displacement boundary conditions on S^2. The integration is performed over the current configuration of the body V, where ρ is the mass density in that configuration, $\ddot{x}^k$ is the acceleration, t^{km} is the Cauchy stress – the stress in the current configuration, f^k is the body force density in the current configuration, and s^k is the surface traction which is acting on S^1. The comma in $x_{k,m}$ denotes covariant differentiation.

The divergence theorem is employed to reveal the differential equations of motion. In anticipation of using the finite element method to generate approximate solutions, the case where $\delta x_{k,m}$ is only piecewise continuous is considered; Jones (1964), Prager (1967), and Key (1971). Interior surfaces where the discontinuities of $\delta x_{k,m}$ occur are denoted by S^0. Only surfaces S^0 which are stationary with respect to the material are considered. The situation is pictured in Figure 1 where n_k is the normal to S^0 and the symbols $+$ and $-$ denote the respective sides of the surface. The result is

$$\int_V (\rho\ddot{x}^k - t^{km}_{,m} - \rho f^k)\delta x_k \, d\upsilon + \oint_{S^0} (t^{km}_+ - t^{km}_-)n_m \delta x_k \, da + \oint_{S^1} (t^{km}n_m - s^k)\delta x_k \, da = 0. \qquad (2)$$

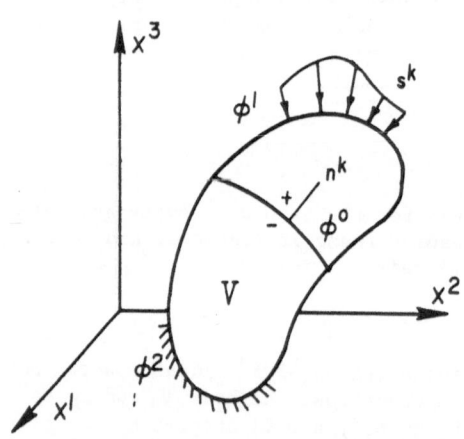

The differential form will vanish if and only if the respective integrands vanish. The resulting expressions are the equations of motion

$$t^{km}_{,m} + \rho f^k = \rho\ddot{x}^k \quad \text{in } V, \qquad (3)$$

the jump condition at a contact discontinuity

$$(t^{km}_+ - t^{km}_-)n_m = 0 \quad \text{on } S^0, \qquad (4)$$

and the traction boundary conditions

$$t^{km}n_m = s^k(t) \quad \text{on } S^1. \qquad (5)$$

The displacement boundary conditions are

$$\chi^i(X^\alpha,t) = \kappa^i(t) \quad \text{on } S^2. \qquad (6)$$

Only initial conditions which are homogeneous in place at time zero are considered. Thus, the initial conditions are given by

Fig. 1. The body V with surface tractions s^k on the boundary S^1 and a prescribed motion on the boundary S^2. An interior boundary S^0 with a unit normal vector n^k is pictured.

$$\chi^i(X^\alpha,0) = X^\alpha \quad \text{in } V, \qquad \dot{\chi}^i(X^\alpha,0) = v^i(X^\alpha) \quad \text{in } V. \tag{7}$$

It is important to realize that these remarks are completely general with regard to the scale of deformations being considered.

While at this point it appears as if the spatial or Eulerian description of the problem is being employed, actual usage is in terms of the material description. That is, all of the quantities are treated as functions of X^α and t. This approach has already been used with good results.

2.2 Kinematics

In finite deformations there are many strain measures which are useful. The majority of them can be computed from the deformation gradient F^k_α defined by

$$F^k_\alpha \equiv \frac{\partial \chi^k}{\partial X^\alpha}(X^\beta,t). \tag{8}$$

The velocity v^k is defined as

$$v^k(X^\alpha,t) \equiv \frac{\partial \chi^k}{\partial t}(X^\alpha,t). \tag{9}$$

The stretching is given by

$$d_{km} = \frac{1}{2}(v_{k,m} + v_{m,k}), \tag{10}$$

with the spin given by

$$w_{km} = \frac{1}{2}(v_{k,m} - v_{m,k}). \tag{11}$$

For use in the invariant time derivatives, the polar decomposition of the deformation gradient

$$F^k_\alpha = R^k_m V^m_\alpha \tag{12}$$

gives the rotation ω defined in terms of the time derivative of rotation R_{kn} provided by the polar decomposition

$$\omega_{km} = \dot{R}_{kn} R^n_m. \tag{13}$$

It is for this setting that suitable finite strain plasticity models are needed.

3. FINITE STRAIN PLASTICITY

Plasticity is the behavior characteristic of ductile metals. Figure 2 shows results which are typical of the behavior of a metal bar loaded first in uniaxial tension followed by uniaxial compression. The straight line representation in Fig. 2 is an idealization of this behavior. This is the approximation which results from the plasticity relations employed here (Goel and Malvern, 1970). Assuming the material to be initially isotropic, they translate uniaxial data into general triaxial behavior by using the notion of a universal hardening curve. A hypoelastic description is used where an invariant stress rate is related to the stretching. Thus,

$$\overset{\triangledown}{t}^{rs} = C^{rsmn} d_{mn}. \tag{14}$$

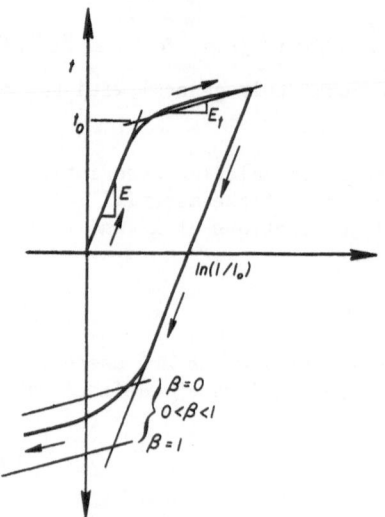

Fig. 2. The typical behavior of a ductile metal bar loaded first in uniaxial tension followed by uniaxial compression. The straight line approximation is characterized as an elastic modulus E, a yield stress t_0, a strain hardening modulus E_t, and a hardening parameter β, where kinematic hardening is obtained with $\beta = 0$, isotropic hardening is obtained with $\beta = 1$, and a linear combination of the two is obtained for β between zero and one.

When no yielding is occurring, C^{rsmn} is the isotropic tensor $\lambda g^{rs}g^{mn} + 2\mu g^{rm}g^{sn}$, where λ and μ are the Lamé parameters. When plastic yielding is occurring, that is, when $\frac{1}{2}\xi'_{ij}\xi'^{ij} - k^2 = 0$ and $\xi'_{rs}\xi'^{rs} > 0$, the moduli are replaced by

$$C^{rsmn} = \lambda g^{rs}g^{mn} + 2\mu(g^{rm}g^{sn} - n^{rs}n^{mn}), \qquad (15)$$

where

$$n^{rs} = \xi'^{rs}/[2k^2(1+H/3\mu)]^{\frac{1}{2}},$$

$$\xi'_{rs} = t'_{rs} - \alpha'_{rs},$$

$$\sqrt{2}k = (2/3)^{\frac{1}{2}}t_0 + \beta \frac{2}{3} H \int_0^t |d^p_{rs}|dt,$$

$$\breve{\alpha}_{rs} = (1-\beta) \frac{2}{3} H d^p_{rs}.$$

The prime denotes deviatoric components and the superscript p denotes the plastic strain. The isotropic hardening description ($\beta = 1$) preserves the approximate symmetry in tension and compression of the true stress versus the logarithmic strain curves. Kinematic hardening is included in a rather obvious way by letting the center of the yield surface α_{ij} move according to $\alpha_{ij} = (1-\beta)\frac{2}{3}Hd^p_{rs}$. When kinematic hardening is introduced ($\beta \neq 1$) the theory departs from a hypoelastic description.

Referring to Eq. 14, two separate invariant time derivatives of the Cauchy stress will be considered. The invariant time derivative of the Cauchy stress is composed of the material derivative of the Cauchy stress plus two additional terms involving a rotation. The first form of the invariant time derivative to be considered is the Jaumann rate. It is based on the spin w_{km}. Until very recently, the Jaumann derivative was considered the appropriate invariant time derivative to use in finite strain plasticity. However, in recent papers by Dienes (1979) and Nagtegaal and de Jong (1981), it was observed that the Jaumann derivative as the basis of a plasticity model possessed an undesirable trait in simple shear. Namely, kinematic hardening resulted in oscillatory shear stresses. The Jaumann derivative is given by

$$\overset{\vee}{t}^{rs} = \overset{\bullet}{t}^{rs} - w^r_m t^{ms} + t^{rm}w^s_m. \qquad (16)$$

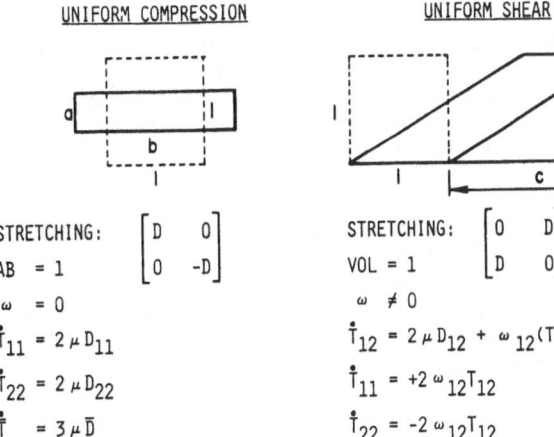

UNIFORM COMPRESSION

UNIFORM SHEAR

STRETCHING:

AB $= 1$

$\begin{bmatrix} D & 0 \\ 0 & -D \end{bmatrix}$

$\omega = 0$

$\overset{\scriptscriptstyle\triangledown}{T}_{11} = 2\mu D_{11}$

$\overset{\scriptscriptstyle\triangledown}{T}_{22} = 2\mu D_{22}$

$\overset{\scriptscriptstyle\triangledown}{T} = 3\mu\bar{D}$

RADIAL LOADING IN PLASTICITY

STRETCHING:

VOL $= 1$

$\begin{bmatrix} 0 & D \\ D & 0 \end{bmatrix}$

$\omega \neq 0$

$\overset{\scriptscriptstyle\triangledown}{T}_{12} = 2\mu D_{12} + \omega_{12}(T_{22} - T_{11})$

$\overset{\scriptscriptstyle\triangledown}{T}_{11} = +2\omega_{12}T_{12}$

$\overset{\scriptscriptstyle\triangledown}{T}_{22} = -2\omega_{12}T_{12}$

NON-RADIAL LOADING IN PLASTICITY

Fig. 3. The homo-
geneous motions of
uniform compression
and uniform shear.

(NOTE: $LN(L/L_0) = 5 \Rightarrow B = 148, C = 10$)

Dienes identified the Green–Naghdi time derivative of the Cauchy stress as a
derivative leading to a far more suitable prediction of material behavior under
finite strains. In simple shear, the Green–Naghdi rate results in a model of
material behavior with monotone increasing shear stress. The Green–Naghdi rate
is given by

$$\overset{\scriptscriptstyle\vee}{t}{}^{rs} = \overset{\bullet}{t}{}^{rs} - \omega^{r}_{m}t^{ms} + t^{rm}\omega^{s}_{m}. \tag{17}$$

4. SIMPLE HOMOGENEOUS DEFORMATIONS

Referring to Fig. 3, the behavior of these two elastic-plastic models will
be examined considering two simple homogeneous deformations, uniform compression
and uniform shear. Both deformations start with a unit square. In uniform compression, as one dimension of the unit square is reduced, the other dimension is
increased. The deformation imposed is volume preserving so that the product of
the length times the height of the deformed rectangle is unity. The horizontal
component of stretching is a constant D, the vertical component of the stretching is -D, and the shear components are identically zero. In this deformation,
both the spin and rotation are identically zero. Figure 3 shows the very simple
differential equations which are obtained for pure elastic behavior. If the
effective stress and effective strain are introduced as scalar measures of the
stress and strain, for elastic behavior they are related through the modulus 3μ.
Since these deformations are volume preserving, the effective strain is a useful
definition. When plasticity is introduced, uniform compression results in proportional loading.

In uniform shear, the top edge of the unit rectangle is moved horizontally
while the bottom edge is held fixed. The top edge and the bottom edge are kept
parallel. The horizontal and vertical components of the stretching are identically zero, and the shear components are given by D. Uniform shear of the unit
square is a constant volume deformation. In this case, neither the spin nor the
rotation is zero. The differential equations describing elastic behavior show
the coupling introduced by the invariant time derivative. When plasticity is
introduced, nonradial loading results.

Deformations leading to logarithmic strains on the order of five will be
considered. In uniform compression, a logarithmic strain of five implies a

horizontal dimension of b = 148.   In uniform shear, a logarithmic strain of five
implies a horizontal displacement of the top edge ten units to the right.

Referring to Fig. 4, in uniform compression with elastic behavior the horizontal stress is linearly related to the horizontal strain with the modulus given
by 2μ. Referring to Fig. 5, in uniform compression with elastic behavior, the
effective stress is related linearly to the effective strain with a modulus 3μ.

Referring to Fig. 6, in uniform shearing with elastic behavior the shear
stress is related to the shear strain in a more complex manner than linear. The
Jaumann time derivative results in oscillatory shear stresses. The Green-Naghdi
time derivative, denoted by the label "Dienes," provides a model of behavior
with monotone increasing shear stress with increasing shear strain. Both material models have an initial slope of 2μ.

For larger values of uniform shear, significant extensional strains develop.
Referring to Fig. 7, in uniform shearing with elastic behavior, the effective
stress versus the effective strain shows a more nearly linear behavior for the
Green-Naghdi time derivative. The departure from a constant slope of 3μ characterizes the departure from linear behavior at large strain. While one would not
expect to find a material with linear elastic behavior at these strain levels, it
is desirable to identify a linear behavior which may be used to characterize the
nature of nonlinearities in more complex models. The material model based on
the Green-Naghdi invariant time derivative approaches this ideal linear behavior.

Referring to Fig. 8, in uniform shearing with elastic-kinematic plastic behavior, the shear stress is again related in a complex manner to the shear strain.
The linear slope of 3.182×10^5 which is tangent to the strain hardening behavior
at low strains is the elastic plastic behavior for uniform compression. Again,
the Jaumann time derivative results in an oscillating shear stress, and the

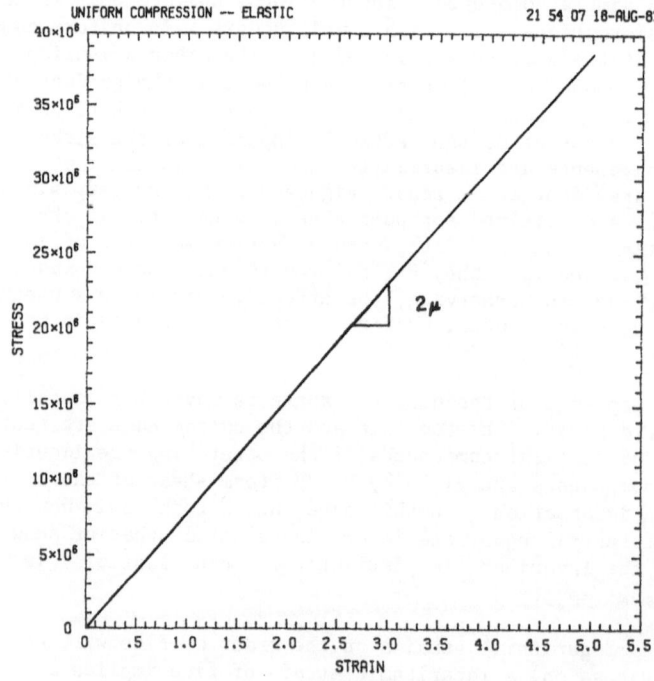

Fig. 4. The stress
versus strain response
of an elastic material
subjected to the homo-
geneous motion of uni-
form compression.

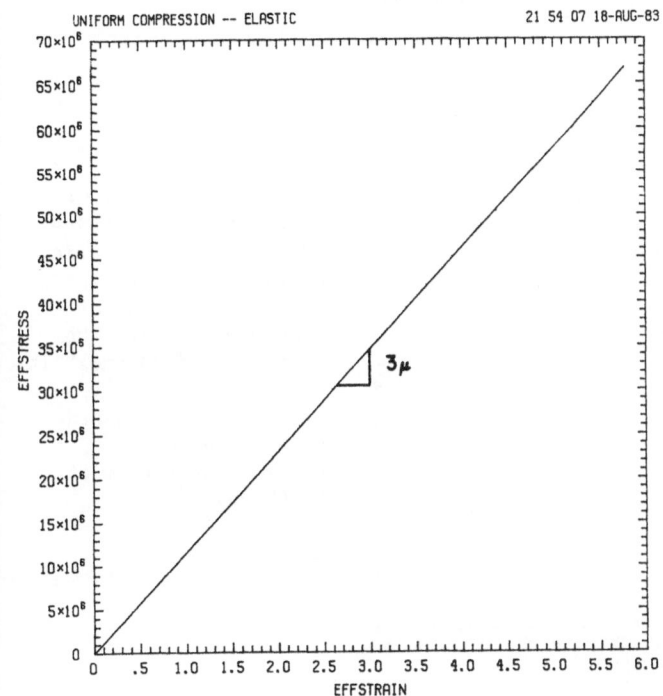

Fig. 5. The effective stress versus the effective strain response of an elastic material subjected to the homogeneous motion of uniform compression.

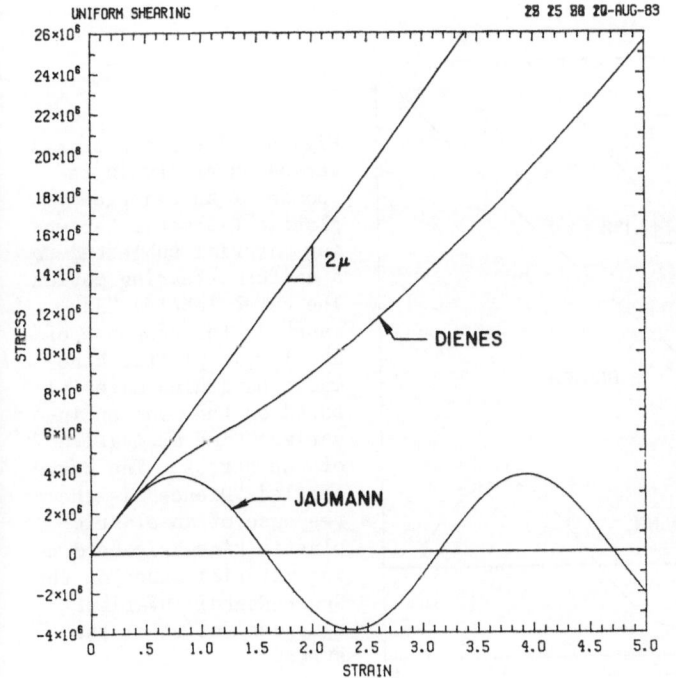

Fig. 6. The shear stress versus shear strain response of an elastic material subjected to the homogeneous motion of uniform shearing. The curve labeled "Jaumann" is the response of a hypoelastic material based on the Jaumann invariant time derivative of the stress. The curve labeled "Dienes" is the response of a hypoelastic material based on the Green-Naghdi invariant time derivative of the stress.

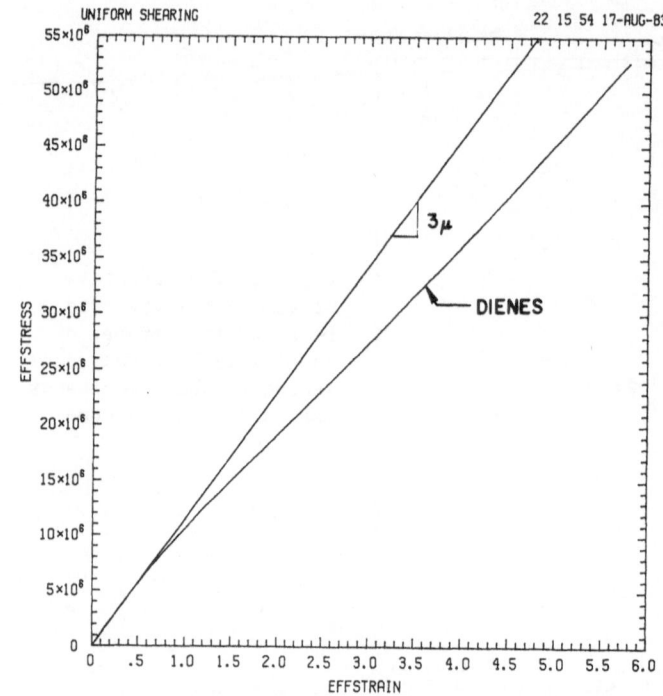

Fig. 7. The effective stress versus effective strain of a hypoelastic material subjected to the homogeneous motion of uniform shearing. The curve labeled "Dienes" is the response of a hypoelastic material based on the Green-Naghdi invariant time derivative of the stress.

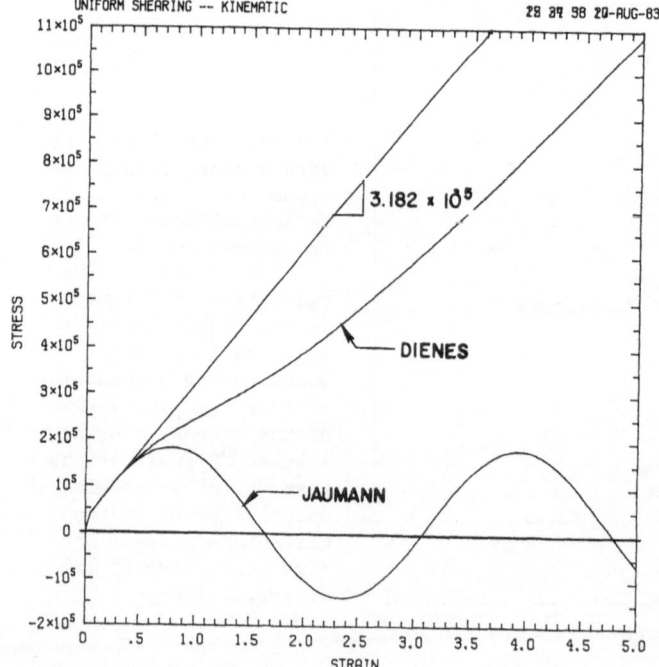

Fig. 8. The shear stress versus shear strain response of an elastic-plastic kinematic hardening material subjected to a uniform shearing motion. The curve labeled "Jaumann" is the response of an elastic-plastic kinematic hardening material based on the Jaumann invariant time derivative of the stress. The curve labeled "Dienes" is the response of an elastic-plastic kinematic hardening material based on the Green-Naghdi invariant time derivative of the stress.

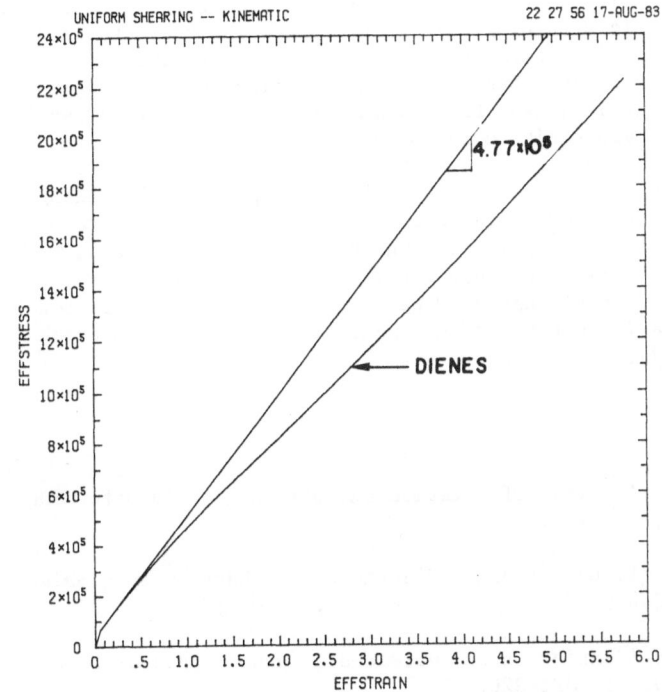

Fig. 9. The effective stress versus effective strain response of an elastic-plastic kinematic hardening material subjected to a uniform shearing motion. The curve labeled "Dienes" is the response of a material based on the Green-Naghdi invariant time derivative of the stress.

Green-Naghdi time derivative results in a monotone increasing shear stress with increasing shear strain. Referring to Fig. 9, in uniform shearing with elastic-kinematic plastic behavior, the effective stress versus the effective strain is nearly a linear behavior.

While one would not expect to find a real material with kinematic hardening behavior that is linear, it is desirable again to have a representation of linear behavior for arbitrarily large strains. Again the nature of the departures from linearity in more complex material models can then be discussed relative to the linear behavior. A material model based on the Green-Naghdi time derivative approaches linear behavior at large strain.

In the case of uniform shearing with elastic isotropic plastic behavior, no oscillatory behavior is observed with either of these invariant time derivatives.

5. CONCLUDING REMARKS

Referring to Fig. 8, while one would not expect linear kinematic hardening behavior for strains on the order of 80%, it is entirely possible in this class of calculations, particularly, for penetration calculations to distort the mesh locally in shear to these levels. Using a material model based on the Jaumann time derivative will result in false indications of strain localization and shear failure. It is for this reason that a change from the Jaumann time derivative should be made. However, with isotropic hardening, the false prediction of strain localization would be avoided.

While the combined isotropic kinematic hardening plasticity model used here is simplistic, it nonetheless foreshadows behavior which will occur in more

sophisticated models. Plasticity, for the most part, centers on the behavior of
the yield surface under plastic straining. Isotropic hardening characterized by
a change in the radius of the yield surface is a first order description of the
size of the yield surface. Kinematic hardening, characterized by the movement
of the center of the yield surface, is a first order description of the translation of the surface. The stress variable which identifies the center of the
yield surface in kinematic hardening is also referred to as the "back-stress"
and is routinely used in the metallurgical literature as a physically meaningful
state variable. The important observation here is that in any finite strain
plasticity model, no matter how sophisticated, which uses back-stress as a state
variable will result in oscillations of this variable if the Jaumann time derivative is used in its evolutionary equation. We would like to suggest that the
Green-Naghdi time derivative is more nearly a canonical derivative with which to
work (cf. Nemat-Nasser, 1983).

6. REFERENCES

Dienes, J.K. (1979), "On the Analysis of Rotation and Stress Rate in Deforming
Bodies," *Acta Mechanica, 32*, 217-232.

Goel, R.P. and L.E. Malvern (1970), "Biaxial Plastic Simple Waves with Combined
Kinematic and Isotropic Hardening," *J. Appl. Mech., 37*, 1100-1106.

Jones, R.E. (1964), "A Generalization of the Direct-stiffness Method of Structural Analysis," *AIAA Journal, 2*, 821-826.

Key, S.W. (1971), "A Specialization of Jones' Generalization of the Direct-
stiffness Method of Structural Analysis," *AIAA Journal, 9*, 984-985.

Nagtegaal, J.C. and J.E. de Jong (1981), "Some Aspects of Nonisotropic Work-
Hardening in Finite Deformation Plasticity," *Proceedings of the Workshop on
Plasticity of Metals at Finite Strain: Theory, Computation and Experiment*,
E.H. Lee and R.L. Mallett (eds.), Division of Applied Mechanics, Stanford Uni-
versity, Stanford, California.

Nemat-Nasser, S., "On Finite Plastic Flow of Crystalline Solids and Geo-
Materials," to appear in *Journal of Applied Mechanics*, 50th Anniversary Issue.

Prager, W. (1967), "Variational Principles for Linear Elastostatics for Discontinuous Displacements, Strains, and Stresses," in *Recent Progress in Applied
Mechanics*, F. Odqvist Volume, Wiley, New York, 463-474.

Truesdell, C. and R.A. Toupin (1960), "The Classical Field Theories," *Encyclopedia of Physics* III/1, Springer, Berlin, 226-793.

GENERAL DISCUSSION

BALADI: Can you define again what you mean by the effective stress and effective strain?

KEY: They are the standard definitions. For example, for the stress, it's the
square root of J_2 for the deviatoric stress with the appropriate constant.

GOUDREAU: In your kinematic hardening example, the yield stress was very small

on the scale which you indicated there. If in fact it goes to zero so you have
zero yield, is that not exactly identical to the elastic case; would you comment
on that? Also, to what extent is the kinematic hardening example with the zero
yield stress equivalent to the hypoelastic example you illustrated first?

KEY: It would be the same. I don't know if those differential equations would
survive the limiting process, but conceptually, it's the same.

HUGHES: We've seen a lot of attention paid to the simple shear example when in
fact I don't think it is possible to do simple shear experiments. Usually what
is done amounts to pure shear, say the torsion of a tube. In that case there's
a lot of experimental data around, and one should perhaps be able to exercise
these models and see how well they compare with that type of data. Due to the
combined shearing and the extension in simple shear, I'm not sure what the
stress-strain curves should look like. I wonder if anyone has any information
about this.

BUDIANSKY: You mean take a tube and just put torsion on it?

HUGHES: Yes.

BUDIANSKY: And do you suppress the axial elongation?

HUGHES: What happens experimentally?

BUDIANSKY: If you leave it free — I just calculated that before I came here. In
the hypoelastic case it elongates without limit, and the diameter shrinks to zero.

HUGHES: I would like to see some of these models now exercised against that
problem. It would seem nice to take a finite element code and run it in a stress
controlled fashion where you would apply pure shear stresses with enough kinematic restraints so that you didn't destabilize the model and then compare the
results with some torsion data. It would seem more valuable than beating this
simple shear case to death.

PHILLIPS: You mentioned about kinematic hardening under large strains. Do you
know anybody who obtained yield surfaces under large strains? It's been my experience that as strains become larger, we get yield surfaces which are smaller
and smaller. There is no such a thing like a yield surface at large strains.
Now, do you know anybody who found them? Can you have kinematic hardening in
such cases?

KEY: You certainly have a point well taken in that kinematic hardening at large
strains is a bit of an artificiality. My only point in driving these equations
so hard into a very unrealistic range is to ensure that they have a core of
linear behavior. When one puts in realistic material behavior that has some
deviation from pure linearity, then the non-linearities seen are governed by the
material non-linearities and not some pathology in the differential equations.
That's my whole point here.

NEEDLEMAN: I have several comments. Maybe I misunderstood Tom Hughes but I
think a simple model for torsion is simple shear, not pure shear. In pure shear
you would not get normal stress effects.
     Another comment is, why linearity? Suppose you were to analyze hyper-elas-
tic material in simple shear. You have to define a strain energy. You can solve

the problem analytically and you get a solution that has a shear stress that peaks. That's for a strain energy function that has a linear effective stress —effective strain relation. The solution is a power term divided by an exponential, and so it peaks and goes to zero. So you don't get linearity in that case, either. Why linearity?

KEY: The golden hope for hypoelasticity back in the 1950's was that you would get a limiting response and have effectively plastic behavior without yield surfaces and the constructions of plasticity.

NEEDLEMAN: I think now one is interested in elastic-plastic material behavior. One wants a computationally convenient representation of the elasticity for small elastic strain. And in that case, I think this is a very reasonable approximation.

　　The restriction for applicability to this situation is that the stress be small compared to any of the elastic moduli. What you are talking about here is stresses on the order of the elastic moduli, and it's not representative of the way this enters into plasticity.

　　In other words, when you write the total strain rate as an elastic part of the strain rate plus a plastic part of the strain rate, and then one writes down a hypoelastic part for the elastic part of the strain rate, one obtains an approximation to a true hyperelastic expression. This approximation is restricted to small elastic strains, which means stresses that are small compared to elastic moduli.

KEY: What I hear you arguing is that you would be perfectly satisfied with either one of these because they're only going to be exercised in terms of elastic behavior down at very low stress levels; for the elasticity part. In that case, be my guest. If you feel comfortable with the potential for oscillating shear stresses, lurking in the background.

NEEDLEMAN: I wouldn't feel comfortable with any of these models at stresses of the order of the elastic moduli. You'd have to do the elasticity right in that case.

KEY: A rigorous hyperelasticity?

NEEDLEMAN: Yes.

HUTCHINSON: I wanted to amplify a little bit of what Tom Hughes said; namely, most of the data at large strain are not for hollow tubes. You can't take the tubes that far; they buckle. It's for solid tubes. I agree with what Tom says. I think it isn't necessarily easy to do, but it's very important to take some of these models and compare them with the data, the experimental data, from torsion of the solid bars at large torque.

PISTER: I'd like to follow up on the conversation about hypo- and hyperelasticity. If you start with the premise that some objective stress rate is determined by a material tensor times a deformation rate (and let's just look at the elastic range of behavior of an elastic-plastic material and get that cleared up first), and if you pose the additional requirement that the material tensor that connects the deformation rate with the stress rate is an isotropic tensor, for a material that is isotropic with respect to a stress-free initial state, you will find the result, namely, that a constant tensor with components λ and μ (the Lamé constants) can't be the material tensor of an isotropic hyperelastic material. However, you can indeed reach a fairly general result that states that λ

and μ, no longer constants, have to be functions of J, the determinant of the deformation gradient. This result gives you a degree of freedom that first of all allows you to linearize in the neighborhood of J = 1 and obtain constant values of λ, μ back again, consistent with infinitesimal elasticity.

Furthermore, for isochoric deformation it results in constant λ and μ. I think this result is germane to what we're talking about here.

The second point, if you relate this to what Tom was talking about this morning, namely, the rate of rotated stress, whether it's Cauchy or Kirchhoff doesn't matter, and make the rate of rotated stress a function of the rotated deformation tensor, you find again that you can't choose the instantaneous elastic modulus tensor arbitrarily. The results stated are based on Bernstein's theorem, which establishes when a hypoelastic material is hyperelastic, and a generalization thereof.

PREVOST: I would like to just comment here, which is going to shock a lot of people, I'm sure, in metal plasticity it is clear that there is a linear regime and after that a plateau where you have some plasticity. But I know of another whole class of materials for which there is not a well-defined yield point. So you think about those materials to be described by plasticity with no specific elastic range.

I want to throw the argument on the table that, in this case, elasticity is in fact introduced in order to regularize the plasticity equations; because the plasticity equation involves the plasticity tensor modulus which is the tensor product of two tensors, and is singular. It cannot be inverted. So you only bring in the elasticity as a matter of convenience: when you form the sum of the elastic and plastic parts, you get something which is regular that can be inverted.

I would like to take this attitude toward elasticity and plasticity, if I may. I think this is something which is fairly adequate for a whole range of materials which I mentioned, which do not have a well defined yield point, below which you can consider they are really elastic. In fact, the definition of the yield point is purely a matter of convenience and is an arbitrary definition!

TING: I'd like to go back to this question of pure shear. Tom asked about the experimental data. I think we have to distinguish that, for large deformation, pure shear in stress does not produce pure shear in strain. Now, the picture you show is a pure shearing strain, so you have to have normal stress added, and thus you don't get a pure shearing stress-state. With added normal stresses, if you twist the hollow cylinder, elongation or shortening can be prevented. What you get probably is a pure shear straining. I don't know how you can get the pure shear stressing.

KEY: Is your concern that the homogeneous motion I picked is impossible to generate in the laboratory?

TING: I don't know impossible, but I don't know how to do it.

KEY: It is difficult to generate in the laboratory. You're obliged, of course, to prevent extensional motion in the gauge section by applying end loads on your torque specimen. It's much easier, of course, to do the torsion experiment where you let the end go free. I'm not sure what relevance it has to what I was doing.

TING: All I wanted to point out is, there are two different kinds of shearing. One is pure shear stressing, and the other one is pure shear straining, and they are not the same.

HUGHES: The torsion data I've seen saturates or thermally softens. It would seem that saturation would be the next increment of improvement in these types of rate models. Again, I don't know what to conclude from simple shear. Maybe Dienes' rate does saturate in pure torsion. I think that would be a worthwhile thing to determine.

BUDIANSKY: Yes, what I would like to say would just underscore what Alan Needleman said. Tom, are you talking plasticity or elasticity?

HUGHES: I'm talking plasticity.

BUDIANSKY: I thought you were, and I think your remarks are well taken. But, you know, a lot of the trouble when you don't have kinematic hardening is only involved in the elastic part of the strain. So let me talk a little bit about this business of a linear relation between some objective stress rate and the strain rate. Does anyone here doubt that this is anything other than just a very convenient thing to do and that nobody believes it's right for strains as large as are shown there? So I don't think it makes any sense, really, to make an issue of defining various kinds of stress rates on the basis of what they will do elastically for very high strains.
   Now, Ras Lee keeps saying, and I guess he's right, that there are problems where you can get large elastic strains in the presence of large plastic strains. And there you really are faced with the issue of really getting the elastic part of the strain right. This was the last of the unsettled issues you put down, Tom: how do you get hyperelasticity into the formulation of constitutive equations in the presence of large strains? I don't know how to do that, but, surely, we're not going to do it with a hypoelastic law, the constants of which never change.

HUGHES: Can I respond to that? There is a paper by Johnson and Bamman which establishes a relationship between hypoelasticity and kinematic-hardening plasticity. In the large deformation case, the results for these models in simple shear are virtually identical. So it's by analogy that one can look at the hypoelastic problem and see problems that one will see in plasticity with kinematic hardening. It's a mathematical analogy. But I still think it's a very valuable thing to do, to upgrade these models to the point where they at least do something like pure shear correctly. Pure shear is a dominant deformational character of plasticity. I think that if curves are not saturating when in fact they do in practice, we should strive to improve these models. They're ad hoc, anyway. Let's at least get them in the right direction. They'll certainly then have a greater range of validity than they do now.

BUDIANSKY: I'm not quite sure we know that the curves will saturate in practice.

HUGHES: I saw several papers full of curves like that.

BUDIANSKY: Nobody has ever tested anything in pure shear.

HUGHES: If you take a solid rod, say, you can solve the problem as a finite-element boundary value problem. And then you can do the comparison.

BUDIANSKY: That's granted.

HUGHES: That seems like a reasonable thing to do.

WILKINS: Well, I find myself agreeing with practically everybody here, but I would like to reinforce some of the comments. I think that the result that

occurs when the rotation terms are included for very large shear is a curiosity.
Other phenomena occur before shear strains of 2 can be realized. Also the
moduli aren't constant. We have done solid bar torsion tests, and hollow cylinder torsion tests. The latter buckle for relatively small strains, as John
Hutchinson said. We must do both tests to get accuracy. The difficulty with
rotation never arose. We monitor rotation terms just to see, by the way, how
large these come compared to the rest of the stresses.

HUGHES: Is that kinematic hardening or isotropic hardening?

WILKINS: Just plain isotropic.

HUGHES: There is no problem there. The problem emanates from kinematic hardening.

WILKINS: What I was going to say is that you asked about the real experiment.
We were able to model the real experiment without having to go to anything fancy
for kinematic hardening. The tests were with aluminum.

PREPARED DISCUSSION

by J. W. Hutchinson

About eight years ago I also served as a discusser of a paper by S.W. Key
presented at a workshop on visco-plasticity here at Northwestern. Now, as then,
I find myself in general agreement with Key's use and formulation of finite
strain plasticity. Eight years ago the central role of the Jaumann-type stress-
rate in generalizing small strain plasticity formulations to large strain was
the point under discussion, and it was roughly at that time that use of the Jaumann-rate became commonplace in the generalizations used in many computer programs. In the intervening period of time, as experience with the generalizations
has accumulated, several problems and limitations with them have appeared.

The particular problem addressed by Key is the unrealistic predictions at
large shears of the most straightforward generalization of the kinematic hardening law using the Jaumann-rate for stress in the flow law and for the back stress
(the stress-like quantity characterizing the center of the yield surface) in its
evolution law. This inadequacy was noted by J.C. Nagtegaal and J.E. de Jong a
little over two years ago, and two papers in addition to Key's have already appeared in the September issue of the *Journal of Applied Mechanics* which help to
clarify the issue (e.g. papers by E.H. Lee, R.L. Mallett and T.B. Wertheimer and
by Y.F. Dafalias).

Key and Dafalias use a generalization of the small strain kinematics hardening law in which the stress-rate and back stress-rate is based on a rate proposed
by J.K. Dienes (and earlier by A.E. Green and coworkers). As opposed to the Jaumann-rate, which uses the vorticity tensor to make the rate objective, the Green-
Dienes-rate uses a material-based spin to enforce objectivity. Lee, et al., use
an alternative material-based spin measure to construct their rate. Both constructions based on a material-based spin eliminate the anomalous oscillations of
stress associated with the generalization based on the Jaumann-rate under large
simple shears. The evolution equation for the material anisotropy, as measured
by the back stress tensor, only makes sense when large shears are involved if the
rate is formed using a material-based spin, as discussed by Lee, et al.

Code developers can refer to the papers by Key, Dafalias, and Lee, et al., for technical details of the alternative rates. Several points should be borne in mind, however, when reading these papers.

1.   There appears to be no essential difference between predictions based on the various generalizations of isotropic plastic flow laws such as that based on the Mises invariant. The distinction between predictions becomes important when the plasticity model includes one or more tensor quantities measuring anisotropy which evolve with deformation.

2.   Even for kinematic hardening, the differences between the Jaumann-rate formulation and the other formulations show up only at fairly large shears. Problems not involving large shearing will not be affected by the issue raised here.

3.   There are sound reasons for using a material-based spin in formulating a rate to characterize the evolution of an anisotropy tensor-measure such as the back stress. One should not lose sight of the fact, however, that there is nothing sacred about the functional form represented by the kinematic hardening yield surface. A balanced approach to improving this law as a representation of finite strain plastic behavior should probably deal both with the choice of objective rate and with the functional form of the relation itself. For example, although the new formulations eliminate the unrealistic oscillations in stress associated with the earlier formulation, it is by no means clear, for example, that the new formulations are better at correlating tension data with simple shear data at finite strain than an isotropic hardening formulation. Further work on the new formulations is needed to gain more insight into their virtues and limitations.

References related to effect of rotation-rate choice in definition of stress-rate:

Dafalias, Y.F. (1983), "Corotational Rates for Kinematic Hardening at Large Plastic Deformations," *J. Appl. Mech., 50, 561-565.*

Dienes, J.K. (1979), "On the Analysis of Rotation and Stress-rate in Deforming Bodies," *Acta Mech., 32,* 217-232. (This reference contains a good list of references to earlier literature.)

Lee, E.H., R.L. Mallett and T.B. Wertheimer (1983), "Stress Analysis for Kinematic Hardening in Finite-Deformation Plasticity," *J. Appl. Mech., 50,* 554-560.

Nagtegaal, J.C. and J.E. de Jong (1981), "Some Aspects of Non-Isotropic Work-hardening in Finite Strain Plasticity," *Proceedings of the Workshop on Plasticity of Metals at Finite Strain: Theory, Experiment and Computation,* E.H. Lee and R.L. Mallett (eds.), 65-102. Published by the Division of Applied Mechanics, Stanford University.

GENERAL DISCUSSION

HUGHES: A point has come up several times today. I think Alan is the one who articulated it the first time. The point is it probably doesn't matter what rate you use in the constitutive equation itself. The critical element here seems to be the equation governing the back stress. It's been mentioned that you could use two different rates. I'd like to point out one pragmatic reason why, if it doesn't matter, you should use the same one because you'll never be able to transform into the simple picture involving one rotating system. You're going to heighten your computational implementation problems considerably if you use different rates. All of those rotations, that I went through some pains to describe, cost money to compute. So if it doesn't matter, I would advocate using one.

HUTCHINSON: Oh, yes, from an aesthetic point of view, I agree. I just mentioned that to illustrate the point that the problem is all tied up in the anisotropic tensor and not in the stress rate itself, although I have not checked this point, but I agree with you completely.

SANDLER: It seems that, in any attempt to get a constitutive behavior appropriate to finite deformations which is to be an extension of constitutive behavior for infinitesimal deformations, one condition ought to be that infinitesimal response from an already finitely deformed state should satisfy the same kind of invariant relations that infinitesimal deformations must satisfy when you're only considering those types of deformations. Geometric nonlinearities should play no role.

To be specific, consider that the material, as it deforms, establishes its own "local coordinate system" (which moves with the material and represents any rotations that it might have undergone). Now it seems that infinitesimal deformations expressed in this special coordinate system ought to be the ones for which material linearity is assumed. In other words, the assumption of incremental linearity, invoked in finite strain models, should be enforced only for such a special coordinate system. So, if you're going to look at linear elasticity as a local description of, let's say, unloading behavior from a finitely deformed state, the linearity ought to be expressible not from some previous or original undeformed configuration, but from the current position taken as a reference state. And done in that way, I think many of the discrepancies between the different tensor rate formulations disappear; you must insist that the constitutive behavior be linearized with respect to a coordinate system which reflects the current "material" configuration.

It seems that it ought to be a reasonable condition, therefore, that if a finite theory is to be an extension of an infinitesimal theory, that infinitesimal deformations from some already finitely deformed state should produce the same kinds of linearity that we require from theories of infinitesimal deformation (in which no distinction is made between "convected" or "spatial" coordinates).

HUTCHINSON: Well, I think these do.

SANDLER: Both of them don't; the one based on Ω does, I think. In this case the incremental material behavior is linear in a coordinate system which has taken account of the deformations of the material, so to speak.

In the other case, when you use the spin tensor (which is always referred to the original undeformed coordinate system), that formulation does not take into account the conversion of the material. In this case, linearity ought not to be imposed because the original coordinate system is related in a geometrically non-linear way to the current convected system. The linearity of the formulation in the original system would seem to have little physical significance. I think this is at the source of the whole problem.

HUTCHINSON: I think there is something to what you're saying, yes.

VALANIS: We are talking about extending small deformation behavior to large deformation behavior. This is just a question. Is there a critical experiment that determines the mix of kinematic anisotropic hardening in real materials?

HUTCHINSON: I don't know of one.

KEY: Certainly, when you unload and begin to yield in reverse, you begin to discover whether the hardening you've been imposing on the material is kinematic or isotropic or some combination of these, or possibly even a little more elaborate than that.

VALANIS: Are you speaking qualitatively and not quantitatively?

KEY: Well, that's the only way to do it. I mean, Prof. Phillips has done an enormous number of experiments doing an initial hardening of the material with proportional loading and then coming back and finding where the yield surface now is by probing in other directions.

VALANIS: But I'm asking for a formula. Is there an equation that tells you that so much of kinematic hardening and so much of isotropic hardening indicates this measurement?

NEMAT-NASSER: Since kinematic hardening is really an assumption, I would like to hear about experimental data that support non-saturation to such large strains.

PHILLIPS: I can say this. Experiments were done and we found that there is no elastic region at large strains. When you go to large strains, then the elastic region becomes zero. There exist elastic strains, but there is no purely elastic region. To talk about kinematic hardening or isotropic hardening or any combination of these, is absurd in light of experimental data.

HUTCHINSON: Let me answer that. Sia, there is experimental data. For example, among others, I know Sig Hecker has compared data at very large strains, in tension, with data in shear. It is a very debatable subject among physical metallurgists whether saturation occurs or not. But it is certainly true that in a number of the materials that he's looked at, he does not yet see saturation, and he goes to very large strains, very large strains. There are some general rules, incidentally, about the relationship of effective stress-strain curves determined in tension as determined in shear. So there's actually quite a bit of finite strain data of that kind.

NEMAT-NASSER: How does it relate to kinematic hardening directly?

HUTCHINSON: Ask Sam.

HUGHES: I want to respond to Kirk's question. In the context of the hardening theory in which you have a linear combination of isotropic and kinematic hardening, you can answer the question of what the proportion is by an explicit formula. This really emanates from Sam's picture. You plastically load, you come down and you hit a point where you reload. That point essentially defines the parameter β. And that gives you the proportion of the two hardening mechanisms. So in the context of a theory, you can answer that.

VALANIS: But only the linear use, because there is kinematic hardening that is not linear.

KRIEG: I don't think anybody has really addressed Budiansky's remark on what difference does the large strain formulations make in the practical case and

Needleman's similar remark. I think we've been satisfied with approximate theories all along. For example, we know that there's good reason to think there are vertices in yield stresses. However, computationally, you don't see us ever use that in practical analysis. The reason is because it's too costly to include them.

In the same way, what we need to do is to find out what is the right formulation. And therefore, if we can find the right formulation, then we can find out how much does it cost. The Green-Dienes rate is essentially computationally free in two dimensions. So we can use it for the same price that it would cost to use the Jaumann rate.

There's one other point, and that's that we need to explore, that is, to find some case in which it's critical that we use a rate of this type rather than the Jaumann rate. In other words, we need one practical case. I don't think it's been found yet, but I think it's a critical issue.

PISTER: I'd like to come back to two things that Kirk Valanis and Ivan Sandler said. I think I must have misunderstood you. You talked about extending linear theory to get a nonlinear theory. You really don't expect to get a covering theory for nonlinear behavior out of a linear theory, right? My point is, it seems to me that no matter where you end up computationally, ideally, if we're going to deal with an elastic-plastic model of a material, there must be an elastic range for that material. I think most people would agree that the elastic range ought to be hyperelastic and not some other kind of "elasticity."

So, no matter where you are, the theory ought to be capable ultimately of saying yes, you're there. If you're around the origin, the identity mapping, you ought to get Hooke's law out of the model. If you don't, you better try something else. Likewise, when you're off in some finitely deformed state of the material when you unload, and this is what Ivan was talking about, you ought to be able to linearize at that state and get the correct elastic tangent modulus for recovery. I contend that you have to do this with a covering nonlinear elastic-plastic theory.

VALANIS: Yes, but you see, the drift of my comment was this. First of all, there ought to be some rigorous way of determining the kinematic hardening law of a material.

PISTER: That's another issue.

VALANIS: Before we even go to finite deformations which are complicated further by objective considerations of the type that we are discussing this afternoon.

PISTER: The whole thrust of Hughes' presentation of finite deformation inelasticity was that beginning with a nonlinear formalism, you can show that the same equations, the same computational algorithm that we had in the infinitesimal case results.

DIENES: Yes, I just wanted to say that I think this large deformation rotation rate that I have been interested in is really a mathematical thing. It's not a physical thing. I think we should think of it in that context. The physics is an entirely different problem. I don't think that it's purely without interest because when you do calculations of the kind I do, which involve large deformations due to impact, you get very large strains and you get very large rotations. And if you're going to use the classical Jaumann rate, the slope of the stress-strain curve is negative, it's unstable and the calculations can blow up. So the utility of the thing and purpose for it was to avoid these instabilities and not to try and produce some better physics. The physics is a separate consideration.

CHAPTER V

SHORT PRESENTATIONS AND OPEN DISCUSSIONS

Chairman: *R. J. Asaro*

Comments by R. J. Asaro

I think we will start off with some general open discussion for about 45 minutes, and then we will start the talks along with a brief comment by Erhard Krempl. Aris Phillips and Rod Clifton will speak on relating experimentation to the development of constitutive laws. Karl Pister will then give a short presentation on rate equations for finite deformation elasticity; Gordon Johnson on material characterization for high velocity impact and explosive detonation; and finally, Mike Gross on the CAVS model for jointed rock. We will then finish off with some more unfinished business.

During this first open discussion we can discuss anything you'd like. This is a time for raising points that there wasn't time to raise earlier; short diatribes are tolerated, if they are not too long. I would like to start off myself on, I think, a positive note.

It is really much too early this first day to think about defining milestones for the meeting. That is, however, something we all should be thinking about for Wednesday afternoon, when we would like to make some statements on what we think we know that perhaps we didn't know yesterday.

The question of kinematic hardening is an important one. It has been discussed quite a bit. It is an important bit of modeling; it is used extensively since it is valuable for modeling a host of phenomena.

One of the primary aims of kinematic hardening theory as it evolved in small strain plasticity, you all remember, was to describe Bauschinger effects; that is, reduced yield stresses following reverse loading. A simple picture for this is just a translating yield surface. The real question, in formulating a kinematic hardening theory, or in extending the small strain kinematic hardening theory to finite strains, is how one chooses to evolve the yield surface center. The center is often interpreted as a back stress to be represented as an offset in the yield surface center. What's really at issue here then is a hardening model. It is really a modeling question, and the way that's going to proceed is by writing down a hardening law for an internal tensorial variable. That law will be written in terms of an objective rate of the yield surface center which is set equal to a function of stress and deformation.

Now, there have been, and rightly so, some issues picked up in connection with this specific model. I think it is really with respect to the kinematic hardening model at finite strain that there is an important question about the choice in stress rates. If in fact the yield surface center is to be interpreted as an internal stress, and if you think of this stress as being caused by some particular microstructural feature, and if we had an electron microscope which could dye this microstructural feature "green," we could follow the green microstructure through the deformation history and, perhaps more to the issue, through any rotations it undergoes. In this way we could follow the rotation of the back stress which represents the yield surface center.

Well, I suspect the simple shear case is the one which dramatizes the effect of rotation. The green microstructure rotates due to simple shear and the most it can rotate is through an angle of 90°, so it isn't surprising that if you choose the Jaumann rate which amounts to rotating the yield surface center, in the simple shear problem, by the constant spin rate of one-half the overall shear rate, then that's not something which was motivated by the microstructural picture. The constant spin rate leads to rotations of more than 90° and in fact causes oscillations in the shear stress-shear strain curve.

Well, that may be a problem, but it is not anything to lose a lot of sleep about. A number of very constructive and sensible suggestions have been made for reformulating a finite strain version of kinematic hardening. One of them is to use the rate $\dot{R}R^T$ where R is the rotation matrix that appears in the polar decomposition theorem in front of the right stretch matrix. In fact, if you work out the component in question of $\dot{R}R^T$ with components on fixed laboratory axes, indeed, you see that rate of rotation defined this way eventually dies down. So, what is accomplished with this rate is to rotate material fibers through a total angle of 90° and, therefore, preclude oscillations. Of course, this isn't the only rotation rate that has this feature; for example, one could simply choose to form a convected rate of the yield surface center which would also preclude oscillations.

One other point I should mention is concerned with the interpretation of Bauschinger effects in terms of kinematic hardening, especially at large strains. I know of no cases in metals and alloys, at least in my own experience, where the magnitude of the Bauschinger effect, measured in simple axial cyclic plasticity tests, does not essentially saturate after 3 or 4% plastic deformation. Certainly, the largest effects accumulate in the first 2 or 3% plastic strains and grow very little at larger strains. So if you are going to use a kinematic model to study fully reversed plastic deformation, or large deviations from proportional loading, you should appeal to experiment and think about an evolutionary law which would very quickly saturate the magnitude of the kinematic effect.

That's all I want to say. I think it is a positive note, because it is an issue of concern, but it is one in which we have identified a number of very sensible methods for improving our models.

* * *

GOUDREAU: Can I just comment on kinematic hardening?

Apart from commenting, dealing with finite strain, the issue was asked earlier how one identified a mix of isotropic and kinematic hardening; and the suggestion was you just unload and identify the reverse yield. But a kinematic model allows an offset of the initial center as well, and I am wondering to what extent people use shear behavior as well as reverse uniaxial yielding to identify a kinematic model?

ASARO: I am not myself aware of anybody who has actually tried to —

GOUDREAU: I am aware of some work in soil to develop multiple yield surfaces, but just within this level of the linear surface, what are people doing for metals? Anybody going to talk about that this week or want to comment now?

ASARO: The question is: Has anybody used combined tension-shear in metals to attempt to match kinematic hardening theory?

GOUDREAU: Right. If you are going to do the linear combination of isotropic/kinematic, as Sam presented earlier today, but assuming the center to be initially at zero, then you can do either the reverse yield or a shear, to identify the parameters. But if you did both, you can fit the third parameter, which would be the shift; you can do the linear combination plus a shift of the center. I'm aware of some of this in soil work, but I don't know who does what in metals.

BUDIANSKY: I am not going to answer that directly, but this leads to a short diatribe. I am very glad you (i.e., Asaro) indicated that one should use the kinematic hardening model with caution, once you get to one or two per cent strain, because it saturates. Then, as far as I know, you can get a reverse Bauschinger effect; in other words, you tend to get an increase in the yield stress in compression as you go up in the strain.

Now I want to add to that and suggest that one should use the kinematic hardening model with caution even up to one or two per cent strain, because as far as I can tell, kinematic hardening was invented because it was an insight, a cute mathematical idea intended as a generalization to polyaxial states of what you could do with the Bauschinger effect in the uniaxial case. But I think it is a bad representation of the Bauschinger effect.

Here is a stress-strain curve (Fig. 1) in tension, $\sigma - \epsilon$. Now if you go up to point Q and back off to R, the conventional model is that if you reload, you will follow the path RQS. But if you continue down past R, the kinematic hardening model says that when you get to Q', where QQ' is twice OP, you will then take off on a curve Q'S' that matches the curve QS.

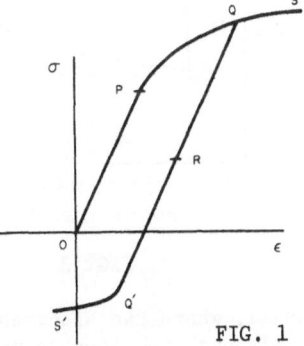

FIG. 1

But this is not the sketch you drew. In fact, since you are a metal physicist, you drew what really happens (Fig. 2). What really happens, as far as I know, is that reverse plastic flow takes off very gradually. The accumulation of plastic strain along RT follows a path very different from the path PQS. So (Fig. 1) is a rotten representation, in that sense, of what really happens.

Now, one hears a lot, especially in recent years, that α, the shift of the center of the kinematic hardening model, represents a back stress. The connection with metal physics seems to end right there. I hear people say, "Yes, it is back stress." But is it really? It sort of looks like it might be, it smells like it a little bit, but has there been any real modeling of back stress that leads to the model?

Now there *have* been calculations, like those John Hutchinson has done, building up from single

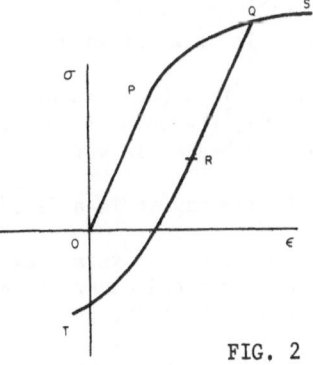

FIG. 2

crystal behavior to polycrystal behavior. What do they find? They in fact find from their calculations that the way the plastic deformation develops *is* more gradual on the reverse loading than on the direct loading. So, yes, back stress is involved in a vague sort of way.

The kinematic hardening model has been justified because it gives Bauschinger's effect and it represents back stress; in fact, I think it does neither very well. I am not suggesting that it not be used anymore, but I think a preoccupation with it has gone too far.

PHILLIPS: To answer the question that has been raised, I only ask the audience to read my papers. They will have both the answer to this question, and the corresponding analytical implementation. In addition to that, this entire work is based on single crystal considerations. I will speak to this when my turn comes.

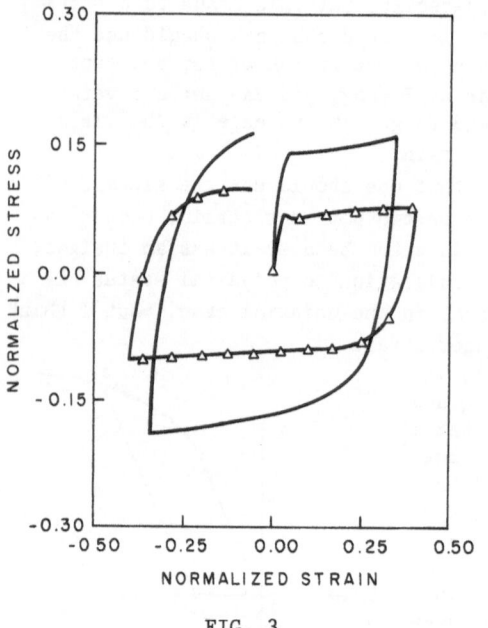

FIG. 3

KREMPL: There was a question whether the Bauschinger effect would exist in proportional loading, combined loading. Fig. 3 shows pertinent experiments on Type 304 stainless steel at room temperature. This figure is part of an ongoing investigation. A fully annealed tube was subjected to simultaneous axial and shear strain controlled cyclic loading. Figure 3 shows that a Bauschinger effect is present in this case. (At the meeting Fig. 5 of a forthcoming paper in *Mechanics of Materials* (1983) by Krempl and Lu was shown. This paper may be consulted for the experimental details.)

PREVOST: This is the kind of behavior which you can observe in testing soil specimens in a uniaxial environment. The loop which has been generated here (Fig. 4) has been obtained in a very special testing condition which is called in soil mechanics, undrained triaxial, where the specimen was sheared at constant volume. You can observe that although the loop is nonsymmetric about the origin, definitely, it is of the same kind as the loop which was shown earlier.

VOICE: Is it a triaxial test?

PREVOST: Yes, but because of the inherent anisotropy, which you observe in any real soil, the loop is very unsymmetrical about the origin. I claim that this kind of behavior is very accurately modelled by using kinematic plasticity.

VOICE: You say it is a fluid saturated material?

PREVOST: Yes. It is a clay which initially was cross-anisotropic because it had been consolidated under the effect of gravity (in the vertical direction), a preferential direction, and with the horizontal plane, a plane of isotropy because of the in-situ conditions.

Now, that clay was taken out of the ground and brought into a testing

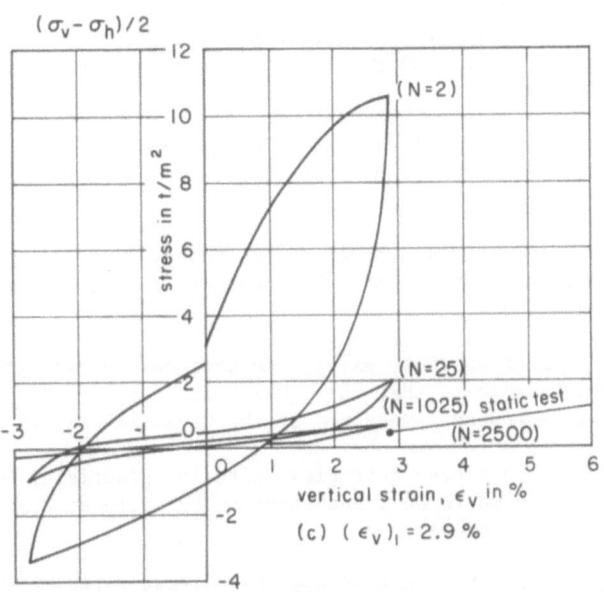

FIG. 4

machine, which is a triaxial testing machine, and tested under undrained loading
conditions, where the fluid was not allowed to escape from the specimen. When
you load in compression — the sign convention is reversed from that you are used
to — then you describe a loop upon loading and unloading, a nonsymmetric loop.

But I think that it is quite interesting to notice that the loops here are
pointed, which clearly identifies a plastic type behavior, rather than a viscoplastic type behavior. For this kind of behavior, there is plenty experimental
and analytical evidence that it can be accurately described by using a very simple and straightforward kinematic type plasticity. So, kinematic plasticity is
a product of our imagination, but it fits pretty well the experiments.

BALADI: In the uniaxial test, are you controlling the axial strain or stress?

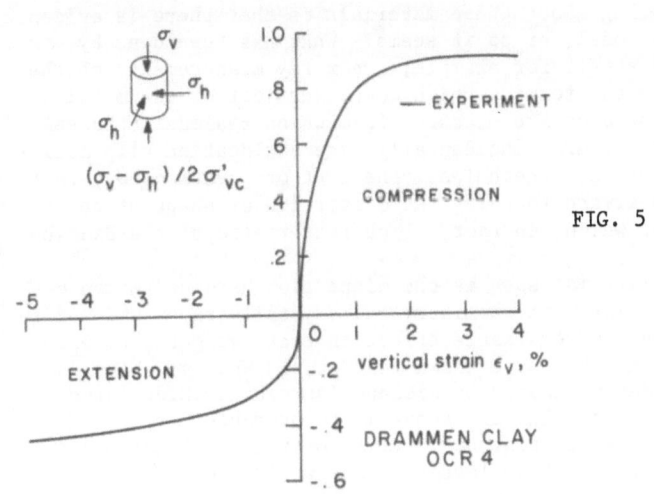

FIG. 5

PREVOST: Well, it doesn't
really matter. But for
this particular experiment,
the test was started from
the initial K_0 condition,
to respect the language of
soil mechanics, and the
cycles of loading were done
under cyclic strain con-
trolled loading conditions.

COMMENT ADDED IN PROOF
(*ASARO*): At this point a
discussion developed which
centered on whether Prof.
Prevost's tests, shown in
Fig. 4, actually displayed
the behavior expected from

a kinematic hardening model. There was some feeling that, although the curve in Fig. 4 displays a Bauschinger effect (see also Fig. 5), the shape of the curve is not what is expected from a simple kinematic hardening model involving a translating yield surface.

GARG: The undrained response of clays will depend upon the rate of loading or the strain rate; the plasticity representation employed by you appears to imply that the rate of loading is unimportant.

PREVOST: That is a subject which is extremely touchy in testing of clays. I can tell you very few experiments have been done to investigate the effect of loading rates in clays.

GARG: I disagree insofar as there exist some data on the hydrostatic response of Shaley sands which indicate that the equilibrium response differs substantially from the short term response measured at the usual laboratory strain rates.

PREVOST: My experience has been that clays with low plasticity — and this one is one of those — are known to be insensitive to the rate of loading, if the test is truly, purely undrained.

ASARO: So, it seems like the issue of possible stress rates, or of saturation of the Bauschinger effect, are the least of our problems. I think that the suggestions about comparing to experiment are really very good ones, but one thing I would caution you, Jerry, about looking at shear, is if you are going to compare shear and tension or compression for the purpose of evaluating kinematic hardening models, then the models that are used to interpret the data must include the effects of deformation induced anisotropy due to, for example, texture.
It seems like a lot of the differences in hardening behavior that are observed in compression, versus large shear, are explainable, at least in part, by textural effects. I am thinking about deformation induced crystallographic texture which has yet to be included in any large strain hardening model.
I wish to make one other comment, based on my own experience, which touches on what Bernie Budiansky was saying about kinematic hardening. I once did some experiments on some particular materials where the Bauschinger effect is most extreme; these are dispersion hardened metals. The simplest case is spheriodized steel, but another case I looked at was a superalloy, Nimonic 80A dispersion hardened with Yittria oxide particles.
What's extremely fascinating about these materials is that there is evidence to support an internal stress model, or so it seems. What has been done by others in dispersion hardened steel, for example, are x-ray measurements of the distorted lattice parameter in the ferrite which comprises most of the matrix. In this way residual elastic strains are measured from which residual stresses can be estimated in the ferrite where, incidentally, the dislocation slip activity takes place. In that material, nonetheless, the sort of phenomenology that Bernie referred to, i.e., the severe rounding and distortion of shape of the reverse loading curve is present which, in fact, is characteristic of the Bauschinger effect in most metals.
Now, when you look at a material such as the dispersion hardened Nimonic 80A alloy — which has a very different chemistry and crystallography and a different slip mode — there you get a remarkable effect in that you get a reverse curve, which actually looks like the curve shown in Fig. 6. What you have in these materials is strain hardening caused by dislocation debris which forms around the particles which in a very literal sense in recoverable.
As for interpreting this effect (Fig. 6) in terms of kinematic hardening, well, I think you can imagine what a translated surface would look like if you

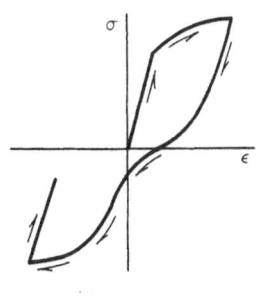

FIG. 6

tried to model the phenomenology sketched in Fig. 6 with kinematic hardening with simple translation of a yield surface. The even more remarkable thing is if you take this material through five or six fully reversed plastic strain cycles it shakes down to classical kinematic hardening.

GOUDREAU: You don't need finite strain to exhibit this phenomenon.

ASARO: This is all for small strains, that is, the experiments of mine that I talked about.

GOUDREAU: What happens if you shear it?

ASASO: I didn't do it and I don't know.

Okay. Let's have the short presentation by Aris Phillips, which should address more of these experimental issues and how they impact the development of constitutive laws. Rod Clifton will quickly follow and then, I think, Erhard Krempl has a quick statement to make right after Rod's talk.

Short Presentation by A. Phillips

Gentlemen, I'm very glad that I came here because I have observed a lot of interest in yield surfaces. We have done some work during the last ten years on this subject. I would like to report on some of this work. I have not prepared slides, because I did not expect to talk here, but my talk will be based on some sketches.

We observed the following in our experiments, Fig. 7. If you load to point a, then you observe not one yield surface, but two surfaces, and there is a very important reason why there exist two surfaces. The outer one, II, behaves like a Mises surface, or approximately so; the inner one behaves as a conventional yield surface. Now there is a difference in the law of growth of the plastic strain, when you are inside of this loading surface II or when you are outside. When you are located inside II, after you have established it at a, then the yield surface is moving around to reach first a; then to reach a', then a", etc. As the yield surface moves around, plastic strains are generated.

Now, if you go beyond the previous loading surface II, then a new loading surface is generated and the yield surface moves accordingly. Outside of II as the new loading surface is generated, a different law of growth of plastic strain is valid.

Now, this is significant, because when you have tests where there is repeated loading, you generate plastic strains which are of the first type, while if you continue to load radially, you generate plastic strains which are of the second type. So you cannot possibly correlate the one law of growth of plastic strains with the other test results; two different laws are valid.

Again, as you move away from the origin, the yield surfaces are becoming smaller and smaller. If you start going back towards the origin, these surfaces become bigger. So, in one sense the material has a memory.

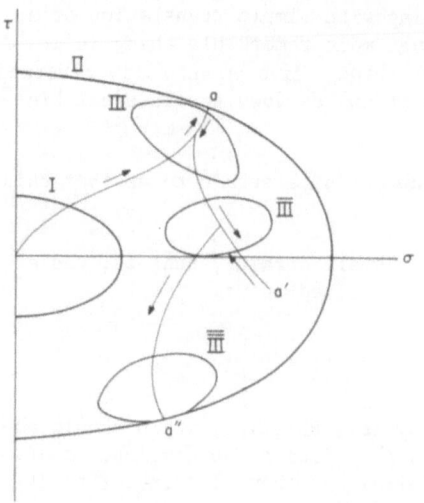

FIG. 7

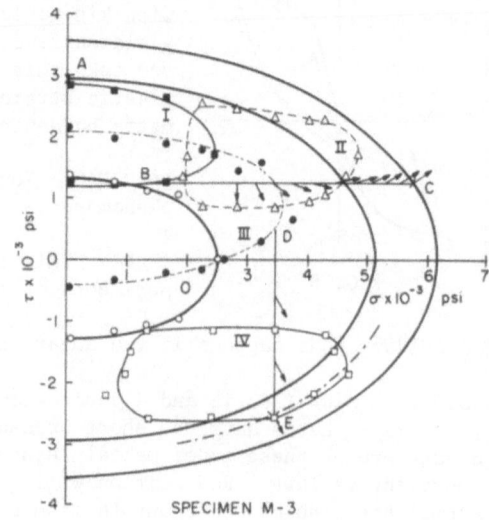

FIG. 8

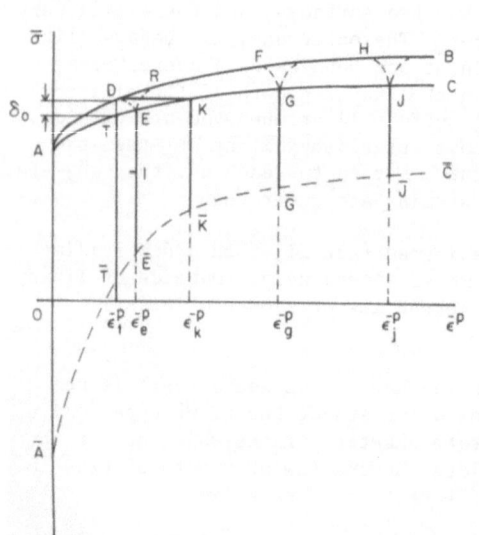

FIG. 9

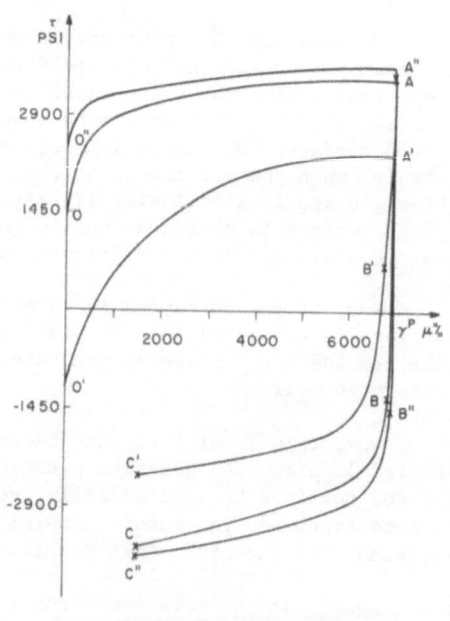

FIG. 10

When the state is located inside the loading surface, the plastic strain increments, as they are generated, are normal to the yield surface, which is very different from the loading surface. As the yield surface tends to reach the loading surface, then the plastic strain increments become normal to both the yield surface and the loading surface; we have a rotation of the plastic strain increments (Fig. 8).

All this is complicated even more because of the strain rate effects. The yield surface does not touch a (Fig. 7). You have to stay at a for considerable time, maybe 24 hours before the yield surface will become tangent to the loading surface at a. The yield surface moves with time; this is a strain rate effect. How important such strain rate effects are depends on the application we consider.

There is also another type of experimental result which has to do with the existence of what we call the equilibrium stress-strain curve. In Fig. 9 $\bar{\epsilon}^p$ and $\bar{\sigma}$ are the equivalent strain and the equivalent stress. If you load with a higher stress rate, then you obtain the curve AB. There exists, however, a curve AC which would be obtained if you had infinite time for loading. The yield surface is represented here by straight lines. As you move to larger $\bar{\epsilon}^p$, the length of this straight line becomes smaller and smaller, until at some distance, this length becomes zero. The two curves, AC and $\overline{AC}$ will meet, and at that place, you have no yield surface.

The same thing happens when you raise the temperature. When you raise the temperature, the yield surfaces become smaller, you have internal curves C and $\bar{C}$ and the intersection occurs at a smaller value of $\bar{\epsilon}^p$. Now, the equilibrium stress-strain curve does not exist only when you increase the strain; it also exists when you decrease the strain. For example, in Fig. 10 if you decrease a strain, then you have curves like A'C and AC'. These are equivalent to the curves introduced previously which are valid for increasing strain.

Now, this is a bare outline of some of the results we have obtained, and if you wish to know more about these results, please write to me and I will send you my papers. These papers have been published since 1965.

***

NEEDLEMAN: I have what I hope is a quick question. Operationally, how do you define the plastic strain?

PHILLIPS: Let me explain. It is a very interesting question because it goes to the heart of the matter.

When we talk about a yield surface, it is important to define what we mean by a yield surface. The yield surface in our definition is the surface which encloses the pure elastic region. The question is: How do you find this pure elastic region? If I had the stress-strain diagram, then I increase the stress until you have the first deviation from proportionality. Now, the question is: How do you find this deviation from proportionality? We allow a deviation from proportionality by 3 micro-inches and then we unload. We find the unloading line to be parallel to the elastic loading line with a parallel shift of 3 micro-inches. Then we backtrack from the highest point reached in order to find the intersection with the elastic line. This gives the proportional limit. Once you do that, you obtain good correlation. If, however, you start using a larger deviation, say 20 micro-inches, or 100 micro-inches or one percent, then you can not get any correlation. So there seems to be, from a practical point of view, some kind of a physical significance to the point we define as a proportional limit.

Short Presentation by R. J. Clifton

I know that many of you are interested in applying these large deformation calculations to problems of very high strain rates, such as occur in the applications that we saw earlier — for example, self-forming fragments. I would like to describe a recent result we have in obtaining data at high shear strain rates.

We use a pressure-shear plate impact experiment shown in Fig. 11. A thin specimen is sandwiched between two hard plates that remain elastic. Impact occurs in such a way that both pressure and shear loading are imposed on the face of the specimen. This is almost the ideal textbook experiment where you simply press a specimen and shear it. The compression wave reflects back and forth in the specimen a few times until essentially hydrostatic pressure builds up — then the pressure stays essentially constant; however, the material continues to be sheared, high shear strain rates being produced. In the early work strain rates of 10^5 sec^{-1} were obtained; I will show some recent work with shear strain rates of 5×10^6 sec^{-1}.

We monitor the wave that passes through the anvil plate. From that wave and the fact that the flyer and anvil remain elastic, we can deduce the shear stress and shear strain rate for the specimen. I say "we can deduce it" in that because the anvil and the flyer remain elastic (shown in Fig. 12 as linear elastic) all states of shear stress and transverse particle velocity in the flyer and the anvil must lie, respectively, on the lines shown for the flyer and for the anvil. Once the shear stress becomes nominally uniform through the thickness of the specimen, the shear stress level can be obtained if you can measure the transverse component of the free surface velocity. This we measure with a transverse displacement interferometer (TDI). Measurement of the free surface velocity and the projectile velocity gives the difference in velocity, Δv, which is used to obtain the average shear strain rate. Thus, measurement of only the free surface velocity gives both the shear stress and the shear strain rate. The shear strain rate can be integrated over the duration of the experiment to obtain the dynamic stress–strain curve.

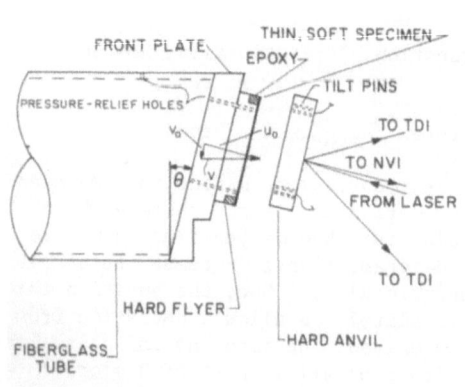

HIGH STRAIN RATE, PRESSURE- SHEAR EXPERIMENT

$> 10^6$ s^{-1}

FIG. 11

We have performed such experiments for a number of materials. The material we have most data on is commercially pure aluminum. These data are summarized by Fig. 13. The shear stress is plotted versus the logarithm of shear strain rate. For reference, quasistatic data would be off the diagram to the left. Results for strain rates in the range of 10^3 sec^{-1} are obtained from the Kolsky Bar experiment. Our early work was at strain rates up to a little over 10^5 sec^{-1}. These experiments are conducted on specimens that are a few tenths of a millimeter thick. Recently, we prepared a specimen by vapor depositing a very thin layer, 18 microns thick, on the flyer plate. If you make the thickness very small, you get a very high shear strain rate because the shear strain rate is the difference in velocity divided by the thickness. The result of this experiment is shown as an open square at a strain rate of 5×10^6 sec^{-1} and a shear stress of 3.2 kilobars. This is a remarkably high flow stress for a very soft aluminum in shear.

I think that measurement of such a large shear stress is important because if you really do have a transition from, say, a thermally activated process which gives a logarithmic dependence of shear stress on strain rate in strain rate regimes below, say, 10^4 sec^{-1} to some other mechanism that is perhaps associated with viscous glide of dislocations through the lattice and which gives a stronger dependence of shear stress on strain rate, then it would seem that in applications, such as the self-forming fragment, you ought to see strong differences between what you predict and what you observe if the predictions are based on the relatively low flow stresses at strain rates of 10^3 sec^{-1}.

The insert in the upper part of Fig. 13 gives an indication of what the same information looks like when plotted on a linear scale of shear stress versus shear strain rate. There are very fundamental questions as to what happens at the higher strain rate; one possibility is a saturation in shear stress level; another possibility is that shear stress will increase proportionally with plastic strain rate; still another possibility entertained by others is a saturation in strain rate. We are trying to do research in the area of very high strain rates in order to resolve some of these matters.

I want to conclude my remarks by asking a question: Those of you who do calculations of very high strain rate problems, do you really use a flow stress corresponding to strain rates of 10^3 sec^{-1} and get the right results?

***

GOUDREAU: Maybe I don't quite understand. From your top view, it seemed to me, you are positing a possibility of a rate saturation. Do I see that as a possible extrapolation that you don't know yet?

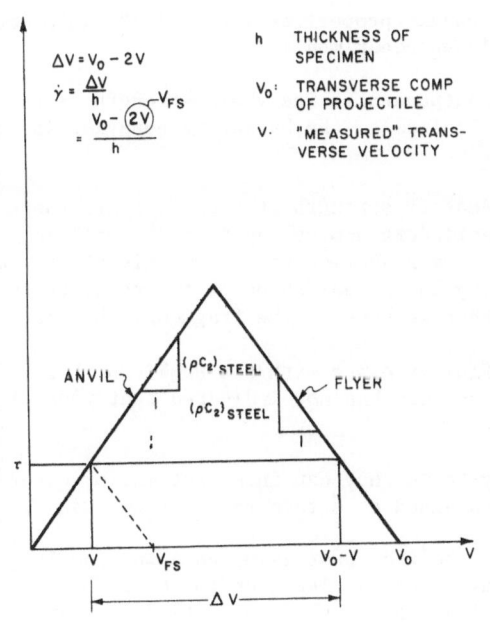

FIG. 12

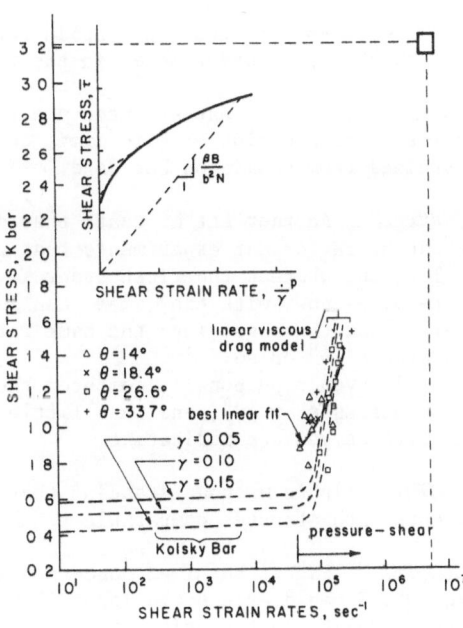

FIG. 13

CLIFTON: I am simply indicating that there are all these possibilities.

GOUDREAU: I am confused by the lower curve, now, the lower set of curves. What am I reading there, shear stress —

CLIFTON: Shear stress versus log strain rate. The open large square is our latest result.

GOUDREAU: Whereas from the top curve, I see at least the possibility of a saturation strain rate reaching a maximum stress, somehow I am having trouble seeing that possibility in the lower set of data. It looks to me like when you look at that 10^5, your stress takes off; it becomes extremely sensitive to strain rate.

CLIFTON: You have to get used to working with logarithmic scales. Actually the sharply increasing dependence of flow stress on shear strain rate that is shown on the lower diagram corresponds to the concave downward curve on the upper diagram.

GOUDREAU: Let me say, briefly, what I think we do. A lot of the high rate properties are measured over in "H" division, but what we have done up to now is to assume a rate independent flow stress, but measured at a high rate, at 10^5 or 10^6.

CLIFTON: But how did you measure that?

GOUDREAU: Well, I have to ask these guys. My concern in the fragment problem was that at late times, the rates slow down to 10^4 or less, and I am concerned about the rate softening that would occur as you go into the late stages of the fragment formation and the rate drops an order of magnitude, and I expect a lower yield stress than what we assumed in the initial early stages of the formation.

VOICE: I am just curious, could you explain about what's done in the calculation?

GOUDREAU: How are the high strain rate strength properties identified? In other words, the first order model is the rate independent model.

MOSS: The models that are used are rate independent. The yield strength is obtained from Hugoniot elastic limit data. Typically, the hardening behavior is obtained from Hopkinson bar data.

GOUDREAU: Another set of tests that do identify strength at high rate are these kinds of Taylor Bar experiments that are cylinders shot up against the wall and splat, and whereas the first order Taylor data reduction gives you only effective stress, we now, with the codes, can identify the actual shape of a stress-strain curve, and these get into the same rates that we have in the fragment, that is, up around 10^5 or so.
     I haven't personally compared those flow stresses with the Hopkinson bar flow stresses at 10^3, and what little I have seen has not indicated that radical a shift as you are indicating.

CLIFTON: It is my understanding that they think they can interpret Taylor experiments with stresses comparable to those obtained at strain rates of 10^3 sec^{-1}.

GOUDREAU: Right, but I am concerned about the fact that as we go down from 10^5, to 10^4, I would expect the lower flow stress, that at least up to now we have not been modeling in the calculation of these fragments. What we do find experimentally is that at late time, when we get into the lower rate regime, they break up.

I'm wondering whether if, in fact, there might be a lower strength at that late low rate?

HERRMANN: The self-forging fragment is an ideal illustration of the complexities that can occur here. Several things that happen, first of all, the plate is soft initially and it is subjected to a high explosive shock wave which reverberates. Measurements we have made in our lab suggest that that introduces a lot of defects and a lot of hardening. The second thing is that the plate undergoes very large deformations, more or less adiabatically.

The calculation of the temperature is an extremely complicated thing; there is, obviously, thermal softening going on. Furthermore, there are very large strains occurring, so that the strain hardening behavior is really not too well defined; and then, on top of this, there is a rate effect, such as you allude to.

The most intelligent work I have seen has attempted actually to back out an average flow stress by comparing calculations with various different constant flow stresses to change of shape. And those suggest that, in fact, the yield strength is pretty high, which would be in consonance with what your observation is here.

But really, it is very difficult to back out anything uniquely about the rate effect, because you must recognize that it is very closely coupled, the temperature effects and very large strain hardening effects.

CLIFTON: Yes, I would say that the temperature here probably had to go up by about 600°C and the maximum strain is four.

GOUDREAU: Again, I would say 600°C is about the kind of temperature changes that these models are predicting. The strains are on the order of 150, 200% in the fragments. If you get into the jets, you know, the shape charge jets, you really can get all the way out to melt strains. But I would ask Dr. Herrmann whether or not you think that essentially yield softening, either because of thermal —well thermally, we kind of know you soften it out —whether rate softening is a phenomenon in the fragment at late time as it starts to slow down? I don't know of anybody that's tried to model that.

HERRMANN: I am really more concerned about the initial explosive environment which suddenly takes these soft copper plates and turns them into something very different, which you don't know.

GOUDREAU: It is only about a 20 microsecond interval.

HERRMANN: Then, from there, you can make some sense.

ASARO: There is another brief comment by Erhard Krempl in connection with experimental methods.

Short Presentation by E. Krempl

I just want to make one comment. The idea came this morning. I want to talk about slow tests. We never talk about boundary conditions when we refer to material properties obtained from such tests. As an example, let us consider macroscopically homogeneous motions only. Fig. 14 shows a typical stress-strain diagram that appears in every textbook. Consider now a test with stress boundary conditions under increasing stress. Then the stress increment has to be positive

and no negative slope is possible. The drops following the upper yield point and the maximum load
are not possible in stress control with increasing
stress. So, therefore, the material property that's
reported in all texts cannot be a material property,
because it can only be observed under a displacement
boundary condition.

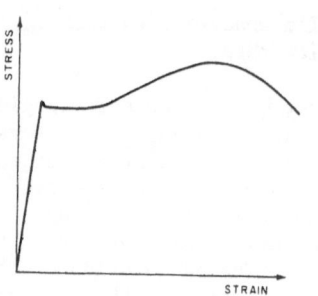

FIG. 14

    A simple, but crucial check for a constitutive
equation is to require the qualitative behaviors
that are observed in tests with displacement or
load boundary conditions. I think that the softening aspect referred to by Sia, and by other people,
this morning is more pronounced in the geological
materials than in metals. My argument applies to the constitutive equations of
these materials as well. A crucial test for the validity of a constitutive equation is to see whether it really can represent the observed difference between
the behavior under stress and displacement boundary conditions in a uniaxial
test. (Experimental results pertaining to this subject are to be found in the
discussor's paper to appear in *Mechanics of Engineering Materials*, C.S. Desai
and R.H. Gallagher, editors, John Wiley (1983).)

***

ASARO: Well, I think there is a question of how measurements of flow stresses
and strain hardening behavior get into the codes versus how estimates of flow
stresses are obtained back out of large scale calculations. This is a very interesting question. I hope we can return to it again at some point, maybe at
the very end of this session, or tomorrow for sure.
    But let's go on with another short presentation by Karl Pister on rate
equations for finite deformation elasticity.

Short Presentation by K.S. Pister

    My remarks are associated with remark number 5 in the section on finite de-
formation inelasticity in Tom Hughes' paper. I want to talk about the constitutive model for materials and not computation with constitutive models, although
clearly they are related. Let's concentrate on a mechanical process for an inelastic material that's associated with some elastic range; in other words, let
the inelastic deformation rate be zero for this process.

    Typically, we have seen today a number of times choice of an objective
stress rate σ^*, defined by some material tensor times the difference between the
total deformation rate and the inelastic deformation rate;

$$\sigma^* = c(d - d^{in}), \qquad c^{ijk\ell} = \alpha g^{ij} g^{k\ell} + \beta(g^{ik} g^{j\ell} + g^{i\ell} g^{jk}). \qquad (1)$$

Let's just concentrate on the elastic part. For computational purposes, as we
have heard over and over again and if you look in the literature, the material
tensor is assumed to be an isotropic tensor — written out here with two functions,
α and β typically taken as constants. The question that I want to pose and simply
make an assertion as to the consequences of this question is the following: What
restrictions does one have to place on c, that is, α and β, for c to characterize a hyperelastic material, and particularly an isotropic elastic material?
That's the question. After all, we are typically dealing with materials with a
hyperelastic range, for lack of material data on anything better. It follows,

and this is an assertion that has been proved, α and β must depend on J (J = det (F)), and the corresponding strain energy function W has a form:

$$W = U(J) + \tfrac{1}{2}\mu I; \quad \mu = \text{constant}, \quad I = \text{tr}(F^T F), \tag{2}$$

where α, β *cannot* be constants. A possible choice for U is

$$U(J) = \frac{\lambda}{2} (\log J)^2 - \mu \log J. \tag{3}$$

Then,

$$\alpha(J) = \frac{\lambda}{J}; \quad \lambda = \text{const.}, \quad \beta(J) = \frac{1}{J}(\mu - \lambda \log J). \tag{4}$$

In other words, the result is a generalization of Neohookean elasticity; it is a compressible version of Neohookean elasticity.

Later on, Dr. Baladi is going to talk about soil inelasticity, and I was interested to see from his experimental results that he has the same sort of result, except he expresses dependence of the material constants on stress.

Now, there is no unique way to characterize the strain energy function, but one possibility is to choose U in the form that I have shown in Eq. (3) for the reason simply that this specializes to the right limit. If you take U this way, then α is λ/J; β is the form shown in (4), and that has a property that if you linearize the deformation about the reference configuration, that is, for an identity map, then α indeed is λ and β is μ, both constant; and more important, for an isochoric motion, which you experience in simple shearing, α and β are constants; everything is fine there.

For finite deformation metal plasticity where there is a tremendous amount of computational experience that ties in with experiments, elastic deformation is infinitesimal typically, and plastic deformations are isochoric; then, approximately, α and β are λ and μ, respectively. Fine. The computational people are right on the mark, but what has been described constitutes a logical reason why they are on the mark and why that's a good statement.

What about the picture that Tom Hughes talked about today and Sam Key mentioned?

$$\dot{\sigma}_R = c_R(d_R - d_R^{in}), \quad \dot{\sigma}_R = R^T \sigma * R. \tag{5}$$

Let's talk about the rotated picture, talk about the rate of rotated Cauchy stress and the rotated rate of deformation which follows from the Green-McInnis generalization of hypoelasticity rather than that of Truesdell which is the earlier version we were talking about.

The rate of rotated Cauchy stress is again some material tensor times the elastic deformation rate rotated. Now, this is the so-called Green-Naghdi rate or whatever name you want to associate it with. After we go through this a few more years, people are probably going to be bowing out as to who did this.

The remark I wish to make here is that it follows that the tensor c_R cannot be the elasticity of a hyperelastic material if it is an isotropic tensor or a constant tensor. So be careful in your computations. Just because c happens to be a convenient isotropic tensor, because that's the first thing that comes to mind, it may not be the material that you think it is.

                                   ***

ASARO: But if it is constant, i.e., if you do use Hooke's law —

PISTER: It just isn't.

ASARO: Not for a hyperelastic material — but it is a first order approximation of one, and if you do that and those are λ and μ, then I think a consistent order representation for an incremental elastic stress-strain law appears using a Jaumann stress rate and rate of deformation. If you need a rigorous finite elastic theory, you would return to basic physics, define a free energy function, and the second derivatives would give you proper elastic coefficients.

PISTER: That's precisely what I did, except I looked at this picture; that's just the big c that you are talking about there and my c are just simple maps of one another.

ASARO: We are in complete agreement, I think.
       The next short presentation — unless there are some questions about what was just said — is by Gordon Johnson on a slightly different topic, viz., "Computations and Material Characterization for High Velocity Impact and Explosive Detonation."

Short Presentation by G. R. Johnson

       I was asked to make some comments on the work we have been doing in computations and material characterization. Being a code developer, I come at this problem from the standpoint of being able to get answers — to increase the *capability* of the codes to include problems of interest. I recognize from the group of people here that this is not necessarily where everyone is coming from.

       What we are faced with in these kinds of problems, high velocity impact and explosion detonation, are some significant computational problems in just being able to obtain a solution. Therefore, I would like to show the kinds of problems we are faced with and then also talk briefly about our material characterization work.

       To begin, a characteristic of the EPIC-2 and EPIC-3 codes is that they use two-dimensional triangular elements and three-dimensional tetrahedral elements. The reason for this is that these elements do not turn inside out under severe distortion, as other elements sometimes do, and we can run these computations for a long time without user intervention.

       To show the effect of triangular elements, here is a DYNA2D solution of Hallquist; a cylinder impacting a rigid surface. It can be seen that quadralaterals are used. If you use triangles and compute an average pressure over every two triangles, you reduce the number of imcompressibility constraints and you can essentially match these results. Therefore, there is not a significant stiffness detriment in this case.

       Here I have used an average pressure, and these elements will in fact turn inside out under some circumstances, just as the quadralaterals will. If we use an individual pressure formulation, the individual triangles will not turn inside out and we can go to very high distortions. Also, if we use a crossed triangle configuration we can almost, but not quite, match these results.

       Yes, triangles are too stiff; it shows up primarily in the centerline

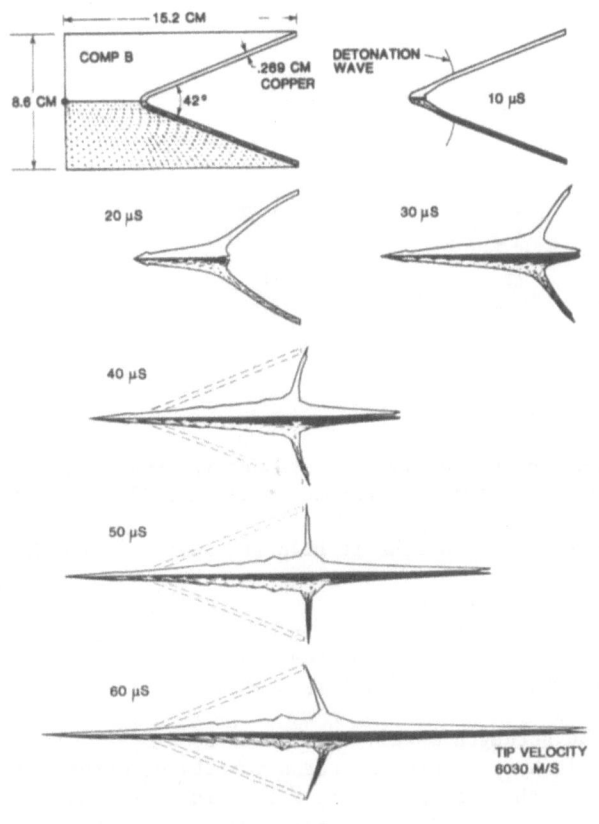

FIG. 15

region. However if we use a crossed triangle configuration, we can give up a small amount of accuracy, but gain added capability in the kinds of problems we are able to run.

Now to show a few examples; this is a shaped charge problem (Fig. 15). Here, we detonate the explosive; it drives the liner material down to the center; it forms a jet and we see extremely high strains, hundreds of percent. The code continues to run and the triangles do not turn inside out, and it gives us a capability to do another class of problem.

Here we have a three-dimensional, self-forging fragment computation. We see the explosive gas expanding out around the edges and we see the fragment begins to form. Self-forging fragments provide a very severe test for material models, because their final shape is a direct function of the material response.

Another kind of problem that we do is impact problems; here is a three-dimensional computation. You can see that we have allowed the material to erode off; essentially, we let the elements disappear when they get very highly strained. The little dots you see are nodal masses whose associated elements have totally failed. Again, this is a very approximate thing to do, because we are letting element volumes disappear. But this problem is really driven by kinematics, and we can show the computed mass and velocity to be in reasonable agreement with experiment.

Another important aspect of these computations are the sliding interfaces. I can't get into the details, except to mention that the master surface is generally considered to be a membrane on the surface of the plate. You can see that the top of the plate stays intact and the copper rod never breaks through the top of the plate. That's okay for thin plates, but, of course, when you get to thick plates, you have to be able to break through that top surface.

Just recently we have put an eroding target capability into the EPIC-2 code. Dr. Fred Stecher should get credit for doing much of this work. Here we have a rod impacting some water (Fig. 16) where we not only allow the rod surface to change and erode materials away, but the initial master surface on the water is also allowed to erode down into the water as the rod penetrates. So now it is possible to perform a Lagrangian computation of a rod perforating a thick plate.

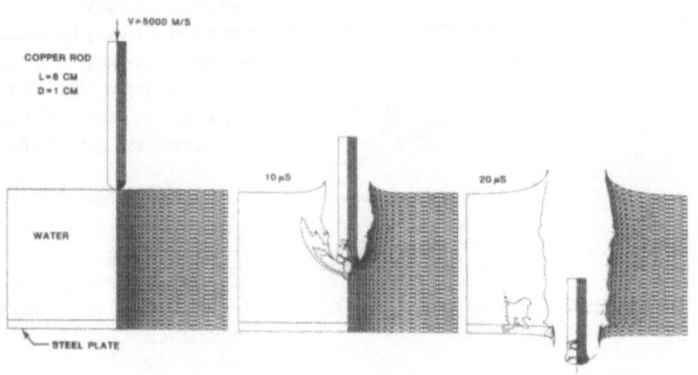

FIG. 16

Again, this is an approximation, and quite frankly, we have not yet evaluated accuracy. However, it does give us an additional computational capability.

I also have some materials work and would like to hold the questions to the end.

Computational material models are driven very much by being able to put them into the codes and effectively use them. Here are the requirements we set out to obtain about two years ago for both strength and fracture models.

I only have time to mention the strength model. It has to be simple. It can't change the code too much and it must be useable for a variety of materials. We should be able to get the data for the constants from a limited number of laboratory tests, and we should be able to evaluate it with some independent tests.

Here, in the strength model, we have a strain effect, a strain rate effect, and a thermal softening effect (Fig. 17). I think you can recognize the terms.

SUMMARY

• THE MODEL HAS THE FORM

$$\sigma = \left[A + B\epsilon^n\right]\left[1 + C\ln\dot{\epsilon}^*\right]\left[1 - T^{*m}\right]$$

$$\underbrace{\qquad}_{\text{STRAIN}} \quad \underbrace{\qquad}_{\substack{\text{STRAIN}\\ \text{RATE}}} \quad \underbrace{\qquad}_{\text{TEMPERATURE}}$$

• DATA OBTAINED FOR 12 MATERIALS

• TESTS PERFORMED TO OBTAIN DATA (A,B,n,C,m)

      −QUASI-STATIC TENSILE TESTS

      −HOPKINSON BAR TENSILE TESTS AT VARIOUS TEMPERATURES

      −TORSION TESTS AT VARIOUS STRAIN RATES

• MODEL AND DATA EVALUATED WITH CYLINDER IMPACT TESTS

• COMPUTED RESULTS SHOW GOOD AGREEMENT WITH CYLINDER IMPACT TESTS

FIG. 17

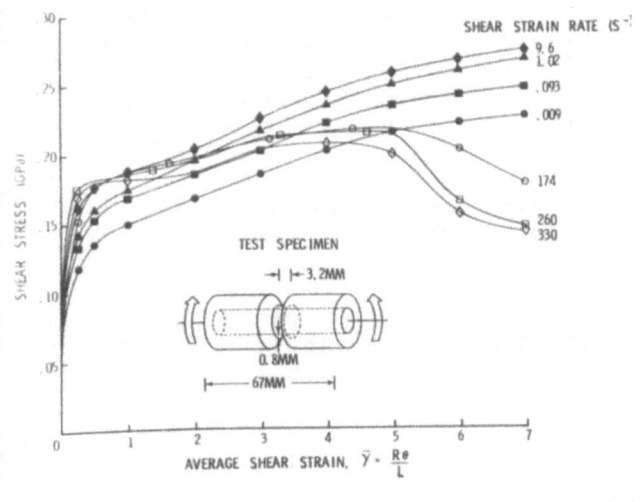

FIG. 18

We have five constants to obtain from quasistatic tensile test data, Hopkinson bar test data at various temperatures, and torsion test data at various strain rates.

Among the interesting things we have obtained are some large strain torsion data for copper, shown as shear stress versus shear strain (Fig. 18). These are large strains, up to 700%. As we increase in strain and strain rate, we see that the stress increases and it does not saturate. When we get up to a strain rate of about 10 and then go to 174, we see a big drop in stress; this is the point at which adiabatic softening takes over. You can see from these data, that there is a strain rate effect in going from low strain rates to high strain rates. There is also a strain hardening effect and a thermal softening effect. We have simulated these tests with the EPIC-2 code, taking into account heat conduction.

The Hopkinson bar data do not go to nearly as large strains. Here we see data at room temperature, data at about 13% of melt and data at 30% of melt.

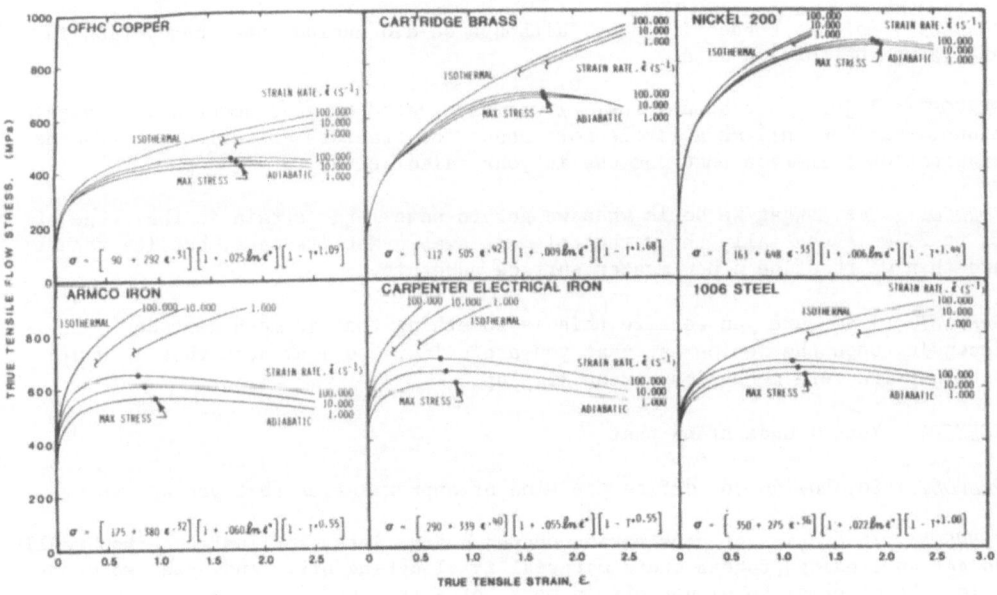

FIG. 19

We use the data to obtain constants for the model and these are the kinds of relationships we obtain (Fig. 19). We have a half a dozen materials here. If we look at the copper, we can see equivalent tensile stress versus equivalent strain; it goes up to a strain of about three. If we ignore the thermal softening, we have an isothermal response. If we look at the adiabatic responses, which take the plastic work and convert it to temperature, we can get an adiabatic stress-strain curve here, which is derived from the basic test data.

Now what we have done is to take these models as they are, and compare computed responses to some cylinder impact test data. So essentially what we are doing is evaluating the model and data with an independent series of tests.

Although the agreement is generally good, we'd like the accuracy to be better. The copper model appears to be a little hard and the Armco iron model appears to be a little soft. But really, we are quite close, and we are able to essentially determine the various effects of the strain, strain rate and temperature.

***

ASARO: These are isothermal calculations?

JOHNSON: No, these are adiabatic calculations here. We assume that the thermal softening is built in with plastic work converted to heat because in these kinds of computations we really don't get any —

ASARO: But the temperature distribution —

JOHNSON: The temperature distribution is directly related to the plastic work distribution.

ASARO: No heat conduction?

JOHNSON: No heat conduction here, although we did include heat conduction in analyzing the torsion data.

PREVOST: I just have a question. I was fascinated by your penetration calculations. Can you tell me a little more about the fracture criterion and fracture propagation criterion that you use in your calculations?

JOHNSON: Yes. What we do is when we get to some large strain in the range of 1½ to 2, we essentially throw that element away. Not its mass, but its volume; and then we redefine a new master surface under it.

PREVOST: I am sure you realize this is something that is mesh dependent; it is depending upon the refinement that you are using, and I am sure that it would make any fracture mechanics person jump up.

JOHNSON: Yes, I understand that.

PREVOST: So, how do you define the kind of approximation that you are using?

JOHNSON: In erosion of penetrating shaped charge jets, for instance, you really do get an eroding process where material is sloughing off. And, yes, we don't allow the material to slough off in terms of a fracture model per se. We can, however, let the deviator and shear stresses be set to zero based on a better fracture model. But where you have a situation where you really are eroding the material away, then you would hope that it would approach the right solution for large values of the eroding strain.

GOUDREAU: I'm impressed by your finite strain torsion test data and your ability to resolve the thermal softening effect. I wonder if you could put back up your equation of your model and indicate, first, how you normalize it to both the torsion and the tensile data at large strain, and second, how you deal with the finite strain rate issue that's been discussed earlier today.

JOHNSON: What we do is to try to find the strain hardening portion of the model for a strain rate of one. We take this to be an average of the torsion data, at a strain rate of one, and the tensile data. This is really a yield stress with strain hardening.

GOUDREAU: Is that the effective stress-strain law?

JOHNSON: Yes. This is effective tensile stress and effective strain. The effective strain rate is in the second set of brackets, and the dimensionless temperature is in the third set of brackets; Fig. 17. We write this equation in a way which we hope fits the experimental data. The isothermal part of it fits copper data very well up to strain rates which get into the adiabatic regime. But when we do get into the adiabatic regime, and when we get large strains, it is very difficult to separate the effects of the strain rate hardening and the thermal softening. This is because they provide opposing effects, and you really don't have a test that lets you know what the effect of each of those variables is.

GOUDREAU: What stress rate formulation are you using for these finite strains? Are you using the Jaumann rate?

JOHNSON: We are using the Wilkins radial return method —

VOICE: J_2 flow theory, isotropic hardening?

JOHNSON: Yes, von Mises isotropic hardening.

GOUDREAU: Oh, isotropic.

VOICE: But you think you get a fit that fits the torsion data and uniaxial data out to these large strains?

JOHNSON: It doesn't fit it real well. We are making a compromise. Generally the tension data strain hardens faster than the torsion data.

NEEDLEMAN: I have a sort of philosophical question. You said one of the criteria you used in choosing material models in these very important, very complicated calculations was how easy the models are to implement into given codes. I have been convinced that these are very complicated, very important computations. Why aren't the best available material models being used whatever it costs?

JOHNSON: I think that we try to do the best job we can in the atmosphere we are working in. In industry, for instance, the kind of work we show here is really a tool to be used to develop better products.

NEEDLEMAN: But Dr. Snowden said earlier that major policy decisions, major economic decisions involving hundreds of millions of dollars are riding on these calculations. You are telling me it is going to cost $10,000 more for a calculation?

JOHNSON: If you look around to the kinds of data and models which are available —

NEEDLEMAN: You are saying then that good materials models for these problems are not available?

JOHNSON: If better constitutive relationships were available, and usable in the codes, then we would put them in. What we wanted to do here was to try to step up one level from a constant flow stress, because realistically, that's the way people are currently doing these problems. Now we can at least phenomenologically see the effects of some of these variables such as strain hardening, strain rate hardening and thermal softening.

So, what we are trying to do is not necessarily to get the exact answer because we know we will never get there. Particularly in industry, we can't spend a long time on most projects, so what we do is to try to make a step upward and to include more than we had before and then build from there.

CURRAN: I'd like to help get Gordon off the hook a little bit. I have admitted to him in private that we use triangles ourselves, and I guess I have admitted it in public now. But, of course, the triangular cells themselves put in, in a sense, a constitutive relation, which is independent of any constitutive model that is used. So, right there, there is a big uncertainty.

The second point is that for many of the cases he showed, the only important part of the constitutive relation is the density. The momentum conservation is almost the only thing governing the problem.

As a taxpayer, I don't want him to spend a lot of money modeling a yield surface if momentum conservation is going to give me the penetration velocity for a particular piece of armor or is going to give me roughly the shape of a fragment. Now, I think what he has told us, though, is that momentum conservation is not the whole story. We are now beginning to see that the constitutive relations will have some influence in many cases.

GOUDREAU: One last comment, and that is that most of the formation is not that sensitive, but late time breakup is an important thing for these long standoff weapons. And I am wondering, in your model, with the reduction of strength as rate drops off, do you ever see in your calculations an indication that the strength has dropped off enough to indicate ahead of time the failure?

JOHNSON: We see the strength dropping off, but in the self-forging fragments computations there does not appear to be enough time for localizations to occur completely. But when we simulate the torsion tests, for instance, where we actually do get failure, then we do, in fact, see the localizations occur.

PREVOST: When you were talking about localization were you talking about explicit calculations?

JOHNSON: Explicit calculations.

PREVOST: Did you see localization in explicit calculations?

JOHNSON: Yes.

PREVOST: That's interesting.

ASARO: Our last short presentation this evening is by Mike Gross on the CAVS model for jointed rock.

Short Presentation by M. B. Gross

I want to talk about an empirical constitutive model we have been developing. This discussion may belong more in tomorrow's talks, but we have been developing a model called CAVS to simulate the response of fractured or jointed rock. CAVS stands for Cracking and Void Strain. It has been developed by one of my colleagues, Don Maxwell.

CAVS is interesting for two reasons. Number one, it is a non-isotropic model which keeps track of the detailed joint and fracture system in a rock mass and allows the joints to slip and possibly separate, depending on the loading on the rock mass; and number two, we try to base CAVS on laboratory data as much as possible. In particular, we have based CAVS on the data due to Nick Barton at Terra Tek. Barton has looked at data from a variety of jointed rock experiments, as well as done some model experiments of his own. Based on those experiments, Barton has developed simple empirical relations for how joints respond in a rock mass. These relations for joint slip and slip-induced dilatancy are the basis for CAVS.

We first began pursuing this approach because of calculations of the Pile Driver experiments where an isotropic constitutive model could not reproduce the observed response; we have subsequently been using CAVS for a variety of other applications, including the explosive fracturing of oil shale and the stability of openings in jointed rock. More recently, we have analyzed tunnel hardness under high overpressures, and we are going to be doing calculations of SRI experiments which explosively load a rock mass.

CAVS has three basic functions. The first function is a detailed bookkeeping system for the fracture system; CAVS postulates a 3-dimensional, orthogonal system of joints, CAVS will keep track of fracture slip, slip-induced dilatancy, and joint stiffness for each joint. The second function of CAVS is to maintain stress and strain compatibility between the fracture system and the host rock. Given the total strain in the rock and joint system, the CAVS logic partitions the total strain into a rock strain and a joint strain that are consistent with Barton's relations for joint slip. The third function of CAVS is the capability to model detailed physical mechanisms.

I will give you a rough idea of what some of these joint properties look like. Figure 20(a) is a plot of joint aperture versus normal stress and slip displacement. This property is physically reasonable in the sense that as you increase the normal stress, you expect the fracture aperture to close down. If you start to slip along a joint, you expect that your aperture will open because you are riding up on the asperities. Finally, if you release the normal stress, you expect to pop back to a larger aperture than originally existed.

Similar comments apply to the sliding friction. Figure 20(b) shows that as

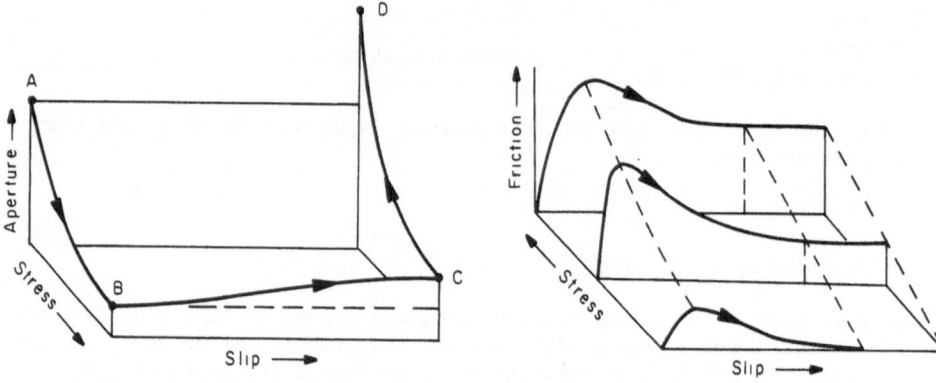

(a) Example path of aperture changes   (b) Slip friction paths at constant effective stress

FIG. 20.   Unified model of fracture stiffness, slip friction, and slip-
            induced dilatancy. See text.

a joint first starts to slip, there is high initial friction, and as the joint
slip wears down the asperities, there is a regime with constant sliding friction.

The CAVS model does require a fair amount of material property data.   For
example, the tensile strength and shear strength of the host rock and the various joint properties, as defined by Barton, are required by CAVS. Over the
years, we have developed default estimates for the major parameters in CAVS, and
these estimates have performed well on a variety of calculations. The program
requires input of the following material parameters: Sample tensile strength,
sample shear strength at zero effective stress, joint slip cohesion, joint rough-
ness coefficient, residual friction coefficient, generalized failure cohesion,
initial generalized failure slope and its damaged value, generalized failure
ultimate strength, stressless-unslipped fracture aperture, and fracture stiffness
modulus.

Let me just quickly show you one computational result with the CAVS model.
Figure 21 presents a tunnel in tuff, and we have postulated a fracture system
that exists throughout the rock. There is also a fault which was postulated
underneath the tunnel. The calculations are performed by increasing the load
on the side of the left boundary until the tunnel collapses and the load at collapse is used as a relative measure of the hardness of the tunnels.

Table 1 presents a series of calculations which analyze the variation of
tunnel strength with joint properties. Table 1 presents four cases: an elastic-
plastic case without joints, a case with strong joint cementation and relatively
high friction, and two cases with no joint cementation and high and low slip
friction. The last column in Table 1 is the load which corresponds to 5% volume
strain in the tunnel; this load is used as a simple measure of the strength of
the tunnel. You can see that the collapse pressure varies rather sharply
throughout the computational sequence, so the strength of a tunnel is extremely
sensitive to the joint properties. These calculations highlight the capability
of the CAVS model to perform a consistent variation of the joint properties in
order to analyze, in a rational way, the effect of joint properties on the geomechanical response of the system.

TABLE 1. Tuff Tunnel Collapse Overstresses
Based on 5% Tunnel Volume Strain

Full Grid Cases	Joint (1) Angle	Cementation	Joints	σ_5 (kbar)
T 1.2		(Elastic-Plastic)		0.73
T 3.2	- 17	Strong	Strong	0.435
T 3.0	- 17	0	Strong	0.33
T 3.1	- 17	0	Weak	0.20
Quarter Grid Cases	Joint (1) Angle	Cementation	Joints	σ_5 (kbar)
Q 1.0	- 45	Strong	Very Strong	0.60
Q 1.01 Q 2.0	- 45 0	Strong	Strong	0.55 0.42
Q 1.1 Q 2.1	- 45 0	0	Strong	0.31 0.315
Q 3.1 Q 4.1	- 45 0	0	Weak	0.15 0.25

***

NEMAT-NASSER: I was wondering if you would amplify on how you calculate the local stresses and local strains, or do you assume them to be the same as the overall ones?

GROSS: The local stresses in each zone are obtained from the local strains in each zone. The local strains are partitioned between rock strain and joint strain to provide stress equilibrium where:
1. The rock stresses depend on the rock strains and the rock model.
2. The fracture stresses depend on the joint strains (apertures and joint slip) and the joint model, which include: a. joint orientation and spacings; b. joint stiffness model; c. joint slip and dilation model.
3. Corresponding stresses (for the rock and the joints) are equal for local stress equilibrium.

NEMAT-NASSER: How do you relate that to the overall farfield stresses?

GROSS: The local stresses are related to the farfield stresses through the code solution for the equations of motion, acting zone-by-zone throughout the grid. For quasi-static cases, this corresponds to both local and global stress equilibrium.

CURRAN: Can I interrupt you? I don't want to put words in your mouth, but I

suspect you do it in a manner very
similar to the way we do it for a comparable model. We divide the total
strain into that produced by the joint
and that by the intact material. So,
basically, it is very similar to a plasticity model; the stress is generated
by the elastic component in the intact
material.

PREVOST: I would like my comment not
be taken as a criticism of Nick Barton's
work whom I know personally and greatly
admire his work —he is the person who
wrote the modeling concepts which are
being used in these kinds of calculations.
    However, there is one concept
which I sort of disagree with when you
treat a rock as a continuum system. You
forget about the discontinuities which
occur along the cracks in such a formulation, that is, you smear out all those
effects. So, my concern is loosing some
special features of the solution which
you could observe in real rocks, i.e.,
in rocks, there are definite fracture
planes. This kind of formulation is
based on a continuum model for rocks
which are definitely fractured systems.

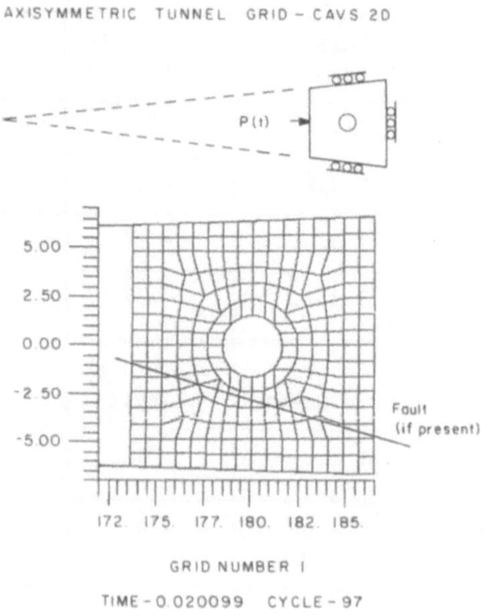

AXISYMMETRIC TUNNEL GRID – CAVS 2D

5.00

2.50

0.00

-2.50

-5.00

172. 175. 177. 180. 182. 185.

GRID NUMBER I

TIME – 0.020099   CYCLE – 97

FIG. 21

GROSS: Well, in fact, I think I do model fracture planes. We have to have a
detailed discussion to define it. I understand what you are thinking when you
say "smearing," but in effect, each zone or particle in the calculation does
keep track of its fracture response. The zones have a minimum of "smearing."
If there is no more than one fracture of a given orientation in a zone, then the
description is discrete in that zone for that direction, i.e., everything about
that fracture is known. If there is more than one fracture of a given orientation in a zone, then the description is ubiquitous in that zone for that direction, i.e., all fractures in that zone for that orientation have identical properties.

PREVOST: Do you have special criteria for doing that? How do you do that?

GROSS: The CAVS bookkeeper keeps track of joint and fracture parameters in great
detail.

PREVOST: In plasticity, you can have localized deformations when you have yield
vertices, but with your formulation, you only have a smooth type of response.

GROSS: The localized deformation (of a zone) can be decoded (through the bookkeeper) to expose the detailed response of each mode of deformation (aperture
squeeze, joint slip, etc. for each joint).

GOUDREAU: Just a brief comment. I have no problem with the averaging over a zone width normal to the crack. My only concern is that these cracks will tend to link up and the propagation criteria at the tip of the crack may be sensitive to smearing across coarse zones.

GROSS: The stress concentration about a crack tip is currently modeled from the Westergaard solution for a sharp crack. Initiation, propagation, and arrest are modeled fairly simply right now, but previous versions (for oil shale) have been quite detailed. This detailed approach is only justified if there are experimental data.

Cracks link up rather realistically in CAVS. This appears to be a natural consequence of the solution of the equations of motion. No explicit "linking" logic is present in the code.

GOUDREAU: I can rephrase my comment by saying sensitivity to the mesh refinement. I doubt if there is that much sensitivity to the mesh refinement normal to the crack, but as you confine your mesh to the tip of the crack you may have—

GROSS: The estimate of the stress concentration of a crack tip to promote propagation necessarily involves the zone size around the tip. This is accounted for in the formulation and provides fracture results that are insensitive to zone size. This feature was developed for application to the explosive fracturing of oil shale.

PREVOST: What is a joint?

GROSS: A joint has no mass in CAVS. The initial forces come strictly from the solid rock mass.

HEGEMIER: Point of clarification. You evidently have a double noding scheme to represent your joints, where you allow slip across the joints?

GROSS: The current version does not account for the conversion of a primary set of joints to a secondary set by slip of an orthogonal joint. The bookkeeper is not that sophisticated at this time.

HEGEMIER: Let me go over this again. I don't think I understand it.

PREVOST: Do you remesh?

GROSS: No, no remeshing.

HEGEMIER: You don't have a double noding system?

GROSS: There is not a double noding system.

HEGEMIER: How do you account, for example, for two joint planes intersecting?

GROSS: Well, the current way it is done is —the model is not that sophisticated in the sense that you do allow slip in one direction; and if you develop sufficient slip in the other direction, it can also go that way.

HEGEMIER: I see. But how do you define "slip"?

GROSS: Slip is accumulated along a joint when the applied shear stress exceeds the allowable value. The "over" shear stress is relaxed to the allowable value with a concomitant increase of joint slip according to linear elastic theory. The current version of CAVS accumulates slip without regard to the sign of the relaxed shear stress, hence slip does not reverse. More detailed versions have accounted for the sign and have allowed slip to reverse. This feature is not thought to be important for tunnel response to a single shock. Of course, it can be important in earthquake cases.

PREVOST: So, if I understand correctly, when you get to a joint, when you have defined some kind of criteria where there is going to be some propagation, you get some of the stresses which are normal to the plane of propagation?

GROSS: Yes, I think that's right.

PREVOST: So, that is a noncontinuum type approach to fracture criterion, isn't it?

GROSS: Yes. It is currently fairly simplified.

PREVOST: So, you rotate the axis, kill the stress and rotate back the axis to get the reference strain?

GROSS: Yes. It is simplified that way.

PREVOST: That's very similar to what people do in mechanics.

GROSS: Probably is the same problem.

CHAPTER VI

MODELING THE BEHAVIOR OF MATERIALS

M. L. Wilkins

University of California, Lawrence Livermore National Laboratory

INTRODUCTION

A first requirement in the calculation of problems in mechanics is a formulation of the material behavior. The material description should include elastic, and elastic-plastic flow. Appropriate yield criteria must be employed. The literature includes many complicated forms to describe material behavior, some of which have been developed to aid the mathematics in the analytic solution of the equation of motion. Numerical techniques can solve the equation of continuum mechanics in two or three dimensions and time with second order accuracy. With these techniques the equations of motion are completely independent of equations that describe material behavior and any mathematical form may be used. The objective of the material models is to provide a theoretical description applicable to a wide class of practical problems, but using simple idealizations of the outstanding features of the real phenomena.

The problems of greatest present interest pertain to metal plasticity. Presented here is the plasticity formulation that has been used for many years at the Lawrence Livermore National Laboratory; Alder et al. (1964).

PLASTIC YIELDING

Most theories of plasticity follow the experimental observation that there is no permanent change in volume due to plastic strain (plastic incompressibility). The description that follows only considers plastic incompressibility which is an appropriate description for ductile metals. The numerical method (Alder et al., 1964) advances the total strain tensor in small increments. The deviatoric stress tensor, s_{ij}, is first calculated with Hooke's law assuming the total strain increment is elastic, $\dot{s}_{ij} = 2\mu\dot{\epsilon}_{ij}$. Here $\dot{\epsilon}_{ij}$ is the deviatoric strain rate tensor and μ the shear modulus. (A correction is made for rigid rotation; see Alder et al. (1975). The dot designates a time derivative along a particle path. The time derivative provides an ordered sequence for the incremental stress/strain relationship. The stress tensor is compared with the condition for plastic yielding. If the condition has been exceeded, each component of the stress deviator tensor is scaled by a constant so that the final stress state satisfies the yield condition. The stress/strain calculations are carried out in the coordinate systems of the equations of motion. To understand the method it is convenient to use principal stress space. The explanation is

schematic assuming proportional loading to clarify the procedure. The actual
equations are in tensor form and contain rigid body rotation terms. Referring
to Fig. 1a, the superscript (n+1)* refers to the equivalent stress resulting
from a strain increment between n and (n+1)*. The von Mises condition, $\sigma_{eq} = Y^0$,
is satisfied by scaling all six components of the deviatoric stress tensor
by m if $\sigma_{eq} > Y^0$.

$$m = \frac{Y^0}{\sigma_{eq}^{(n+1)*}}, \qquad \sigma_{eq} = \sqrt{\frac{3}{2}}\sqrt{2J_2}; \tag{1}$$

J_2 = second invariant of the deviatoric stress tensor = $\frac{1}{2}(s_1^2 + s_2^2 + s_3^2)$.

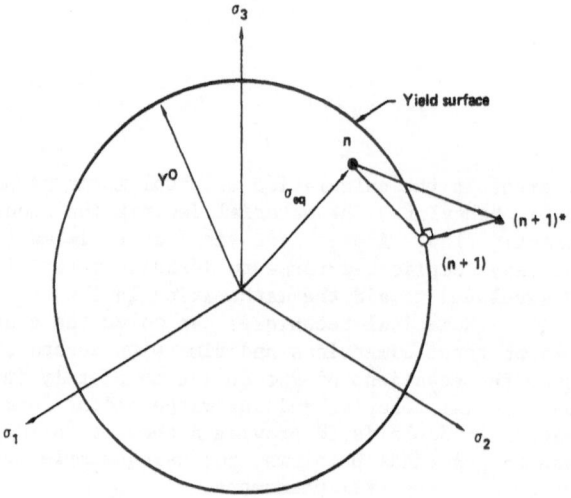

FIG. 1a. Intersection of von Mises yield surface with plane $s_1 + s_2 + s_3 = 0$.

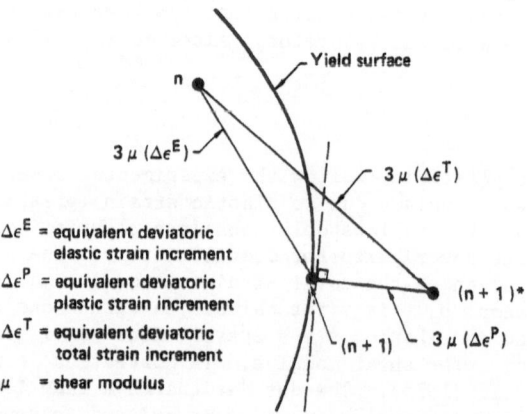

FIG. 1b. Schematic of the stress scaling procedure for elastic-plastic behavior.
From n to (n+1)* represents the stress calculation, assuming complete
elasticity. From (n+1)* to (n+1) represents the portion of the elastic
stress that is set to zero. The corresponding strain increment is
counted as plastic strain. From n to (n+1) represents the actual increase in the elastic stress for the strain increment, $\Delta\varepsilon^T$.

The total deviatoric strain increment $\Delta\varepsilon^T$ is considered to be the sum of elastic, $\Delta\varepsilon^E$, and plastic, $\Delta\varepsilon^P$, components.

$$\Delta\varepsilon_i^T = \Delta\varepsilon_i^E + \Delta\varepsilon_i^P . \tag{2}$$

In terms of the principal deviatoric stresses, s_i, the procedure gives:

$$\Delta\varepsilon_i^P = \frac{1}{2\mu}\left[\frac{1}{m}-1\right]s_i, \qquad i = 1,2,3; \ \mu = \text{shear modulus}. \tag{3}$$

Equation (3) is obtained from Eq. (2) with

$$\Delta\varepsilon_i^T = \frac{s_i^{(n+1)*} - s_i^n}{2\mu} ; \quad \Delta\varepsilon_i^E = \frac{s_i^{n+1} - s_i^n}{2\mu} ; \quad \text{and} \quad s_i^{n+1} = m s_i^{(n+1)*}.$$

Thus, implicit in the method is the result that an increment of plastic strain is related to the corresponding deviatoric stress by a positive constant, namely, $\frac{1}{2\mu}\left[\frac{1}{m} - 1\right]$ where $m \leq 1$. It is seen that the sum of the plastic increments $\Delta\varepsilon_i^P$ is zero (since the sum of the deviatoric stresses is zero), implying plastic incompressibility. The increments of plastic strain are integrated to give the equivalent plastic strain ε^P:

$$\varepsilon^P = \int d\varepsilon^P \quad \text{where} \quad d\varepsilon^P = \sqrt{\frac{2}{3} d\varepsilon_{ij}^P d\varepsilon_{ij}^P} , \quad d\varepsilon_{ij}^P = d\varepsilon_{ij}^T - d\varepsilon_{ij}^E. \tag{4}$$

The increments of total strain, $d\varepsilon_{ij}^T$, are given by the equations of motion. The increments of elastic strain, $d\varepsilon_{ij}^E$, are obtained from Hooke's law using the stresses that have been scaled to satisfy the yield condition. When it is not required to know the components of plastic strain a saving in computation time is realized by calculating the equivalent plastic strain ε^P directly. The ratio of an increment in equivalent stress to an increment in the equivalent strain deviator is 3μ. Referring to Fig. 1b

$$\text{i)} \qquad \sigma_{eq}^{(n+1)*} - \sigma_{eq}^{n+1} = 3\mu\Delta\varepsilon^P$$

$$\text{ii)} \qquad \sigma_{eq}^{n+1} = m\,\sigma_{eq}^{(n+1)*} = Y^0 \tag{5}$$

$$\text{iii)} \quad \Delta\varepsilon^P = \frac{\sigma_{eq}^{(n+1)*} - Y^0}{3\mu} .$$

The equivalent strain ε^P is obtained as before by summing the increments $\Delta\varepsilon^P$.

The incremental plastic work ΔW^P is calculated from the product of components of the deviatoric stress tensor with the corresponding components of the incremental plastic strain tensor.

$$\Delta W^P = [s_1\Delta\varepsilon_1^P + s_2\Delta\varepsilon_2^P + s_3\Delta\varepsilon_3^P]/\rho. \tag{6}$$

ρ is the local mass density. The incremental plastic work can also be determined from the product of the equivalent stress with the increment in equivalent plastic strain.

The plastic work is always non-negative since from Eq. (3) the stress appears squared for each component of Eq. (6). Thus, the increment of plastic work is zero or positive for a loading or unloading cycle.

In principal stress space (σ_1, σ_2, σ_3), the yield surface is a cylinder. The scaling procedure does not change the ratio of the stress deviators and since the scaling is along the radius of a cylinder the normality condition is always satisfied. The result corresponds to an associated flow law, i.e., the plastic strain increments are associated with the yield surface.

Strain hardening is included by expanding the yield surface as a function of equivalent plastic strain, $Y^0 = Y(\varepsilon^P)$. (We are only considering directionally independent yield surfaces.) When strain hardening occurs the yield surface is not constant during a given strain increment. The conic surface representing a material that strain hardens is approximated by a cylinder with an incrementally increasing diameter as shown in Fig. 2. The error accrued has been found to be negligible (Bradley, 1973) as would be expected since the increments of strain are small. (The increments of strain are controlled by the stability conditions for integrating the equations of motion; see Alder et al. (1964).)

It is noted that the yield surface is located unambiguously by the scaling method. Any yield condition which can be expressed in terms of invariants of the stress tensor can be used. The yield surface may have any shape including corners and cusps. With this radial scaling method the relationship between the incremental plastic strains and the corresponding stresses, Eq. (3), remains unchanged for any yield surface. The plastic work is always positive or zero.

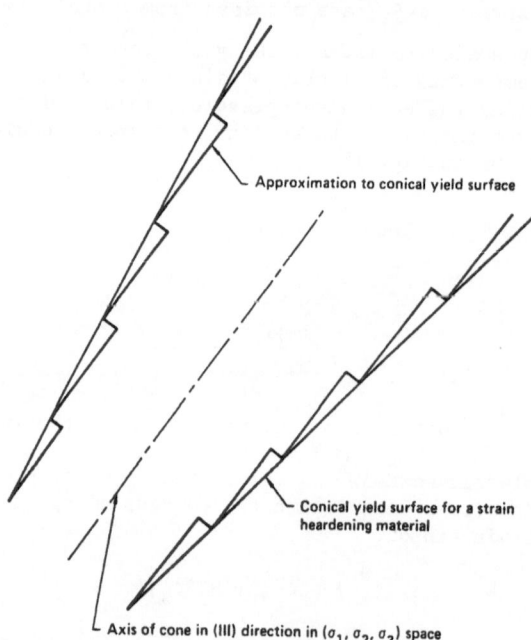

FIG. 2. True yield surface of a material that strain hardens and the approximate yield surface used in the stress scaling computational scheme (pictured in (σ_1, σ_2, σ_3) space).

When the yield surface is not normal to the radius vector as occurs for example with strain hardening, a nonassociated flow law is implied by the scaling procedure. The above results become clear when it is noted that the radial scaling method locates a von Mises surface that intersects the actual yield surface. An implied plastic potential flow rule is applied to the von Mises surface by the scaling procedure. The fact that the increments of plastic strain are not normal to the yield surface is of no consequence. There is no experimental evidence to support this requirement. (In fact experimental data from Flügge (1973) indicate the contrary.) It is, of course, required that the plastic flow process be dissipative which introduces irreversible thermodynamics into the mathematical formulation of plastic flow. This result follows by the manner the total strain is partitioned into elastic and plastic components. The stress increments corresponding to part of the total elastic strain increments are set to zero by scaling the stresses. The corresponding strains are counted as plastic strains, Eq. (2). The remaining stress increments correspond one to one with the elastic strain increments from the linear stress strain relationship. This process is obviously dissipative since the capacity to do work for part of the total strain has been lost. The dissipation is quantitatively described by Eq. (6).

As an example of the versatility of the method consider the Tresca yield condition:

$$s_1 - s_3 = c^0 = \text{constant.} \qquad (7)$$

The principal stress deviators s_i are assumed to be strictly ordered,

$$s_1 \geq s_2 \geq s_3 \qquad (8)$$

after advancing the strain tensor to state (n+1)*, Fig. 3, calculate

$$c^{(n+1)*} = s_1^{(n+1)*} - s_3^{(n+1)*}. \qquad (9)$$

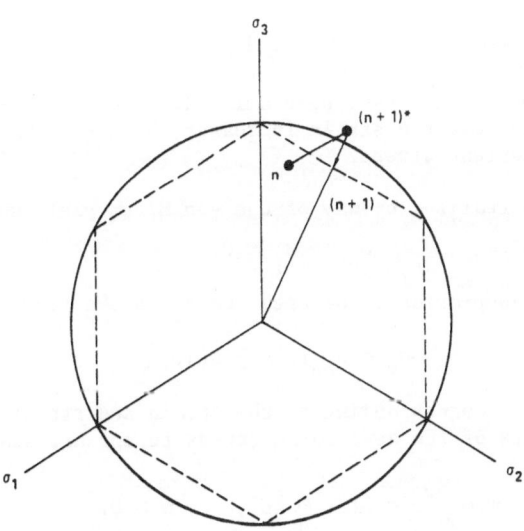

FIG. 3.   Tresca and von Mises yield assumptions in principal stress space.
           Hexagon: Intersection of the Tresca yield surface with plane
                     $s_1 + s_2 + s_3 = 0$.
           Circle: Intersection of the von Mises yield surface with the plane
                     $s_1 + s_2 + s_3 = 0$.

If $c^{(n+1)*} \le c^0$, the stresses are left as they are and tagged with the superscript $(n+1)$. If $c^{(n+1)*} > c^0$ scale all of the components of the deviatoric stress tensor by $m = c^0/c^{(n+1)*}$, i.e., $s_{ij}^{n+1} = ms_{ij}^{(n+1)*}$. The plastic strains are calculated as before, Eqs. (2) and (3).

In summary (1) the requirement for the plastic flow process to be dissipative can be satisfied without the requirement for normality and (2) when a non-associated flow law is used any shaped yield surface, even those containing corners or cusps, can be utilized as required by experiments.

VISCOUS MODELS

The above formulation of the plasticity problem provides the framework for introducing easily other models of material behavior. A simple viscous model that has a direct physical interpretation is the Maxwell solid. In this model the total strain rate $\dot{\varepsilon}_{ij}$ is the sum of two components: the elastic component $\dot{\varepsilon}_{ij}^E$ and the viscous component $\dot{\varepsilon}_{ij}^V$. We will assume that the viscous behavior applies only to the deviatoric stresses s_{ij}.

$$\dot{\varepsilon}_{ij} = \dot{\varepsilon}_{ij}^E + \dot{\varepsilon}_{ij}^V = \frac{\dot{s}_{ij}}{2\mu} + \frac{s_{ij}}{2\eta}. \tag{10}$$

The strains are the deviatoric strains, μ the shear modulus as before, and η the coefficient of viscosity. A dot over a symbol represents a material time derivative.

Rewriting Eq. (10) gives Eq. (11), that describes the stress relaxation of a Maxwell solid:

$$\dot{s} = 2\mu\dot{\varepsilon} - \frac{\mu}{\eta} s. \tag{11}$$

For clarity the tensor indices have been omitted. Equation (11) states that the stress increases as the elastic strain increases, but is relaxed by an amount proportional to the current stress.

Equation (11) is realized by way of the von Mises yield condition.

$$\sigma_{eq} - \sigma_y \le 0. \tag{12}$$

Here σ_{eq} is the same as previously defined and σ_y is the flow stress with

$$\sigma_y = \sigma_{eq}[1 - \frac{\mu}{\eta} \Delta t]. \tag{13}$$

Δt = the time increment corresponding to the strain and stress increments. When $\sigma_{eq} > \sigma_y$ the components of the deviatoric stress tensor are scaled by

$$m = \frac{\sigma_y}{\sigma_{eq}} = [1 - \frac{\mu}{\eta} \Delta t], \qquad m \ge 0. \tag{14}$$

The physical interpretation is σ_y represents the material strength. Maximum and minimum boundary values placed on σ_y can be interpreted as changes in the coefficient of viscosity η.

$$Y_{min} < \sigma_y < Y_{max}. \tag{15}$$

Stresses falling below Y_{min} correspond to $\eta = \infty$, i.e., no relaxation with the result the material is described by Hooke's law in this stress region.

Following dislocation theory the maximum stress the material may attain is considered to be the actual applied stress minus the component of stress corresponding to plastic flow. Equation (16) gives the stress relaxation according to dislocation theory.

$$\dot{s} = 2\mu\dot{e} - 2\mu\dot{\gamma}. \tag{16}$$

Here the stress increases as the elastic strain increases, but is relaxed by the plastic strain rate $\dot{\gamma}$, which is given by dislocation theory. Equation (16) is realized by defining the flow stress σ_y of Eq. (12) as:

$$\sigma_y = \sigma_{eq}[1 - \frac{3\mu\dot{\gamma}}{\sigma_{eq}} \Delta t]. \tag{17}$$

The scale factor becomes

$$m = [1 - \frac{3\mu\dot{\gamma}}{\sigma_{eq}} \Delta t], \quad m \geq 0. \tag{18}$$

It is seen from Eqs. (12) and (17) that the stress increment corresponding to the strain increment $\dot{\gamma}\Delta t$ is set to zero by the scaling method. The plastic strain rate $\dot{\gamma}$ can be defined as

$$\dot{\gamma} = b\ NW, \tag{19}$$

where b = Burgers vector. The mobile dislocation density N and the average dislocation velocity W are functions of the plastic strain and the applied stress.

In terms of dislocation theory, the dislocations in motion and hence the plastic strain rate $\dot{\gamma}$ can be postulated to be zero until the applied stress exceeds some minimum value. This condition is introduced by the model used to describe the dislocation velocity W as a function of applied stress. The applied stress is taken as the equivalent stress σ_{eq}. Equation (16) reduces to Hooke's law for stresses below the given minimum value. With the Maxwell solid model, when the coefficient of viscosity η in Eq. (11) is assumed to be infinite for stresses below a minimum value and zero for stresses above this value, the familiar elastic-plastic behavior is obtained. The corresponding result with the dislocation model (Eq. (16)) is obtained by postulating that the plastic strain rate is zero for stresses below a minimum value and infinite for stresses above this value.

In conclusion, the plasticity formulation described here offers a numerically simple way of introducing a wide range of material behavior into a calculational program.

ACKNOWLEDGMENT

Work was performed under the auspices of the U.S. Department of Energy by Lawrence Livermore National Laboratory under contract #W-7405-Eng-48 and partially supported by Defense Advanced Research Projects Agency.

REFERENCES

Alder, B., S. Fernbach, and M. Rotenberg (editors) (1964), *Methods in Computational Physics*, Vol. 3, Academic Press, New York and London. See also: HEMP3D in UCRL-51574, Rev. 1, Lawrence Livermore National Laboratory, May 30, 1975.

Bradley, W.L. (1973), "Strain Hardening in the HEMP Code," UCID-16328, Lawrence Livermore National Laboratory, July 25, 1973.

Flügge, S. (1973), *Handbuch der Physik*, Band VIa/1, Festkörpermechanik I, C. Truesdell (ed.), Springer-Verlag, Berlin-Heidelberg-New York, p. 681.

GENERAL DISCUSSION

HERRMANN: I'd just like to clarify, Mark, if — on your radial release, you have used kinematic hardening; and if so, would you release to the center of the yield surface (the back stress) or to zero stress?

WILKINS: I haven't —it's been a long time since I experimented with things like that. Now I'm working with compaction of porous models.

PHILLIPS: You state that there is no experimental evidence to support the requirement of normality. This is completely untrue. There is a clear normality. What I understand is that you have taken the Mises surface and then, because there is no normality with respect to the Mises surface, then you say there's no normality whatsoever. Well, you take the wrong surface, a surface which is not valid for the material you use, and then you proceed to say that you have no experimental evidence.

WILKINS: The example I showed used a von Mises surface for the plastic potential, but an independent yield surface. The yield surface selects the von Mises surface. Any yield surface can be used as long as it can be represented by invariants of the stress tensor. The yield surfaces are determined by experiments. We're doing close-coupled experiments and calculations.

   Now, if I had information from experiments that said I must be normal to the surface, then I would restructure the model to include this experimented result. I've searched the literature and I could never find the experiment that gave this result. In fact, in the course of searching, I found something in a book by Truesdell that gave experimental results to the contrary. I have only considered incompressible plasticity appropriate for modelling the behavior of metals.

PHILLIPS: We're talking about metals.

KREMPL: I want to bring up a question with regard to your viscosity model, the Maxwell body. On the one hand you showed very good results of your computation. On the other hand, you have in your model linearity, and we all know that metals are non-linear. Further, the model can only represent secondary creep. It doesn't show transient creep. It appears that the results are not very sensitive to these parameters because you have shown that the shapes of the projectile are very well reproduced. How can this be explained?

WILKINS: I'm not sure I follow that —

KREMPL: I'm saying that the model is linear. We all know that metals are nonlinear in their behavior. Secondly, the model represents only secondary creep, no transient creep or primary creep.

WILKINS: Are you talking about the Maxwell solid model?

KREMPL: Right.

WILKINS: My point for showing that was to demonstrate the ease of introducing a model into the calculational procedure. I'm not trying to represent it as being a physical model. It's a model that's in the literature. It's been known for a hundred years.

CURRAN: You weren't using it for creep anyway.

WILKINS: I wasn't using it for creep.

KREMPL: But you showed good results.

WILKINS: No, but I said by chance it looks like what we see experimentally. I was not claiming that that was a valid representation.

CURRAN: It wasn't creep, either. That was a plate slap.

WILKINS: That was a plate slap, yes. These are at high strain rates.

NEMAT-NASSER: I just want to first make a comment and then ask a question. I hope that you are not citing Truesdell as an experimentalist authority. I'm sure you are not. I am wondering if you feel that dissipation and stability of solutions are synonymous. It seems to me that there is, at least maybe an inadvertent, implication of that kind in the comments that you made.

   We know that most processes can be made to be dissipative even though they may be unstable and highly sensitive to small perturbations in data. In this sense and only in this sense, it seems to me that if you have anything other than a sphere or a circle for the yield surface, which is the Mises condition, then you do introduce by the radial-return method a tangential plastic strain rate component which changes the structure of the differential equations at certain stages in such a profound manner that the solution (if the solution method itself is sensitive enough) becomes highly sensitive to small perturbation in data. And this is not; that is, what I'm suggesting here is not a contrived problem. These can be seen in problems of necking and shear bands in pulled bars or in buckling problems. These problems all involve dissipative processes. But the corresponding solutions become very sensitive to small perturbations in data if we include the tangential plastic strain rate component.

WILKINS: I wanted to start off stating that I was working from experiment, modeling the outstanding features of the real phenomena. The yield surfaces are not theoretical yield surfaces. They are derived from experiment. The experimental results must be in a form that's correct physically and mathematically by using invariants of the stress tensor. If the real process is unstable, then the calculation will show that it's unstable. For example, in some of my powder compaction experiments, a cylinder is imploded by lateral burning explosive. When a low detonation velocity explosive is used, oscillations start in the tube; and the tube goes unstable before the detonation has swept over it. The experiment is sensitive to the mechanical properties of the material. In other words, there is an instability.

   A calculation can simulate a physical condition that is unstable. The calculation can be correct but the instability does not occur because there is nothing in the calculation to trigger it. For example, we could super cool this room to a condition that is below the dew point. The condition can be unstable, but nothing would happen unless there was a piece of dust or some other perturbation present. We do try to watch for instabilities that can come up. The method we use is to put a Gaussian distribution on some material parameters. A random number generator selects values for each zone in the computational grid.

As I said earlier, we do develop a model to describe the experiment at hand, then we try to predict the next experiment. If we can identify from a calculation a situation that's really going to explore the model, then we'll try to perform that experiment. But it's always back and forth between calculations and experiment.

CURRAN: I think in order to keep on schedule we'll postpone the rest of the questions until after Alan has given his presentation.

PREPARED DISCUSSION

by A. Needleman

The radial return algorithm described by Wilkins is an appropriate method for integrating the constitutive relation of a Mises solid, as demonstrated by the solutions obtained using this approach by Wilkins and coworkers, e.g. Wilkins (1978), Norris et al. (1978) and as also shown by Krieg and Krieg (1977). The question arises, however, as to how to proceed when deviations from the Mises idealization, such as yield surface vertex effects and anisotropic hardening, are to be incorporated into the analysis. Wilkins suggests using the radial return method in conjunction with various shaped yield surfaces, including those with corners. Such a procedure can rule out significant phenomena.

In order to illustrate this, a framework for describing plastic constitutive relations needs to be specified. A (rate-independent) plastic constitutive consists of a yield condition demarking the stress combinations that permit plastic response, a flow rule giving the plastic strain rate when inelastic behavior occurs and a workhardening relation determining how the yield condition and flow rule evolve. Furthermore, there are a few basic principles that any plastic constitutive relation must satisfy, such as invariance under rigid body rotations and the requirement that the plastic dissipation be positive. However, these basic physical principles do not greatly restrict the prescription of plastic response.

A workable theory stems from consideration of detailed mechanisms of plastic flow for specific materials or from general principles abstracted from such micro-mechanical models. In structural metals the predominant mechanism of plastic deformation is dislocation slip; see Asaro (1983) for a discussion of the micromechanics of crystalline slip and its incorporation into continuum plasticity. When viewed within the continuum slip framework, plastic deformation by dislocation slip involves shearing along certain crystal lattice planes in certain lattice directions. The combination of slip plane and slip direction defines a slip system. With $\underset{\sim}{n}^i$ the normal to the slip plane, $\underset{\sim}{s}^i$ the slip direction and $\dot{\gamma}^i$ the shear rate on the ith slip system, the plastic strain rate $\underset{\sim}{\dot{\varepsilon}}^p$ is given by

$$\underset{\sim}{\dot{\varepsilon}}^p = \sum_i \frac{\dot{\gamma}^i}{2} (\underset{\sim}{n}^i \underset{\sim}{s}^i + \underset{\sim}{s}^i \underset{\sim}{n}^i) \tag{1}$$

where the sum extends over all active slip systems.

If slip activity is taken to be governed by Schmid's law, i.e.

$$\tau^i = \underset{\sim}{n}^i \cdot \underset{\sim}{\sigma} \cdot \underset{\sim}{s}^i \tag{2}$$

then this model for plastic flow, i.e. τ^i = constant defining the yield surface, leads directly to the maximum plastic work inequality (Bishop and Hill, 1951, McClintock and Argon, 1966)

$$(\underset{\sim}{\sigma} - \underset{\sim}{\sigma}^*) : \underset{\sim}{\dot{\varepsilon}}^p \geq 0 \tag{3}$$

where $\underset{\sim}{\sigma}$ is the stress giving rise to $\underset{\sim}{\dot{\varepsilon}}^p$ and $\underset{\sim}{\sigma}^*$ is some other stress in or on the yield surface.

Furthermore, within each crystallite of a polycrystalline metal there are a discrete number of slip systems. This discreteness of slip systems implies a vertex on the polycrystal yield surface (Hill, 1967). The significance of a vertex does not lie in the details of the yield surface shape, rather it lies in the implication of the vertex structure for the plastic flow rule.

Maximum plastic work implies convexity of the yield surface and normality of the plastic strain rate to the yield surface where the yield surface is smooth. When the yield surface has a vertex at the current stress point, maximum plastic work permits the plastic strain rate to fall within the "forward cone of normals." Therefore, at a vertex the plastic strain is not restricted to a direction fixed by the current state but can follow the stress rate. Thus, a vertex gives rise to a relatively "soft" response to an abrupt change in loading path. Using the radial return expression for the plastic strain increment (Eq. 3 of Wilkins' paper) in conjunction with a yield surface with corners does not give this softening effect.

The softening effect of a vertex manifests itself in the solution to boundary value problems (as has been appreciated, at least within the context of structural mechanics, since the late 1940's) and can lead to a solid with a vertex on its yield surface exhibiting a plastic flow field qualitatively different from that occurring in a solid with a smooth yield surface. Recently, finite element solutions for large plastic strains have been obtained based on a plasticity theory (Christofferson and Hutchinson, 1979) designed to model the features of a polycrystalline yield surface vertex. These solutions exhibit the development of localized shear bands, for example, Tvergaard, Needleman and Lo (1981) and Needleman and Tvergaard (1983). Such shear bands are observed to play a prominent role in limiting tensile ductility and occur in circumstances in which the material continues to strain harden. Thus, these localizations are not due to softening as arising from micro-rupture phenomena or thermal effects (although there are, of course, also circumstances when these effects induce localization). By way of contrast, solutions to identical boundary value problems based on smooth yield surface flow theory lead to smoothly varying deformation fields. Employing a yield surface with corners in conjunction with the radial return expression for the plastic strain rate can rule out shear band formation in strain hardening solids since the flow rule does not incorporate the vertex softening effect.

Drucker (1951) introduced a definition of a "stable" plastic solid which is equivalent to defining a stable plastic solid as one for which the inequality (3) holds. As Drucker (1954) and others have noted, maximum plastic work is not a general restriction on plastic constitutive relations but defines a class of plastic solids. For structural metals deforming by crystallographic slip, its validity is tied to that of Schmid's law. Deviations from Schmid's law lead to crystal plasticity relations for which maximum plastic work, (3), does not hold (Asaro and Rice, 1977). Since convexity, which follows from the maximum plastic work principle, has strong implications for uniqueness of solution of boundary value problems, it is hardly surprising that materials with flow rules which fail to satisfy this inequality are prone to instabilities (Rice, 1977, Needleman and Rice, 1978).

These comments are meant to emphasize that in going beyond the Mises idealization seemingly subtle features of the plastic constitutive relation can have major implications for the character of boundary value problem solutions. It is important to develop plastic constitutive formulations which mirror the essential features of the physical mechanism of plastic response for the class of materials being considered.

REFERENCES

Asaro, R.J. (1983), "Micro-Mechanics of Crystals and Polycrystals," *Adv. Appl. Mech.*, *23*, 1.

Asaro, R.J. and J.R. Rice (1977), "Strain Localization in Ductile Single Crystals," *J. Mech. Solids*, *27*, 445.

Bishop, J.F.W. and R. Hill (1951), "A Theory of the Plastic Distortion of a Polycrystalline Aggregate under Combined Stress," *Phil. Mag.*, *42*, 414.

Christoffersen, J. and J.W. Hutchinson (1979), "A Class of Phenomenological Corner Theories of Plasticity," *J. Mech. Phys. Solids*, *27*, 465.

Drucker, D.C. (1951), "A More Fundamental Approach to Stress-Strain Relations," *Proc.*, First U.S. Nat. Cong. Appl. Mech., ASME, 487.

Drucker, D.C. (1954), "Coulomb Friction, Plasticity and Limit Loads," *J. Appl. Mech.*, *21*, 71.

Hill, R. (1967), "The Essential Structure of Constitutive Laws for Metal Composites and Polycrystals," *J. Mech. Phys. Solids*, *15*, 79.

Krieg, R.D. and D.B. Krieg (1977), "Accuracies of Numerical Solution Methods for the Elastic-Perfectly Plastic Model," *J. Pres. Vessel Piping*, *3*, 510.

McClintock, F.A. and A.S. Argon (1966), *Mechanical Behavior of Materials*, Addison-Wesley, New York.

Needleman, A. and J.R. Rice (1978), "Limits to Ductility Set by Plastic Flow Localization," *Mechanics of Sheet Metal Forming*, D.P. Koistinen and N.-M. Wang (eds.), Plenum Publ. Co., New York, 237.

Needleman, A. and V. Tvergaard (1983), "Crack Tip Stress and Deformation Fields in a Solid with a Vertex on its Yield Surface," *Elastic-Plastic Fracture Mechanics: Second Symposium*, Vol. 1, *Inelastic Crack Analysis*, C.F. Shih and J.P. Gudas (eds.), ASTM STP 803, 80.

Norris, D.M. Jr., B. Moran, J.K. Scudder and D.F. Quinones (1978), "A Computer Simulation of the Tension Test," *J. Mech. Solids*, *26*, 1.

Rice, J.R. (1977), "The Localization of Plastic Deformation," *Proc.*, 14th Int. Congr. Theoret. Appl. Mech., W.T. Koiter (ed.), North Holland, Amsterdam, 207.

Tvergaard, V., A. Needleman, and K.K. Lo (1981), "Flow Localization in the Plane Strain Tension Test," *J. Mech. Solids, 29*, 115.

Wilkins, M.L. (1978), "A Method for Determining the Work Hardening Function to Describe Plasticity of Metals," *Formability: Analysis, Modeling and Experimentation*, S.S. Hecker, A.K. Ghosh and H.L. Gegel (eds.), AIME, New York, 111.

GENERAL DISCUSSION

CURRAN: As before, you may address your questions to either speaker or commentator.

HEGEMIER: Actually, my question is for Mark Wilkins. On one hand, I can appreciate the radial return procedure from the standpoint of computational efficiency. On the other hand, it's very confining in the sense that you're restricted to von Mises-type flow rule. For example, if you have the J_3 involved, then you have a non-circular yield surface but yet you've got, as you've indicated, a Mises-type flow rule.

   Now my question. Have you attempted to extend the procedure to cases where in fact you would have normality or some other type of flow rule, associated, non-associated, but perhaps more general?

WILKINS: Yes, we have worked out the complete equations of classical plasticity theory; it is a natural thing to do. For models assuming plastic incompressibility and path independent yield surfaces we have found the requirements for normality and convex surfaces are not necessary and limit the usefulness of the theory. There are cases, however, where we do use normality. Some materials when they fracture, for example, dilate. However, we always start from concepts of classical plasticity, i.e. normality and convexity and then depart from this base. So the answer is yes, but my goal here was just to show a method for solving elastic-plastic flow in its simplest form as a base reference point. More complicated models of material behavior are included as required by experiment.

   I want to make it clear that I'm not wedded, as you say, to any one model. But I want to have the truth come out of a model with the simplest number of statements. If I can model simply the physics of material behavior, then I want to do it simply. If I can't, then I have to add more complexity. As required by experimental information, to do dilatational modeling I have to have a little bit more complexity.

VALANIS: I have two comments. First, if one plays fast and loose with what is known as the Drucker inequality or the positive plastic work, as you put it, one might run into serious problems with questions of uniqueness of solution of the initial value problem.

   And number two, there was a very interesting paper at the recent Canadian Congress of Applied Mechanics where it was shown that you can take a monotonic stress strain curve for a metal, for instance, and model the material either by pure isotropic hardening or by pure kinematic hardening or by a combination of both. And when you did monotonic tests in the sense that you never unloaded, i.e., you traced a path in the strain space without unloading, you found really very little difference in the calculations. There were differences, maybe ten per cent, maybe fifteen per cent, something of that order, but there were not chaotic differences in the results.

On the other hand, if you went into unloading, then differences became really large, I mean of the order of seventy-five per cent, of the order of a hundred per cent, of that kind of order.

So yes, it does matter what you do sometimes, and sometimes it doesn't matter what you do. But I think that you ought to know when it matters and when it doesn't matter. You can just not leave the question open like that and let people fend for themselves.

CURRAN: Does either the speaker or the commentator want to make a reply before we move on?

WILKINS: Well, I mean, you certainly have to be aware of instabilities. But the idea as I tried to point out from the beginning, the way to proceed is to make a model that describes what you see and predicts what you haven't done.

We use the computers to explore the consequences of what we've just described. If we can find an interesting spot to explore, then we'll call for an experiment there. We're really interested in experimental plasticity, not just theory alone. The theory is a means to an end to understand the experiments. After performing calculations and experiments a number of years, one gets an appreciation of what's important and what isn't important. Otherwise, one can thrash around and do a lot of things that do not keep one channeled to achieve the desired good.

NEEDLEMAN: I agree with what Dr. Wilkins said and I also feel that it is important to know when one has instabilities and when one doesn't. I think it's fair to say that there's been a considerable amount of effort in the past five or ten years in plasticity in trying to delineate these circumstances.

SCOTT: I'm, I guess, getting increasingly puzzled by the role of experiment in general at this meeting. I hear Dr. Wilkins say that he has done what I take to be immense numbers of calculations involving different models by putting in various different assumptions into the calculations. Then he says when he wants to refer to experiments, he searches the literature for evidence of certain experimental results.

I don't really understand why there are not more experiments done, experiments which involve both loading and unloading paths, single-element experiments compared to multi-element experiments. In the latter you have to write a finite element code to describe the geometry of the test. In a single-element test, you actually get the material response telling you what it's doing. Whereas, if you do a very complex boundary value problem which you then try to simulate on a computer, most of the time a very large smoothing goes in the test which obscures the unit material behavior. In other words you cannot discriminate differences in individual element responses when you're looking at an overall boundary value problem.

I would like to hear somebody say something more about experiments. Why aren't there more single-element experiments? If there are, why are they not being reported here?

CURRAN: I'm going to let Mark say something, but I can't restrain myself at this point. I've been very quiet. It's my belief that there is no such thing as a single-element experiment. And with that statement, I will turn it over to Mark.

SCOTT: But before you do, there are experiments which are more single-element than other experiments, and that's really what I'm getting at.

I recognize there is a wide spectrum of these tests, but I think just to take a chunk of metal and do something to it including everything in the metal from small strain linear elastic material, say, in the core, to very large plastic strains out at the boundary is a very different test from actually trying to construct something where stress and strain are more or less the same everywhere.

CURRAN: Yes, I think some of us who have actually done computer modeling of a round bar tension test have received very unpleasant surprises, if you think that's a single-element test; but I really should get out of this.

WILKINS: Well, since you referred to searching the literature, you're working with trying to describe material behavior because you have an end use in mind in research on fracture or research on designing some particular application. You don't want to be forced to do experimentation that has already been done. So you actively are, in the literature, trying to find out everything that's known on the particular material.

What you'll find out, the experiments done on engineering tests are very crude. They'll use strain gauges that span two inches. Well, all I have to do is do a calculation and find out they haven't measured anything. There's no way you can use that information. I see engineers even in modern time using the tension test, putting a strain gauge on it. There is no way they can understand what's happening.

And it is very essential to understand at what plastic strain the material fails — and you can do this with a tension test by measuring the cross-section of the neck. You get a very good measure of the plastic strength. You get a good enough measure to recognize two blocks of the so-called same material.

You can — you order some material; you've got a specification on it. And you do the tension test and say, yep, it has the right work hardening; it has the right hardness; it has the right density. And all I have to do is do an experiment and find out it's a different material and go back and do the right experiment, which they don't do, and show it's not the same material.

CURRAN: Would you comment on what the right experiment is?

WILKINS: I've been searching for the right experiment; and I can find some experiments that are a lot better than the others. It turns out that the tension test is really a pretty good experiment for obtaining the plastic work function and for getting the fracture initiation stress/strain state. If this test is backed up with the flat plate plane stress test, supporting information can be obtained.

A simple experiment is the Taylor test where the flow stress is obtained from the impact of cylinders on rigid boundaries.

But there isn't any one experiment that tells everything we want to know. A series of experiments is required. And you have to make the right measurement on the experiment.

NEMAT-NASSER: Yes, I think perhaps you may be talking about different things than, say, Aris Phillips here would, when you are talking about "experiments"! I wish to make a comment about this and please correct me if my memory is failing me.

I recall one of your papers, Dr. Wilkins, where you had data on shear and data on tension, and you fitted each by a curve which was, I think, a power law of some kind. These curves in the stress-space met each other at a cusp. Now, it seems to me — and please correct me if I'm wrong — that is what you refer to as the experimental evidence.

On the other hand, when I look at what Aris Phillips does, I see that he uses many specimens; he puts them in the testing machine; and he does many loadings and unloadings in a certain defined way so that anyone can go to the lab and reproduce (almost) the same results. In this manner he then defines a yield surface.

Now, are these the same — first of all, did I comment on your paper properly, and secondly do these two procedures, in your view, define the same "experiment"?

WILKINS: Well, I would say so if you're developing a yield surface. Experimental data is used to suggest the form of the yield surface which is then expressed in terms of invariants of the stress tensor. Next a calculation is performed to reproduce the experiment. If we don't succeed, then there is something wrong with the model which must be corrected.

NEMAT-NASSER: May I just comment? I think that perhaps here is the point that must be clarified. I think it would be a very good idea if Aris Phillips and Dr. Wilkins have a long discussion in private because I'm sure that here is where the difference lies.

Some of us define experiment in a very, very different way than just taking a few samples — or even a lot of samples — and fitting them with some kind of curve and then saying that this is experimental evidence. This is much more than the experimental evidence.

PHILLIPS: What I want to mention is that you said that you put a particular equation and receive the typical data. That is what you do? Did I say something wrong?

WILKINS: No.

PHILLIPS: Well, if you put the wrong equation assumption in, of course it will not fit with the data.

CURRAN: I would suggest at this point that we encourage that this discussion be continued afterwards.

I'd like to make also a comment: I think Scott's point that we must try to find experiments that have as much as possible the proper boundary conditions is very important. I think that's a very key problem for the experimentalists to discuss with the analysts. We find when we try to do calculational simulations of even the round bar tension test, we have significant problems in that each particle in that rod is seeing a different stress-strain history despite the fact that we would like to believe that there is some sort of a constant path followed by each particle. There are, unfortunately, gradients that are significant.

WILKINS: Can I reply to that one question that you raised? If the experimentalist gives me the experimental results, it will be in some stress-strain space that he's measured. What I must do is generalize the data into three-dimensional space, using invariants of the stress tensor. I should be able to exactly reproduce what you measured. Otherwise, I haven't generalized correctly your experimental data.

CURRAN: One more comment and then we'll have to postpone further discussion.

VARDOULAKIS: Working along the lines of late Prof. Roscoe and trying to develop element tests, we have been confronted with the problem of bifurcation of element tests. I have used the theory of bifurcation in order to estimate properties which are non-observable during homogeneous deformations. Because element tests can give us at least some information about the incipient properties which are non-observable during the homogeneous part of the deformation. And my question to Prof. Needleman is, do you have surface instabilities that will occur before shear banding is taking place in your calculations?

NEEDLEMAN: Yes, the shear bands occur after the bifurcation point for surface waves.

CHAPTER VII

CALCULATION OF PENETRATION

Kent D. Kimsey

US Army Armament Munitions and Chemical Command
Ballistic Research Laboratory
Aberdeen Proving Ground, Maryland

ABSTRACT

This paper provides an overview of the calculation of penetration. Particular emphasis is placed on numerical simulation of penetration. The predictive capability of current wave propagation codes for impact problems as well as requirements for improving their predictive capability are discussed.

1. INTRODUCTION

The mechanics of penetration and perforation of solids has long been of interest for military applications and is currently being applied to a number of industrial problems such as the integrity of nuclear reactor pressure vessels, crashworthiness of vehicles, protection of spacecraft from meteoroid impact and explosive forming and welding of metals. Thorough reviews of the fundamentals of penetration and perforation and their application to practical problems have been prepared by Goldsmith (1960), Johnson (1972), Backman and Goldsmith (1978), and Zukas (1982).

Impact phenomena can be characterized by the impact angle, the geometric and material characteristics of the target or projectile, or the striking velocity. A short classification of impact as a function of striking velocity, V_s, and strain rate, $\dot{\varepsilon}$, is given in Table 1. The velocity ranges depicted in Table 1 should be considered only as reference points. The velocity ranges are extraordinarily flexible since the displacement field under impact loading depends on numerous parameters in addition to impact velocity.

Impacts at velocities in excess of 1 km/s excite the high frequency modes of the colliding solids. The response is confined to a localized region (typically 2-3 projectile diameters) and is characterized by the presence of shock waves and high hydrodynamic pressures which, on contact, can exceed the material strength by an order of magnitude. For ordnance velocity impacts (1-3 km/s) the pressures decay rapidly due to the presence of free surfaces and the effects of material strength and, except for the interface, oscillate at values comparable to the material strength. Under hypervelocity conditions (4-12 km/s) hydrodynamic pressure dominates the behavior of the solids throughout the bulk of the penetration process. Material strength effects become

165

Table 1.   Impact response of materials (Zukas, 1982)

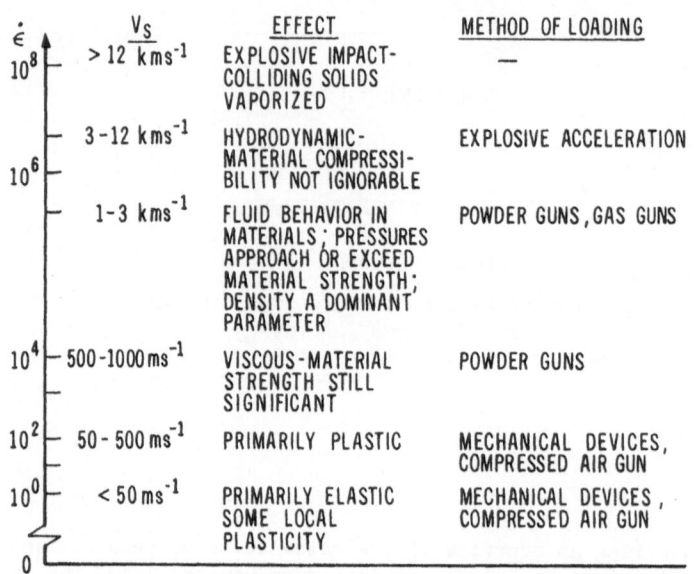

significant only in the very late stages of the process. Superimposed on these
are extensive plastic deformation (plastic strains > 60%), large localized heating and material failure due to a number of mechanisms (i.e., petalling, plugging, brittle fracture, fragmentation, spall, adiabatic shear), the mechanism(s)
activated depending on geometry, loading history and material constitution.
Strain rates exceeding $10^5 s^{-1}$ at the impact interface and $10^2 - 10^3 s^{-1}$ elsewhere
are not uncommon. Penetration and perforation processes are therefore formidable physical problems and it is not surprising that the bulk of the research in
this area has been experimental in nature.

Analytical approaches have tended to fall in three categories:

1. Empirical or quasi-analytical: algebraic equations are formulated
based on correlation with a large number of experimental data points and these
are used to make predictions to guide further experiments. See Backman and
Goldsmith (1978) for a survey of empirical correlations of ballistic data.

2. Approximate analytical methods: these concentrate on one aspect of the
problem such as plugging, spall, crater formation, etc., by introducing simplifying assumptions into the governing equations of continuum mechanics in order to
reduce these to one- or two-dimensional algebraic or differential equations. Simple models of penetration have been described in recent surveys by Backman and
Goldsmith (1978), and Jonas and Zukas (1978).

3. Numerical methods: finite element and finite difference methods are
used to obtain solutions to the governing partial differential equations. Previous reviews of one- and two-dimensional codes for wave propagation and impact
have been prepared by Mescall (1974), Von Rieseman et al. (1974), Herrmann (1975)
and Belytschko (1975). A review of three-dimensional wave propagation codes for
impact problems is given in Zukas et al. (1981).

While empirical and approximate analytical methods are quite useful for
developing an appreciation for the dominant physical phenomena, they are limited

in scope. The terminal ballistician must be prudent in applying these models to a given impact situation, taking care not to violate the simplifying assumptions introduced in their derivation or exceed the data base from which their empirical constants are derived.

2. NUMERICAL SIMULATION OF IMPACT

A complete description of the dynamics of impacting solids must account for the geometry of the interacting bodies, elastic, plastic, and shock wave propagation, hydrodynamic flow, finite strains and deformations, work hardening, thermal and frictional effects, and the initiation and propagation of failure in the colliding solids. Recourse must be made to numerical methods to obtain the complete solution to high velocity impact situations. Numerical simulations in two dimensions have been performed routinely for a number of years. Today, with the advent of fourth generation computers, array processors, and vector processors, three-dimensional simulations are frequently performed for oblique impacts where a three-dimensional stress state is dominant.

A number of two- and three-dimensional wave propagation codes exist and have been successfully applied to problems in terminal ballistics and warhead dynamics. Before discussing the characteristics of these codes, it is worthwhile to classify the main features of the three problem areas of interest to the terminal ballistician:

(a) *Kinetic energy penetration* - the behavior of both inert (solid) projectiles and barriers (often in the form of plates) at impact velocities of 0.5 - 2 km/s is dominated by inertia with material failure as an added complication. Since the problem is momentum-driven, the key parameters are the equations of motion and the descriptions of material failure. Large plastic flow is highly localized and is typically accounted for with an incremental elastic-plastic relationship. Experience has shown that this is adequate for many practical problems. For good correlation with experiments, it is crucial that material parameters be determined from dynamic (wave propagation) experiments.

(b) *Hypervelocity impact* - hydrodynamic pressure dominates the behavior of solids for impact velocities between 4 - 12 km/s. Hence, the equations of motion and a high pressure equation of state are the key descriptions of material behavior. Material strength is only significant for the very late stages of this energy-driven problem which can often be adequately treated with a simple elastic perfectly-plastic relationship with an appropriate value of yield strength obtained from dynamic experiments.

(c) *Warhead formation* - the collapse of an explosively loaded metal to form a short (L/D between 2-3) slug or long jet moving at velocities of 2-8 km/s results in a striker under extremely high pressures (0.2 - 1 megabars), with metal temperatures ranging from one-half to just near the melting temperature. Computations of liner collapse phenomena have had to assume failure strains in excess of 100% in order to obtain decent correlations with experiments. This problem is dominated by very large plastic flows under conditions of extreme temperature and pressure. The adequacy of incremental plasticity theories for such applications has been questioned. Clearly, this is an area for fruitful research.

The predictive capability of current wave propagation codes for impact studies is dependent on the material model and properties that are used in the calculation. It is common in existing wave propagation codes for the study of high velocity impact phenomena to divide the deformation behavior of metals into volumetric and shear (deviatoric) parts. The volumetric behavior is described by a high pressure equation of state. Considerable equation of state data exist for various metals and additional data can be readily obtained. Consequently, the state of the art for metallic equations of state is adequate for present needs.

An incremental elastic-plastic formulation is used to describe the shear response of metals in present finite-difference and finite element codes. Most follow the description first given by Wilkins (1964). The plasticity descriptions are usually based on an assumed decomposition of the velocity strain tensor into elastic and plastic parts. Generally, the plasticity model in wave propagation codes is a simple elastic, perfectly-plastic description. Minor modifications have been introduced to this basic description by allowing the yield stress to vary with the amount of plastic work, temperature, strain rate or some combination thereof.

Zukas and Ringers (1980) have applied the EPIC-3 code to the study of the dynamic buckling of yawed rods. Their results demonstrate the excellent agreement with experiments that can be achieved for an ordnance velocity impact using an elastic perfectly-plastic material model with a dynamic flow stress. The specific problem considered was the impact of a 2024-T3510 aluminum rod (length 5.6 cm, diameter .635 cm) of rolled homogeneous armor at a velocity of 550 m/s. Experimental results were obtained from reverse ballistic tests in which the rod was suspended by tungsten wires in a plastic frame at an angle of 45 degrees.

Figure 1 shows a comparison of experimentally recorded deformation profiles with those obtained with EPIC-3 at 48 μs. The agreement is generally excellent. At this late time it can be seen that the curvature of the deformed portion of the rod does not quite match that shown in the framing camera record. This is due to inadequate numerical resolution in that portion of the rod, which experiences severe stress gradients. Repeating the calculation with a finer grid produced almost exact agreement with experiment.

A principal limitation of wave propagation codes for high velocity impact studies is the uncertainty in the material response description with regards to failure. In dynamic fracture a range of damage is possible. Damage grows as a function of time and applied stress. As damage grows, material stiffness decreases so that even incipient damage levels are important. Despite the fact that material failure is a time-dependent process, most production calculations are performed with simple time-independent criteria based on maxima or minima of field variables (i.e., maximum tensile stress, maximum shear strain, maximum plastic work, relative volume). In certain impact calculations, excellent results have been obtained through their use. However, since failure can occur by a variety of mechanisms it can be expected that different criteria will be appropriate for different applications. Moreover, failure models are not available for all conditions likely to be encountered, and there is little guidance for analysts as to the appropriate choice of failure model for a given condition.

One of the earliest time-dependent failure criteria to be successfully applied to intense impulsive loading is the Tuler and Butcher (1968) model. Failure is assumed to occur instantaneously when a critical value of the damage, K, defined by the integral

$$K = \int_0^T (\sigma - \sigma_0)^\lambda \, dt \qquad (1)$$

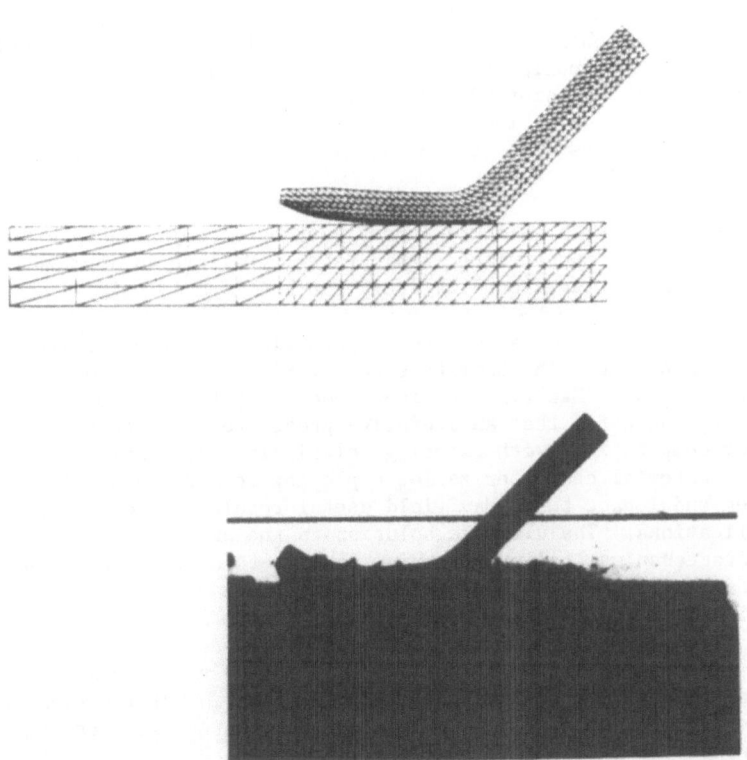

FIG. 1.   Experimental and Computational Results of Oblique Impact at 48 μs
          (Zukas and Ringers, 1980).

is reached. Here σ(t) is a tensile stress pulse of arbitrary shape, σ_0 a thres-
hold stress below which no significant damage will occur regardless of stress
duration, and λ is a material parameter.

Failure criteria in which the damage accumulation is a function of damage
as well as field variables such as temperature, strain rate, and pressure-stress
ratio, have been devised by Davison (1977) and Johnson (1982).

Micromechanical behavior has been incorporated in a continuum-damage model
by researchers at SRI International; see Seaman and Shockey (1975), Seaman et al.
(1976), Erlich et al. (1980), Curran (1982) and the proceedings of an Army Research Office conference on dynamic failure (1982). Ductile failure damage is
initiated when the average stress exceeds a tensile-pressure criterion. Brittle
failure is initiated whenever the maximum normal stress exceeds a tensile thres-
hold. Shear banding occurs when the maximum plastic-shear strain exceeds a
critical value. After initiation, voids, cracks, or shear bands nucleate and
grow according to experimentally determined rate equations.

Computer simulation of ordnance impacts (projectile masses ranging from a few grams to 5 kilograms at striking velocities from 0.5 to 2 km/s) using an incremental elastic perfectly-plastic material model can serve as a first order approximation provided care is taken to select an average dynamic flow stress. Excellent results have been obtained with an elastic perfectly-plastic material model in conjunction with a simplistic description of material failure. This approach is appealing from the point of view that the degree of dynamic material characterization required is quite low. Also, many high-strength alloys show little variation in flow-stress with strain rate and relatively low rates of strain hardening.

Computations involving high velocity impact must be accompanied by efforts to characterize materials at appropriate strain rates. Failure to do this can lead to results which are not even qualitatively correct.

Problems in accounting for dynamic material behavior and fracture are difficult but not insurmountable. The Committee on Materials Response to Ultrahigh Loading Rates of the National Materials Advisory Board (NMAB) recommended a practical approach. They suggested that an iterative procedure of successive refinements involving computations with existing relatively simple failure descriptions, dynamic material characterization employing relatively simple and standard techniques and impact tests may yield useful results for design.purposes in many applications. The ultimate solution to the problem requires development of micromechanical failure theories through theoretical and experimental research such as the nucleation and growth models under development by researchers at SRI International and Sandia National Laboratories.

Hypervelocity impacts (4 - 12 km/s) readily lend themselves to computations. Here, the principal factor in characterizing material behavior is the high pressure equation of state. Since high pressures and temperatures exist throughout the penetration process, strength effects are negligible except at late stages of the penetration. The problem is dominated by energy deposition. Density is a key parameter in determining the amount of penetration. The plasticity model is a secondary consideration.

An accurate plasticity model including the effects of large strain (up to 100%), high pressure (0.2 - 1 mbar) and high temperature (between 60 -100% of the melt temperature) could prove very valuable for the study of explosive-metal interactions such as self-forging fragment formation and the early stages of jet formation. A serious problem here would be the primitive state of experimental techniques which may not be sufficiently sensitive to permit discrimination between different constitutive models. Analytical work in this area currently involves use of semi-empirical one-dimensional relationships for various aspects of the formation problem together with two-dimensional computer simulations with simple incremental elastic-plastic models.

Computational investigation of explosive-metal interaction such as self-forging fragment devices and the formulation of shaped charge jets has been successfully simulated in both two- and three-dimensions (e.g., Johnson, 1981, Hallquist, 1980, Weickert and Kimsey, 1983). Lagrangian codes with an elastic perfectly-plastic material model are frequently used to conduct these analyses. Eulerian analyses have also been used to successfully predict deformation fields and tip velocities. Close agreement for displacements and peak velocities with experimental data can be achieved provided that the code can model arbitrary two-way sliding between adjacent materials and the use of a material model which varies the yield strength of the material as a function of temperature and pressure. Since strain rates are typically on the order of 10^5 s^{-1} strain rate enhancement is considered saturated and has negligent effect on the yield strength. The Steinberg and Guinan (1978) material model is an example of such a model which assumes that the elastic shear modulus and the yield strength of an isotropic material depend on pressure and internal energy.

A major limitation with wave propagation codes, particularly for three-dimensional solutions, is long run times. One-dimensional problems on computers such as the CDC 7600 can be performed in a few minutes. Two-dimensional problems can require from 15 minutes to 2 hours. Typical run times for three-dimensional problems, however, are from 4 - 20 hours or more. This limitation is in the process of being circumvented by developments in both hardware and software. The trend in computer hardware has been toward increased capability at ever-decreasing cost.

The localized nature of the high velocity penetration problem makes it an ideal candidate for the application of both subcycling and reduction techniques which have been successfully used for transient analysis of structures and brought about significant economies. Briefly, subcycling involves integrating different parts of the computational grid with different time steps, so that the bulk of the effort is spent on regions undergoing violent change while quiescent portions of the grid are updated only when needed. Reduction methods can be entirely numerical or combined numerical analytical procedures. They likewise take advantage of the localized nature of the response and can considerably reduce both computer storage requirements and run times.

Coupled Eulerian-Lagrangian techniques make use of the advantages of both methods in different regions of the problem. Eulerian computations can be performed in regions undergoing large distortions since they are ideally suited for such problems. They can then be interfaced to a Lagrangian region where distortions are not so large but boundaries of material histories need to be tracked accurately. This approach has been successfully applied to study the penetration of spaced plates by kinetic energy projectiles (Matuska and Osborn, 1981).

A most promising technique to extend the capability of Lagrangian codes to deep penetration and spaced plate perforation problems is the concept of eroding slidelines. It is being actively investigated at a number of centers. The original Lagrangian codes required that sliding interfaces specified at the beginning of the problem remain unchanged throughout. This requirement was imposed not from physical considerations but to simplify the programming of the interface logic. Its effect was to prohibit total failure of material dictated by the physical problem (i.e., front-face spall), resulting in either unrealistic distortions of the computational grid leading to large truncation errors or to such low values of the time step that continuing the calculation beyond the early stages of impact was economically prohibitive. A first attempt at dynamically redefining sliding surfaces in the presence of total element failure was made by Johnson (1979). This approach, however, has several limitations and restrictions (i.e., only obliquities of 45° or less can be treated and users must specify a priori the extent of target damage) so that it has not been used extensively. Snow (1982) implemented logic to dynamically redefine the master surface as element failure occurs in the EPIC-2 code. The approach retained the requirement that the master surface remain continuous and employed an asymmetric interface treatment. Nonetheless the technique permitted deep target penetration calculations to be performed with a Lagrangian code. During the 1978-1983 time frame, similar refinements of interface treatments were being developed by Massman,

Scharpf, Poth, and others at Industrieanlagen-Betriebsgesellschaft, Munich, West Germany, and have been incorporated in the DYSMAS family of codes, which are capable of analyzing both penetration phenomena as well as dynamic structural response (Scharpf, 1983).

As an example, consider developments at BRL where this technique has been extended to include discontinuous master surfaces and a symmetric interface treatment. The procedure is based on discrete master segments, removing the requirement for a continuous master surface. Furthermore, the interface treatment is symmetric. Initially, nodal points on one side of the interface are designated slave and the nodal points associated with the opposite side of the interface are designated master nodes. The steps to advance in time are briefly stated below:

1. Apply equations of motion to all master nodes.

2. Apply equations of motion to the ith slave node.

3. Identify the discrete master segment, formed by two master nodes, which encompasses the ith slave node within a specified search radius.

4. Perform vector cross-product to test if the ith slave node has penetrated the master segment. If penetration has not occurred, repeat steps 2 - 4 until all slave nodes have been processed.

5. If penetration has occurred, assign new velocities to the slave and master nodes to satisfy conservation of angular and linear momentum.

6. Place the ith slave node onto the discrete master segment. Repeat steps 2 - 6 until all slave nodes have been advanced in time.

7. Interchange master and slave node designations and repeat steps 3 - 6.

8. Identify all elements which have exceeded the user specified failure criteria.

9. Remove master node designation for all master nodes which no longer lie on the surface as a result of element failure.

10. Identify all newly exposed surface nodes as new master nodes.

11. Interchange designation of slave and master nodes and repeat steps 8 - 10.

Figure 2 shows results obtained with this method for the penetration of a steel plate by a 65 gram, hemispherically-nosed steel rod with a striking velocity of 1103 m/s. Figure 3 shows similar results for a plane strain simulation at an obliquity of 60° and striking velocity of 1647 m/s. Both computations were performed using an elastic perfectly-plastic material model with a dynamic flow stress of 1.8 Gpa and an equivalent plastic strain failure criteria of 1.7. Table 2 shows a comparison of computed residual masses and velocities with those obtained experimentally from radiographic data. Agreement is quite good. Additional work is still required to refine and optimize the technique. It appears, however, to be a very promising technique which extends the capabilities of Lagrangian codes.

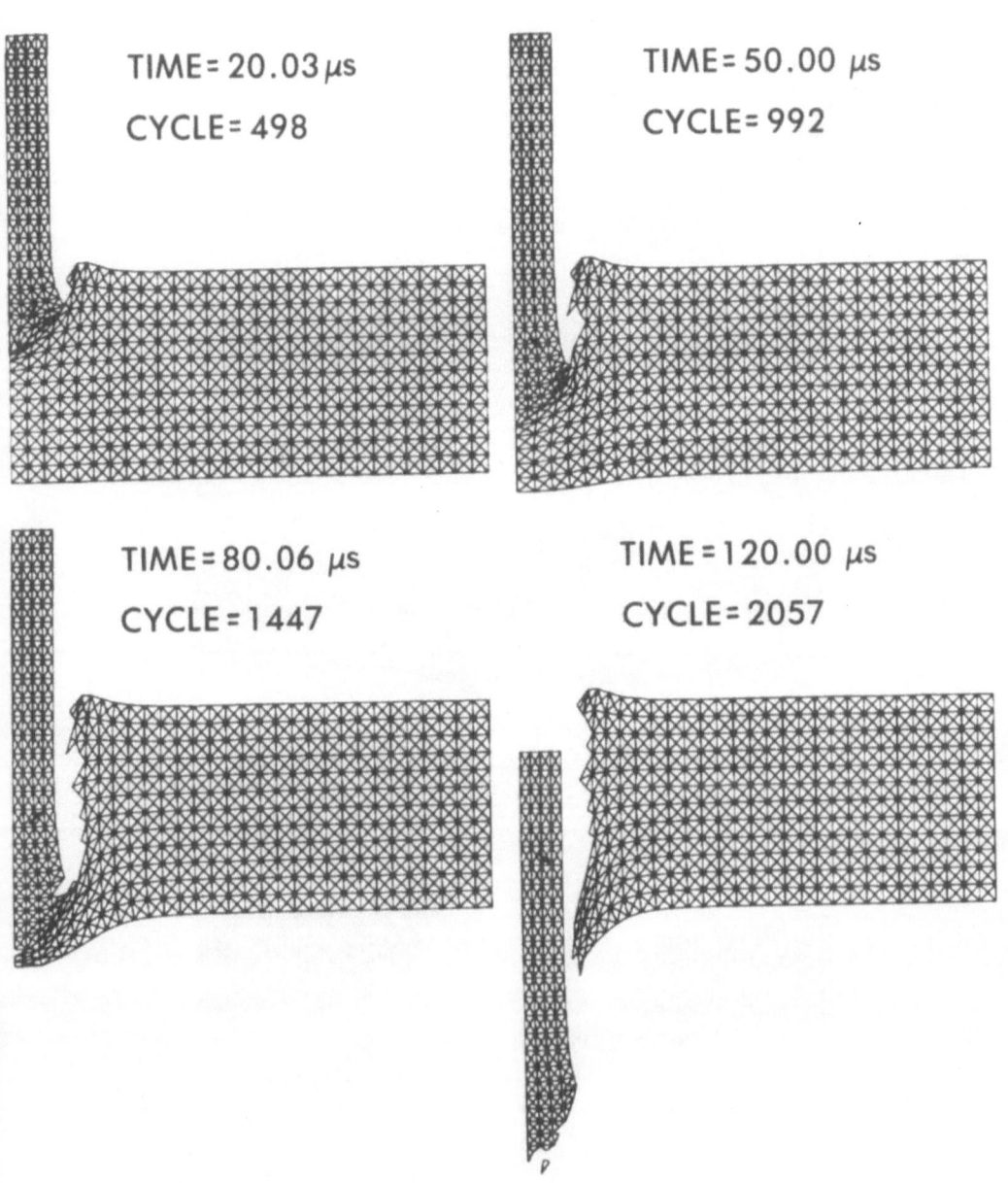

FIG. 2. Eroding Contact Surface for an Antisymmetric Impact.

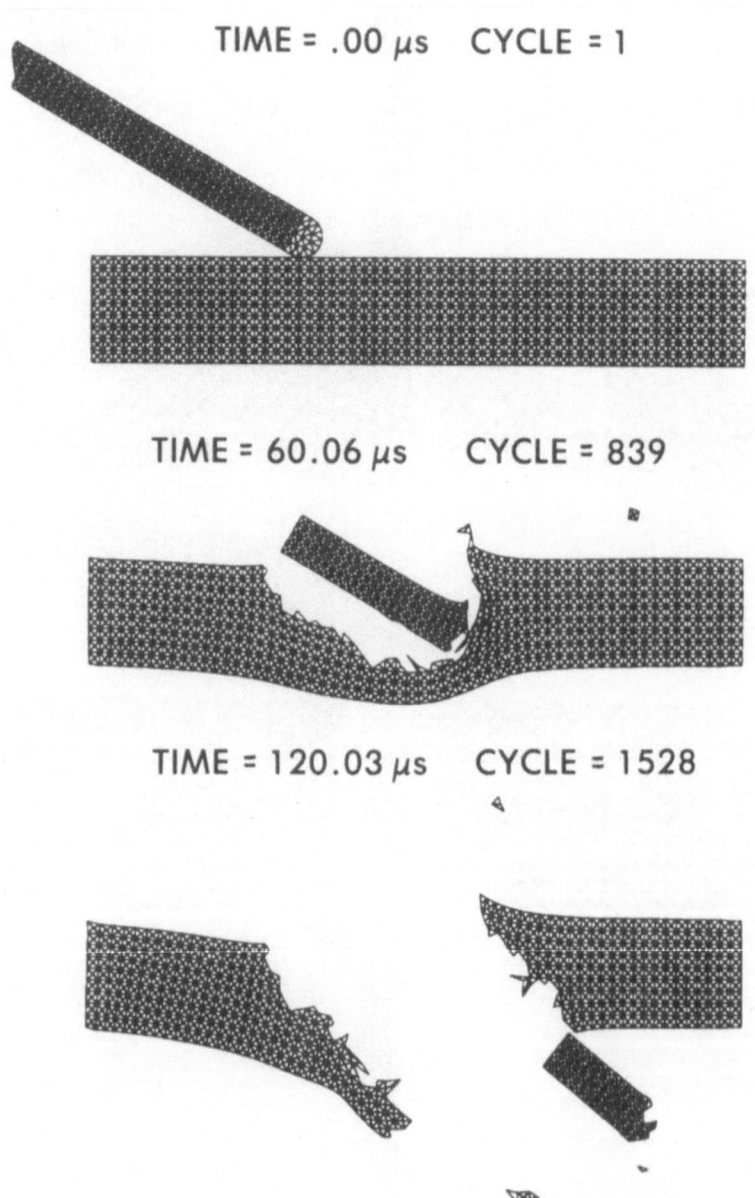

FIG. 3. Eroding Contact Surface for an Oblique Impact.

Table 2. Comparison of calculated and measured residual parameters.

θ	V_s (m/s)	Residual Velocity (m/s)		Residual Mass (g)	
		Calculated	Measured*	Calculated	Measured**
0°	1219	925	910	34.5	39.1**
0°	1103	709	690	32.1	32.7
60°	1647	1202	1145	22.9	16.8

*Lambert (1978).   **Estimated from radiograph.

3. SUMMARY

Uncertainties in dynamic material properties and failure models still exist. However, the NMAB iterative approach allows terminal ballisticians to solve practical problems. However, wave propagation codes for the calculation of penetration need ongoing research in the areas of:

a) dynamic data for high strength steels and high density alloys,

b) time-dependent failure models, and

c) accurate plasticity theory for strains up to 100% in the presence of temperatures at or near the melt temperature and pressures in the high kilobar to megabar region.

ACKNOWLEDGMENTS

The author is indebted to Dr. J. A. Zukas for his suggestions and criticisms which have markedly improved the paper. The author also expresses his gratitude to Mary P. Hubble for her diligent preparation of the manuscript in final form.

REFERENCES

_____(1980), "Materials Response to Ultra-high Loading Rates," National Materials Advisory Board, NMAB-356, Washington, D.C.

_____(1982), Workshop on Dynamic Fracture: Modeling and Quantitative Analysis, May 17-19, 1982, Baltimore, Maryland, Ballistic Research Laboratory, Army Research Office and Army Research Office— Europe.

Backman, M.E., and W. Goldsmith (1978), "The Mechanics of Penetration of Projectiles into Targets," *Int. J. Eng. Sco., 16,* 1-99.

Belytschko, T. (1975), "Nonlinear Analysis —Description and Stability," *Computer Programs in Shock and Vibration,* W. Pilkey and B. Pilkey (eds.), Shock and Vibration Information Center, Washington, D.C., 537-562.

Curran, D. (1982), "Dynamic Fracture," Chapter 9 in J.A. Zukas et al., *Impact Dynamics,* Wiley-Interscience, New York.

Davidson, L., A.L. Stevens and M.E. Kipp (1977), "Theory of Spall Damage, Accumulation in Ductile Metals," *J. Mech. Phys. Solids, 25,* 11.

Erlich, D.C., D.R. Curran and L. Seaman (1980), "Further Development of a Computational Shear Band Model," Army Materials and Mechanics Research Center Final. Report AMMRC-TR-80-3.

Goldsmith, W. (1960), *Impact,* Edward Arnold, London.

Hallquist, J.O. (1980), "User's Manual for DYNA2D — An Explicit Two-Dimensional Hydrodynamic Finite Element Code with Interactive Rezoning," Lawrence Livermore Laboratory, UCID-18756.

Herrmann, W. (1975), "Nonlinear Transient Response of Solids," *Shock and Vibration Computer Programs Reviews and Summaries,* W. and B. Pilkey (eds.), Shock and Vibration Information Center, Washington, D.C.

Johnson, G.R. et al. (1979), "Three-Dimensional Computer Code for Dynamic Response of Solids to Intense Impulsive Loads," *Int. J. Eng. Sci., 14,* 1865-1871.

Johnson, G.R. (1981), "Recent Developments and Analyses Associated with the EPIC-2 and EPIC-3 Codes," *1981 Advances in Aerospace Structures and Materials,* S.S. Wang and W.J. Renton (eds.), AD-01, ASME, New York.

Johnson, G.R. (1982), "Status of the EPIC Codes, Material Characterization and New Computing Concepts at Honeywell," *Proc. of the Army Research Office Workshop on Computational Aspects of Penetration Mechanics,* J. Chandra and J.E. Flaherty (eds.), Springer-Verlag, New York.

Johnson, W. (1972), *Impact Strength of Materials,* Crane, Russak, New York.

Jonas, G.H. and J.A. Zukas (1978), "Mechanics of Penetration: Analysis and Experiment," *Int. J. Engng. Sci., 16,* 879-903.

Lambert, J.P. (1978), "The Terminal Ballistics of Certain 65 Gram Long Rod Penetrators Impacting Steel Armor Plate," Ballistics Research Laboratory, ARBRL-TR-02072.

Matuska, D.A. and J.J. Osborn (1981), "HULL/EPIC3 Linked Eulerian/Lagrangian Calculation in Three-Dimensions," Ballistic Research Laboratory Contract Report ARBRL-CR-00467.

Mescall, J.F. (1974), "Shock Wave Propagation in Solids," *Structural Mechanics Computer Programs,* W. Pilkey, S. Saczalski, and H. Schaeffer (eds.), University of Virginia Press, Charlottesville.

Scharpf, F. (1983), Industrieanlagen-Betriebsgesellschaft, Munich, West Germany, private communication.

Seaman, L., D.R. Curran and D.A. Shockey (1976), "Computational Models for Ductile and Brittle Fracture," *J. Appl. Phys., 47,* 4814.

Seaman, L. and D.A. Shockey (1975), "Models for Ductile and Brittle Fracture for Two-Dimensional Propagation Calculations," Army Materials and Mechanics Research Center Final Report AMMRC-CTR-75-2.

Snow, P. (1982), "KEPIC-2," Kaman Sciences Corporation Report No. K82-46U (R), August.

Steinberg, D.J. and M.W. Guinan (1978), "A High-Strain-Rate Constitutive Model for Metals," Lawrence Livermore Laboratory, Livermore, California , UCRL-80465.

Tuler, F.R. and B.M. Butcher (1968), "A Criterion for the Time Dependence of Dynamic Fracture," *Int. J. Fract. Mech., 4, 431.*

Von Rieseman, W.A., J.A. Stricklin and W.E. Haisler (1974), "Nonlinear Continua," *Structural Mechanics Computer Programs,* W. Pilkey, S. Saczalski and H. Schaeffer (eds.), University of Virginia Press, Charlottesville.

Weickert, C.A. and K.D. Kimsey (1983), "Three-dimensional Computer Simulation of a Linear Self-Forging Fragment Device," *Proc. Seventh International Symposium on Ballistics,* Royal Institution of Engineers (KIvI), The Hague, The Netherlands.

Wilkins, M.L. (1964), "Calculation of Elastic-Plastic Flow," *Method in Computational Physics, 3,* B. Alder, S. Fernbach, and M. Rotenberg (eds.), Academic Press, NY.

Zukas, J.A., et al. (1981), "Three-dimensional Impact Simulations: Resources and Results," *Computer Analysis of Large-Scale Structures,* AMD, *49,* K.C. Park and R.F. Jones (eds.), The American Society of Mechanical Engineers, New York.

Zukas, J.A. and B.E. Ringers (1980), "Numerical Simulation of Impact Phenomena," *Proc. 1980 Summer Computer Simulation Conference,* Simulation Councils, Inc., La Jolla, California, 307-310.

Zukas, J.A. et al. (1982), *Impact Dynamics,* Wiley-Interscience, New York.

GENERAL DISCUSSION

CLIFTON: Let me start things off by making sure that we all understand some of the terminology. You used the phrase "reverse ballistic technique" for getting dynamic data. Could you explain that for us, please?

KIMSEY: In the reverse ballistics technique you have a plastic frame, and the projectile is suspended from tungsten wires to accurately control the yaw. The target is launched from a light gas gun.

CLIFTON: Now, the outcome of this experiment would be the values of the yield stress for the two materials that you would use, then, in the other calculations?

KIMSEY: The experiments were conducted to look at the effect of yaw on impact. This particular experiment was chosen in order to compare computational results with experimental results for which we had very good camera records.

CLIFTON: The point I'm trying to understand is: is this experiment the way that you got the yield stress values used for the calculations?

KIMSEY: No, the yield stress value for the calculation came from free-flight impacts performed by Bell at the Johns Hopkins.

HEGEMIER: You've talked entirely about the case where the target is metal. What about the case where the target is something like concrete? First of all, what do you do modeling-wise; and second, do you get any agreement at all between the modeling and the experiments?

KIMSEY: We have looked at penetrations in concrete. The concrete model used was the model in the EPIC 3 code and the Hull code. The model allows the concrete to go through a crush-up phase and then continue to increase in pressure. When the concrete unloads, it comes back to a fully compacted state. We've simulated bomb impacts into concrete. But we have not been able to compare the results with experiments primarily because the computations ran so long that we couldn't continue them to late response times.
    This was a low-velocity impact. Once the initial stress state is set up in the bomb it would be more computationally efficient at that point to then use an implicit method to go out to later times.

NEMAT-NASSER: I have actually two questions, and let me ask them both at the same time because I think they relate to each other. In one of your earlier slides you indicated that since failure modeling is incorrect, then constitutive formulations should be done most cost effectively; meaning as simple as possible.
    Noting that simplicity does not necessarily go synonymous with cost effectiveness, my two questions are: what do you really mean by agreement, and are you willing to extend your "principle of cost effectiveness" to its logical conclusion and include some of these complicated numerical calculations, as well?

KIMSEY: Yes, I think we would include some of the advanced failure models in the codes for our particular applications, providing they do not require an extraordinary amount of material characterization.

NEMAT-NASSER: No, you didn't understand my question. I meant the principle that since the modeling is not completely correct, we must do the simplest type of constitutive modeling.
    I'm asking, are you willing to extend this principle into the area of computation, as well?

KIMSEY: No, a crude mesh will not do an adequate job for impact problems. You cannot make sacrifices in the mesh.

NEMAT-NASSER: Then please tell us, what do you mean by "agreement"?

KIMSEY: If we can get within say ten per cent of the values obtained in experiments, I feel we then have fairly good correlation with experiments. That's probably within the error of the material properties used in the computations.

GARG: Since, I believe, you get very high temperatures in these experiments, what do you do for the material properties at those temperatures? I think you alluded to thermal softening. I don't quite understand what you mean by that.

KIMSEY: Thermal softening means that as the internal energy of a particular element goes up, its associated flow stress is reduced according to some empirical correlation based on the large number of experiments. Dr. Gordon Johnson has some torsion data in which he looked at temperature effects.

HEGEMIER: That is for a rate-independent material model?

KIMSEY: Yes, although there are rate dependent models that account for different strain rates.

CLIFTON: I have Walt Herrmann next. Perhaps you could include in your discussion, Walt , whether the iterative procedure envisioned by NMAB is the type of procedure that you see here.

HERRMANN: I really wanted to make one other comment first, and then I'd be glad to address that. Could you put on your first slide again, please [see Table 1 of Kimsey's text]. It relates to the comments by Dr. Garg. There seems to be a real misunderstanding here. The slide showed at strain rates of 10^8 a velocity of 12 kilometers per second, explosive impact occurred and colliding solids are vaporized. At 3 to 12 kilometers per second hydrodynamic material response occurs, that is, compressibility is all important.

I think that hydrodynamics includes the notion that material strength is not important. At one to three kilometers per second the slide showed that fluid material pressures approach or exceed material strength.

Now, colleagues of mine at Sandia Labs are currently engaged in doing experiments in which they're directly observing the yield strength of materials like aluminum, copper, tungsten, and other materials, above a megabar. They're doing recovery experiments and looking at the micro-structure. One of the things that's very, very obvious is that you have a large number of effects which can occur at the same time. It relates to Garg's question.

First of all, there is heating. However, the material is compressed. The material is well above its critical point and does have strength. There are a large number of defects introduced. The material is very strongly shock hardened, and the material strength goes up by a large amount because of that particular factor. That is not negligible at megabar pressures.

We have a lot of application calculations for complicated convergent flows in metals in which there is no hope of a hydrodynamic, strengthless calculation reproducing the observations. I don't have time to go into what those are, but there result large discrepancies if you don't get the right strength in at stresses above a megabar.

Now, there's thermal softening going on that competes against hardening, and there are large strain rate effects which also affect strength. We had a comment previously that said, it's very difficult to disentangle all those effects by very simple tests. And that is true here. It's really very difficult to disentangle all those things. Nevertheless, in many cases I think you can make some engineering progress in applications by taking an average flow stress under the proper conditions.

Now, the iterative process that I think the NMAB suggested was that first of all, the NMAB recommended very urgently, particularly in the range of material failure, that there be a very intense research effort which begins by looking at the micro mechanics of failure in order to come to grips with what the right kinds of material failure criteria and mechanisms ought to be and how those ought to be incorporated.

Now, in the process of doing the research, which takes some time, you can do some interim things. You can utilize the basic models that are available today and compare those, as I think you suggested, with experiments such as you showed to see whether, in fact, existing models, or maybe other simpler models, would give you the major features that you're looking for.

Now, I'm very concerned with the actual engineering realm where one is attempting to design things. It's very important to distinguish what material behavior is actually crucial to the particular application. It's very easy, for example, to try to include all kinds of effects, such as those I've mentioned. In material failure you can try to follow the growth and coalescence of cracks in a great amount of detail, based perhaps even on micro-mechanical models. It is very expensive to do the calculations, and extraordinarily expensive to do all the material tests that disentangle all these effects. If and only if, in fact, simpler models work, then I would say it's not necessary to do all that. I can suggest to you that you don't do a great deal of work if you have a simple technique that gives you satisfactory results for your particular application. If your calculation shows discrepancies, then you try to build in more realism into your model and try to distinguish what features of the material behavior you really have to build in.

So in a sense, this is a kind of iterative process. You need to carry on the research very urgently, I think, to enhance your understanding so that you know what next to build in. But at the same time, I don't think you wait for all of that to be completed. You can begin by doing calculations of a simpler nature and then build in more complications as time goes by and you find the need for it.

VALANIS: I was really going to ask a question but Dr. Herrmann already brought it up. Since here you have a very high temperature involvement in the vicinity of the impact and failure, wouldn't it make more sense to consider material properties at the temperature of the environment in which the event is actually happening than to consider properties at room temperature?

KIMSEY: Yes, it would. The problem is obtaining the dynamic material properties at elevated temperatures. Some of this data is only now starting to surface.

ASARO: Following the comments that were just made, let me ask a specific question. With regard to the examples of penetration that you showed with oblique projectiles and which seem to show reasonable agreement, at least in the pattern of penetration and deformation, I have two questions. One, how sensitive would the predictions have been to any of the details of the constitutive models used; what I'm getting at is, how much of that phenomenon is, say, dominated by inertial effects rather than material constitutive properties? And secondly, in the cases where penetration involves very localized failure modes, for example one or two catastrophic shear bands — and you must have examples where you have that sort of phenomena — how well are you able to model the occurrence and details of such processes? These sorts of processes, I think, would not necessarily be dominated by inertial effects.

KIMSEY: The capability to model shear banding has been developed through work conducted by Barbara Ringers at the BRL in which she has introduced a concept of node splitting in the EPIC 2 code. The technique allows a node to split, and then track a shear band as splitting continues. The technique is used primarily to look at plugging problems.

HEGEMIER: As I understand it, your thermal softening amounts to strain softening. Is that true?

KIMSEY: No, thermal softening amounts to strain softening. I mean — not strain but stress. That is, one reduces the yield stress as a function of internal energy or temperature.

HEGEMIER: Now, that worries me from a computational standpoint. Have you done any studies where you've changed the mesh size and looked at your results? I realize these things are enormous computations.

KIMSEY: We have looked at different mesh sizes, but only to assess its effect on the deformation. We have not looked at its effect on thermal softening. For impact problems if one uses a very coarse mesh, the mesh tends to be overly stiff, and you won't get as severe deformations as observed in experiment.

BELYTSCHKO: Strain softening is of concern if the stress decreases with strain. In other words, if you have a direct relationship between the stress and strain. If there is another state variable, such as temperature, that governs the change in the yield stress, you will not have that kind of instability.

In one case you have a loss of the hyperbolicity of the basic equations. But if you have another parameter that changes your yield stress, for example strain rate effects, you don't necessarily have localization.

PREPARED DISCUSSION

by Werner Goldsmith
University of California, Berkeley

(Presented by Marvin E. Backman)

This summary of work by the author and associates describes the use of complex computer programs for analyzing penetration, with emphasis on kinetic energy projectiles, hypervelocity impact and warhead applications. Observed macroscopic phenomena are embedded in the formulation with dominance of certain patterns listed for specific ranges of striker velocity. The versatility of this tool is emphasized as exemplified by interfacing of various codes based on finite difference or element discretization which, jointly with explicit integration algorithms, have attained substantial maturity. Associated difficulties are also noted, including the high operational cost and current areas of uncertainty regarding dynamic material properties, high-temperature effects on plastic deformation and time-dependent failure mechanisms. Two other analytical approaches are also mentioned: (a) Empirical equations as found in JTCG/ME (1977) and Backman and Goldsmith (1978), and (b) Models embracing engineering approximations. The first is correctly labelled as applicable to a narrow, often unknown range of parameters. The second is described as descriptive only for simplified material and stress distribution characterization and, further, incorrectly in the view of the discusser, labelled as concentrating but upon a single deformation mechanism.

Large-scale numerical codes constitute a very powerful and widely applicable tool for calculating the field variable history in penetration problems (Wilkins, 1964, 1978). However, as in the case of phenomenological modelling, the validity of the end result depends upon an *a priori* understanding of all the significant mechanisms as well as on a sufficiently accurate material characterization. They can serve as a criterion to check the accuracy of simpler theories and can be used to assess the effect of parametric changes within a reasonable band when acceptable agreement with experimental results has been obtained within this range without the need for additional expensive experimentation. Since no analysis can reproduce unincorporated behavior patterns, these codes are no exception; they do not properly predict the two types of adiabatic shear bands, nucleated at local discontinuities that are often found in plugging failure of steel plates due to projectile impact (Backman and Finnegan, 1973; Olson et al., 1980). Thermal softening provisions in the computational program (Sedgwick et al., 1978) designed to account for this feature are apparently not sufficiently accurately represented or localized.

Although huge data banks store the high-pressure behavior of many materials, the stated need for further information at high strain rates and temperatures as

well as for improved failure models suggests some discontent with the code, at
least under certain conditions. The constitutive equations employed, the Tillot-
son or Mie-Grüneisen relations and deviatoric stress behavior obeying a Mises or
Tresca yield criterion, with some adjustments for hardening, compressibility and
thermal effects have not been changed for metallic substances in two decades.
The discusser believes that it is necessary to prescribe a complete interactive
thermo-mechanical field and constitutive equation including rate effects in order
to solve problems involving extreme environments. This enormous task will require many years and intensive additional theoretical and experimental effort
even for an approximate solution, but is mandatory if observed features such as
solid phase changes, vaporization and impact explosions are to be included in
the capability of these large-scale codes.

The accuracy of the behavioral description is very difficult to verify
since (a) strain-rate effects depend upon the specific experiment employed, and
(b) measurement of transient temperatures is several orders of magnitude slower
than that for strain. When coupled with the need to consider inhomogeneity and
anisotropy or multi-component systems such as fiber-reinforced plastics, the
number of material characterization parameters, which must be found empirically
by tests, needs to be severely limited to permit their practical utilization.
As a result, it may even be necessary to relax some of the foundational concepts
of continuum mechanics such as invariance requirements and the relation between
yield surfaces and flow rules in plasticity theory. This problem may be further
aggravated if it should turn out that all relevant mechanisms cannot be completely characterized by continuum concepts. The latter purport to encompass
the aggregate of all microstructural effects, but a complete demonstration of
this equivalence has not yet been successfully executed.

Some of the local phenomena, such as nucleation and growth rate of voids or
microcracks or else shear banding have been included in computational efforts to
evaluate macroscopic damage although a complete constitutive theory has not yet
been delineated (Seaman et al., 1976; Shockey et al., 1975).   Zukas et al. (1982)
state that an analysis of void mechanics requires too complicated a material representation for use in current hydrocodes which employ very simple failure models.
By extension, such an argument should also logically be used for the shear band
representation which is treated like a giant dislocation. However, such a position is inconsistent with the present author's remark that engineering models are
too limited in scope when, in principle, large-scale computers exhibit comparable
limitations. Inclusion of other microstructural characteristics, such as dislocations, would undoubtedly necessitate a plethora of assumptions concerning their
distribution and velocity whose implementation would tax the capacity even of the
next series of computer generations.

Large computer codes are useful and have often been employed for plate perforation analysis or penetration of semi-infinite solids. They are less suitable
for the examination of structural response due to impact and projectile penetration near the ballistic limit where dysfunction of a target due to large displacements might be sought. Initial velocities well above the ballistic limit
produce only local effects without destruction of the overall structural integrity (Backman and Goldsmith, 1978).

Engineering penetration models can also only reproduce the mechanisms included, but changes in the relative importance or addition of other phenomena
observed in tests can be executed more rapidly and inexpensively. A useful criterion for the application of the simpler model can be a stipulated degree of
accuracy required for the prediction of a particular parameter. The sequel of

this discussion will be primarily concerned with the phenomenological modelling of plate penetration, an approach which currently or in the recent past has been pursued at very few centers in the world both at institutional and government laboratories.

Analyses of this type utilize a momentum, energy or a combined approach, occasionally also incorporating some needed dimensions based on test data or *a priori* assumptions. A widely cited energy balance for plugging at normal or oblique incidence includes the ballistic limit for the system (Recht and Ipson, 1963). Wave propagation modelling of a deformable cylinder normally striking a plate or a deformable half-space leads to an evaluation of the mass loss produced by shear and the magnitude of uniaxial dynamic yield shear strength and work-hardening coefficient (Recht, 1978). Hydrodynamic representations of long-rod penetration have delineated the wave propagation and fracture processes in the rods and the strain fields in the targets (Tate, 1967, 1978). Projectile-induced plugging at normal incidence in metallic plates and penetration in a semi-infinite solid represented by compression of an equivalent cylinder have been described by means of a factor defining the constraint to flow by the surrounding target material (Woodward and DeMorton, 1976; Woodward, 1982).

A series of models combining shear and compression in plates have been developed for normal incidence, with a simple cosine factor attempting to account for obliquity (Awerbuch and Bodner, 1974, 1977); plugging is developed by means of an adiabatic shear zone. This has recently been augmented by inclusion of the formation and propagation of a bulge in a viscoplastic target and use of the upper bound theorem of plasticity (Ravid and Bodner, 1983). A membrane model of the normal perforation of an infinite plate has also been proposed that utilizes a critical-strain failure criterion and a quantitative penetration resistance factor (Dienes and Miles, 1977).

Combined analytical and experimental investigations of the normal impact and perforation of soft aluminum and carbon steel plates in the velocity range from 50 - 950 m/s have recently been carried out by the discusser and his associates using conical- and hemispherical-tipped as well as blunt cylindrical strikers. The theories applied have been based on one or more observed deformation mechanisms. An energy balance was utilized for the sharp projectile upon normal impact with an intact plate, specifying crack propagation and subsequent petal bending by hinge rotation so as to permit projectile passage; concurrent plate bending response was considered by means of a semi-empirical relation (Landkof and Goldsmith, 1983). Excellent agreement was found between predicted and measured terminal striker velocities for the intact plate, except near the ballistic limit. When the projectile was fired into the center of a predrilled hole smaller than striker, the velocity drop was found and predicted to be greater than for the intact target when the ratio of hole to projectile diameter was less than 1/3 due to a change of the perforation mechanism from petalling to extrusion.

Below the ballistic limit, a simple mass-dashpot system provided an excellent prediction of the terminal geometric and kinematic parameters for hemispherically-tipped strikers; somewhat poorer agreement was found in the perforation region upon use of a dual mass-dashpot model (Levy and Goldsmith, 1983). All system constants can be explicitly evaluated from available material and geometric properties. The correlation of theory and experiment is depicted in Fig.A. While this simplistic approach is obviously unsatisfactory for any even highly circumscribed description of such an event, it might be possible to deduce the approximation of such a situation as a special case from a more comprehensive

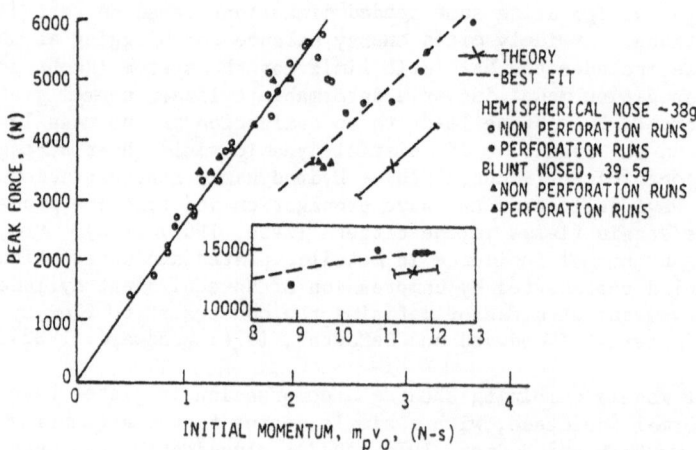

FIG. A.   Peak Force as a Function of Initial Momentum
for the Normal Impact and Perforation of Hemispherically-Nosed Cylindrical Steel Projectiles
of 12.7 mm Diameter against a 1.27 mm Thick
2024-0 Aluminum Target Clamped on a 119.7 mm
Diameter.

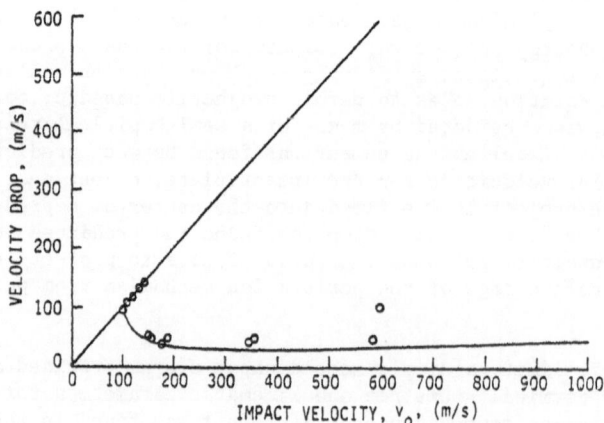

FIG. B.   Experimental and Predicted (solid curve) Velocity
Drop as a Function of Normal Impact Velocity of a
40g Rigid Steel Projectile for a 6.4 mm Thick
2024-0 Aluminum Plate.

continuum theory. A cap initiated by shearing is separated from the ductile, initially plane target in the region of maximum curvature change coinciding with the greatest cross-sectional change.

The analysis of the normal impact of hard, blunt projectiles on softer plates was performed by characterizing the process as a 5-stage series involving indentation, plug initiation by shear, slipping and separation, and post-perforation deformation (Liss, Goldsmith and Kelly, 1983). Dynamic target response is included by energy transfer from a plastic shear hinge. The system is divided into a limited number of rigid bodies separated by moving plastic wave fronts and hinges. Excellent correspondence was found between predicted post-perforation parameters and corresponding observations, as illustrated in Fig. B. Still better correlation would be expected upon inclusion of target bending, neglected here, which will always absorb a significant amount of energy near the ballistic limit.

The documented results are intended to contrast the levels of effort and expenditure for comparable degrees of correlation with data for predictions by simple models and large-scale computers, respectively.

ACKNOWLEDGMENT. This work was sponsored by the Army Office of Research under Contract DAAG 29-80-K-0052.

REFERENCES

Awerbuch, J. and S.R. Bodner (1974), "Analysis of the Mechanics of Perforation of Projectiles in Metallic Plates," *Int. J. Solids Struct.*, *10*, 671.

Awerbuch, J. and S.R. Bodner (1977), "An Investigation of Oblique Perforation of Metallic Plates by Projectiles," *Exp. Mech.*, *17*, 147.

Backman, M.E. and S.A. Finnegan (1973), "The Propagation of Adiabatic Shear," *Metallurgical Effects at High Strain Rates*, R.W. Rohde et al. (eds.), Plenum, New York, 531.

Backman, M.E. and W. Goldsmith (1978), "The Mechanics of Penetration of Projectiles into Targets," *Int. J. Engng. Sci.*, *16*, 1.

Dienes, J.K. and J.W. Miles (1977), "A Membrane Model for the Response of Thin Plates Ballistic Impact," *J. Mech. Phys. Solids*, *25*, 237.

Joint Technical Coordinating Group for Munitions Effectiveness (Anti-Air), "Penetration Equations Handbook for Kinetic-Energy Penetrators, *61* JTCG/ME-77-16.

Landkof, B. and W. Goldsmith (1983), "Petalling of Thin Metallic Plates during Penetration by Cylindro-Conical Projectiles," *Int. J. Solids Struct.* (in press).

Levy, N., and W. Goldsmith (1983), "Normal Impact and Perforation of Thin Plates by Hemispherically-Tipped Projectiles, I. Analytical Considerations," submitted.

Liss, J., W. Goldsmith and J.M. Kelly (1983), "A Phenomenological Penetration Model of Thin Plates," *Int. J. Impact Engng.*, *1*, 321.

Olson, G.B., J.F. Mescall and M. Azrin (1980), "Adiabatic Deformation and Strain Localization," *Shock Waves and High-Strain-Rate Phenomena in Metals, Concepts and Applications*, Ch. 14, M.A. Meyers and L.E. Murr (eds.), Plenum Press, New York, 221.

Ravid, M. and S.R. Bodner (1983), "Dynamic Perforation of Viscoplastic Plates by Rigid Projectiles," *Int. J. Engng. Sci., 21,* 577.

Recht, R.F. and T.W. Ipson (1963), "Ballistic Perforation Dynamics," *J. Appl. Mech., 30,* 384.

Recht, R.F. (1978), "Taylor Ballistic Modelling Applied to Deformation and Mass Loss Determination," *Int. J. Mech. Sci., 16,* 809.

Seaman, L., D.R. Curran and D.A. Shockey (1976), "Computational Models for Ductile and Brittle Fracture," *J. Appl. Phys., 47,* 4814.

Sedgwick, R.T. et al. (1978), "Investigations in Penetration Mechanics," *Int. J. Engng. Sci., 16,* 859.

Shockey, D.A., L. Seaman and D.R. Curran (1975), "A Computational Model for Shear Bands," SRI Poulter Laboratory Technical Report 003-75.

Tate, A. (1967), "A Theory for the Deceleration of Long Rods after Impact," *J. Mech. Phys. Solids, 15,* 387.

Tate, A. (1978), "A Simple Hydrodynamic Model for the Strain Field Produced in a Penetration of a High Speed Long Rod Projectile," *Int. J. Engng. Sci., 16,* 845.

Wilkins, M.L. (1964), "Calculation of Elastic-Plastic Flow," *Methods in Computational Physics,* B. Alder et al. (eds.). *Fundamental Methods in Hydrodynamics.* Academic Press, New York, 211.

Wilkins, M.L. (1978), "Mechanism of Penetration and Perforation," *Int. J. Engng. Sci., 16,* 793.

Woodward, R.L. and M.E. DeMorton (1976), "Penetration of Targets by a Flat-Ended Projectile," *Int. J. Mech. Sci., 18,* 119.

Woodward, R.L. (1982), "Penetration of Semi-Infinite Targets by Deforming Projectiles," *Int. J. Mech. Sci., 24,* 73.

Zukas, J.A. et al. (1982), "Three-Dimensional Impact Simulations: Resources and Results," *Computer Analysis of Large Scale Structures,* K.C. Park and R.F. Jones, Jr. (eds.), New York, ASME, AMD *49,* 35.

GENERAL DISCUSSION

CURRAN: I'm sure that some of you who know me would be surprised if I didn't leap up when someone mentions shear bands, and so I'm going to make a very brief comment about them. I think they're very interesting to this group because they are plastic instabilities, and we've been talking about plastic instabilities. We've seen examples where they're very important.

In the impact of a long rod on a thick armored plate, they govern both the ballistic limit, that is, the velocity required to penetrate the target, and the fragmentation, because it's intersecting shear bands that cause the fragmentation. I'm also responding to a previous question by Bob Asaro when he asked how well present shear band models can predict the ballistic limit and the formation of the plug.

I put my name down on the list to talk a little bit about these things tomorrow, so I won't say much about them now, but I wanted to make a very quick response to a few things that came up.

First, the shear band models that my colleagues and I have developed are internal state variable models that fit into the constitutive relation family that we've been talking about at this meeting. They treat damage as an internal state variable and describe the formation of the bands as plastic instabilities. Now, the key property that we think governs their nucleation is a trade-off between work hardening and thermal softening. In most applications, we're talking about the thermal softening that's caused by adiabatic heating, the thing that Ted talked about.

BELYTSCHKO: My statement was wrong, incidentally, because in the case of adiabatic heating, stress is related to plastic strain, as Goudreau has just shown me.

CURRAN: Jim Rice and his colleagues, Asaro and Rudnicki and company, have worked out a lot of the cases where these instabilities can occur. Marv's example of mild steel (which shows what I suspect is a shear band plugging in one case, and ductile petaling in the other) is a good example where such models can help, because mild steel is notoriously nonresistant to shear banding because it doesn't have enough work hardening in it, whereas aluminum is notoriously resistant to shear banding because it has a lot of work hardening and lots of ductility in it. It fights back.

HUTCHINSON: Well, following along this same thought, I thought Dr. Backman made a very pointed remark when he said that he felt that the shear banding should come out naturally from the flow law and the calculation itself. To a large extent, I think he's right. It shouldn't have to be put in as an artificial failure mechanism. The question I'd like to address to you is, do you have any feel for whether the reason apparently these aren't coming out is because of the thermal flow stress, however you want to put it, modeling, or is it possibly because of the yield stress modeling?

Let me remark that we know from lots of examples now that there are in general multi-axial stressing histories that if you use isotropic hardening, Mises characteristically, it tends to suppress any kind of shear banding, even if it does want to occur. So do you have any feel for what is lacking in the modeling here such that it doesn't come out naturally?

BACKMAN: I think it probably is material properties that are restricting the solution. It is a very fine process, too. I have seen a region of the body developing the kind of shape that is eventually bounded by shears, and then that disappears as the calculation proceeds, so that I have presumed that something within the material modeling that's going on in that particular code is suppressing the shear bands.

JOHNSON: I'd like to make a comment on the shear banding we've observed with a number of materials we've torsion tested. I think we have pretty good evidence that we do in fact see shear banding, where the thermal softening overtakes the strain hardening. Also, from experimental observations, if we look at stress vs. strain, as soon as the stress begins to decrease, it localizes very quickly. Moreover, in terms of doing the computations, we can show the instabilities beginning to form.

But I think the problem for these large-scale computations is that at very

high strain rates the shear bands get extremely thin. As a result, we can't
model the shear bands by putting many elements in the band area and include heat
conduction and other effects. It just isn't compatible with the whole model.
But I would suggest that one of the things we might be able to do is to find
those conditions which will lead to the onset of shear banding. Maybe, when the
stress begins to decrease, although there's still some argument along those
lines, or other critical events begin, then we could say this represents the onset of shear banding. Then we could take the appropriate measures to either
split the elements, as Kent made reference to Barbara Ringers' work, or to let
the element fail in some other manner. To automatically include shear banding
and heat conduction and other effects in the context of a big problem, just isn't
realistic.

HUGHES: I'm kind of surprised to hear some of these comments because it seems
to me that one should be able to model shear banding even with a fairly crude
mesh. It will be spread over an element; I've done calculations like that, and
I know other people here have done calculations like that. If your elements are
performing correctly, you should be able to get shear banding with no particular
gridding whatsoever, no biasing, no refinement. It's the same thing as in, say,
gas dynamics. You have shock capturing schemes. You don't have to have a real
fine mesh, but you capture a shock over so many elements. It's just a natural
part of the overall computational process.

BACKMAN: I know that John Mescal has used large-scale representations of continuum processes and gotten shear bands that agree with experiments. But you
don't see them just happening, at least in the penetration codes that I know of.

NEEDLEMAN: Just a comment about Gordon Johnson's comment. I'd be rather pessimistic about the effectiveness of imposing a critical strain criterion for the
onset of shear banding or failure, because it's a plastic process and inherently
path dependent. If you had a critical strain one would need a class of critical
strains depending on the stress state and the history of how one got there.

BACKMAN: Let me make another statement on this. I have a biased view on this
because I looked at a lot of cross-sections of craters in which one sees these
shear bands going in recognizable patterns, not always regular, but essentially
axisymmetric. When you actually look at the crater from above they're not axisymmetric, but more like sections of spiral, and they're three-dimensional
things. I don't know whether that's a reason of why computations don't reproduce them, but that is certainly a non-axisymmetric property that one can easily
miss.

NEMAT-NASSER: I want to make a comment. I'm glad that Ted Belytschko corrected
himself. It was a mistake to say that if you include the effect of temperature
you do not get a softening effect. You do get — I did the calculation here —
and you can get a negative slope for the stress-strain curve. Therefore, going
back to John Hutchinson's comment, the question that I think the numerical analysts, the numerical codes, have to address is that why doesn't this naturally
emerge from the calculations? Why don't the codes predict at a certain stress
level when the adiabatic heating makes the curve to slope down, a loss of hyperbolicity? I think this question is directed to Mr. Kimsey.

KIMSEY: I think Gordon addressed this earlier.

WILKINS: I was going to support what Tom Hughes said where one should be able
to model this in the programs. For example, one can model shear fracture in one

zone, which is analogous to shear banding.

CLIFTON: Let me ask Sia's question to you just directly, Dr. Wilkins. If you do a calculation where you are accounting for thermal softening, do you see localized deformation?

WILKINS: Well , I don't — I wouldn't expect to see localized deformations because the loading may not be in that particular direction. But I think if you had a yield surface that responded to what your stress state is, and you have thermal softening, then I think you'd see shear banding. This is essentially the same as what we do in fracture, where we use some kind of cumulative damage model. You will always get brittle fracture with the model if you say fracture occurs perpendicular to the tensile load.

But if you've done experiments and you realize that the fracture can occur at lower strains for certain types of loading, for example for shear types of loading, you can get the fracture to occur at lower strains in your model. The physical explanation could be that the holes can form in a group for a shear type of load. When you include that in your model, then you'll start to get shear fracture. And this is comparable to shear banding as far as the mechanics is concerned. The code should be able to reproduce shear banding if you make the constitutive model correct.

HUTCHINSON: I just want to very quickly emphasize again that there is a difference between failure like ductile hole growth or brittle fracture which you do have to add separately to the plastic flow law, and the plastic flow. A decent plastic flow law should lead to these instabilities without having to add anything extra to it. It's a very big difference.

SANDLER: The question is one of loss of hyperbolicity due to softening occurring from straining versus softening occurring from temperature effects. The loss of hyperbolicity in strain softening is due to the fact that the stress and the strain (through the displacement or acceleration) are coupled in the equation of motion, in which acceleration is directly related to stress gradient. If changes in strain directly produce changes in stress with a negative modulus, you get a loss of hyperbolicity through the equation of motion. Softening due to temperature effects, while it certainly leads to a decrease in stress, may not necessarily lead to a change in the character of the partial differential equation. The wave speeds may remain real if you have stress decreasing, depending upon the details of the thermal behavior and the way the energy equation is handled.

Another form of "softening" is that due to viscoelastic effects, which is also called stress relaxation. By means of a viscoelastic formation, one arrives at a perfectly properly posed dynamic problem. Wave speeds remain real. The problem is that if you introduce softening strictly through the strains, the wave speeds become complex. The governing system of partial differential equations is no longer hyperbolic and forward time marching is not possible. For this reason there can be a fundamental difference mathematically between temperature softening and strain softening.

CHAPTER VIII

ON THE EXPERIMENTAL DETERMINATION OF CONSTITUTIVE EQUATIONS

W. Herrmann

Sandia National Laboratories
Albuquerque, New Mexico 87185

1. INTRODUCTION

The widespread and economical use of computers to solve partial differential equations by finite element or finite difference methods has made it possible to consider nonlinear problems in thermomechanics representing large deformations and complex material behavior. In fact, it is relatively simple to formulate some finite difference and finite element methods in such a way that the constitutive equations can be inserted in subroutines without disturbing the remainder of the computer code implementing the method. As a result, there has been an increase in interest in the development of constitutive equations purporting to represent "realistic" material behavior in the nonlinear range. In some cases empirical equations have been proposed which have been fitted in some, usually restricted, way to experimental data. These have been inserted without further ado into the codes, and numerical solutions for specific and often very complicated initial-boundary value problems have been obtained as a result of many hours of computer time. Little attention has been given to the character of the theory resulting from the constitutive formulation, or to validation of results against exact solutions or experiments representative of the problems being solved.

The relaxed attitude with which come practitioners approach the subject is understandable in the context of the simplicity of insertion of constitutive equations of almost arbitrary complexity into the codes, the rather forgiving nature of the codes themselves (if the solution is ragged, increase the artificial viscosity), and the fact that the procedures by which the classical theories like elasticity and hydrodynamics, plasticity and ideal compressible fluids were derived are taken for granted, seldom taught, and all but forgotten in many present curricula.

Fortunately there has been a resurgence of interest in the foundations of nonlinear theories of thermomechanics. There is also renewed interest in the mathematics of nonlinear partial differential equations, especially with regard to their stability. To this should be added an interest in the methodology for the deduction of constitutive equations from experiment.

It is impossible to consider the constitutive equations apart from the conservation laws and boundary conditions, which together constitute a mathematical

theory. Consequently, the next section will briefly review the requirements of
a complete mathematical theory. With this background, the methodology of construction of constitutive equations will be illustrated by means of a number of
specific case histories.

2.   GENERAL REQUIREMENTS

The construction of a complete theory involves all of the following elements:

i. *Postulation of the proper balance laws.* Theories involving microstructural variables, for example, may involve internal degrees of freedom governed
by balance laws in addition to the usual mass, momentum and energy conservation
equations for a single constituent material. Their omission produces an inconsistent theory. This is a particularly active area of research at present, because the proper formulation of the extra balance laws is not always self
evident.

ii. *Postulation of the proper entropy principle, or other stability condition.* Once again, the proper form of this condition is not clear in many cases.
For example, the proper stability condition which places restrictions on the
elasticity tensor in finite deformation elasticity remains unresolved, and the
proper formulation of the Clausius-Duhem inequality for certain types of mixture
theories remains a topic of debate.

iii. *Choice of the general functional form of the constitutive equations.* The
variables to be included in the constitutive equations, and the general form of
the dependency of the material response on these variables, together with the
balance laws and stability conditions, determine the nature of the theory and
the kind of phenomena exhibited in its solution.

iv. *Analysis of the theory to ensure its consistency.* This includes proper
kinematical invariance to choice of coordinate system, limitations on the constitutive equations by the assumed stability or entropy condition, limitations
arising from assumed material symmetries, and limitations arising from assumptions that the material response should be frame-indifferent, in particular to
rotations of the material with respect to the observer.

v. *Postulation of the class of general initial-boundary conditions to be
considered.*

vi. *Proofs of existence and uniqueness of solutions to the theory represented
by the above assumptions.* More properly for non-linear theories, this involves
an analysis of domains of existence of possibly none, one or multiple solutions
and their stability.

vii. *The investigation of the general behavior of solutions to the theory.* In
some cases, exact solutions to particular initial-boundary value problems can be
found without knowledge of the specific functions representing a specific material. Such solutions give insight into the question whether the theory is useful in modeling observed behavior in specific circumstances.

viii. *Construction of particular functions within the general functional forms
of the constitutive equations in order to represent a specific material.* Some-

times the functional forms may be suggested by a consideration of the micro-mechanical processes by which the material is expected to respond.

ix. *Fitting of the parameters in the constitutive functions to experimental data obtained in "material property" tests.* It is sometimes possible, and extremely advantageous, to deduce some parameters directly from measurements if the test configuration corresponds to an initial-boundary value problem for which an exact general solution is known.

The classical theories of thermomechanics usually meet these requirements. Many current attempts to construct constitutive equations in fact fall within the general form of an existing classical theory, and all of the mathematical structure of the theory is available.

To give a particularly simple example, many of the attempts to describe the behavior of metals fall within the theory of infinitesimal plasticity. Provided that the restrictions of the theory are observed, and in particular those arising from Drucker's quasi-thermodynamic stability postulate, all is well. Trivial homogeneous solutions of the theory apply (at least approximately) to tests such as the tension-torsion-inflation test, which allow the deduction of "stress-strain" laws from experiment. Of course, tests generally provide incomplete information for constructing a complete constitutive relation for all possible stress states, and it is necessary to introduce assumptions to complete the description. These may be based on arguments regarding the micromechanical deformation mechanisms which are thought to be active in particular materials. The extension of incomplete experimental data by different assumptions may result in different constitutive descriptions for the same material, each providing a satisfactory fit to the same experimental data. Each may, in fact, be useful in problems which exercise the material in a range in which the constitutive equation gives a good description of the behavior. With the assurance that unique solutions exist and are in some sense well behaved for the class of problems to be solved, it is possible to attack the problem of constructing particular solutions. If it is necessary to use numerical methods, then, since exact solutions are known to exist, the construction of numerical algorithms and the investigation of their convergence to the exact solution can be tackled with some confidence.

When an attempt is made to extend the classical plasticity theory to large deformations by introducing a nonlinear strain measure in an arbitrary fashion, the mathematical structure of the classical theory may no longer apply. If an attempt is made to fit the stress-strain relation to data beyond the point at which strain localization or buckling occurs in the specimen, it is clear that the trivial homogeneous solution which underlies the reduction of load-elongation data to stress-strain information no longer applies. More seriously, fits to the data beyond the ultimate load may not satisfy the stability postulate, the theory may imply instability and bifurcation of solutions perhaps representing strain localization, or nonexistence of solutions perhaps representing failure. However, it is just these situations which may be of most practical interest. In fact, many problems require "post-failure" analyses, which inherently involve large deformations and instabilities such as strain localization or localized cracking. Forcing such problems into the context of a "well-posed" classical theory may exclude precisely the physical phenomena which are crucial to the problem.

It is, of course, not feasible to carry out the entire program outlined above for most non-linear theories before they are used. Nevertheless, it is

important for the practitioner to be able to recognize when he is dealing with
a constitutive equation which falls into a theory for which at least some of the
mathematical structure has been worked out, and to make use of this structure in
the experimental evaluation of his equations. It may be even more important for
him to recognize when his constitutive equations fail to satisfy the stability
or consistency requirements of known theories. In many cases, the failure may
be intrinsic to the physics, but the practitioner is at least alerted to the
fact that the resulting theory may not behave the same as well known classical
theories.

3. STRESS WAVE PROPAGATION IN POLYMERS

This example illustrates a case where an existing non-linear theory could
be used for which some of the mathematical structure already existed to aid in
the evaluation of constitutive functions. Schuler, Barker, Nunziato, et al.*
performed impact experiments using PMMA in which they were able to resolve the
detailed profiles of strong plane stress waves, and from which they intended to
deduce constitutive equations for use in calculations of dynamic loading problems. They observed acceleration waves (in which the first derivatives of the
velocity, etc. are discontinuous), which either decayed or grew depending on
their amplitude, Fig. 1.

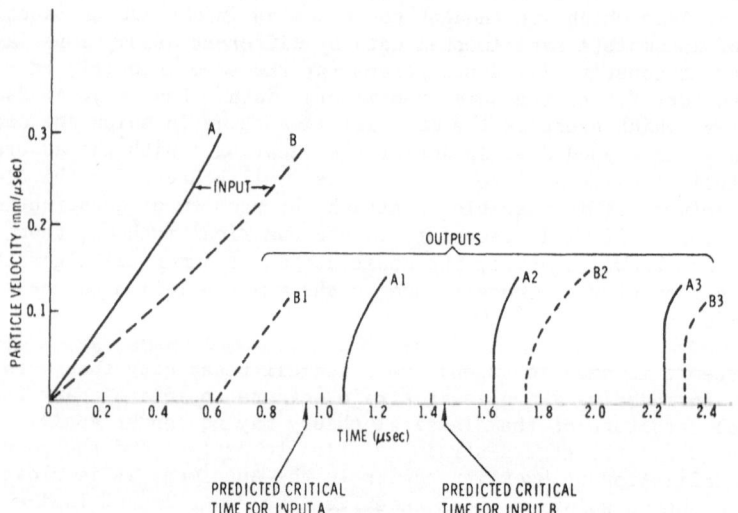

FIG. 1. Experimentally Observed Acceleration Waves in PMMA (from Nunziato et
         al. (1974), *Encyclopedia of Physics*).

They also observed shock waves (in which the velocity, stress, etc. are
discontinuous), generally followed by continuous waves which changed shape as
they propagated, appearing to settle down to steady propagating waves after
some time, Fig. 2. Ultrasonic data were also available.

---

*See Ref. Nunziato et al., (1974), *Encyclopedia of Physics*.

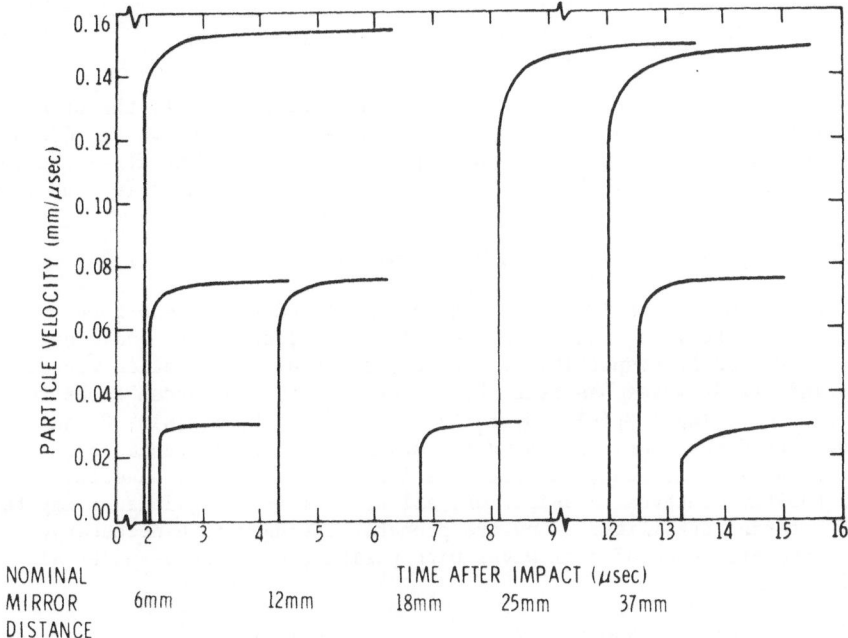

FIG. 2.   Experimentally Observed Shock Waves in PMMA (from Nunziato et al. (1974), *Encyclopedia of Physics*).

Coleman and Noll* had developed the theory of materials with fading memory, generalizing concepts of linear viscoelasticity, and several properties of the theory had been worked out. In particular, Coleman, Gurtin and Herrera* had shown the existence of acceleration and shock waves under particular conditions on the constitutive functions, and derived equations for their growth and decay. Pipkin and Greenberg* showed the existence of steady waves, and exhibited their general properties. These general results qualitatively agreed with the observed behavior, suggesting that the theory might be useful in describing the polymer.

Restricting the present discussion to one dimension for simplicity, Coleman and Noll* argued that the stress σ at a material point X at the present time t is a functional of the entire strain history $e^t(s) = e(t-s)$ for all past times $(t-s)$; $0 \leq s < \infty$

$$\sigma = \mathcal{F}(e^t). \tag{1}$$

The concept of fading memory was made precise by smoothness assumptions on $\mathcal{F}$ and by introducing an influence function h operating on the strain history which is smooth and monotone decreasing, so that strains in the distant past have a smaller influence on the behavior than recent strains. This provided enough mathematical structure to investigate the general properties of the theory for simple strain histories such as creep and stress relaxation. They also constructed sequences of approximations to the functional in terms of integral expansions, and showed that the first approximation corresponded with classical linear viscoelasticity in the case of infinitesimal deformations. For large strains the first approximation is

---

*See Ref. Nunziato et al., (1974), *Encyclopedia of Physics*.

$$\sigma(t) = \sigma_I(e) + \int_0^\infty G'(e,s)e(t-s)\{2 - e(t-s)\}ds \tag{2}$$

where $\sigma_I(e)$ is an instantaneous response function corresponding to the stress immediately following a strain jump to e, and $G' = dG(e,s)/ds$ where $G(e,s)$ is a relaxation function. The stress corresponding to a constant strain history for all past time is given by the equilibrium response function $\sigma_E(e)$ and it is easily shown that

$$\sigma_E(e) - \sigma_I(e) = \{G(e,\infty) - G(e,0)\}(2 - e). \tag{3}$$

First and second derivatives of σ_I and σ_E exist under the smoothness assumptions, and are the instantaneous and equilibrium first and second order moduli E_I, $\tilde{E}_I$, E_E and $\tilde{E}_E$ respectively. The existence of acceleration waves, shock waves and steady waves has been shown under the further assumptions that $\sigma_I(e) > \sigma_E(e) > 0$, $E_I(e) > E_E(e) > 0$, $\tilde{E}_I(E) > 0$, $\tilde{E}_E(e) > 0$, $G(e,s) > 0$ and $G'(e,s) \leq 0$. The first four correspond to convexity of the instantaneous and equilibrium response functions, and with the last two, they are connected with concepts of positive internal dissipation, all of them relating in some way to concepts of material stability. They are probably stronger than necessary, guaranteeing the existence of real waves with positive wave speeds under all conditions.

Steady wave analyses show the existence of several classes of steady waves, depending on the conditions. One of these corresponds to the type of waves seen in the experiment, Fig. 2. If the steady wave speed is V, then the analysis shows that

$$\rho_0 V^2 = \frac{\sigma_I(e^-)}{e^-} = \frac{\sigma_E(e_\infty)}{e_\infty} \tag{4}$$

where ρ_0 is the density of undisturbed material ahead of the wave, e^- is the strain immediately behind the leading shock wave, and e_∞ is the asymptotic strain attained far behind the wave. Thus, without specifying the particular functions σ_I, σ_E or G, measurements of steady wave velocity, strain behind the leading shock, and asymptotic strain behind the wave in different experiments can be converted directly to points on the $\sigma_I(e)$ and $\sigma_E(e)$ curves, which can then be fitted with empirical expressions.

In a steady wave, the motion is described by a single parameter $\zeta = (t - X/V)$. Continuity and momentum conservation imply that $\sigma(\zeta) = \rho_0 V^2 e(\zeta)$ so that the general constitutive equation becomes

$$\rho_0 V^2 e(\zeta) = F(e(\zeta - s)). \tag{5}$$

Solutions of this functional equation provide the steady wave profile $e(\zeta)$. Since $\sigma_I(e)$ and $\sigma_E(e)$ are known, information about $G(e,s)$ should be obtainable from measured steady wave profiles. In some cases this can be done directly. Alternatively, a particular form can be assumed for G, the steady wave equation can be solved explicitly, and the parameters in G can be chosen to fit the steady wave data. The latter course was followed here, by assuming that G could be expressed in terms of a single relaxation time

$$G(e,s) = \{G(e,0) - G(e,\infty)\}e^{-s/r} + G(e,\infty). \tag{6}$$

Average values of the three parameters in this expression $G(e,0)$, $G(e,\infty)$ and r could be evaluated directly from data at only three points on the steady wave profiles.

Since the theory corresponds to classical linear viscoelasticity for infinitesimal strains, the usual analysis of plane acoustic waves holds, and the relaxation function at zero strain can be found by the usual procedure from ultrasonic dispersion and attenuation data. While it was found for this material that a relaxation spectrum with 20 terms was needed to fit the entire range of (time-temperature shifted) frequencies over 20 orders of magnitude, a single relaxation time sufficed to fit the data over the three orders of magnitude or so spanning the impact experimental conditions. This relaxation time agreed well with that obtained from steady wave analysis, suggesting that, at least for the range of conditions of interest, the time dependence of G might be obtained from ultrasonic experiments which are considerably easier to perform. Of course the resulting description is then limited to the particular time regime of the tests, and cannot easily be extrapolated to other conditions.

We note that the theory yields expressions for the growth and decay of unsteady shock and acceleration waves, which also depend on the relaxation function. Since all the parameters of the theory had been determined from steady wave data, these expressions could be used as consistency checks. In fact good agreement was noted between predictions of the theory and shock and acceleration wave growth or decay data.*

Finally, the constitutive theory was incorporated into a finite difference code, and predictions were compared with data from various more complex impact experiments to further validate the theory. While the final constitutive equation is one dimensional, the general theory is not so restricted. Plausible assumptions about material isotropy lead in a natural way to a properly invariant description for arbitrary motions.

While the theory outlined above was appropriate for modeling PMMA under the conditions of interest, the integral evaluation in (2) is difficult and time consuming when used with present finite element or finite difference methods. Rearrangement and differentiation of the previous equations allows them to be recast in the general form appropriate to a material with internal state variables (Nunziato, Schuler and Hayes, 1974),

$$\sigma = \sigma(e,\underset{\sim}{\alpha}) \qquad \dot{\underset{\sim}{\alpha}} = \underset{\sim}{h}(e,\underset{\sim}{\alpha}) \qquad\qquad (7)$$

where $\underset{\sim}{\alpha}$ and $\underset{\sim}{h}$ are vectors with as many components as terms in the relaxation spectrum, and the functions σ and $\underset{\sim}{h}$ are related to the functions σ_I, σ_E and G used previously. Implementation in this form in finite element and finite difference methods is considerably more efficient. While the theories are equivalent in the present special case, this is not true in general. The theory was also extended to include thermal effects; see Nunziato et al. (1974), and Coleman and Gurtin (1967).

Schuler et al.* found that, at stress levels above about 6 kilobars, the $\sigma_I(e)$ and $\sigma_E(e)$ curves appeared to undergo an inflection, perhaps suggesting a phase change or plastic transition of some sort. Since such an inflection violates the convexity conditions under which the above wave analyses hold, the present theory is inapplicable and the data for higher stress tests could no longer be reduced using them.

---

*See Ref. Nunziato et al. (1974), *Encyclopedia of Physics*.

4.   STRESS WAVE PROPAGATION IN METALS

It is tempting to extend the viscoelastic constitutive description of the last section to viscoplastic phenomena. Continuing in one dimension for simplicity, this has been done by adding the simple elastic-plastic strain rate (velocity strain) decomposition

$$\dot{e} = \dot{e}_e + \dot{e}_p \tag{8}$$

where $\dot{e}_e$ and $\dot{e}_p$ are the elastic and plastic strain rates, respectively, given by

$$\dot{\sigma} = E\dot{e}_e \qquad \dot{e}_p = g(\sigma, e, \underset{\sim}{\alpha}). \tag{9}$$

Here, E is an elastic modulus while g is a relaxation function. The $\underset{\sim}{\alpha}$ are again internal state variables governed by evolution equations of the form (7_2). These equations have a general form similar to that discussed in the last section for a material with internal state variables, but the appearance of the stress in (9_2) prevents exact correspondence in general. Moreover, plasticity theories usually introduce a yield surface in general six-dimensional stress space $\Phi(\underset{\sim}{\sigma}, \underset{\sim}{\alpha}) = 0$. If the current stress is on or within the yield surface $\Phi \leq 0$, then the relaxation function g is assumed to be zero. If the stress state lies outside the yield surface $\Phi > 0$, then the plastic strain rate is assumed to be normal to the yield surface $\dot{e}_p = \zeta d\Phi/d\underset{\sim}{\sigma}$. These features render the previous theory for a viscoelastic theory inapplicable. For large deformations, no theoretical structure exists comparable to that for non-linear viscoelasticity. In particular, general constraints on the shape and motion of the yield surface have not been worked out.

In one dimension, the complications of the shape of the yield surface and the normality condition are absent if the material is initially isotropic. If we consider only the simplest case of perfect plasticity, then the motion of the yield surface may be ignored. The theory for this case has received sufficient attention to give some basis for analysis. A summary with references has been given by Herrmann (1976). The properties of the theory are very similar to those for a viscoelastic material. In particular, under suitable inequalities on E and g, plane shock, acceleration and steady waves exist which are qualitatively similar to those described in the last section, except that they are preceded by an elastic precursor shock.

One of the earliest studies of strong plane waves noted that the precursor decayed with propagation distance, Figs. 3, 4. If the strength of the shock a is represented by the jump in stress across it, then its rate of change turns out to be given by

$$\dot{a} = \left[\frac{2LE}{L+3E}\right] g - \left[\frac{2(L-E)}{L+3E}\right] \dot{\sigma}_2 \tag{10}$$

where $\dot{\sigma}_2$ is the stress rate immediately behind the shock, and $L = \rho V^2$ where V is the shock wave speed.

Early studies of shock wave attenuation assumed that, since precursors were generally of low enough amplitude to assume that the material was nearly linearly compressible, $L \simeq E$ and the second term of (10) could be neglected. Then (10) reduces simply to $\dot{a} = \frac{1}{2}g$, and the stress dependence of the relaxation function can be determined directly from fits to the shock attenuation data. Most studies actually began with specific models of dislocation motion, with terms for dislocation multiplication and velocity, which could be used to suggest the functional form of (9_2). The coefficients were then evaluated from

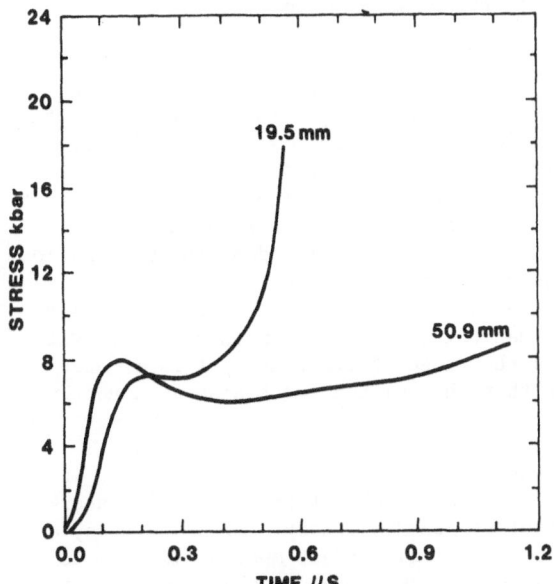

FIG. 3. Precursor Wave Profiles in Armco
Iron (from Herrmann, 1976).

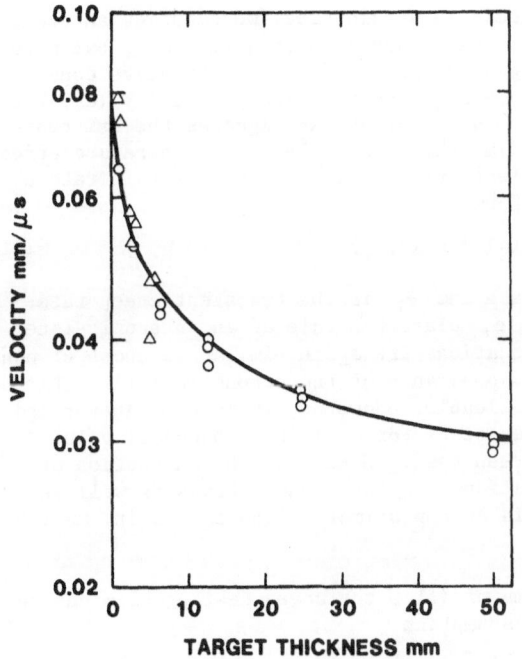

FIG. 4. Precursor Wave Decay in Armco Iron
(from Herrmann, 1976).

the shock wave decay curve. The fact that a good fit was obtained was generally taken as "proof" that the assumed dislocation mechanism was in fact operative.

Unfortunately, different dislocation mechanisms appeared to fit the various sets of data. In fact it was found that most assumed mechanisms led to two parameter equations which fitted given sets of data equally well, and fits to the data could not be used to distinguish among them. Moreover the initial dislocation density which had to be assumed in most cases was much too high. A detailed study (Herrmann et al., 1971) of shock attenuation in single crystal copper, for which very good data existed, showed that the nonlinear term in (10) could not in fact be neglected. The exact solution had to be used. Consequently, all of the previous work was called into question. A very careful study (Asay et al., 1975) of shock wave decay in single crystal lithium fluoride using the nonlinear relation showed that none of the previously assumed mechanisms fitted the data, and new mechanisms involving heterogeneous nucleation of dislocations were invoked, mechanisms which have subsequently also been questioned.

Steady waves in metals have also been analysed (Herrmann, 1974). In this case, it was possible to deduce stress, strain and strain rate along steady waves for different experiments directly from the data. These data could then be used to test various relaxation functions. Once again it became evident that the data could not be used to discriminate between different dislocation mechanisms; several mechanisms could be fitted to the data equally well.

It has become evident that, while considerations of the micromechanical mechanisms of deformation are very useful in suggesting forms for the constitutive functions, it is often found that forms derived from several different mechanisms will fit given sets of data equally well, and it is not possible to assert with any assurance that a particular mechanism must be the only one that is operative on the basis of limited comparison of theory with data over a restricted range of conditions. It is more helpful if the dominant deformation mechanism is identified by direct observation of the microstructure, and this is being attempted; see Grady et al. (1982). Fits of the relevant theoretical expressions to the measurements of mechanical quantities can then be made with some confidence. Extrapolation beyond the range of experiment still is uncertain, however, unless it can be shown that the dominant mechanism does not change.

While progress has been made in viscoplastic plane wave propagation, it must be noted that its generalization to other stress states is highly uncertain, because of the lack of a satisfactory general theory. Even the study of unloading is uncertain, since motion of the yield surface must be invoked to describe anisotropic hardening, which seems to be a feature observed in most metals. Generalization to other than one dimensional motions is, of course, even more difficult and uncertain without the guidance of a general theory.

5.   CREEP

Jumping about 20 orders of magnitude in strain rate, we consider an example of creep modeling, specifically in rock salt (Herrmann et al., 1982), but more generally in ductile materials. The general forms of the constitutive equations for creep are often taken to be the same as those for viscoplasticity (8,9), but with a reinterpretation of the terms. The simplest case ignores the existence of a yield surface. Again using one dimension and neglecting temperature effects for simplicity, one approach to transient creep takes the creep strain rate $\dot{e}_p$ to be given by a function of general form

$$\dot{e}_p = \dot{e}_s(\sigma) + \dot{e}_t(\sigma, e_t) \tag{11}$$

where $\dot{e}_s$ is the steady state creep rate and $\dot{e}_t$ is the transient creep rate. Thus the total transient creep strain e_t plays the role of an internal state variable. In the present case, the equations are again similar to those of non-linear viscoelasticity except for the appearance of the stress in (11). This fact renders the previous theory inapplicable. However, it is straightforward enough to solve the one-dimensional equations for an initial-boundary value problem corresponding to a uniaxial creep test. Note that the assumption of one-dimensional motion is an approximation. In fact, end effects as well as lateral strains and barreling make this an approximation whose validity must be investigated.

For various assumed specific forms of (11), the creep test problem can be solved explicitly. For example, the assumption

$$\dot{e}_t = \xi(e_\infty - e_t) \tag{12}$$

where ξ and e_∞ together with $\dot{e}_s$ are parameters which depend on the stress, leads to the solution

$$e = e_0 + \dot{e}_s t + e_\infty \{1 - \exp(-\xi t)\}. \tag{13}$$

This expression has in fact been used extensively to fit high temperature creep test data in metals, and was found to fit relatively low temperature creep data in salt as well, Fig. 5. Once again, a variety of two parameter expressions was found to fit the data equally well, so that no firm deductions about the active creep mechanisms could be made.

Attempts to fit the stress dependence of the parameters ξ, e_∞ and e_s over a relatively wide range of conditions to expressions motivated by various dislocation mechanisms suggested different regions in which different dislocation mechanisms seemed to dominate. However, the usual lack of discrimination among mechanisms on the basis of the mechanical data prevented determining with any assurance the mechanisms in each region, or in fact the boundaries of the regions. Consequently, studies are now underway, in which the dislocation structures are being observed directly. With a firm knowledge of the dominant mechanisms in each region and the region boundaries, the constitutive functions can be assigned with more confidence. In fact, it already appears that the incorrect mechanisms were used in the cited study (Herrmann et al., 1982), and that the constitutive descriptions will have to be revised.

Once again, the absence of a satisfactory general theory complicates the extension of the one-dimensional theory fitted to uniaxial creep tests to general stress states. In the absence of a yield surface, this problem is not as severe as in the case of unified creep plasticity theories (Krieg, 1980) which do introduce a moving yield surface. The above constitutive description was generalized using an assumption of material isotropy, inserted into a numerical solution scheme, and used to obtain solutions to initial-boundary value problems corresponding to closure of a salt mine for which data existed. Satisfactory agreement served to provide some confidence that the model is a useful one for the prediction of mine closure, at least within carefully prescribed conditions. Numerical solutions were also obtained for the two-dimensional deformation of the specimens in creep tests to heuristically bound the error involved in the interpretation of the tests.

It is interesting to note that the examples we have chosen all involve some variation of the theory of materials with internal state variables. One of the requirements of a theory stated at the outset was an investigation of the existence and uniqueness of solutions. This has so far been ignored in the present discussion. Passman and Trucano (1983) have begun an investigation of these questions, taking a very simplified constitutive model of this type, and restricting attention to homogeneous deformation idealizing one-dimensional creep elongation. They have found a very

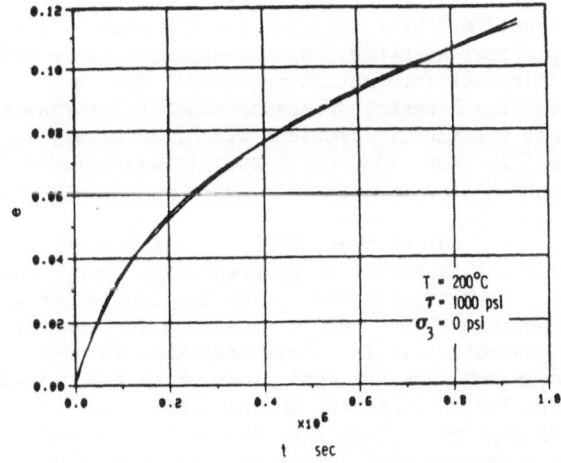

FIG. 5.   Data and Exponential Fit for High Temperature Creep of Rock Salt (from Herrmann et al., 1971).

rich stability theory, even for this very simple case, identifying regions in which no, one, or more solutions exist. The existence of multiple solutions for certain ranges of material parameters and test conditions suggests material instabilities. These may not be physically realistic, in which case they may suggest stability inequalities limiting the constitutive functions. Nonexistence of solutions implies that homogeneous solutions cannot be found under the relevant conditions, and may suggest that some other deformation mode may occur, for example strain localization. While this work is of a preliminary nature, it suggests some of the complications which must be understood before such constitutive theories can be used with confidence.

6. DISCUSSION

The examples of previous sections were given in order to illustrate a number of points, which will be summarized here.

First of all, if a relatively complete general constitutive theory exists, and some of the mathematical structure following from it has been worked out, then there may exist the basis for constructing specific constitutive functions, perhaps based on models of micromechanical deformation mechanisms (if these are known). The theory provides limitations, usually in the form of inequalities, which the constitutive functions must obey to satisfy invariance, symmetry and stability. Solutions to the theory for initial-boundary value problems representing test configurations provide the basis for evaluating coefficients in the constitutive functions, particularly if they are exact closed form solutions. Moreover, it is then usually possible to tell which tests provide information about which coefficients. Several tests may give information about the same coefficients, thus providing compatibility checks on the applicability of the theory, and perhaps allowing a choice of test method (based on sensitivity or cost) for routine "material property tests."

Depending on the application of the theory, it is often unnecessary (and is usually impossible or prohibitive) to evaluate the constitutive equations for all possible motions and conditions. The theory may suggest which coefficients need careful evaluation for the proposed application. For example, uniaxial test methods may be available, but the theory is required to predict behavior in spherical symmetry. Do the uniaxial tests provide sufficient information, or is additional information required? The theory will often be useful in suggesting additional tests, or assumptions, to provide missing information. If only spherically symmetric problems are of interest, it is obviously not necessary to proceed to generalize the theory to all motions, or to evaluate coefficients which do not affect the problem in spherical symmetry.

Any set of constitutive functions and their evaluation for a particular material represent an approximation to the real behavior. The range of satisfactory approximation is always limited, since it is impossible to describe, in a tractable theory, all of the complications in behavior of which a real material is capable. It is always desirable to ensure that "material property tests" cover the range of conditions which the material is expected to encounter in the intended application, and to bound the range of applicability of the constitutive description. It is also desirable to establish the applicability of the theory by comparing predictions with experimental observations for the actual application. If this can be done, then the theory may be used with some confidence for that application. By the same token, it is unnecessary (and usually prohibitive) to extend material property tests beyond the range of conditions of interest in the intended application, or to include in the description mechanisms which will not be exercised.

Conversely, it is not always possible to validate the theory over the full range of conditions of interest. The uncertainty inherent in extrapolation must be recognized. The uncertainty may be minimized if the constitutive functions are based on observed micromechanical deformation mechanisms, and there is sound reason to assume that new mechanisms will not come into play under the extended range of conditions.

It is obvious that the desirable state of affair described above is not always attainable before predictions are needed for some applications. Progress is still possible, even if the mathematical structure of the theory has not been worked out fully. However the increased uncertainties must be recognized.

Often, exact solutions are unavailable for "material property test" configurations, and approximations must be used. Such approximations must be justified carefully, usually heuristically by using the evaluated constitutive equation to find solutions to the problem representing the test configuration. Least desirable is the use of numerical solutions of relatively complicated tests, since numerical uncertainties are added to other sources of error. The use of complicated test configurations is sometimes unavoidable if the material cannot be fabricated into a simple test configuration without altering its properties. (For example, a filament wound composite will have altered properties if it is fabricated into a flat test coupon.) In such cases, some progress may be possible if a careful "material identification" process is used to deduce coefficient values in the constitutive functions from a large number of measurements. The process of iterative adjustment of a large number of material coefficients, matching numerical solutions with experimental observations on complex configurations without the benefit of a rational statistical optimization technique, is least desirable and unlikely to yield useful results.

If the restrictions on the constitutive functions, arising from invariance, stability, etc. are not understood, then the opportunities for generating a spurious theory are enhanced. When numerical solution methods must be used, then it is especially difficult to detect any inconsistencies in the theory by examining solutions, since difficulties in the theory may be masked by additional problems attendant on the numerical approximation, especially if it is difficult or impossible to perform a reasonable analysis of convergence and stability of the numerical method. It is especially necessary, in such cases, to scrutinize numerical solutions using extensive parameter variations in an attempt to isolate numerical effects from behavior of exact solutions of the theory. The usefulness of a constitutive theory under such conditions is extremely limited, and predictions must be viewed with extreme caution.

7. RECOMMENDATIONS

A number of general recommendations will be ventured, regarding areas for research in constitutive theory. Since it is impossible to consider the construction and evaluation of specific constitutive functions from the theory as a whole, no such distinction will be made here.

    i. While several general constitutive theories have been developed, and others are under development at this time, there are areas which merit increased activity.

Among these, theories of damaging materials merit special attention. Heterogeneous processes of micro shear banding or micro cracking, macroscopic shear localization (faulting) or crack propagation are so ubiquitous in problems of current interest that they need to be incorporated into rational constitutive descriptions. Extensions of classical plasticity theories appear to be inappropriate, particularly for brittle materials like ceramics and rocks.

Another area of particular interest concerns theories of porous materials and heterogeneous mixtures. The thermomechanical response of granular or cracked mixtures, often accompanied by phase changes or chemical reactions, is of importance in many present applications. Frequently, the microstructure of these materials must be accounted for to provide an adequate description.

ii. The central problem in constitutive theory today is the investigation of stability properties. It is crucial, for a given nonlinear constitutive theory, to identify the conditions under which solutions exist, and to investigate their general behavior. These conditions include restrictions on the constitutive equations, which must be known in order to construct specific constitutive functions and evaluate their coefficients. Moreover, existence of solutions, their bifurcation, and possible non-existence must be known in order to construct solution methods, especially numerical methods.

iii. It is important to obtain solutions, particularly exact closed form solutions, to as many problems as possible. These solutions explore the behavior of the theory, and can serve to indicate whether a constitutive theory may be useful for a particular application. These solutions may also provide the basis for "material property tests" from which constitutive functions may be evaluated.

iv. Finally, it is important to develop solution techniques (usually numerical) for problems representing the particular applications to be solved. The development of numerical solution techniques for a general constitutive theory requires appropriate analysis of numerical accuracy, convergence and stability.

REFERENCES

Asay, J.R., D.L. Hicks and D.B. Holdridge (1975), "Comparison of Experimental and Calculated Elastic-Plastic Wave Profiles in LiF," *J. Appl. Phys.*, *46*, 4316.

Coleman, B.D. and M.E. Gurtin (1967), "Thermodynamics with Internal State Variables," *J. Chem. Phys.*, *47*, 1625.

Grady, D.E. et al. (1982), "Microstructure and Mechanical Properties of Precipitation Hardened Aluminum under High Strain Rate Deformation," in *Proc. 29th Sagamore Army Materials Research Conf.*, Sagamore, NY.

Herrmann, W. (1976), "Some Recent Results in Elastic-Plastic Wave Propagation," in *Propagation of Shock Waves*, E. Varley (ed.), ASME AMD-*17*.

Herrmann, W., D.L. Hicks and E.G. Young (1971), "Attenuation of Elastic-Plastic

Stress Waves," in *Shock Waves and the Mechanical Properties of Solids*, J.J. Burke and V. Weiss (eds.), Syracuse Univ. Press.

Herrmann, W. (1974), "Development of a High Strain Rate Constitutive Equation for 6061-T6 Aluminum," *Sandia National Labs.* SLA-73 0897.

Herrmann, W., W.R. Wawersik and S.T. Montgomery (1982), "Review of Creep Modeling of Rock Salt," in *Intl. Conf. Constitutive Modeling*, Tucson, AZ.

Krieg, R.D. (1980), "A Unified Creep-Plasticity Model for Halite," *Sandia National Labs.* SAND80-1195.

Nunziato, J.W., E.K. Walsh, K.W. Schuler and L.M. Barker (1974), "Wave Propagation in Nonlinear Viscoelastic Solids," in *Encyclopedia of Physics, VIa/4*, S. Flügge (ed.), Springer-Verlag, Berlin, 1-108.

Nunziato, J.W., K.W. Schuler and D.B. Hayes (1974), "Wave Propagation Calculations for Nonlinear Viscoelastic Solids," in *Proc. Intl. Conf. on Comp. Methods in Nonlinear Mechanics*, J.T. Oden (ed.), U. Texas at Austin, 489-498.

Passman, S.L. and T.G. Trucano (1983), "Stability, Instability and Localization in Materials with Damage," *Intl. J. Numerical and Analytical Methods in Geomechanics* (submitted).

GENERAL DISCUSSION

HUGHES: In the beginning about localization, you mentioned that you thought perhaps triangles were too stiff and perhaps quadrilaterals were too loose. I think one has to be a lot more precise with statements like that. I believe, if you have triangles in arbitrary gridding modes, that statement is correct. However, if you use the so-called cross triangle pattern, that element is rank sufficient, yet it flows incompressibly correctly and people have done localization studies with it, Alan being an example.

NEEDLEMAN: These were all crossed triangles.

HUGHES: Even though they're drawn as quads, they're really crossed triangles.
    Another point with regard to quads themselves: Not everyone uses one-pointed quadrature on quads. In fact, the localization calculations that Jean Prevost and I did were done with full integration of the deviator. Consequently, there were no hourglass or keystone modes, and yet we still see very pronounced, clear localization. So it's just not triangles and quads; it's the particular triangle, the particular quad, and how you treat them.

HERRMANN: Yes, thank you for balancing out that statement. I think the numerical solution of situations such as that can be extremely useful and can help you understand those kinds of instabilities very well. A number of people, as you suggest, have done very careful analyses in which those kinds of uncertainties are made to be minimal.
    I think you would agree, however, that if you had a nonlinear stability analysis in the first place, that would help guide your numerics. Conversely, those numerics, carefully done, can certainly illuminate many of the issues of the stability question. However, what I think really is very difficult is, in a calculation of ballistic impact, to include even a reasonable description of, say, shear banding if you don't already know a whole lot about the shear banding proc-

ess. You're just not using higher order elements or four-node quadrilaterals.
You can't afford the resolution, I don't believe, so that shear banding falls
out naturally. Perhaps, if you know about the shear banding problem ahead of
time you could arrange your mesh and your methods so that those would fall out
if you thought you wanted to do that.

HUGHES: Let me just follow up on that statement. Again, the calculations that
Alan has done and Jean Prevost and I have done use regular meshes. This was intentionally done not to try to bias the calculation, and the bands were forming
at correct angles. In fact, in the case that we studied, we had analytical information and precisely reproduced that numerically.

BELYTSCHKO: I would like to comment in support of Walt Herrmann's point. I
think that localization problems are something that's really on the forefront
of numerics today. It can be said without any reservations that we can accurately compute problems, particularly in a post-failure mode, which show localization. For example, in discussions I've had with Jean Prevost, he tells me
that he has gotten to the point of incipient localization, but he has not really
calculated to the point where you really have a lot of slip, and it's in that
regime where things become very difficult. We cannot perform with reliability
localization calculations where we do not know the direction in which this
localization is going to take place.

NEEDLEMAN: A couple of comments about what Ted said. Bob Asaro and I have done
some computations where we take localization out to very large accumulations of
slip, so it has been done. But I think it's extremely helpful and very important
in these calculations, which are relatively simple problems, to have gone through
the analytical theory and use that information in designing meshes, in knowing
what's going to happen. For arbitrary problems, it's a very difficult task. But
having the analytical information available before doing the numerics has been a
significant aid.

VALANIS: Could you describe precisely the problem that you've done? I mean,
what are the geometry, boundary conditions and such?

NEEDLEMAN: The work was done with Bob Asaro and Dan Pierce on localization in a
single crystal. A rate-dependent material model is used with slip on discrete
systems. Uniaxial extension of a single crystal is considered with traction
free on the sides and applied displacement boundary conditions at the ends.

VALANIS: Uniform displacement on the outer boundaries? How about the transverse
displacement? Are you keeping that rigid?

NEEDLEMAN: It's shear free at the end.

VALANIS: Shear free — and you are getting localization?

NEEDLEMAN: Yes. This end is shear free.

CLIFTON: Alan, there are a couple things you didn't say. One, you didn't say
that you put a little perturbation on the free boundaries!

NEEDLEMAN: There are several things that I didn't say. One is that we've put a
little perturbation on the surface and we've done computations where we put in
the actual shape of the grips.

VALANIS: So in the case where you are getting localization, you have some per-turbation of the boundaries.

NEEDLEMAN: Yes, the perturbation picks out the place. In the crack problem it is stress gradients that pick out the location. There are no geometric pertur-bations there.

VALANIS: What is the perturbation that you used in your specific case?

NEEDLEMAN: One is a sine. The imperfection is something like 10^{-3} times the thickness times a linear combination of sine X and sine KX. And I don't know whether K is 2 or 3 or 4.

VALANIS: So your surface is then sort of corrugated.

NEEDLEMAN: The surface has a corrugation. That's it. The surface is stress-free and has a corrugation.

GOUDREAU: Is the period in terms of grid size?

NEEDLEMAN: No.

CLIFTON: What is the period relative to the grid size?

NEEDLEMAN: There are two or three wiggles over the whole length, so that it's much, much larger than the grid size. It's much larger than the localization.

HERRMANN: This is undoubtedly a matter of great interest, but let me go to the other extreme. We're trying to calculate a ballistic impact. You'll find maybe 30 or 40 of the shear bands that Marv Backman was talking about. It turns out that you have to introduce a softening behavior because the shear banding is a preferred mode rather than homogeneous deformation. Unless you introduce the softening due to the shear bandings, you get the depth of the penetration wrong.

Now, I would maintain that it's very difficult to include explicit shear bands in actual application calculations. The fact that Prof. Needleman's cal-culations are extremely useful I don't contest in the least, and they need to be done very badly.

PREPARED DISCUSSION

by Erhard Krempl

1. INTRODUCTION

The presence and continuing increase of computing power in the design office offers almost unlimited possibilities of treating complicated structural engineer-ing problems. Most importantly, when judiciously used, computers will constitute the key tool with which reliability and predictability of performance of engineer-ing structures can be assured in the design stage. Before this is possible, the limiting states of materials must be captured in appropriate constitutive and failure (life prediction) laws. These are, with a few exceptions, nonlinear and therefore inherently difficult. Appropriate strategies must be formulated to develop these laws and to educate potential users.

Herrmann has given some of these strategies which concentrate mostly on wave propagation and geological materials. The following will emphasize slow processes in metals and multiaxial aspects.

2. GENERAL REQUIREMENTS

Until recently nonlinearity was considered outside the realm of engineering interest. Many methodologies and approaches which are presently used are rooted in a linear world and may have to be modified for nonlinear conditions.

There is a tendency to concentrate on modeling a specific test with considerable emphasis to fit a few experimentally obtained curves. As an example, the creep test is cited. Usually no consideration is given to relaxation or strain (stress)-rate sensitivity in monotonic and cyclic loading which are found in materials that exhibit creep. The perfectly fitted "constitutive equation" may fail to give the proper *qualitative* response when applied to these different conditions (Krempl, 1974). For this reason it is important to characterize material response in various test conditions qualitatively, i.e. by bounds on derivatives (Krempl, 1975). These qualitative properties can then be compared with the qualitative solution properties of the proposed model.

In the above it is implied that even the trivial solutions cannot be obtained on closed form, i.e. those for which the equations of motion are satisfied on account of the macroscopically homogeneous state of stress and of slow (accelerationless) motions. This is indeed the case for many old and newly developed nonlinear theories for rate-dependence and rate-independence: Bodner and Partom (1975), Chaboche (1977), Hart (1976), Liu and Krempl (1979), Miller (1976), Rhode and Swearengen (1980), Valanis (1980). Closed-form solutions may not be obtainable for these models, even in the uniaxial case. The situation is similar to the one encountered in Krieg's paper listed in Herrmann's references. Before numerical methods are employed, the qualitative properties such as asymptotic solutions, Cernocky (1982), Cernocky (1982), Cernocky and Krempl (1979), and stability of critical points, Cernocky and Krempl (1983), must be ascertained.

The discussor is in full agreement with the author who stresses the importance of obtaining a good understanding of the constitutive theory before resorting to numerical schemes. There are other points raised by the author which have the full support of this discussor.

- A "universal" nonlinear constitutive equation is not obtainable presently. Each constitutive equation has a range of validity which must be specified.

- Micromechanics can help in the formulation of constitutive equations. Unfortunately the help is not as strong as one would like due to the simultaneous presence of many mechanisms and the uncertainties involved in "translating mechanisms into mathematical expressions."

- Material data are advantageously obtained directly from tests on specimens representing a homogeneous state of deformation. Indeed these are the tests from which we learn about the macroscopic behavior of real materials.

- Differential formulations of constitutive equations are preferable to integral formulations for numerical applications. Since all nonlinear problems will ultimately resort to numerical schemes, a differential formulation appears to be a prerequisite for the engineering use of a constitutive equation.

3. NEED FOR IMPROVED MATERIAL MODELS

The author appears to have de-emphasized the need for new material models which do not use traditional concepts and have capabilities beyond those of the classical theories of plasticity and creep. Multiaxial tests for determining material behavior are not mentioned.

The discussor has a different viewpoint. In recent years the same technology that revolutionized the state of computing has drastically changed the mechanical testing. Feedback principles and computer control together with servohydraulics has brought about new capabilities. They are:

- Enforcement of either load or displacement boundary conditions. Especially the latter was not possible with conventional testing.

- Arbitrary control of loading rates (displacement or load) within the frequency characteristic of the load frame-specimen assembly.

- Independent control of each axis in multiaxial tests such as an axial-torsion test.

- Repeatability of the same test history on different specimens through computer control.

- Test results in the form of digitized data which permit repeated analyses using different "philosophies."

These new capabilities have already given rise to various "surprises."

- The stress-strain diagram with upper and lower yield points cannot be a material property. It cannot have this shape under linearly increasing stress (stress boundary condition), Krempl (1983).

- In contrast to generally held beliefs, inelastic deformation of ductile engineering alloys was found to be significantly rate(time)-dependent at room temperature, Krempl (1979), Krempl (1983), Krempl and Lu (1983), Kujawski and Kallianpur and Krempl (1980), Kujawski and Krempl (1981).

- In biaxial axial-torsion tests, out-of-phase loading was shown to produce much stronger cyclic hardening than in-phase loading, Lamba and Sidebottom (1978), Lu (1983) and McDowell (1983).

- Cross hardening or latent hardening is observed when axial (shear) tests are made after prior shear (axial tests, Kallianpur (1982), Lu (1983). This hardening exceeds that predicted by the classical theories of kinematic hardening and isotropic hardening. (The modelling of such results requires deformation induced anisotropy.)

These are just a few results obtained with modern testing capabilities. They would undoubtedly have given rise to other and improved material idealizations if they would have been available at the conception of the theories presently used in nonlinear analysis.

Multiaxial testing using servocontrol has just begun. An improved understanding of material behavior will result from such testing and will ultimately improve the constitutive equations used in nonlinear analyses.

4. RECOMMENDATIONS

While nonlinear constitutive equations are important, they are not sufficient to predict failure, lifetime or reliability of performance of engineering structures. Failure or damage accumulation laws must be adjoined that tell which combinations of stress and strain lead to failure. (Failure must be appropriately defined; see Stouffer et al. (1980), Krempl (1980).) At a given location and under a given operating history, the result of a nonlinear analysis is envisioned to be lifetime rather than stress and strain. Lifetime, of course, is of direct interest to the designer.

Based on the present view of the discussor and with the ultimate goal of computing lifetime, the following items deserve further study. (Because of space limitations, only items related to constitutive behavior are addressed.)

- Continued development of "unified theories" and others which do not use a loading/unloading criterion or a yield surface. These theories may not distinguish between plastic and creep strain. Interactions between micro- and macro-approaches can be very fruitful here.

- Performance of systematic uniaxial and multiaxial experiments at various temperatures. These experiments should be geared to the constitutive equation development but should provide data in such a form that they can be used by all. Data should not be presented so that they can only be interpreted by a specific theory.

- The unified theories result in systems of nonlinear differential equations which are stiff. Special integration schemes must be developed together with variational principles. It would be desirable to have a minimum principle and efficient and stable solution methods.

- Links between constitutive theory and damage (failure) laws must be established.

5. ACKNOWLEDGMENT

This research was supported by the Solid Mechanics Program of the National Science Foundation.

6. REFERENCES

Bodner, S.R. and Y. Partom (1975), "Constitutive Equations for Elastic-Viscoplastic Strain-Hardening Materials," *J. Appl. Mech., 42, 385.*

Cernocky, E.P. (1982), "Comparison of the Unloading and Reversed Loading Behavior of Three Viscoplastic Constitutive Theories," *Int. J. Non-Linear Mechanics, 17,* 255.

Cernocky, E.P. (1982), "An Examination of Four Viscoplastic Constitutive Theories in Uniaxial Monotonic Loading," *Intl. J. Solids and Structures, 18,* 989.

Cernocky, E.P. and E. Krempl (1979), "A Nonlinear Uniaxial Integral Constitutive Equation Incorporating Rate Effects, Creep, and Relaxation," *Intl. J. Non-Linear Mechanics, 14,* 183.

Cernocky, E.P. and E. Krempl (1983), "Evaluation of a Uniaxial, Nonlinear, Second-Order Differential Overstress Model for Rate-Dependence, Creep and Relaxation," *Intl. J. Solids and Structures, 19,* 753.

Chaboche, J.L. (1977), "Viscoplastic Constitutive Equations for the Description of Cyclic and Anisotropic Behavior of Metals," *Bull. de l'Acad. Polonaise des Sciences, Série Sc. et Techn., 25,* 33.

Hart, E.W. (1976), "Constitutive Relations for the Nonelastic Deformation of Metals," *J. Eng. Matls. and Tech., 98,* 193.

Kallianpur, V.V. (1982), Ph.D. thesis, Rensselaer Polytechnic Institute, Troy, NY.

Krempl, E. (1974), "Cyclic Creep. An Interpretive Literature Survey," *WRC Bulletin No. 195,* Welding Research Council, New York.

Krempl, E. (1975), "On the Interaction of Rate- and History-Dependence in Structural Metals," *Acta Mechanica, 22,* 125.

Krempl, E. (1979), "An Experimental Study of Room-Temperature Rate Sensitivity, Creep and Relaxation of Type 304 Stainless Steel," *J. Mech. and Phys. of Solids, 27,* 363.

Krempl, E. (1980), "Inelastic Constitutive Equations and Phenomenological Laws of Damage Accumulation for Structural Metals," *Phys. Non-Linearities in Struc. Anal., Proc. IUTAM Symp.,* Senlis, France, 117.

Krempl, E. (1983), "Viscoplasticity Based on Overstress. Experiment and Theory," Ch. 19, pp. 369-384 in *Mechanics of Eng. Materials,* C.S. Desai and R.H. Gallagher (eds.), J. Wiley and Sons, to appear.

Krempl, E. and H. Lu (1983), "The Rate(Time)-Dependence of Ductile Fracture at Room Temperature," to appear *Engineering Fracture Mechanics.*

Kujawski, D., V. Kallianpur and E. Krempl (1980), "An Experimental Study of Uniaxial Creep, Cyclic Creep-and Relaxation of AISI Type 304 Stainless Steel at Room Temperature," *J. Mech. and Phys. of Solids, 28,* 129.

Kujawski, D. and E. Krempl (1981), "The Rate(Time)-Dependent Behavior of Ti-7Aℓ-2Cb-1Ta Titanium Alloy at Room Temperature under Quasi-Static Monotonic and Cyclic Loading," *J. Appl. Mech., 48,* 55.

Lamba, H.S. and O.M. Sidebottom (1978), "Cyclic Plasticity for Nonproportional Paths. Parts I and II," *J. Eng. Matls. and Tech., 100,* 96.

Liu, M.C.M. and E. Krempl (1979), "A Uniaxial Viscoplastic Model Based on Total Strain and Overstress," *J. Mech. and Phys. of Solids, 27,* 377.

Lu, H. (1983), Work in progress at the Mechanics of Materials Laboratory at Rensselaer Polytechnic Institute, Troy, NY.

McDowell, D.L. (1983), "On the Path Dependence of Transient Hardening and Softening to Stable States under Complex Biaxial Cyclic Loading," *Proc. Intl. Conf. on Constitutive Laws for Eng. Materials,* C.S. Desai and R.H. Gallagher (eds.), Tucson, AZ.

Miller, A.K. (1976), "An Inelastic Constitutive Model for Monotonic, Cyclic and Creep Deformation, Parts I and II," *J. Eng. Matls. and Tech., 98,* 97.

Rhode, R.W. and J.C. Swearengen (1980), "Deformation Modeling Applied to Stress Relaxation of Four Solder Alloys," *J. Eng. Matls. and Tech., 102,* 207.

Stouffer, D.C. et al. (1980), Workshop on Continuum Mechanics Approach to Damage and Life Prediction, (May 4-7, 1980), Carrollton, KY.

Valanis, K.C. (1980), "Fundamental Consequences of a New Intrinsic Time Measure. Plasticity as a Limit of the Endochronic Theory," *Arch. Mech., 32,* 171.

GENERAL DISCUSSION

TRULIO: I've heard of only one other laboratory where this kind of thing was being attempted. That's at Terra Tek. You may know of what they're doing. I'm not sure. Where servomechanisms are being used to follow specified strain histories. I wanted to ask you what particularly your range of control is over strain rate and either strain amplitude or stress amplitude are with the servomechanisms.

KREMPL: As far as the rates are concerned, the frequency characteristic of the hydraulic system and the characteristic of the frame are limiting. Typically, the limit is something like one per second in these machines. Stresses are determined by the size of the load frame and the material.

KRIEG: There is some work going on at Sandia in which the load that they put on the specimen is primarily due to the gravity load, but incremental loads are added with a magnetically driven head. In this case, the feedback is much quicker. I think we have something like 20 hertz, somewhere in that range.

KREMPL: But isn't that a special development that you personally have constructed? But that's not a commercially available machine!

KRIEG: That's right. This is specially built for doing stress drop and stress rise tests.

HEGEMIER: I just want to make a comment. I agree with you, we need more closed loop, servo-type testing. But Jack, with respect to your comment, there are other programs going on in the country where closed loop servo-equipment is being used. In my lab, for example, at UCSD off-campus, we've got up to 21 closed loop servos going through a mini computer system on a huge biaxial test. It's a pain in the neck, but it's feasible; there are programs where this sort of thing is going on.

KRAJCINOVIC: I'd like to argue with you about the micro-models and their use. I think we have enough macroscopic theories that in some unexplained way tell you what to do and how the things behave by adding another term or so, when needed.
      Consider, for instance, a uniaxial compression of unconfined rock specimen. As we know, the softening happens first due to nucleation and growth of cracks, and then eventually you get a fault of some type and the thing breaks. Now, if we didn't know that, we would, of course, do what people do now in concrete: they take a curve and they match some polynomial of 4, 6 or 8 order depending on how many points they have and formulate the theory that cannot be used for any other response. I don't think that that's what you advocated.

KREMPL: No.

KRAJCINOVIC: But then you said that micromodels are not useful.

KREMPL: Not helpful, I said. But I qualified that later. What you're saying is, you can find out from these studies that the motion is no longer homogeneous.

KRAJCINOVIC: That's correct.

KREMPL: But I didn't mean this type of help. I meant a predictive help. Somebody is around and says dislocations move in a certain way, and therefore you have to add this particular term to your constitutive equation. It's usually done the other way, that people do macroscopic tests and then they rationalize them in terms of dislocation motion.

KRAJCINOVIC: That's because the plasticity was developed before electron microscopes. I mean, the experimental signs were not introduced at a time when the plasticity was developed, but that's not a proof that it really should happen that way.

NEEDLEMAN: I don't know all the details, but as an example, there was quite a bit of work done based on a micro-model, the slip model, due to Taylor, by Gilbert Chin at Bell Labs and others, for predicting textures. This turned out to be extremely useful for predicting deformation-induced textures.

VALANIS: I'd like to add to that. For instance, in a paper that Gilman gave at San Antonio about ten years ago, from considering the notion of dislocation — he comes to the conclusion that rate sensitivity, for instance, in metals has to be described through a plastic strain rate and not a total strain rate. I think that that's an important conclusion because I submitted a proposal to the National Science Foundation on strain rate in which I did indeed suggest the plastic strain rate as the proper mechanism of strain rate sensitivity, and all the reviewers jumped on me because I didn't use the total strain rate and I used the plastic strain rate. But in fact, if you look at the dislocation theories, they suggest that the plastic strain rate is the proper vehicle for instituting strain rate effects in metals. So that all these things are somehow being ignored.

KREMPL: You would not put it into the elastic relations because in the ranges we're talking about, elasticity has to hold. So strain rate sensitivity cannot come into the elastic relations, so what else is left but plastic strain?

VALANIS: I didn't quite get that.

KREMPL: I'm saying that if I look for a vehicle to produce strain rate sensitivity, I cannot use the elastic relations because I want them to produce elasticity. Therefore, from that alone I have to put it into the inelastic strain rate. I agree with you, but I don't think a micromechanical consideration is needed.

VALANIS: Let me give you a counterexample. Let's take linear viscoelasticity for example. If I increase the strain rate, I get closer and closer to an elastic response. And for all strain rates, as a matter of fact, the initial slope of the stress-strain diagram is the elastic slope. So there I have elastic response, which is not violated by using the total strain rate as a vehicle for strain rate sensitivity. So that's a counterexample.

KREMPL: You can write that also. We have to talk about that.

HERRMANN: I just wanted to rebut this point in general. My statement was that

I thought micromechanical theories were most helpful. But in fact, my argument is that the folks that start with the micromechanical theory and then find agreement with, say, the wave data that I showed, and then conclude that this is the mechanism that's going on in the material are concluding too much. In the creep literature, for example, we find dislocation climb at high tempera- ture creep. It turns out that the expressions for dislocation climb fit the low-temperature creep in salt just beautifully. But in fact, when you go back to the microscope, you observe wavy slip bands all over the specimen. Wavy slip bands suggest a different functional form for constitutive equations. For one, you get a power law and for the other one, you get an exponential, and the ex- ponential fits the data just as well, but it extrapolates differently.

My argument is, if you're going to use a micromechanical model, you need to be open to the possibility that the model might not be unique, and that you may be in deep trouble by extrapolating unless you have a direct observation of the micro-structure.

If, for example, in the application of salt creep to mine closure, you dis- cover that samples from the walls of salt mines that have been closing for the past 250 years show the same wavy creep bands that you see in the lab at three orders of magnitude higher rates, then you may have some reason for extrapolat- ing that data to lower rates, but not otherwise.

NEMAT-NASSER: In relation to micromechanics versus macromechanics, I think it's very important to keep in sight that we do like to understand the physics. When we talk about micromechanics, we are saying that we hope to understand what is the most essential feature that gives rise to certain macroscopic response.

Now, we understand that real problems are complicated: Loading regimes change and therefore the micro-events that give rise to certain macroscopic aspects may also change, and there may be coupling of many micro-events! But that does not say that we should not try to understand them. I think the whole exercise, this intellectual exercise, is to be able to understand the basic physics. For an eventual engineering application, of course, we would like to be able to put this understanding in terms that can be used. But I'm sure that neither Erhard nor Walter or anybody else here suggests that we should stop understanding physics. For this understanding, micromechanics is most essential.

CHAPTER IX

VERIFICATION OF THE EFFECTIVE STRESS
AND AIR VOID POROSITY CONSTITUTIVE MODELS

J. T. Cherry* and N. Rimer

S-CUBED, P.O. Box 1620
La Jolla, California 92038

1. INTRODUCTION

Of all the elements required for a numerical simulation of seismic coupling from a nuclear explosion, the most important and the most uncertain are the physical (constitutive) models of each nonlinear process induced in the medium by the propagating stress field, and the material properties required by the constitutive model which makes the simulation site specific. These two elements cannot be validated by numerical experiments. In fact, there is a trade-off between the two, with an incorrect constitutive model requiring incorrect material properties in order to produce results which agree with measured ground motion data.

Our approach to resolving this uncertainty has been to base the development and formulation of each constitutive model on the results of quasi-static laboratory tests on rock samples, and then to apply the model to the simulation of explosion induced ground motion. If the application appeared promising in terms of its ability to match ground motion data with the constitutive model using independently measured material properties, then the model was retained for use in future ground motion simulations.

This approach has shown that two specific constitutive models severely affect seismic coupling. They are: the irreversible collapse of air-filled porosity and the modification of material strength due to pore fluid pressure, i.e., effective stress. Therefore, the validation and testing of these models are issues of continuing importance and concern.

In this report, we present a rather severe test of these models by using them to simulate experiments in which small scale explosions were detonated in grout spheres. High quality, reproducible, particle velocity data were obtained from these experiments. We show that an effective stress law coupled with the irreversible collapse of air-filled porosity provide a very simple, straightforward explanation of these data. These results have served to reenforce our confidence in both the validity of the effective stress and porosity constitutive models and the importance of these models for determining seismic coupling.

---

*Present address: Science Horizons, Inc., 710 Encinitas Blvd., Encinitas, CA 92024.

2. CONSTITUTIVE MODELS FOR POROSITY AND EFFECTIVE STRESS

2.1 Porosity

The first laboratory tests on the irreversible collapse of air-filled porosity in rock samples were reported by Stephens and Lilley (1970) in which they
summarized their work on rocks of interest to the Plowshare Program. Based on
this work, a constitutive model for porosity collapse was formulated (Cherry and
Peterson, 1970), refined (Cherry et al., 1973), and used to match ground motion
data from explosive sources (Cherry and Peterson, 1970; Riney et al., 1973).

These results were encouraging enough to suggest that we should use the
model to determine the effect of air-filled porosity on teleseismic magnitudes.
Cherry, Rimer, and Wray (1975) performed the parameter study and found that the
introduction of porosity causes a severe reduction in magnitude with magnitude
changing by 0.17 units when porosity changes from zero to one per cent.

The porosity model developed by Cherry et al. (1973) has essentially re-
mained unchanged for our applications. In the model, the hydrodynamic component
of the stress tensor, P, is dependent on air-filled porosity via a parameter, α,
given by

$$\alpha = \frac{V_r + V_\omega + V_P}{V_r + V_\omega} = \frac{v}{\hat{v}} \tag{2.1}$$

where V_r, V_α, and V_p are the respective rock, water and air pore volumes, and v
is the specific volume of the partially saturated mix.

The material properties required by the model are:

1. $\hat{P}(\hat{v})$, the pressure/volume relation for the rock with air-voids removed.

2. P_c, the pressure at which all air-voids are removed.

3. P_e, the pressure above which irreversible removal of air-voids begins.

4. ϕ_0, the initial air-void volume fraction, where ϕ_0 and α_0 are related by

$$\alpha_0 = \frac{1}{1 - \phi_0} . \tag{2.2}$$

5. k_0, the zero pressure bulk modulus of the porous mixture.

Given the above material properties, then

$$P = \frac{1}{\alpha} \hat{P}(v/\alpha) \tag{2.3}$$

where

$$\alpha_0 \geq \alpha \geq 1 \tag{2.4}$$

and the material properties (P_c, P_e, ϕ_0, and k_0) determine the variation of α
with v under the constraints for pore collapse assumed by Cherry et al. (1973).

As noted earlier, this model has been used successfully in a variety of
ground motion simulations since its development. In this report we present additional evidence that the model adequately matches observed ground motion in rock
materials whose air-void porosities range between 1.5 per cent and 13.4 per cent.

2.2 Effective Stress

The development of constitutive models for an adequate treatment of material strength has been an issue of debate and uncertainty among the ground shock calculation community for over ten years. The major issue may be stated as follows: "Given laboratory measurements of the static strength of rock, what are the appropriate values to be used in simulating the response of rock to an underground nuclear explosion?" The uncertainties associated with answers to this question arise because laboratory measurements of strength vary greatly depending on whether the sample is wet or dry, fractured or intact, and on the strain rate at which the test is conducted.

However, there is a minimum strength for a rock mass; i.e., the limit when the rock is completely broken into blocks. In this case Byerlee (1979) has shown that the strength is determined by the frictional strength, which is

$$\tau = 0.85 \ \sigma_\eta \tag{2.5}$$

where τ is the shear stress and σ_η the normal stress applied to the frictional surface. Equation (2.5) is applicable at low normal stresses ($0.2 \ Kb < \sigma_\eta < 2.0$ Kb). At higher normal stress, the applicable law is

$$\tau = 0.5 + 0.6 \ \sigma_\eta \tag{2.6}$$

expressed in kilobars. Equations (2.5) and (2.6) have been shown to be independent of lithology, temperature and scale, and to exhibit a weak rate dependence in which the frictional strength decreases a few per cent per decade of increase in sliding velocity (Scholz and Engelder, 1976; Dieterich, 1978).

When the rock contains a fluid within its pore structure at an internal pressure, p_f, it has been shown (Garg and Nur, 1973) that the strength of the rock does not depend only on the externally applied stress, but on an effective stress ($\sigma_\eta - p_f$). For example, when fluid is present in the rock, equations (2.5) and (2.6) should be modified to be

$$\tau = 0.85(\sigma_\eta - p_f) \tag{2.7}$$

and

$$\tau = 0.5 + 0.6(\sigma_\eta - p_f) \tag{2.8}$$

and the strength is said to obey a "law of effective stress."

Even the synthesis of a large amount of data as represented by Eqs. (2.7) and (2.8) involves a large variation in strength, τ, depending on the variation of pore pressure, p_f, with the stress state, σ. One might assume that the issue could be resolved by testing rock samples obtained from core taken at the site of the event. In general, this has proven to be an unsatisfactory procedure because the rock sample is not representative of the average water content or fracture density of the site, and because the laboratory induced pore pressure does not correspond to that generated *in situ* during the passage of the shock wave.

Therefore, due to the wide range of strength values available from laboratory tests, the only way to decide what is important is to try to match ground motion data from explosive sources, compare the strength required to match the data with that from laboratory tests, and then attempt to draw general conclusions concerning the types of laboratory tests most appropriate for the nuclear test rock environment. That is the approach we have taken.

The conclusions we have reached regarding the strength required to simulate explosive induced ground motion in low porosity brittle rocks (crystalline igneous rocks, crystalline metamorphic rocks, and well-consolidated sedimentary rocks) and high porosity brittle rocks (poorly consolidated sedimentary rocks and vesicular igneous rocks) are as follows:

1. The strength assumes a "high" value during the initial portion of the stress pulse and then relaxes to a "low" value.

2. The "high" value appears to be that obtained from laboratory measurements on dry rock samples. However, there is some evidence which indicates that for high strain rates ($10^4 - 10^6$/sec), a strain rate effect should be added to the static dry strength measurements.

3. The low value appears to be that obtained by assuming that an effective stress law is operative and that the pore fluid pressure, p_f, is equal to the pressure of the externally applied stress, P.

Apparently the first to recognize the importance of effective stress for matching explosively induced ground motion data were Cherry and Peterson (1970). They assumed that the wet strength was the "equilibrium" strength, added a strain rate dependence to the dry strength, used a Maxwell solid formulation to relax between the dry and wet strengths, and succeeded in matching the stress history data obtained from small scale experiments in blocks of grout. Figure 2.1 shows the differences in measured strength characteristic between dry and wet grout. Figure 2.2 shows the comparison between the measured and calculated stress history, 6.5 centimeters from the charge, in grout with 10 per cent air-void porosity, where the calculation was performed assuming relaxation between the strain rate dependent dry strength and the wet strength.

This model was revised by Cherry and Rimer (Bache et al., 1975) where they used an effective stress law and coupled the Maxwell solid stress relaxation from the dry to the saturated strength with the removal of all air-void porosity from the rock matrix.

The last modification to the strength constitutive model has involved replacing the Maxwell solid relaxation with a direct calculation of the change in pore fluid pressure as a function of the parameter α. In the model, we assume that the effective pressure, $P - p_f$, varies as follows:

$$P - p_f = P \qquad\qquad \alpha > \alpha_e \qquad\qquad (2.9a)$$

$$P - p_f = \left(\frac{1 - \alpha}{1 - \alpha_e}\right)P \qquad \alpha_e \geq \alpha \geq 1 \qquad\qquad (2.9b)$$

$$P - p_f = 0 \qquad\qquad \alpha = 1 \qquad\qquad (2.9c)$$

where α and P are defined by Eqs. (2.1) and (2.3), respectively, and

α_e is the value of α at P_e.

The modifications given by Eqs. (2.9) were introduced into the model both to improve the physics in the model so that effective stress could be used directly in the calculation of strength, and to remove the strain rate dependence of the Maxwell solid which obviates the cube root scaling of ground motion measurements with explosion yield. The material strength is given by

$$Y = Y(\bar{P} - p_f) \qquad\qquad (2.10)$$

with the strength function determined from laboratory tests on dry ($p_f = 0$) rock samples. Equation (2.10) is intended to be a generalization of the effective

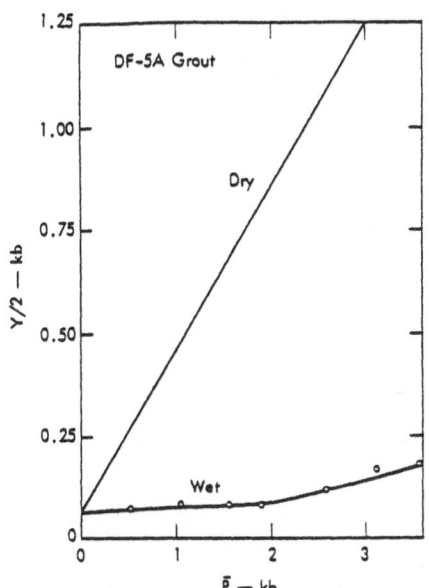

FIG. 2.1. Strength of DF-5A grout (Cherry and Peterson, 1970). In this
figure and in the remainder of this report, Y corresponds to the difference between the maximum and minimum principal stresses at failure
under triaxial stress conditions and $\bar{P}$ is half the sum of these
stresses. For a definition of these parameters in terms of stress
invariants, see Cherry and Peterson (1970).

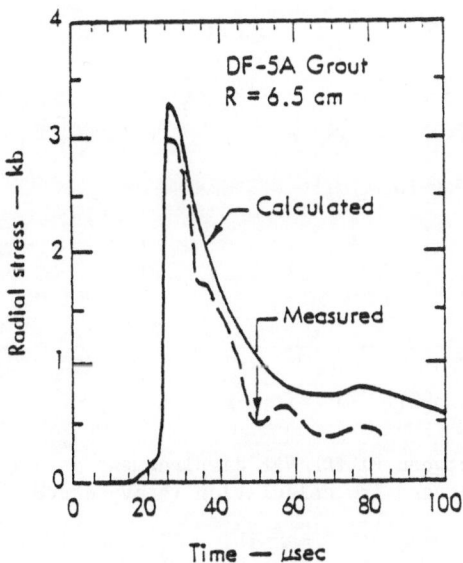

FIG. 2.2. Stress history in DF-5A grout 6.5 cm from high explosive
detonation (Cherry and Peterson, 1970).

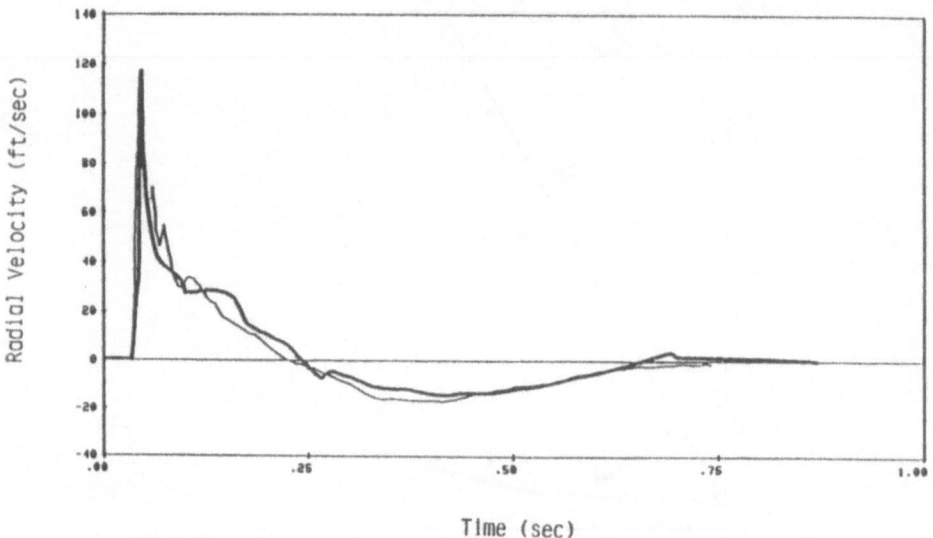

FIG. 2.3.   Comparison between PILEDRIVER particle velocity data (light
curve) and the effective stress law calculation (heavy curve)
204 meters from the source.

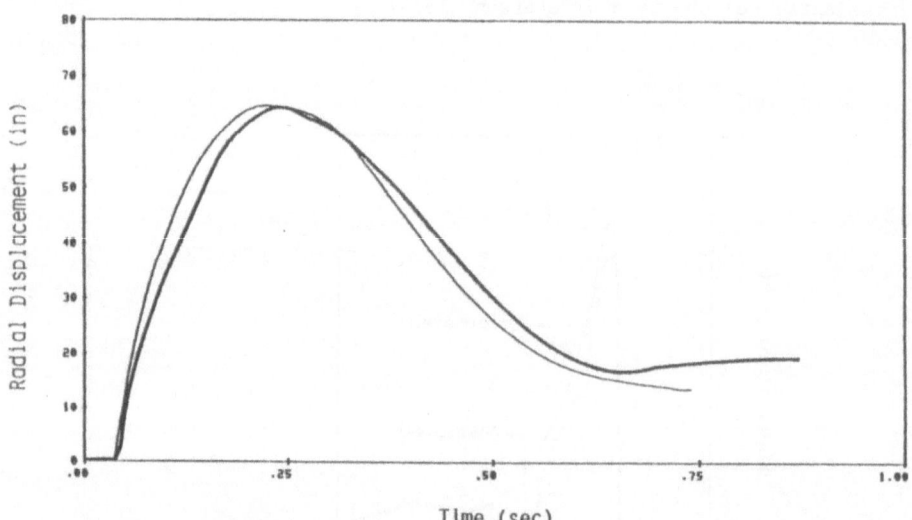

FIG. 2.4.   Comparison between PILEDRIVER displacement data (light curve)
and the effective stress law calculation (heavy curve) 204 meters
from the source.

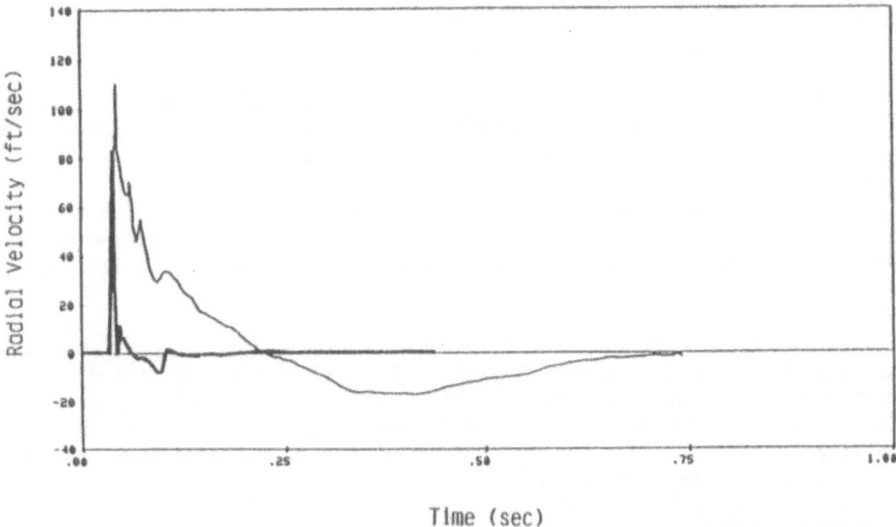

FIG. 2.5. Comparison between PILEDRIVER particle velocity data (light curve) and the noneffective stress law calculation 204 meters from the source.

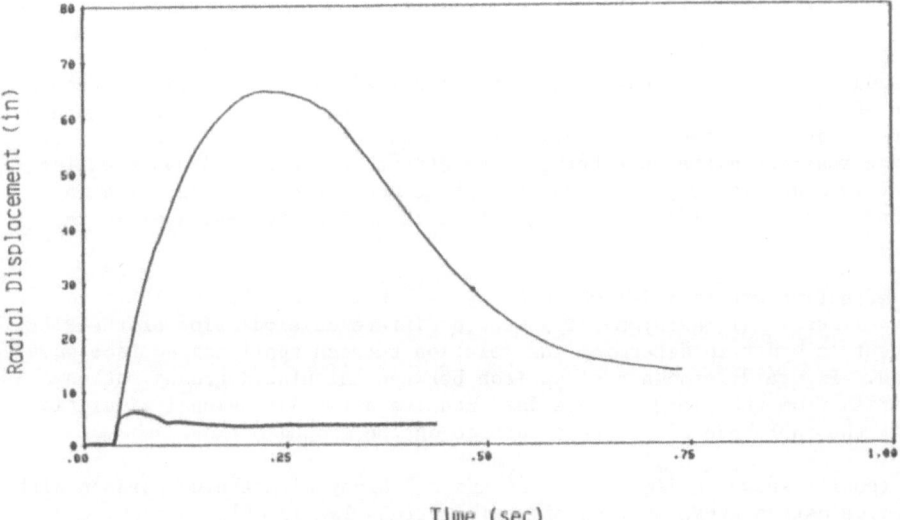

FIG. 2.6. Comparison between PILEDRIVER displacement data (light curve) and the noneffective stress law calculation 204 meters from the source.

stress law given by Eqs. (2.7) and (2.8). The reduction in strength due to p_f is introduced into the calculation of strength via the effective stress constitutive model given by Eqs. (2.9) and (2.10).

It is worth emphasizing that the effective stress constitutive model makes simple and straightforward requirements on material properties data, with the critical material properties being the dry strength, $Y(\bar{P})$, and the pressure, P_c, at which all air-voids are removed. Even if site specific measurements of these material properties are not available, enough laboratory tests have been conducted on a variety of rock types so that in most cases reasonable estimates of these properties can be made, e.g., Stephens and Lilley (1970), Heard (1970), Byerlee (1979).

Our first application of this model involved a simulation of ground motion data recorded from the 60 KT PILEDRIVER event. The material properties used for the simulation were obtained from Heard (1970) for the strength of dry, fractured granite, and from Brace (1965) for the pressure (P_c = 1 Kb) at which all void space (fracture porosity) is closed. In addition, we assumed that the site contained 0.1 per cent air-void porosity.

Figures 2.3 and 2.4 compare the calculations with radial ground motion data recorded at a horizontal range of 204 meters. Figures 2.5 and 2.6 show a comparison between the data at 204 meters and a calculation in which the effective stress law was not used, i.e., the pore fluid pressure, p_f, in Eq. (2.10) was assumed to be zero for all strain states and, therefore, the dry granite strength was not influenced by effective stress. The difference between these two calculations is quite significant, a factor of approximately 15 in peak displacement for instance, with the effective stress model providing good agreement with the data.

In addition, the final cavity radius calculated by the effective stress model was 40 meters, also a good agreement with PILEDRIVER cavity measurements. The cavity radius from the noneffective stress calculation was 26 meters. It is interesting that the difference between the effective stress and noneffective stress cavity radii is very close to the difference between NTS cavities in granite and those reported by the French from their granite test area in the Sahara.

The effective stress model was used by Day, Rimer and Cherry (1981) to perform a two-dimensional (axisymmetric) finite difference simulation of the PILEDRIVER event in order to determine the relation between spall and surface wave generation. Figure 2.7 shows a comparison between calculated ground motion and data recorded from the event. These data require a two-dimensional simulation because of the influence of the free surface on the ground motion.

The results shown in Figs. 2.3, 2.4 and 2.7 along with the comparison with cavity radius data suggest that an effective stress law is able to provide a satisfactory explanation of the PILEDRIVER ground motion data. Unfortunately, this type of data from underground nuclear tests is not frequently obtained. In fact, the PILEDRIVER data is unique in terms of the large amount of data available and the extent of both surface and subsurface coverage provided. Therefore, if further verifications of both the air-filled porosity and the effective stress constitutive models are to be obtained, it would seem logical that we use less expensive, small scale explosive tests. Data from such tests have recently become available. In the remainder of this report, we show the results of applying these models to these data.

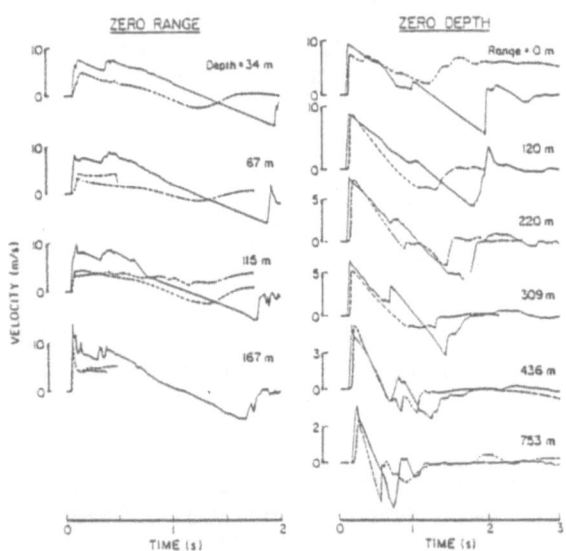

FIG. 2.7. Vertical velocity time-histories (positive up) obtained from
the two-dimensional simulation of Day, Rimer and Cherry (1981) compared with recorded velocities for PILEDRIVER.

3.   THE GROUT EXPERIMENTS

As part of the Defense Nuclear Agency (DNA) late-time containment program
for underground nuclear testing, SRI International has, since 1976, been conducting small scale laboratory experiments to study explosion phenomenology related to the containment of the cavity gas. The initial experiments involved
casting a 12 inch diameter sphere of either 2C4 or low density LD2C4 grout
around a lucite-encased sphere of high explosive (3/8 gm of PETN), placing this
grout sphere in a pressurized water tank to simulate overburden pressure, and
detonating the PETN. While maintaining overburden pressure, the sphere was then
hydrofractured from the explosively formed cavity using a tube emplaced preshot.
Hydrofracture breakdown pressures from these tests are compared with breakdown
pressures from precast (unexploded) spheres to obtain an estimate of the magnitude of the explosively formed residual stress fields. Figure 3.1 shows the experimental apparatus. A detailed description of these experiments and a discussion of the experimental results may be found in Cizek and Florence (1981).

Recently, SRI has begun to emplace particle velocity gauges inside the grout
spheres (Cizek and Florence, 1981). These gauges consist of concentric circular
current-carrying loops of wire cast symmetrically about the charge. A magnetic
field is generated normal to the plane of the loops by passing current through a
coil which surrounds the sphere. Charge detonation produces radial motion of
the loops that cut the magnetic flux lines. In accordance with Faraday's law,
the voltage induced in each conducting loop is proportional to the particle velocity. Records of radial particle velocity versus time obtained from these
gauges have proved to be reproducible from shot to shot.

In addition to these experiments, DNA has supported the measurement of material properties for 2C4 and LD2C4 grout. These measurements have been made at

TABLE 3.1

MATERIAL PROPERTIES FOR 2C4 ROCK MATCHING GROUT
AND LD2C4 LOW DENSITY GROUT

Property	2C4	LD2C4
Density (gm/cc)	2.205	1.921
P-Wave Velocity (m/s)	3318.0	3128.0
S-Wave Velocity (m/s)	1792.0	1629.0
Air-Filled Porosity (%)	1.5	13.4
Unconfined Strength (bar)	270 ± 30	220 ± 30
Tensile Strength (bar)	36 ± 10	31 ± 10

SRI and Terra Tek, Inc. (Cooley et al., 1982). They include static load–unload tests under uniaxial strain conditions, hydrostatic compression tests, triaxial compression tests on both virgin samples and on samples damaged by hydrostatic compression to 4 Kbar and subsequent decompression, ultrasonic velocity tests and porosity, water content and density measurements.

Table 3.1 and Figs. 3.2, 3.3, and 3.4 summarize the results of these tests. The data given in the table and in the figures were used in the effective stress simulation of the experiments. However, the strength data of Fig. 3.4 were modified to account for the fact that the triaxial tests were not conducted either on dry samples or in an environment in which the pore fluid pressure was monitored during the test.

In the absence of strength tests on dry grout samples, we assumed that the dry failure surface was simply an upward extension of the sloped portion of the intact failure surface shown in Fig. 3.4,

$$Y = a + b\bar{P}. \tag{3.1}$$

For 2C4 grout, we obtained the best agreement with the particle velocity data using

$$a = 0.09 \text{ Kb} \qquad b = 1.6 \tag{3.2}$$

while for LD2C4 grout, we used

$$a = 0.055 \text{ Kb} \qquad b = 1.5. \tag{3.3}$$

Both the slopes and the intercepts given by Eqs. (3.2) and (3.3) are within the constraints imposed by the data shown in Fig. 3.4. The grout strength is now modified by pore fluid pressure according to

$$Y = a + b(\bar{P} - p_f) \tag{3.4}$$

where p_f is calculated using Eq. (2.9).

4. SIMULATION OF EXPERIMENTS IN 2C4 and LD2C4 GROUT

In this section, we present three simulations of high explosive experiments in grout and compare the results of these simulations with particle velocity measurements at distances of 1.27 cm, 1.9 cm, 2.54 cm and 4.0 cm from the charge center.

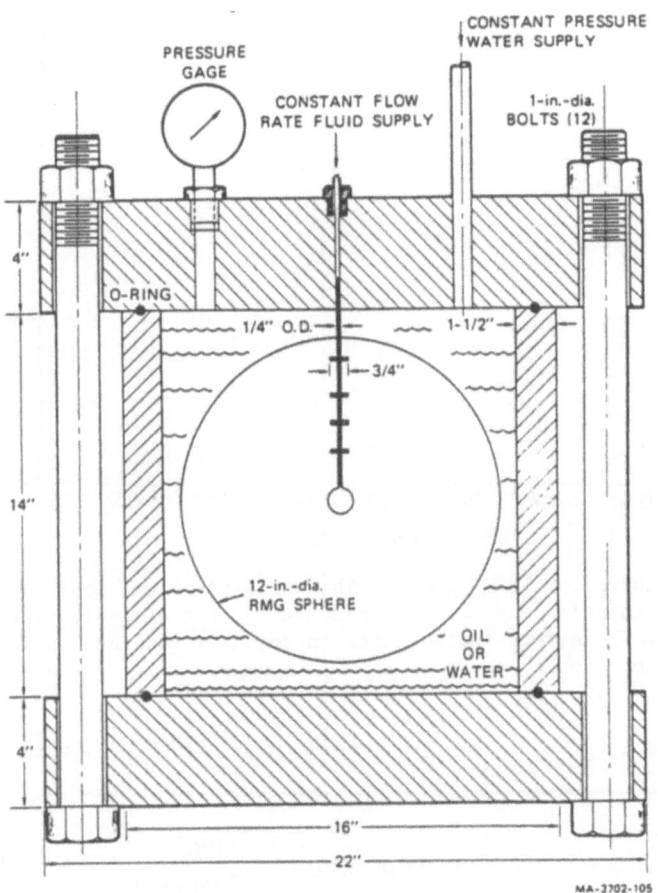

FIG. 3.1. Grout spheres experiment apparatus (Cizek and Florence, 1981).

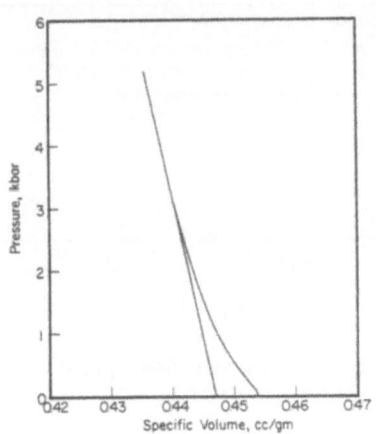

FIG. 3.2. Crush curve for 2C4 grout.

J. T. Cherry and N. Rimer

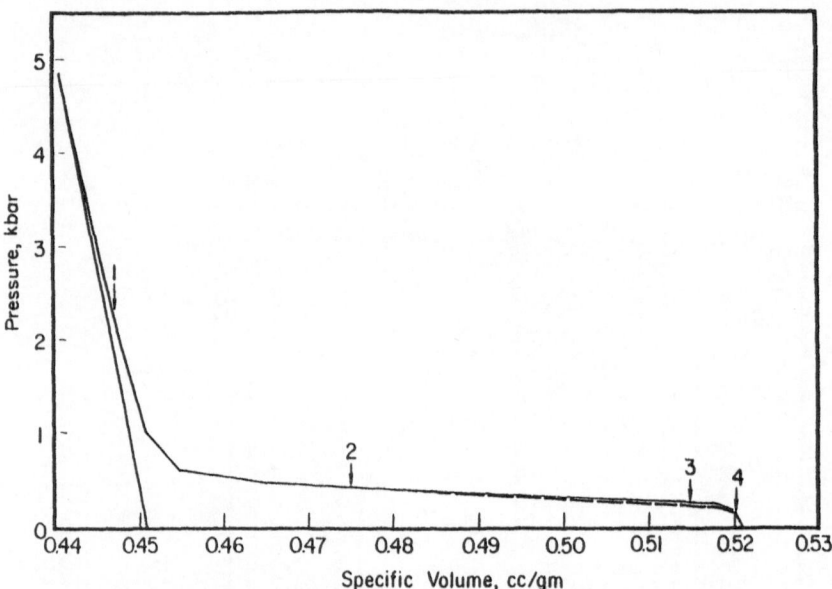

FIG. 3.3. Crush curve for LD2C4 grout. The solid curve was used for the RDD model calculation and the dotted curve for the effective stress calculation. The arrows indicate the approximate loadings corresponding to the velocity peaks at the four gauges.

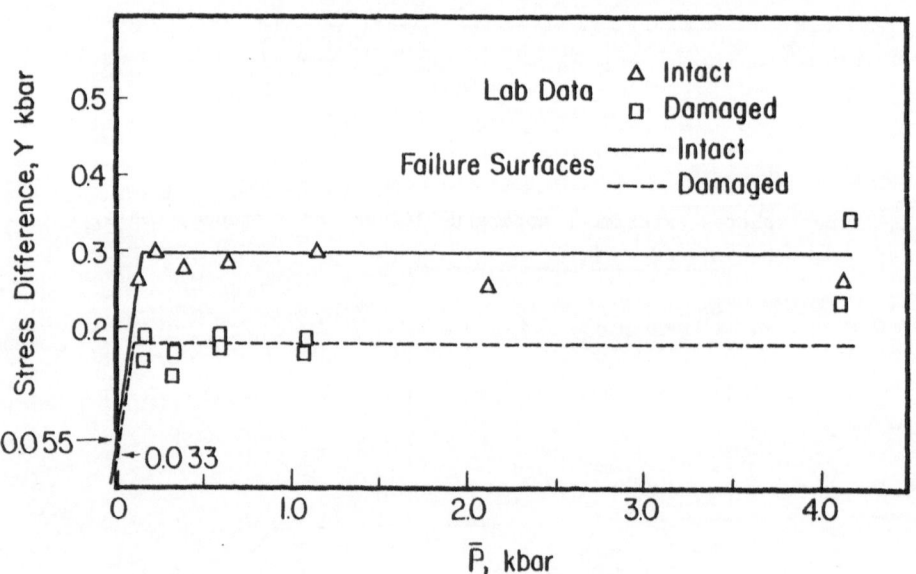

FIG. 3.4. Failure surface for 2C4 rock matching grout.

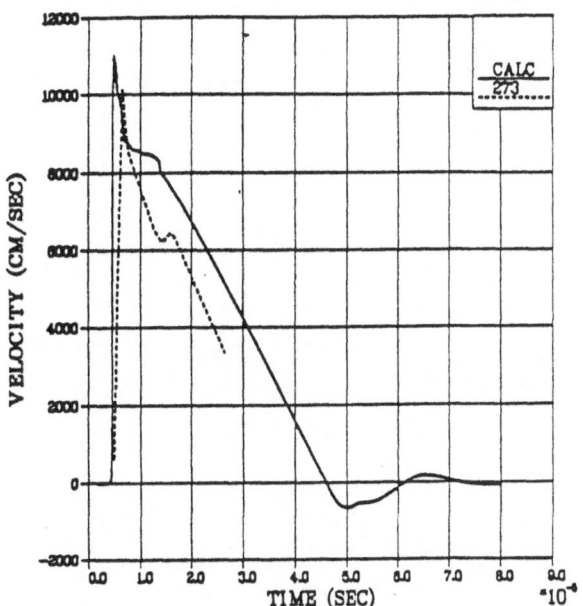

FIG. 4.1.   Comparison between measured velocities in 2C4 grout and
those calculated from a noneffective stress law simulation at
1.27 cm.   In this and all subsequent comparisons the solid
curves are calculated, and dashed curves are measured velocities.

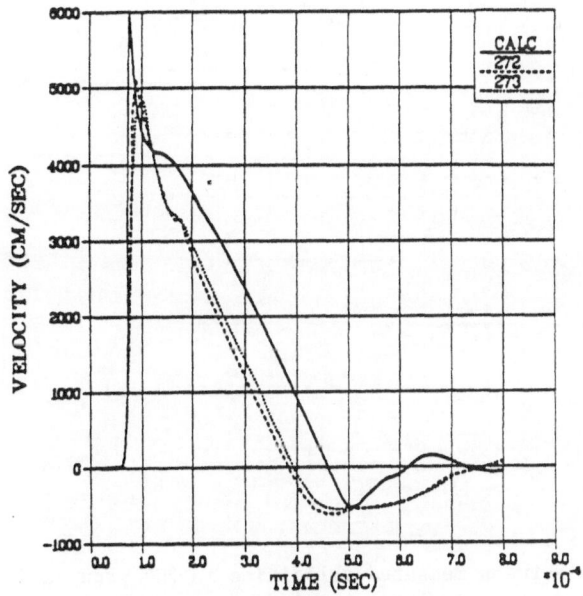

FIG. 4.2.   Comparison between measured velocities in 2C4 grout and
noneffective stress law simulation at 1.9 cm.

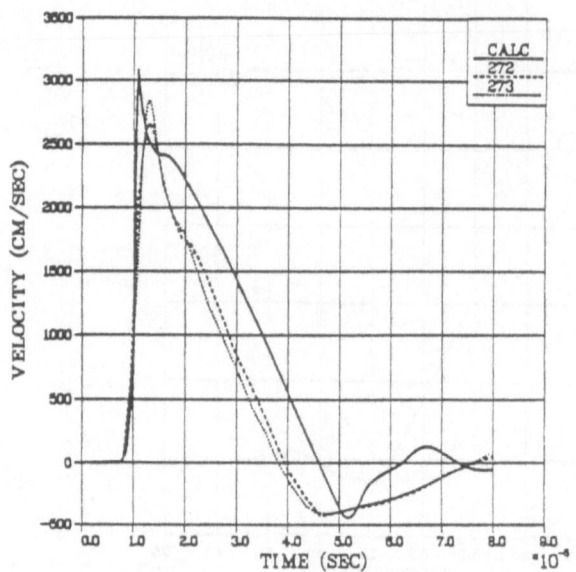

FIG. 4.3. Comparison between measured velocities in 2C4 grout and
noneffective stress law simulation at 2.54 cm.

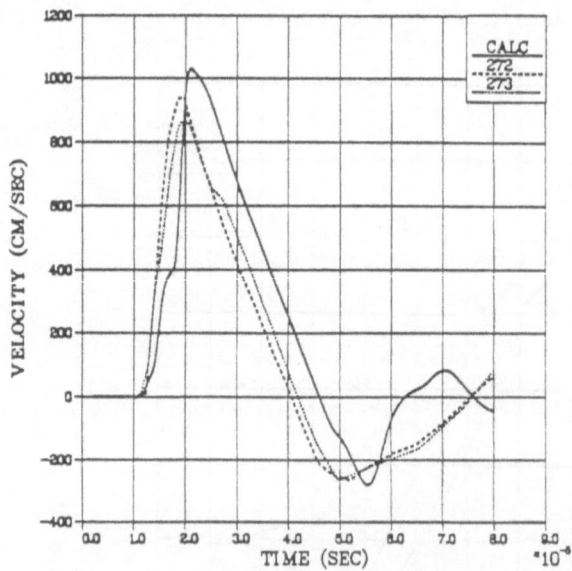

FIG. 4.4. Comparison between measured velocities in 2C4 grout and
noneffective stress law simulation at 4.0 cm.

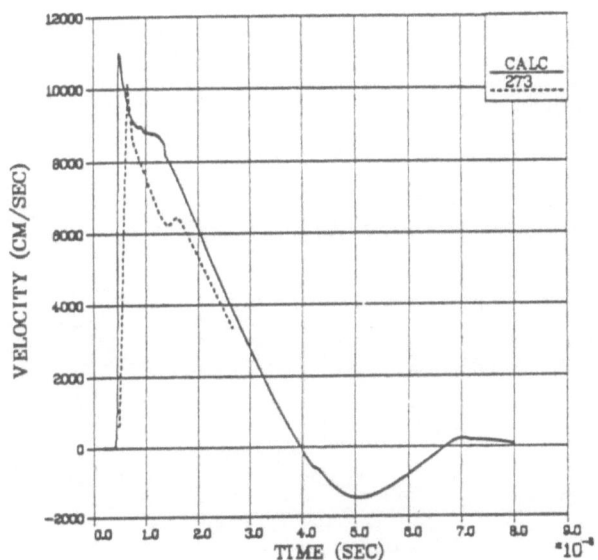

FIG. 4.5.   Comparisons between measured velocities in 2C4 grout and
effective stress law simulation at 1.27 cm.

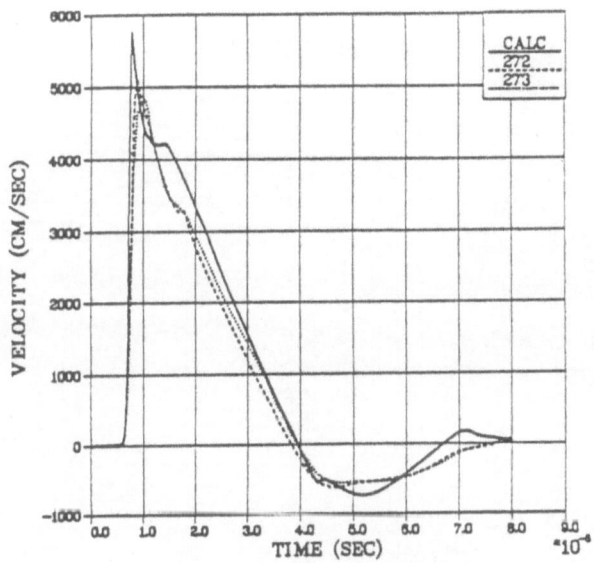

FIG. 4.6.   Comparisons between measured velocities in 2C4 grout and
effective stress law simulation at 1.9 cm.

J. T. Cherry and N. Rimer

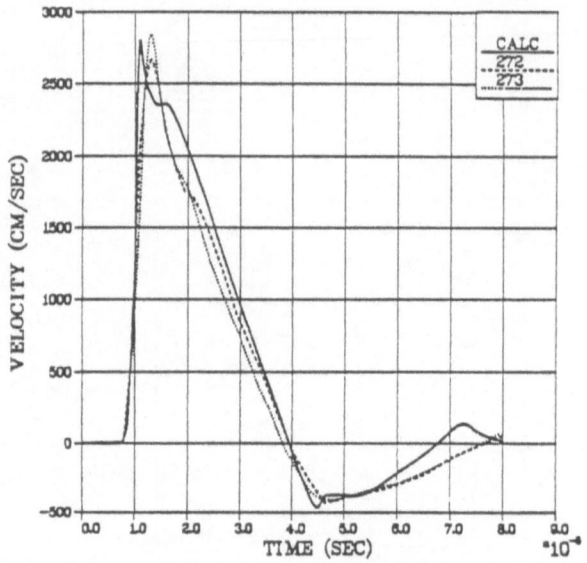

FIG. 4.7. Comparisons between measured velocities in 2C4 grout and
effective stress law simulation at 2.54 cm.

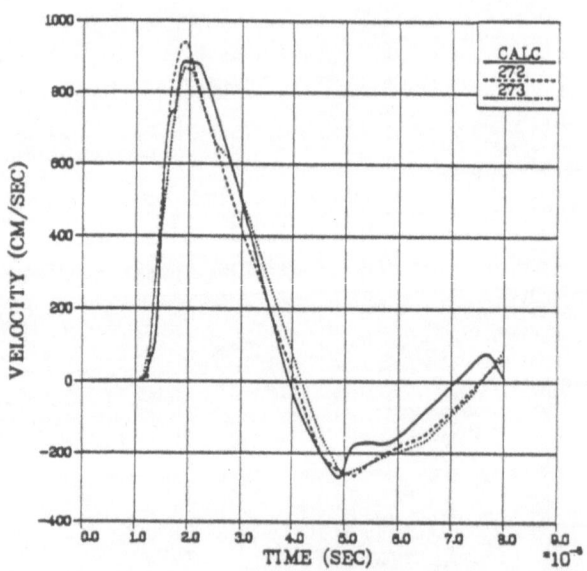

FIG. 4.8. Comparisons between measured velocities in 2C4 grout and
effective stress law simulation at 4.0 cm.

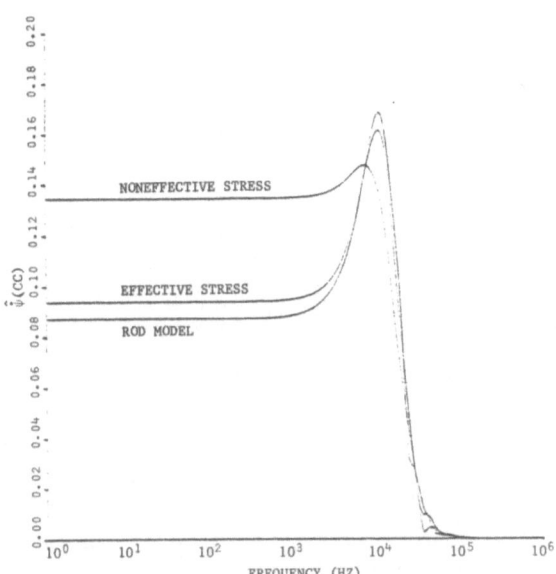

FIG. 4.9. Spectra of the reduced velocity potentials from the noneffective stress and the effective stress simulations. Also
included is the spectra from the RDD model (Rimer and Lee, 1982).

The first simulation used the material properties for 2C4 grout but no effective stress law, and the strength used in the simulation was that denoted in
Fig. 3.4 as the "intact failure surface." Figures 4.1, 4.2, 4.3, and 4.4 compare the particle velocities calculated from the simulation with those measured
during the experiment. At all ranges, the calculated velocity pulse is too wide
during the positive phase and too narrow during the negative phase.

The second simulation used the material properties for 2C4 grout and included the modification of material strength due to pore fluid pressure (effective stress) given by Eq. (3.4). The comparison between calculated particle
velocities and those measured is shown in Figs. 4.5, 4.6, 4.7 and 4.8. At all
ranges, we obtain excellent agreement with measured data from effective stress
law simulation.

The consequences of these two simulations (with and without effective
stress) on seismic coupling is interesting and is shown by the RVP spectra in
Fig. 4.9. The RVP spectrum with effective stress is peaked, with the spectral
peak about a factor of two larger than the spectral amplitude at low frequencies.
The RVP spectrum without effective stress is flat and greater than the effective
stress spectrum at low frequencies by a factor of 1.5.

Peaking of the RVP spectrum is a direct consequence of the width and amplitude of the negative portion of the velocity pulse, and the frequency at which
the peak occurs is approximately equal to the period of the pulse. The effective
stress model produces the correct shape for the velocity pulse during both the
positive and negative phases of the pulse by providing high material strength
during the initial stages of pore collapse and then lowering the strength during
the final stages. The high strength retards the velocities during the positive
phase while the low strength broadens the negative portion of the pulse.

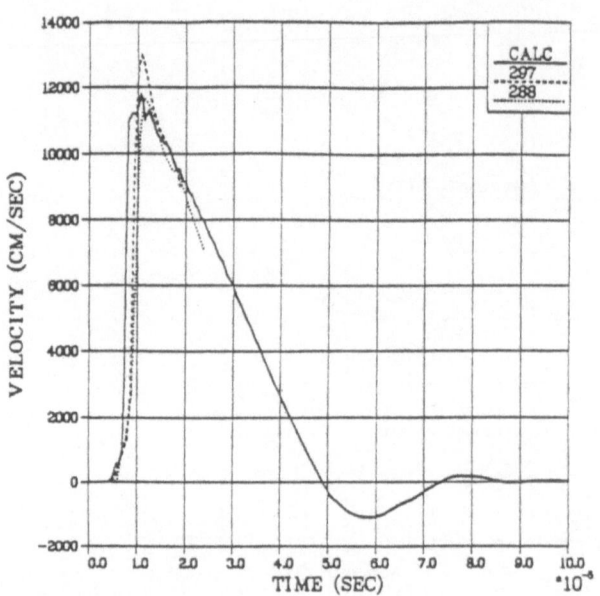

FIG. 4.10. Comparisons between measured velocities in LD2C4 grout
and effective stress law calculation at 1.27 cm.

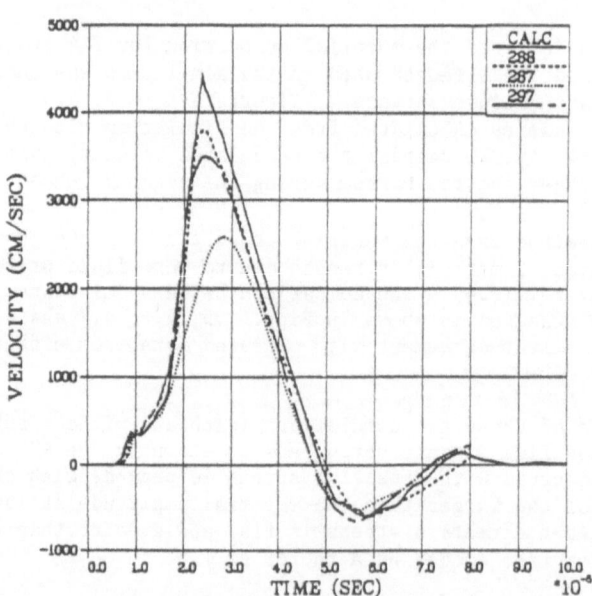

FIG. 4.11. Comparisons between measured velocities in LD2C4 grout
and effective stress law calculation at 1.90 cm.

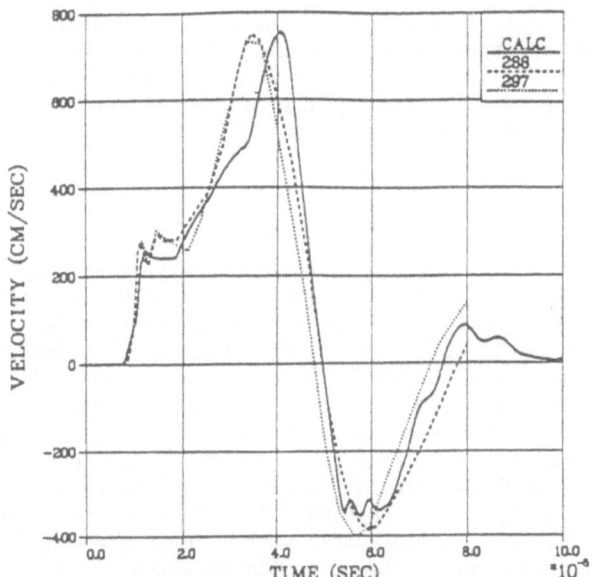

FIG. 4.12. Comparisons between measured velocities in LD2C4 grout and effective stress law calculation at 2.54 cm.

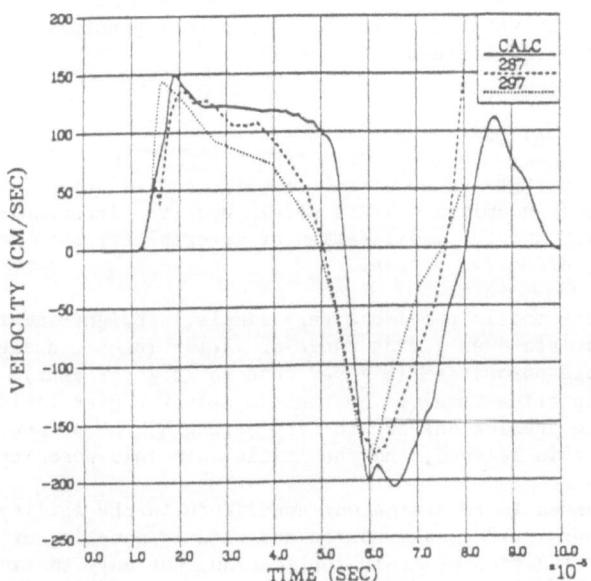

FIG. 4.13. Comparisons between measured velocities in LD2C4 grout and effective stress law calculation at 4.0 cm.

The third simulation used the material properties for LD2C4 grout and included the effective stress constitutive model. The comparison between the calculated particle velocities and those measured during the grout experiments is
shown in Figs. 4.10, 4.11, 4.12, and 4.13. The agreement between the calculated
results and the data is acceptable, indicating that the effective stress model is
applicable to a wide range of air-void porosities.

5. UNIQUENESS

Other constitutive models can be constructed which provide as good (or better) a fit to the data as the effective stress law results shown in Section 4.
Rimer and Lie (1982) have developed such a model. They discuss in detail a complicated rate dependent and damage dependent set of constitutive models (referred
to as the RDD model) which they used to simulate the SRI experiments in 2C4 and
LD2C4 grout. The RDD model has three important features: a failure surface which
is strain rate dependent upon loading; a rate dependent stress relaxation from
this failure surface to a damaged failure surface upon unloading; and a reduced
shear modulus upon loading. Both the rate of deviatoric stress relaxation and
the failure strength of the damaged grout are made functions of the amount of
damage, as is the shear modulus. The deviatoric stress relaxation to the damaged
failure surface is accomplished using a Maxwell solid approach.

This model successfully matches the SRI data in 2C4 and LD2C4 grout (Figs.
5.1, 5.2, 5.3, 5.4, 5.5, 5.6, 5.7, and 5.8), and the seismic coupling is approximately the same as that from the effective stress law (Fig. 4.9). However, the
complicated RDD model requires many assumptions regarding material properties
which are certainly not easily verified (and indeed may be impossible to verify)
from laboratory material properties tests.

6. CONCLUSIONS AND RECOMMENDATIONS

We have reviewed the current status of two specific constitutive models
which severely affect seismic coupling. These models were the irreversible collapse of air-filled porosity and the modification of material strength due to
pore fluid pressure, i.e., effective stress.

We have shown how these models provide a very simple, straightforward explanation of ground motion data over a wide range of yields (61 KT, nuclear to
3/8 gram, PETN) and air-void porosities (0.1 per cent to 13.4 per cent). The
most critical material properties required by these models for near field ground
motion and seismic coupling predictions are the dry strength, the pressure at
which all air-void porosity is removed, and the initial air-void porosity.

These results have served to reenforce our confidence in the ability of
these models to predict near field ground motion and seismic coupling effects.
We recommend intensive continuation of their application, not only to explosion
sites which have preshot measurements of material properties and/or near field
ground motion measurements, but also to sites where explosion yield can only be
estimated from far field seismic measurements.

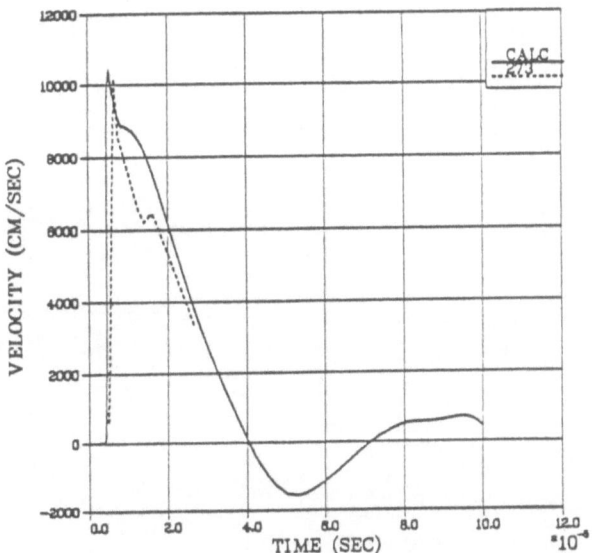

FIG. 5.1.   Comparison between measured particle velocities from SRI 2C4
grout test 273 and results of RDD model calculation at a radial
range of 1.27 cm.

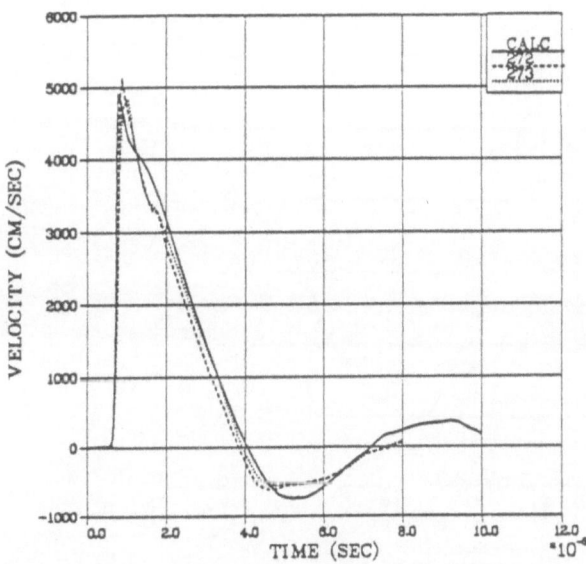

FIG. 5.2.   Comparisons between measured velocities from 2C4 grout
tests 272 and 273 and results of RDD model calculation at 1.9 cm.

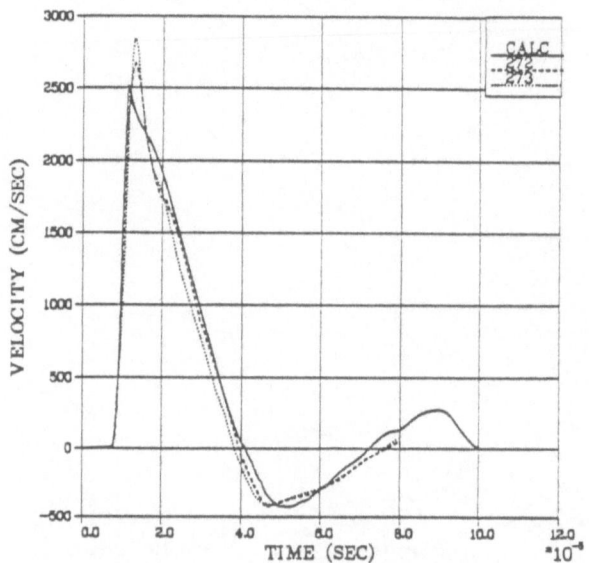

FIG. 5.3. Comparisons between measured velocities in 2C4 grout and
RDD model simulation at 2.54 cm.

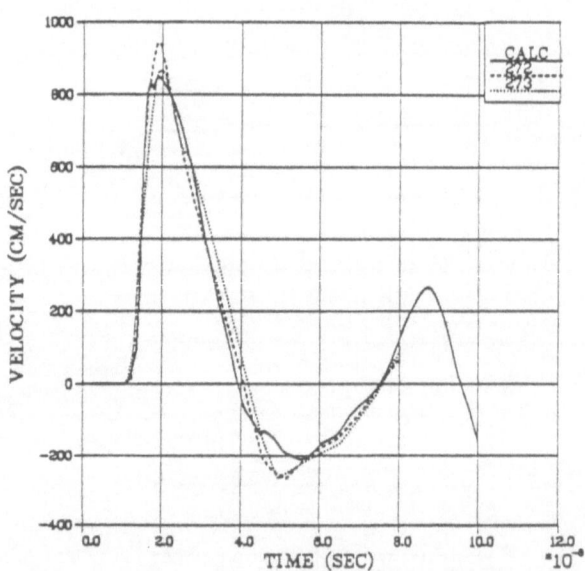

FIG. 5.4. Comparisons between measured velocities in 2C4 grout and
RDD model simulation at 4.0 cm.

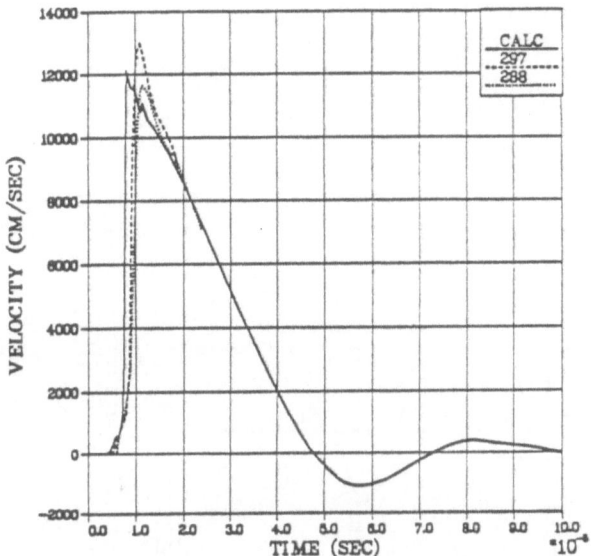

FIG. 5.5. Comparisons between measured particle velocities from SRI
    LD2C4 grout tests 288 and 297 and results of RDD model calculation
    at a radial range of 1.27 cm.

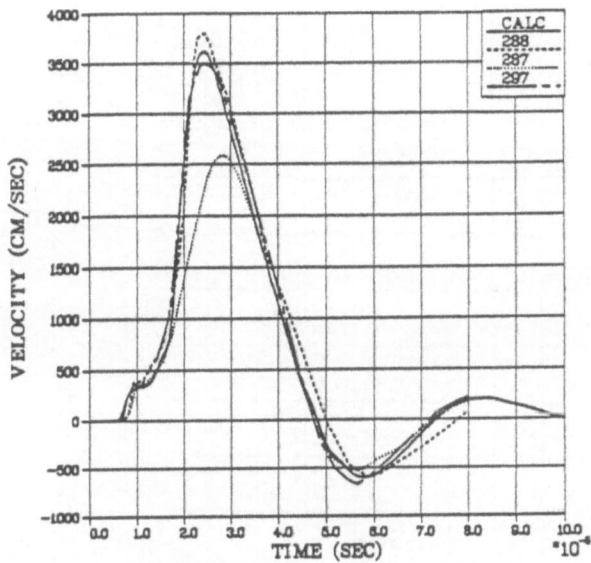

FIG. 5.6. Comparisons between measured velocities in LD2C4 grout and
    RDD model simulation at 1.9 cm.

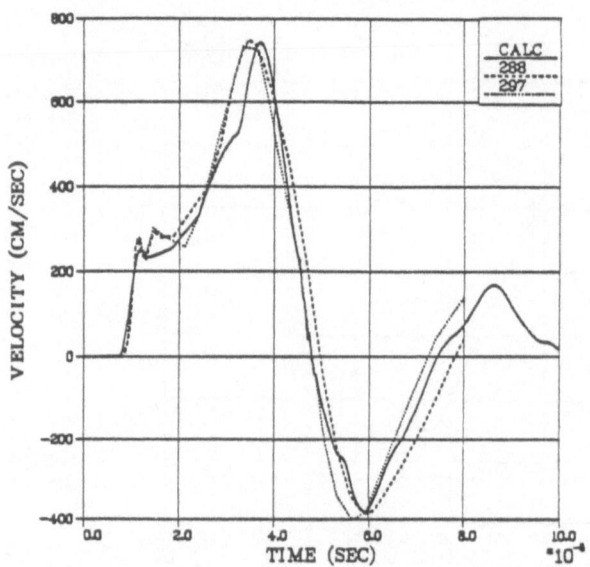

FIG. 5.7. Comparisons between measured velocities in LD2C4 grout
and RDD model calculation at 2.54 cm.

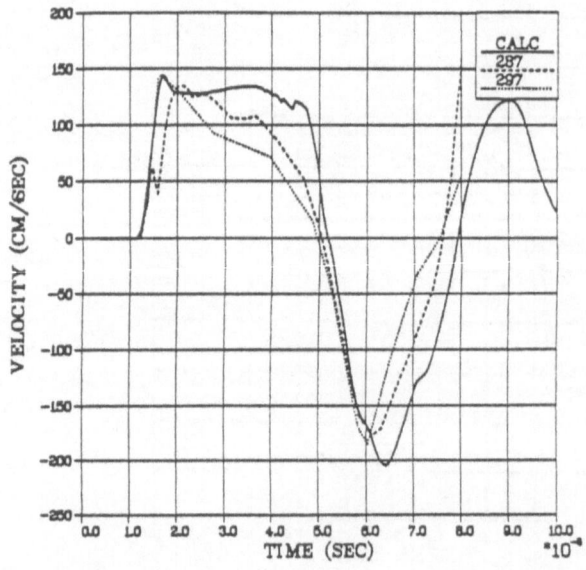

FIG. 5.8. Comparisons between measured velocities in LD2C4 grout
and RDD model calculation at 4.0 cm.

ACKNOWLEDGMENT

    This research was supported by the Advanced Research Projects Agency of the
Department of Defense and was monitored by AFTAC/VSC, Patrick Air Force Base,
Florida 32925, under Contract No. F08606-79-C-0008.

    The authors wish to acknowledge the contribution to this work of Dr. Carl E.
Keller, Chief, Containment Division, DNA Field Command. He not only made the
grout data available to us but had the foresight and tenacity to insist on its
acquisition.

REFERENCES

Bache, T.C., T.G. Barker, T.R. Blake, J.T. Cherry, D.G. Lambert, N. Rimer, and
J.M. Savino (1975), "An Explanation of the Relative Amplitudes of the Teleseismic
Body Waves Generated by Explosions in Different Test Areas at NTS," Systems,
Science and Software Final Report submitted to Defense Nuclear Agency, DNA 3958F.

Brace, W.F. (1965), "Some New Measurements on the Linear Compressibility of
Rock," *JGR*, *70*, 391-398.

Byerlee, J.D. (1979), "Experimental Study of Rock Friction with Application to
Earthquake Prediction," *Pure Appl. Geophys.*, p. 117.

Cherry, J.T. and F.L. Peterson (1970), "Numerical Simulation of Stress Wave Propagation from Underground Nuclear Explosions," in *Engineering with Nuclear Explosives*, Vol. 1. Available from the Clearing House for Federal Scientific and
Technical Information, National Bureau of Standards, Springfield, Virginia.

Cherry, J.T., C.B. Archambeau, G.A. Frazier, A.J. Good, K.G. Hamilton, and D. J.
Harkerider (1973), "The Teleseismic Radiation Field from Explosions: Dependence
of the Seismic Amplitudes upon Properties of Materials in the Source Region,"
Systems, Science and Software Final Report submitted to Defense Nuclear Agency,
DNA 3113Z.

Cherry, J.T., N. Rimer, W.O. Wray (1975), "Seismic Coupling from a Nuclear Explosion: The Dependence of the Reduced Displacement Potential on the Nonlinear Behavior of the Near Source Rock Environment," Systems, Science and Software Technical Report submitted to VELA Seismological Center, SSS-R-76-2742, September.

Cizek, J.C. and A.L. Florence (1981), "Laboratory Investigation of Containment
in Underground Nuclear Tests," SRI International Final Report for 1980 submitted
to Defense Nuclear Agency, DNA 5731F, February.

Cooly, C.H., R.H. Smith and J.F. Schatz (1982), "Properties of Tuffs, Grouts and
Other Materials," Terra-Tek Final Report submitted to Defense Nuclear Agency,
TR-82-05, January.

Day, S.M., N. Rimer, and J.T. Cherry (1981), "Surface Waves from Underground Ex-
plosions with Spall: Analysis of Elastic and Nonlinear Source Models," Systems,
Science and Software Topical Report submitted to VELA Seismological Center,
VSC-TR-82-11.

Dieterich, J.H. (1978), "Time Dependent Friction and the Mechanics of Stick-Slip,"
Pure Appl. Geophys., *116*, 790-806.

Garg, S.K. and A. Nur (1973), "Effective Stress Law for Fluid-Saturated Porous Rocks," *JGR, 78,* 5911-5921.

Heard, H.C. (1970), "The Influence of Environment of the Inelastic Behavior of Rocks," in *Engineering with Nuclear Explosives,* Vol. 1. Available from the Clearing House for Federal Scientific and Technical Information, National Bureau of Standards, Springfield, Virginia.

Rimer, N. and K. Lie (1979), "Spherically Symmetric Calculations of the SRI Grout Spheres Experiments for Four Different Laboratory Configurations," Systems, Science and Software Topical Report, SSS-R-80-4240, November.

Rimer, N. and K. Lie (1982), "Simulation of the Velocity Records from the SRI Grout Spheres Experiments," Systems, Science and Software Topical Report, SSS-R-82-5580.

Riney, T.D., G.A. Frazier, S.K. Garg, A.J. Good, R.G. Herrmann, L.W. Morland, J.W. Pritchett, M.H. Rice, and J. Sweet (1973), "Constitutive Models and Computer Techniques for Ground Motion Predictions," Systems, Science and Software Report submitted to Defense Nuclear Agency, DNA 3180F.

Scholz, C.H. and J.T. Engelder (1976), "The Role of Asperity Indentation and Ploughing in Rock Friction, I," *Int. J. Rock Mech. Min. Sci., 13,* 149-154.

Stephens, D.R. and E.M. Lilley (1970), "Loading-Unloading Pressure-Volume Curves for Rocks," in *Engineering with Nuclear Explosives,* Vol. 1.

GENERAL DISCUSSION

SCOTT: Do you happen to have a slide that would show, for example, measured stress or particle displacement histories at the same radial distance from a shot, but at different angular or azimuthal positions? That is, I'd be very interested to know what the azimuthal variation was in a random geological material. This would give a better feel for the correlation of measured vs. calculated behavior.

CHERRY: The one that I showed was the shot in alluvium and tuff, and these were the site characteristics that were available pre-shot from logging techniques that were conducted in the hole, in the inplacement hole or very near to the inplacement hole. On-shot depth was 445 meters and the site geology that we could gleen from well cuttings and logging data said there was a layer of alluvium, density 1.85 and 19 per cent air-voids, and so on down the line...

BACHE: I think his question was: If you moved that hole three meters, how different would the properties be?

SCOTT: A little further than that.

CHERRY: We had pre-shot for one hole. We actually made a one-dimensional model code, and we found that, indeed, as far as the attenuation of the shock wave was concerned, most of the attenuation was occurring because of the alluvium layer, having an air-void content of 6.4 per cent. When we then went to the two-dimensional model — and by the way, we had to put a free surface into the model or into the calculation — what we did was simply replace the whole column of alluvium by a single material whose P wave velocity varied with the position in the hole, but which had a fundamental density of 1.89 or an ambient density of 1.89

and air-voids of 6.4 per cent. That's all that was needed.

SCOTT: By two dimensional, you mean axisymmetric?

CHERRY: Axisymmetric, that's correct.

SCOTT: Was the array of pickups aligned along more than one radial direction?

CHERRY: No, they weren't. These were all in a single hole. However, in the granite calculation that I showed, there were a large number of gauges available, distributed at various distances and azimuths from the shot. We had gauges also along the surface, starting from directly above the shot to 753 meters along the horizontal surface.

SCOTT: What are the azimuths for these gauges?

CHERRY: I don't know.

SCOTT: But they are all at different azimuths.
Let me make my point. The point is: he gets extraordinarily good correlations, which I am always deeply suspicious of in any geologic material, between the calculations and the results. All I would like to see is what is the comparison between a measurement at the site at one azimuth at a particular distance with a measurement at another azimuth at the same radial distance to see if those two correlate as well as his calculations do at one place.

CHERRY: They vary, they do vary. The point is, though, when you look at all of the data from an environment that is full of gauges, one sees agreements about this well. Some gauges match very well, some you have questions about.

KELLER: I have looked very closely at a lot of the field data, and generally speaking, I have a fairly strong opinion that azimuthal variations seen in field data are more the problem of the quality of the gauge measurements than they are with real variations. We have some recent results measured in a very controlled environment near a thousand pound TNT charge in a massive bed. We see "azimuthal variations," but we have two gauges at the same range, on different azimuths and with different measurement techniques, that have records which are nearly identical. I believe those two records.
I suspect that azimuthal variations that have been invoked in a lot of past data interpretation have been more a problem, or a fault, of gauging. We see a variety of records from the gauges that we have, and there is no statement from the experimenter that some of them are indeed very peculiar and probably not valid. If you don't have enough field measurements in order to sort the good from the bad, you blame it on azimuthal variations.
The reason we went to very small tests is that we have much better control. We have measured particle velocities in tuff, alluvium and grouts. In fine grain tuffs, we see very little evidence of azimuthal variation.

CHERRY: It is an average measurement around the azimuths.

KELLER: We measured velocities very recently in a coarse grained tuff. You would expect to get azimuthal variation, since the occlusions were of the size of the charge. We do see a very slow rise time and a rounded peak which is exactly the kind of thing you would expect to see on our gauges, if you have an asymmetry in the shock wave. The point is, I am very suspicious of much of the field test data.

CHERRY: My approach, again, though, was to believe the data. That's all I could do — to see what the calculations gave when I put all of the data together.

BALADI: A few months ago I performed code calculations against the MISERS BLUFF II-1 event in which we have several duplicated measurements along the same axis. I used a two-phase elastic-plastic isotropic hardening type model for calculating pore pressure and effective stresses. I performed two calculations, one using a single-phase model and one two-phase model, and the result could match one gauge or the other, it depends on which gauge you choose. The difference between the duplicated measurements is as much as the difference between the result of the two models.

CHERRY: The gauge that you have trouble matching, is that in the saturated layer?

BALADI: Both the saturated and the dry top layer.

CHERRY: I will tell you what my conclusion is: there is no such thing as a dry rock: the rocks are always partially saturated. And there is no such thing as a saturated rock, that rocks always have some air in them. I would very much like to look at your data at the ground level.

PREPARED DISCUSSION

by L. W. Morland

1. INTRODUCTION

The authors' aim is to confirm a constitutive model proposed to describe the response of partially water saturated porous rock to large amplitude stress waves. Recognizing that ground motion data from explosion tests may well correlate with a variety of distinct models by suitable adjustment of the parameters, their approach is to develop a form of constitutive model based on quasi-static laboratory tests, with parameters determined by the measurements, and then compare the predicted wave propagation with ground motion data. This procedure has led to a conclusion that irreversible collapse of voids and the effect of pore fluid pressure on the rock strength both have a significant effect on the dynamic loading-unloading response during a wave passage. The paper adopts a model in which inelastic pressure-density and shear strength-effective pressure relations are the key ingredients, and predicts the spherical wave motion for comparison with controlled small scale laboratory explosions in grout spheres. The predictions for both low and high initial air content grouts correlate well with the experimental results, while predictions based on a model ignoring the pore fluid pressure in the strength relation are widely different. It is noted that a much more complex model has been developed which fits as well, or better, the same experimental results, but requires many assumptions about material properties which are not measured or verified by laboratory tests. I support the approach to model development which incorporates the minimal ingredients found necessary to match (qualitatively) observed response, adding to the ingredients only as predictions prove incompatible with further observations.

2. PRESSURE-DENSITY RELATION

The rock, fluid, and air mixture is treated as a single material with total pressure p related to the relative volume $\hat{v}$ of the rock-fluid composite and the air volume fraction ϕ by

$$p = \frac{1}{\alpha} \, \hat{p}(\hat{v}) = \frac{1}{\alpha} \, \hat{p}\left(\frac{v}{\alpha}\right), \qquad \alpha = \frac{1}{1-\phi}, \tag{1}$$

where v is the specific volume of the total mixture and $\hat{p}(\hat{v})$ is the pressure-volume relation for the rock-fluid composite with air removed. Irreversible removal of air starts at $\alpha = \alpha_e$, $p = p_e$ and is completed at $\alpha = 1$, $p = p_c$ during hydrostatic loading, with loading-unloading reversible above p_c and $\hat{p}(\hat{v})$ given by p(v) for the subsequent unloading response ($\alpha \equiv 1$). A load (above p_c)-unload cycle determines $\hat{p}(\hat{v})$ and $\alpha(v)$. A constitutive theory for a mixture incorporating intrinsic constituent properties, developed by Morland (1972), allows the simple form (1) to be derived as an approximation. First, the restriction of rock and fluid response to a single composite function $\hat{p}(\hat{v})$ implies the neglect of water diffusion through the rock, as noted by Garg (1983), who estimates diffusion lengths behind a sharp front to infer that diffusion will be negligible for low permeability rocks in the wave front vicinity, but not so for high permeability rock on laboratory wave length scales. Denoting partial pressure carried by the (assumed) rock-fluid composite by p^m and that carried by the air by p^a, then (Morland, 1972)

$$p = p^m + p^a, \qquad p^m = (1-\phi)^E p^m, \qquad p^a = \phi^E p^a, \tag{2}$$

where $^E p^m$ and $^E p^a$ are intrinsic constituent pressures which define the forces per unit area of constituent (not mixture). Let ρ be the mixture density and ρ^m, ρ^a the partial densities of composite and air, then

$$\rho = \rho^m + \rho^a, \qquad \rho = 1/v, \qquad \rho^m = (1-\phi)^E \rho^m, \qquad \rho^a = (\phi)^E \rho^a, \tag{3}$$

where $^E \rho^m$ and $^E \rho^a$ are intrinsic densities. By the previous definition of $\hat{p}(\hat{v})$, the unloading response when $\phi = 0$,

$$^E p^m = \hat{p}(1/^E \rho^m). \tag{4}$$

Hence the relation (1) follows from (4) only if

$$p^a = 0 \quad \text{and} \quad ^E \rho^m = 1/\hat{v}, \tag{5}$$

where the latter implies that the air mass is negligible, which is true. The zero air pressure condition supposes that the air diffuses freely during void collapse induced by the stress wave loading, as assumed in the quasistatic loading test (Morland and Hastings, 1973), but if highly compressed trapped air pockets remain on the wave passage time scale, then the model should incorporate an air pressure contribution to the total pressure, just as the rock-fluid partial pressure is the sum of rock and fluid partial pressures.

3. STRENGTH RELATION AND PORE FLUID PRESSURE

A failure criterion

$$\tau = a + b(\bar{p} - p_f) \tag{6}$$

is adopted, where τ is the maximum shear stress, $\bar{p}$ is the normal total pressure on the failure plane, and p_f is the pore fluid pressure (intrinsic fluid pressure); $\bar{p} - p_f$ is termed the effective pressure. As the fluid pressure increases, the failure shear stress decreases. It is postulated that $p_f = 0$ for $\alpha > \alpha_e$ $(p < p_e)$, then p_f increases proportionally with $(\alpha_e - \alpha)p$ until $\alpha = 1$, $p_f = p$ $= p_c$. The model therefore assumes connected pores with fluid and air at zero pressure until α decreases to α_e, then (since it is assumed that air pressure remains zero) the fluid pores disconnect from the air pores so that fluid pressure increases, reaching the total pressure just as the air content becomes zero $(\alpha = 1)$. The latter appears rather arbitrary if fluid and air pores are disconnected. Accepting zero air and fluid pressures until $p = p_e$, and assuming that both rock and fluid compression is linearly elastic, with intrinsic compressibilities κ_s, κ_f, and that the fluid volume fraction ϕ_f in the rock-fluid composite undergoes infinitesimal reversible changes only, Morland's (1978) analysis of undrained response shows that

$$p_f = \frac{1}{\phi_f} \frac{\bar{\delta} - \kappa_f}{\kappa_f - \kappa_s} \left(E_p^m - E_{p_e}^m \right) = \frac{1}{\phi_f} \frac{\bar{\delta} - \kappa_s}{\kappa_f - \kappa_s} (\alpha p - \alpha_e p_e), \qquad (7)$$

where $\bar{\delta}$ is the undrained compressibility and air pressure is assumed to remain zero. This dependence is quite different from the postulated proportionality to $(\alpha_e - \alpha)p$, and does not predict $p_f = p$ when $\alpha = 1$. Garg (1983) deduces a proportionality to αp assuming that the fluid pores are disconnected from the air pores throughout the loading, equivalent to $p_e = 0$ in (7), to exhibit an upper bound to the fluid pressure, but with a different coefficient based on relations incorporating the Biot symmetry (Garg and Nur, 1973). Since fluid pressure has a significant role in the failure response, this aspect of the model should be examined further to assess the relevance of the above inferences.

4. WAVE PROBLEM

The paper does not present the nature of the material response during failure. By inference the failure criterion is "non-hardening," but a "flow rule" for the irreversible deviatoric strain is required to complete the model for general stress loading. The assumed spherically symmetric motion with purely dilatational waves determines the circumferential strain independent of a flow rule, but compatibility with a postulated model response should be demonstrated. The question of associated or non-associated flow rule with the failure criterion becomes crucial when combined shear and dilatational waves are considered. In the most elementary case of plane uni-axial waves in an infinitesimal strain elastic, perfectly plastic, theory with non-associated flow rule, it has been demonstrated that neither existence nor uniqueness of the motion of an interface between loading and unloading zones is guaranteed (Morland, 1977). The physically expected existence and uniqueness are obtained only if the four wave speeds interlace, which follows only for an associated flow rule. Hence failure criterion and flow rule must be formulated together, with care given to the implications for dynamic response under general stress loading.

The above remarks reflect personal experiences of model development and wave propagation analysis, and are not intended to detract from the achievements of the combined theoretical and experimental programme described.

REFERENCES

Garg, S.K. (1983), "On the Effective Stress and Air Void Porosity Constitutive Models," S-Cubed Memorandum.

Garg, S.K., D.H. Brownell, J.W. Pritchett and R.G. Herrmann (1975), "Shock-Wave

Propagation in Fluid-Saturated Porous Media," *J. Appl. Phys.*, *46*, 702.

Garg, S.K. and A. Nur (1973), "Effective Stress Laws for Fluid-Saturated Porous Rock," *J. Geophys. Res.*, *78*, 5911.

Morland, L.W. (1972), "A Simple Constitutive Theory for a Fluid-Saturated Porous Solid," *J. Geophys. Res.*, *77*, 890.

Morland, L.W. (1977), "Plane Wave Propagation in Anisotropic Jointed Media," *Quart. J. Mech. Appl. Math.*, *30*, 1.

Morland, L.W. (1978), "A Theory of Slow Fluid Flow through a Porous Thermoelastic Matrix," *Geophys. J. Roy. Astr. Soc.*, *55*, 393.

Morland, L.W. and C.R. Hastings (1973), "A Void Collapse Model for Dry Porous Tuffs," *Eng. Geol.*, *7*, 81.

GENERAL DISCUSSION

TRULIO: I would like to ask: When the compressibility of the dry skeleton of a rock is measured, and found to be much lower than that of water, how can one conclude from effective-stress theory that only its cohesive strength will remain if one saturates it?

MORLAND: I think Dr. Cherry should discuss the basis of his 3-phase model.

CHERRY: As I understand it, the law of effective stress, which is really due to Garg and Nur for rocks —

GARG: No, Ted, the effective stress law was first proposed by Terzaghi.

CHERRY: Anyway, it is the only paper that I have read on effective stress. It is basically an Archimedes principal type of law that says that if you have water between two surfaces or any fluid between two surfaces, a pressure builds up in that fluid that counterbalances the normal stress that's acting on that surface. So, the total normal stress that one uses to estimate the strength of the material is the difference between the normal stress that's applied on the surface and the pore fluid pressure that's in the crack.

TRULIO: But the skeleton has a compressibility. For the material's frictional resistance to shear not to increase in Coulomb fashion, the liquid somehow has to bear any increased compressive load; that it will do, if the skeleton is more compressible than the liquid. But there is no rock that I know of — no hard rock with a few tenths of a percent total porosity which is more compressible than water. If you measure the strength of such a material in the laboratory and find that somehow it *is* reduced to cohesion, then that result would appear to contradict effective stress theory, not confirm it.

GARG: First, let me address the comments on the effective stress law.
   There is no such thing as a unique effective stress law. There is one effective stress law that applies to fracture or strength of geologic materials, which is simply the confining stress minus pressure. This is an empirical law, and has nothing to do with the rock compressibility or water compressibility.
   The volumetric strain is, however, governed by a different effective stress law , and that's where fluid compressibility and rock compressibility play a role.

The volumetric strain response involves four compressibilities, that is, (1) the drained compressibility of the rock material, (2) water compressibility, (3) compressibility of the rock grain, and (4) the undrained rock/fluid compressibility. Of course, only three of these four compressibilities are independent.

The question I have for Prof. Morland pertains to air pressure in partially saturated porous media. I think what Prof. Morland said would be all right if water and air were immiscible, but they are not immiscible. On an application of a confining pressure, air would tend to go into water and one won't get into the kind of contradiction Prof. Morland was talking about.

MORLAND: I will accept that I had not considered that physical situation.

McFARLAND: I was going to address Jack's question. I can't explain the micro-modeling that accounts for the decrease in strength, but this has been shown by Brace at MIT, Heard and Schultz, people at S-CUBED, Green, and people at Terra-Tek. It is well documented in the mechanic's literature.

TRULIO: I have seen the results and some of them were shown here. The question was of reconciling these with effective-stress theory.

McFARLAND: I have got a comment, if I may, relative to Ted's presentation. The results he showed were calculations as opposed to predictions; the calculations always agree with data, the predictions not so frequently.

CHERRY: Yes. Unfortunately, I ran out of time. We have done one prediction with this model, and since I ran out of time, I won't be able to show it to you. The agreement was not marvelous, but it wasn't bad either. Assuming that you accept effective stress as the mechanism that causes the strength relaxation, the next fundamental problem is to go to a site and to characterize it in terms of the material properties that are required for this model or indeed for any other model. And that's something that we haven't even talked about.

NEMAT-NASSER: I really enjoyed Leslie Morland's presentation. The ordering of the velocities, as you have indicated, in a work hardening state, where all the velocities are indeed real, relate, it seems to me, to the essential structure of the constitutive relations. I think it pertains to some of the comments we have been making all throughout this workshop about the "well-posedness" and you have presented another effective way of looking at this issue, which should not escape attention.

NEEDLEMAN: Just a comment, it is about non-normality and the sensitivity to initial data. I don't know whether it is entirely relevant, but something that has been found in the metal plasticity context when dealing with constitutive relations which give rise to instabilities, I presume the constitutive laws you refer to are rate independent?

MORLAND: Yes.

NEEDLEMAN: Then if one includes material rate dependence in conjunction with the other effects, the problem can become well posed even if the rate independent problem is not. It doesn't mean the instabilities go away, they show up, but the problem is well posed and perhaps analyzable.

MORLAND: I'm sure that's true. I have only looked at rate independent theories in coming up with these conclusions.

HERRMANN: I got the impression from Ted's talk that the field was cored, the core was taken to the lab, lab tests were done, the material properties were fitted, calculations were done, and then afterwards, calculations came out fine by comparison with the field data. And I think it was just suggested that that was not so. Could you tell us what, if anything, was tweaked to make those results come out right, or was it really the way I said?

CHERRY: Let me talk about the last thing that we have done, since that's always supposed to be the best. It was the result of an experiment that DNA, Carl Keller and Sandia, and SRI conducted, along with Terra-Tek, and S-CUBED was sort of in the middle, and I didn't run any of the calculations.
    Carl, would you like to explain what you all did?

KELLER: I can tell what was done, but I won't tell what the results were.

CHERRY: I will tell them what the results were and you tell them what was done.

KELLER: Terra-Tek and S-CUBED have both been doing modeling work for DNA. We wanted to test how well we actually could calculate a real world situation. There were two, what I'd call, real world situations.
    We went to the Nevada Test Site in Rainier Mesa, G Tunnel, and we emplaced one ton of TNT. This was done by Sandia. From that site, twelve-inch core was taken to SRI. The core was sawn in half, and the gauges were emplaced. SRI drove that tuff sample with a 3/8 gram charge. Sandia drove the in situ tuff medium with one ton of explosives.
    We took core to Terra-Tek and measured its properties any way that the calculators wanted it to be measured. Using the models we developed and the Terra-Tek data, we calculated the SRI results at the 3/8 gram scale. We also calculated the Sandia results at the one ton scale. Then, we looked at the SRI results, and we said that we will allow ourselves to make this data part of our material properties measurement. We adjusted the model and recalculated the one ton situation. Then we compared the one ton predictions to the actual one ton results.
    In summary, we compared directly 3/8 gram measurements to one ton measurements, and we compared calculations to both measurements. We also compared the calculations to each other.

CHERRY: You have already seen the results of the comparison with the SRI small shots. They look pretty good.

VOICE: Did you get the arrival time right or is it adjusted?

CHERRY: No, there is some problem with the arrival time; the arrival time was not done properly either. And that's going to have to be looked at. As I understand it, we did not have any in situ measurements of compression wave velocity in this material, and that's unfortunate. But at any rate, the prediction is a lot better than it would have been two years ago. We then went back and reran the calculation with 1.6 per cent porosity, and we got a little better results.
    We haven't talked about site characterization. It is an absolutely critical problem. Air-filled porosity is really important; the cohesion of the material is really important. Those of you that have any capabilities to measure air-filled porosity in situ and can find compressive strength in situ, ought to do it, because that would be worth a lot.

HERRMANN: What did you tweak to make the SRI test come out right?

CHERRY: I don't know whether we have made it come out all right. There was a range of variation in the hydrostat that was available to us from the strength test, and we just played around within that range until we got a good agreement. This mattered some, but it didn't seem to matter very much within the constraints placed on us by the test data provided from Terra-Tek. At any rate, that's a predication, the first one we've done. We are subjecting the core to the in situ stress data to see if the amount of air-filled porosity is going to decrease from 2.4 per cent to whatever.

CHAPTER X

STRAIN-PATH MODELING FOR GEO-MATERIALS

John G. Trulio

Applied Theory, Inc.
930 South La Brea Avenue, Los Angeles, California 90036

ABSTRACT

The explosions that most concern us all have effects known mainly by calculation. Since the calculations' credibility influences the cost and scope of deterrence, reliable numerical prediction of explosion-effects is a long-sought goal. For free fields, flaws in stress-strain relations limit the accuracy of predictions: Such relations are hard to establish for specific geo-materials and admit a very wide range of fields. However, the kinds of deformation occurring in free fields depend more on burst geometry than on material properties. In particular, for contained, nearly spherical bursts —the basic events in nuclear monitoring —the paths traced in strain space by deforming material elements ("strain paths") have shapes and orientations ("patterns") that vary little with medium. Also, a) the patterns form a simple set, b) at a given strain amplitude they are not diverse, and c) hardware is at hand for stress measurement along them in the laboratory and (probably) in situ. Hence, it now appears feasible to obtain by *measurement* the stress-strain curves needed for reliable prediction of seismic sources.

A subset of the same curves also prevails on a sizable, downgoing part of the field produced by a surface burst. Further, while paths are more complex and variable near the surface, they may still form a set small enough for stress-strain measurement — though that will not be so if full control over plane strain, including axis-rotation, is needed. Moreover, in surface-burst fields, paths of spherical type give way to plane-strain paths on a transition region for which measurement of stress-strain curves looks impractical. Thus, for surface bursts, mechanistic models must be asked to extrapolate from measured curves — but to a far smaller extent than at present.

Extrapolation with mechanistic models is to be avoided because i) their record of ground-motion prediction is poor, and ii) the predicted effects of their mechanisms on the stress-strain curves traced in explosions have yet to be confirmed. Also, despite its glitter, postdiction is almost useless for model-validation, because infinitely many models will reproduce any given field of motion. However, free-field prediction can be made more reliable —now —by means of strain-path models, in which mechanistic models are used freely for interpolation among measured stress-strain curves. The same curves are needed to validate mechanistic models, whose development (the ultimate goal) is therefore aided by strain-path methods —all the while minimizing unvalidated use of the models for extrapolation.

1. BACKGROUND AND SUMMARY: OVERVIEW

To compute the response of the earth to explosions below or near its surface, discrete analogs are written of equations that govern continuum motion.
Among them are analogs of constitutive relations that define the mechanical behavior of geologic solids. From the time computers became powerful enough to
make such geo-modeling practical (~1955-1960), it has differed in one major
respect from other parts of numerical continuum mechanics: The accuracy of
model-predictions is limited mainly by defects in constitutive relations. At a
time (for example) when computed fields of viscous compressible fluid motion
were approaching wind-tunnel accuracy (Kitchens, 1972), critical ad hoc assumptions had to be made about material behavior to account for far simpler motion
observed during bursts in material as ideal as dome-salt (Trulio et al., 1975).
They still do.

Typically, differences between computed and observed motions have been explained as due to physical processes not included in a given model; typically
also, experiments have not been performed to verify such explanations independently of the observations that evoked them (they remain ad hoc). There has
thus accumulated a set of at least ten major model-elements (e.g., nonlinear
elasticity; Appendix A), each offering modelers wide latitude in matching known
motions ("postdiction"); at the outset, there were perhaps four. The many post-
dictions made to date have, by and large, succeeded — but the reverse is true of
pre-shot predictions for the handful of explosive events in which they have been
tested against measured motions.

The course being taken to model explosive events is evidently open to criticism, and some of that appears below. However, the aim of this paper is not to
carp, but to map a more promising course — one that, in a few years, can either
yield reliable predictions of explosively-driven motion or tell us that, in all
likelihood, no such predictions can be made.

By way of explanation, suppose that in each prediction-test conducted to
date, a given model had proven accurate — correct, say, to within 10-20% in both
peak velocity and displacement throughout the regions instrumented. Even then,
the model's reliability would be in doubt for a burst differing in gross ways
from the test-events (e.g., in burial depth, charge distribution and/or yield);
after all, extrapolations commonly disclose the limits of validity of theories
(and nature's surprises). However, doubt would largely vanish if the stress-
strain curves traced by material in the new (computed) field closely matched
those of the previous test-events. Again, if stress-strain curves unlike those
seen earlier occurred widely in the new field, they still might prove to be followed closely in the laboratory or — better — in situ; confidence in the predicted field would then approach certainty. On the other hand, the measured
curves could differ greatly from those calculated. The causes of the differences
would then have to be found by testing one or more of the model's elements under
conditions set by the wayward curves — and, after similar tests of their validity, equations might have to be added to the model to account for physical processes previously ignored.

As matters stand, model-predictions rarely approach the accuracy just presumed. Also, i) the standard data-base for prediction comes from a few laboratory tests in which the stress-strain curves traced run far from those of the
fields predicted, and ii) infinitely many stress fields (and hence models) permit exact postdiction of a given field of motion. Hence, standard practice has
taken us but a small step toward model validation of the kind just described.

Yet, such validation (or equivalent) is vital if conventional methods are to yield models that can be counted upon for accurate prediction of explosively driven ground motion. For those methods require us not just to make correct predictions, but to identify each important physical process triggered by explosions, write equations that correctly describe it, and combine the resulting equation-sets in a theoretically sound way (e.g., in accord with the entropy principle). In short, if the predictions of models like those used now are to be trusted, they must be right for the right mechanistic reasons.

Ultimately, correct mechanistic models are the objects sought. How long it will take to find them can only be guessed, but it must be admitted that validating all important model-elements for geo-materials is a major enterprise (Appendix A). Yet, reliable predictions are needed now. To get them, another approach can be taken to modeling which, due to its very empiricism, may also speed the process of mechanism-validation. In particular, study of the stress-strain curves on which attention must center in certifying mechanistic models, has shown that the kinds of deformation material undergoes are a) determined more by burst geometry than constitutive properties, and b) dominated by a few fairly simple kinds of strain. To be sure, there are limits to the truth of a) and b); however, for deep, nearly spherical bursts those limits are more conjectural than real — and even for surface bursts they lie well beyond the range of validity of present computational models.

Findings a) and b) form the basis of strain-path modeling, whose hallmarks are these: Most of the stress-strain curves needed for accurate prediction of motion at a given site are obtained by measuring stress along the kinds of strain paths already known to dominate explosively driven fields. Mechanistic models are then required to reproduce the measured curves, if they can. If they cannot, then the stress-strain data needed for accurate prediction (having been measured) are forced into them in openly ad-hoc ways. At the same time, their authors are presented, in stress-strain form, with specific conflicts between actual data and their constitutive equations. Any improved understanding that results translates at once into better models for predicting ground motion due to explosive loading —a result that plainly benefits organizations whose missions call for such predictions. A similar benefit arises in dealing with differences between material in the laboratory and in situ, which pose a basic problem for all modeling: For most strain-paths of the dominant set, stress gauges may finally be at hand with which to quantify those differences, so that sample-and-lab-test procedures can be revised accordingly.

A designer of footings (for example) may see little hope for help in such a program, but it could also be held that better knowledge of the mechanisms of soil response will have across-the-board benefits. Indeed, developing valid special-purpose models of geo-materials could prove an efficient way to arrive at all-purpose models, and strain-path methods may find fruitful application in other branches of materials science. However that goes, the fact is that even well-tested mechanistic models can fail if used for long extrapolation. Hence, when the jump is made to predicting the effects of a nuclear explosion at a given site, it must also be verified experimentally that the stress-strain curves traced by material in the calculated field really do apply to the material at that site. Fortunately, the job may not have to be done for more than a few sites; the evidence is already strong that strain-path shapes are not greatly altered by a change in source-type from chemical to nuclear (burst geometry outweighs that change). The economic stakes alone easily suffice to justify the cost of such verification.

2. PREDICTION AND POSTDICTION

On occasion, calculated free-field motion is made available for a given event before it occurs. In that case — if the event does not merely copy an earlier one — a model's predictive power can be tested. A prediction of that sort for the Hudson Moon HE Experiment (Bjork, 1972) proved as accurate as any yet made. The calculated field was spherical, and for practical purposes the actual field probably was also. Predicted and actual pulses from the event are shown in Fig. 1 for the smallest and largest radii of measurement (1.05 m and 3.82 m), and an intermediate one. The outermost gauge may well have been placed at a smaller radius than intended (Bjork, 1972, p. 23), but even then calculated pulse-peaks drift below those measured, as radius increases. Indeed, though uncommonly accurate, the calculated pulses as a whole tend to decay faster with range than those measured, from mid-range or less (~1.68 m) outward.

Harder to assess is the decision of the shot's designers to measure radial stress, σ_r; that task, whose subtlety was not well understood in 1971, is only now becoming feasible for spherical fields in geo-solids. Nevertheless, *these* measured stresses could be substantially correct, especially at the smaller gauge-ranges, because they generally exceed by far the material's failure limit, Y_0: $|\sigma_r - \sigma_\theta| \leq Y_0 \approx 35$ MPa, in the laboratory [$\sigma_\theta \equiv$ hoop stress]. Thus, over most of the space-time domain of measurement, the medium was more a fluid than a solid. Principal stresses were of course measurable in fluids before 1971, and calculations of fluid motion were quite reliable by then; as already noted, mazy fluid motions have long been predictable using computational models like the ones applied to ground motion problems. Further, while radial stresses at the two greatest gauge ranges were only about as large as Y_0 much of the time, motion there was driven by the predictable quasi-fluid field found at somewhat smaller ranges.

The main goal (and problem) in modeling geo-materials is to describe the effects of shear accurately when stress gradients have strong deviatoric components. The match between predicted and measured pulses in the field of Fig. 1, which is especially close at the smaller gauge-ranges, would be clear evidence of such power if octahedral shear stresses at gauge-ranges were about as large as mean stresses. Actually, for the most part, they were much smaller (though a little shear was astutely observed to have sizeable effect on mean stress (Bjork, 1972, p. 5)). On the other hand, for the Diamond Dust event (also in tuff; Sauer and Kochly, 1971), the maximum possible shear stress, $\frac{1}{2}Y_0$, was greater than a third of the mean stress at even the smallest gauge-range. Also, pulse-amplitudes varied by almost twice the factor seen in Fig. 1 (radial velocity was measured, as usual, rather than stress). Moreover, in the model used to predict Diamond Dust motion, the dependence of mean stress on compression was found from hydrostatic tests alone — whereas a major result of the later Hudson Moon HE Experiment was that uniaxial strain data are more germane to nearly spherical bursts. Despite these drawbacks, the pulses predicted for Diamond Dust were about as accurate as those of Fig. 1 (Trulio and Perl, 1974, pp. 48-53). Computed peak velocities did depart from the observed power-law decrease with range, and, at the farther ranges, predicted pulses decayed more slowly than those measured — but the errors were not larger than in Fig. 1. Still, when all the experimental and modeling procedures followed for Diamond Dust were used again for the Diamond Mine event (Sauer and Kochly, 1972), the predicted Diamond Mine pulses were much less accurate than for Diamond Dust (Trulio and Perl, 1974, pp. 96-107), although the two bursts took place only 190 m, and 7 weeks, apart. Indeed, even the grout columns that housed the gauges again ran radially outward from shot-point.

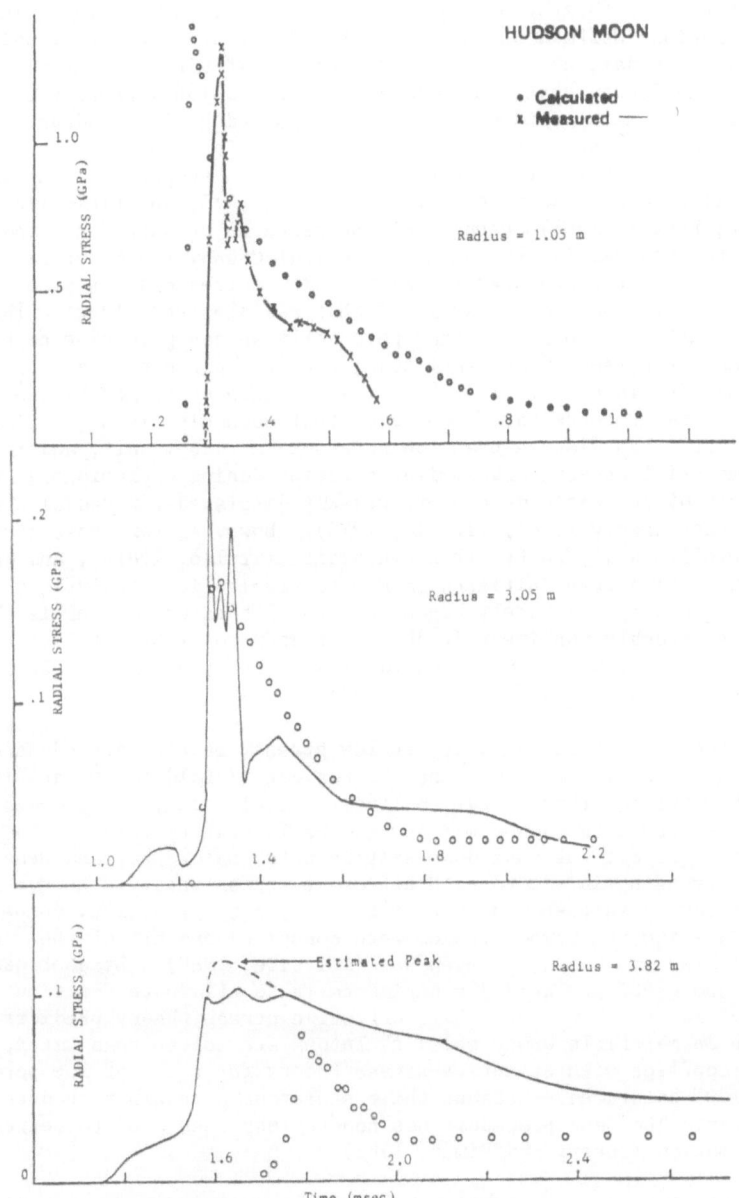

FIG. 1. Hudson Moon HE Experiment: Spherical Radial
          Stress vs. Time, Measured and Predicted at Three
          Distances from the Burst Point.

The Piledriver event in Climax granite (1964) came too early in the models' development to make their predictions (if any) informative. Yet, with regard to Piledriver motion (Hoffman and Sauer, 1969), one truly noteworthy point has emerged: To this day, no models can account for it without critical ad hoc assumptions. The Cowboy shots in dome-salt (Murphey, 1960) present an even more striking case in which postdiction is not supported by independent measurements of the host medium's properties. Specifically, before 1975, the models' verdict was that dome salt would behave elastically once it stopped failing in shear. Also, according to all laboratory (Heard et al., 1975) and field (Terhune and Glenn, 1977; Perret, 1967) measures of the strength of salt (then and now), no shear failure occurred in the roughly spherical Cowboy fields at radial stresses below 35 MPa. Yet the measured radial-velocity pulses made it clear that deformation was far from elastic at 1% of that radial-stress level. [Note: Early work on the salt problem showed that pore collapse can give rise to highly inelastic decay of pulses of any amplitude, even for porosities as low as dome-salt's (~1%) (Trulio et al., 1975, pp. 41-68; since confirmed by others). While no pertinent measurements have been made, that mechanism is not a likely cause of the inelastic behavior seen at low stresses in Cowboy salt, and in almost every geo-material observed at similar stresses during explosions.] Later, the large effects of inelastic decay were starkly displayed for radial stresses > .05 MPa (the lowest Cowboy level) (Trulio, 1978). However, for those studies, the key issue was linearity rather than elasticity (Trulio, 1981a), and laboratory and in situ testing were initiated in 1981 to resolve it. Analysis of the resulting in-situ data has barely begun, but the laboratory answer is this: Deformation is strongly nonlinear in dome salt down to .1 MPa of live stress, and weakly nonlinear (slowly tending toward elastic behavior) as far below that as the data reach (~.001 MPa) (Tittmann, 1983).

Given the fact of inelasticity at low stress, models were adjusted: To better match measured Cowboy motions, Y_0 was cut 17-fold to 40 bars (Rimer and Cherry, 1982; the resultant break to linear elastic decay at $\sigma_r \approx 4.5$ MPa remains at odds with both Cowboy motion and the laboratory results above). The rationale for the cut was that dome-salt is water-saturated, and hence has no frictional strength, in accord with effective-stress theory. However, all analyses of Cowboy salt showed that it was very dry; also, many competent persons familiar with the Cowboy medium were consulted and all of them rejected the idea that it was saturated, or even wet (ATI file, 1980). By contrast, saturations well above 90% are needed experimentally to eliminate frictional strength in soils and rocks. Moreover, while effective-stress theory predicts such a loss of strength in materials whose solid skeletons are softer than water, the same loss would conflict with effective-stress theory for rocks of low porosity (≤1%) —even if 100% saturated —because their skeletons have always proven much stiffer than water. The same procedure has nonetheless been used to better match Piledriver motion (Cherry and Rimer, 1982).

For surface bursts, prediction began very poorly with events Middle Gust (Wright, Sandler and Baron, 1973) and Mixed Company (Ialongo, 1973), on clay/shale and siltstone/sandstone media, respectively. However, in perhaps the only other documented surface-burst prediction, accuracy approached that of Fig. 1 for the few near-surface pulses computed and reported (Misers Bluff Event II-1; Thomas, 1979). Yet, two caveats are in order, besides the dearth of comparisons between predicted and measured pulses: In the first place, prediction accuracy fell with distance from ground zero, and the extent to which fluid motion was observed out to the two stations of high accuracy is an open question. Secondly, motions were known beforehand for many relevant shots fired earlier on desert

soils (five at the Misers Bluff site); what can be left to predict after several
similar fields have been measured, is a sticky question that keeps growing in
importance (it might best be answered by basing a purely empirical prediction on
those fields). Nevertheless, the field predicted for Misers Bluff Event II-1
was much more accurate than those for Middle Gust and Mixed Company. In addition
motion is said to have been forecast well at the edge of the crater dug by a
burst on desert alluvium (Mill Race; 1981), and on a grout/wet tuff medium (Mini
Jade; 1983); as yet, results are not available in print for either field, but the
medium in each was quite likely quasi-fluid from ground zero to almost all gauge
stations.

As for postdiction, modelers may have produced an accurate account of motion
for Middle Gust Event 3 a few years ago, after i) minor model-adjustments, and
ii) sample-and-lab-test procedures were modified, giving rise to major changes in
the material properties data-base (the new model and postdiction are not yet published). However, while defects of a model are disclosed by its postdiction
errors, accurate postdiction is all but useless as a measure of its correctness.
To see why, let $\sigma_{ij}(\underline{x},t)$ denote stress in a given field at position $\underline{x}$ and time t,
where $\underline{x}$ lies on a region enclosed by some surface S; also, let $\sigma_{ij}^e(\underline{x},t)$ be any
time-ordered sequence of stress fields i) that satisfy the equation of static
stress equilibrium $[\sigma_{ij,j}^e = 0]$, and ii) whose tractions vanish all over S. It
then follows from the equations of continuum motion that the sum of the two
stress fields $[\sigma_{ij}(\underline{x},t) + \sigma_{ij}^e(\underline{x};t)]$ will produce exactly the same motion as
$\sigma_{ij}(\underline{x},t)$; neither the momentum nor density of any material element is affected
by the addition of $\sigma_{ij}^e(\underline{x};t)$ to the stress field. [Note: Temperature is affected,
but has not been measured; while internal energy and motion do not generally suffice to determine stress, they might in spherical fields.]

For example, exactly the same spherical motion is produced by two stress
fields (σ_r,σ_θ) and $(\hat{\sigma}_r,\hat{\sigma}_\theta)$ that differ as follows on the interval $r_1 \leq r \leq r_2$:

$$\hat{\sigma}_\theta = \sigma_\theta + f(r;t) \tag{1}$$

$$\hat{\sigma}_r = \sigma_r + (2/r^2) \int_{r_1}^{r} x\, f(x;t)dx \tag{2}$$

where f(r;t) is any function such that

$$f(r;0) = 0 \quad \text{and} \quad \int_{r_1}^{r_2} x\, f(x;t)dx = 0. \tag{3}$$

The first of Eqs. (3) insures that the two fields evolve from the same initial
(t = 0) conditions; the second dictates that $\hat{\sigma}_r = \sigma_r$ at $r = r_2$ [as at $r = r_1$; Eq.
(2)]. The number of functions f(r;t) satisfying Eqs. (3) is strongly infinite,
and remains so even if we require i) f(r;t) to be continuous in r, t and ii)
$\hat{\sigma}_\theta$ to equal σ_θ at r_1 and r_2 $[f(r_1;t) = 0 = f(r_2;t)]$.

For general fields, components of σ_{ij}^e are readily found as products of functions of the single Cartesian variables x, y and z:

$$\sigma_{11}^e = A_{11}fg''h'', \quad \sigma_{22}^e = A_{22}f''gh'', \quad \sigma_{33}^e = A_{33}f''g''h,$$

$$\sigma_{12}^e = A_{12}f'g'h'', \quad \sigma_{23}^e = A_{23}f''g'h', \quad \sigma_{31}^e = A_{31}f'g''h', \tag{4}$$

where f(x;t), g(y;t), h(z;t) are any functions twice-differentiable with respect to x, y and z, respectively, and primes (') denote their spatial differentiation; also

$$\sigma_{ji}^{e} = \sigma_{ij}^{e}; \quad A_{ji} \equiv A_{ij}; \quad A_{11} + A_{12} + A_{13} = 0. \tag{5}$$

Initial conditions will be unaffected by σ_{ij}^{e}, and its tractions will vanish over the faces of a rectangular parallelepiped RP at $x = \pm a$, $y = \pm b$, $z = \pm c$, if we require that

$$0 = f(x;0) = g(y;0) = h(z;0) \tag{6}$$

$$f = 0 = f' \quad \text{at} \quad x = \pm a; \ g = 0 = g' \quad \text{at} \quad y = \pm b; \ h = 0 = h' \quad \text{at} \quad z = \pm c. \tag{7}$$

If, in addition, $f'' = 0$ at $x = \pm a$, $g'' = 0$ at $y = \pm b$ and $h'' = 0$ at $z = \pm c$, then stress (not just traction) will be continuous at the surface of RP. Thus, for any spatial region in which RP can be embedded (and hence all regions of practical concern), the number of stress fields that will produce identical motions under the same initial and boundary conditions is again strongly infinite.

More general static equilibrium fields are obtained if the arbitrary space-time function $W(\underline{x};t)$ replaces f g h in Eqs. (4); components of σ_{ij} then become weighted partial derivatives of W ($\sigma_{11}^{e} = A_{11}W_{yyzz}$, $\sigma_{12}^{e} = A_{12}W_{xyzz}$, etc., where subscripts denote partial differentiation). However, only causal stress-strain relations are strictly admissible, and any stress field consistent with a given field of motion must be tied to the strain field implied by that motion. The added constraint of causality is too broad to discuss here, but enough progress has been made with it to show that there is no limit to the number of causal stress-strain fields consistent with a motion-field.

There are at least two special motions (uniaxial and nonradial-cylindrical) that uniquely determine a single stress component. Otherwise, forcing a constitutive model to reproduce a given field of motion makes it just one of an infinite set of models with that property. Within said set, departures from the stress-strain curves actually followed in the field can be small, moderate or large. Hence, a constitutive model that does not reproduce the motion observed will generally be more accurate than many models that do (the accuracy of a constitutive model is set by its stress-strain curves, not by the motions those curves imply). The fact that correct postdiction of a field of motion flows from certain mechanistic assumptions therefore cannot rationally be taken as evidence of their validity; the job of validation is not that easy.

3.   STRAIN PATHS IN NEARLY SPHERICAL FIELDS

The criteria for model-validation are clear (section 1), but only a detailed account of present models and their growth can show how formidable the task itself has become. As an example, recall that flow rules (just one of the models' major elements) have evolved from Prandtl-Reuss form, to the rule "associated" with yield surfaces (via normality of plastic strain increments), to recent rules that forgo normality — slowly growing complexity whose experimental basis has been very difficult to establish, even in the laboratory. Thus, three hard facts make it unlikely that mechanistic models will soon yield reliable predictions of motion driven by explosions in geo-solids: i) Their record of prediction is poor. ii) Postdiction is unacceptable as a means of validating them. iii) Validation of model-elements one by one for typical materials is a difficult, long-range task that has hardly begun. The ongoing need for reliable predictions — despite these facts — forced us to ask how one might get them even without improved

theoretical bases for models. Accordingly, about ten years ago, we re-examined
the role of constitutive models in ground-motion prediction.

In calculations of ground motion, constitutive relations serve simply to
supply the stress increments produced in material elements as they deform. To
the extent that the resulting stress-strain curves do not come from mechanistic
models (i.e., from theory), only experiments can furnish them. Of course, the
idea of measuring most of the stress-strain data needed in a computational model
for reliable prediction seemed wholly impractical. Yet, explosions clearly subject host media to highly specialized loads, and the resulting stress-strain
fields should reflect that limitation; stress states in soil under the corner of
a building (say) are not produced by a contained, nearly spherical burst in the
same medium. We were thus led to ask just what stress-strain data would suffice
for the sole purpose of predicting explosively driven ground motion.

An answer was sought at first in terms of the paths traced by material elements in stress space ("stress paths"), mainly because attention in modeling was
then focused on yield surfaces. The spectrum of calculated paths was surveyed
in terms of path-shape and orientation, or "pattern," since those properties —
and not amplitude — tell us what kind of loading a path represents [for example,
states of stress are hydrostatic not because of their amplitudes, but because
all three principal stresses are equal in such states (regardless of amplitude),
and hence lie in stress space on a straight line whose direction cosines equal
$3^{-\frac{1}{2}}$ along normal-stress axes]. The study was halted, however, when it became

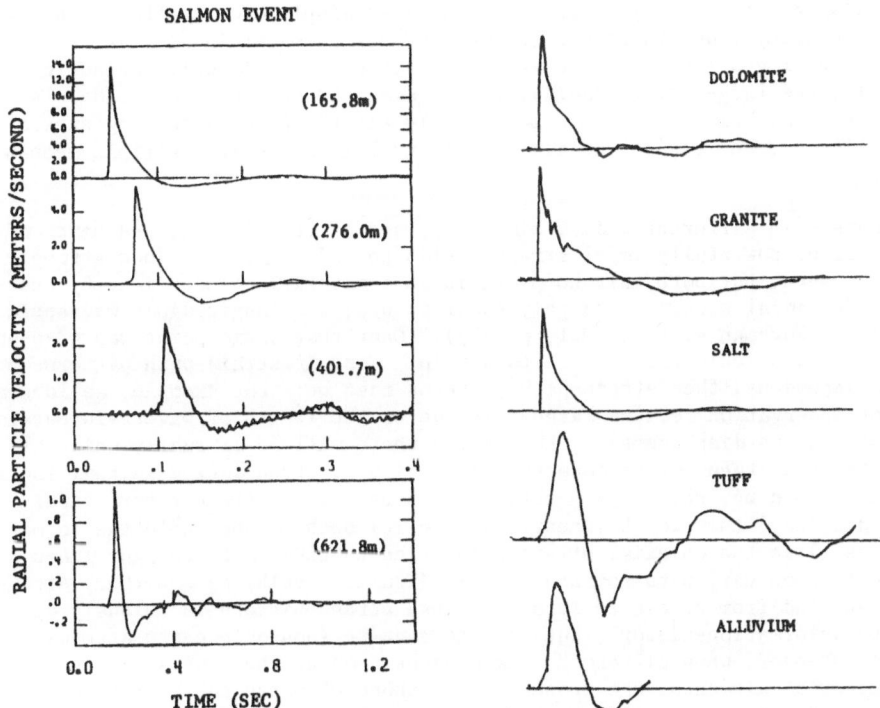

FIG. 2. Measured Wholespace Motion: Radial Particle Velocity vs. Time in
         Fields Thought to be Spherically Symmetric. Left: pulses at various
         distances from the center of the Salmon burst in dome-salt. Right:
         one pulse from each of five buried bursts in different media.

evident that strain paths, not stress paths, held the answer sought (Trulio, 1975). Once comprehended, the telltale clue to that finding was the similarity of velocity pulses as functions of range, from one deep burst to another, whether measured or calculated, and regardless of medium. Velocity pulses from the Salmon event (Fig. 2; Perret, 1967) illustrate the slowly changing waveform typical of spherical motion from buried bursts; also shown in Fig. 2 are measured pulses of material velocity for bursts in five different media ranging from hard rock to dry soil.

To account for the similarity of the pulse shapes in Fig. 2, we note first that compact energy sources in roughly homogeneous media create nearly spherical fields in which material is first driven outward from the source. Soon thereafter, strong velocity divergence (implicit in spherical symmetry) gives rise to inward accelerations, but inertia carries each material element beyond its final position before its outward movement is arrested. Then, following some inward motion, spherical convergence leads to outward acceleration, etc. In elastic media, the resulting radial oscillations about final positions are heavily damped by the loss of radiated energy; even stronger damping occurs in the inelastic media nature offers. Thus, radial velocity pulses from deep shots consist of an initial positive lobe (outward motion), a negative lobe of lower amplitude, and not much else.

An increase in negative-lobe size with radius, relative to that of the positive lobe (Fig. 2; Salmon pulses), is also quite general. In elastic media, velocity pulses (being transients) decay in amplitude in proportion to $1/r$, while permanent displacements decay as $1/r^2$; the opposed displacements effected by each of the two velocity lobes therefore become more nearly equal in size as range increases. In real geo-materials, outward displacements are largely inelastic where strains are large (small radii), heightening the effect in question — and the tendency of materials to become more nearly elastic with decreasing stress-strain amplitude makes the effect almost universal (despite its multiple mechanistic causes).

For bursts in different media, similar pulses of velocity, U, and hence of displacement, D, powerfully impel strain fields toward similarity (how strongly will soon be seen, but note that hoop strain in a material element is very nearly D/r, while radial strain is roughly equal to U/c; $c \equiv$ longitudinal wavespeed (Trulio, 1977; Workman et al., 1981, p. 23)). Once that basic point was grasped, emphasis shifted quickly from stress to strain. For, if strain-path patterns are not medium-dependent, then stress-path patterns must be: For example, as long as a radius of observation remains large relative to the length of wavetrain outside that radius (and to displacement), the strain there will be virtually uniaxial. However, stress is then not so constrained or simple. Even during initial shock loading and unloading, stress-path patterns can be quite different for linear elastic media (say) than for bilinear (hysteretic) ones — and while the former stress paths (like the uniaxial strain paths tied to them) are straight lines, their orientations vary with Poisson's ratio (Trulio, 1981b, pp. 36-10). Attention also shifted from stress to strain because stress-history is uniquely determined by deformation-history, but not the reverse (non-uniqueness attends yielding); moreover, when different strain paths produce the same stress path in a set of material elements, the properties of those elements can thereafter differ greatly. In short, specifying stress paths is not adequate to determine the stress-strain curves followed by material in a given field.

Since strain-path patterns do not recur precisely in different media, no strict theorem can underlie the physical plausibility (above) of their near

A. Paths from Measured Motion   **B. Calculated Paths**

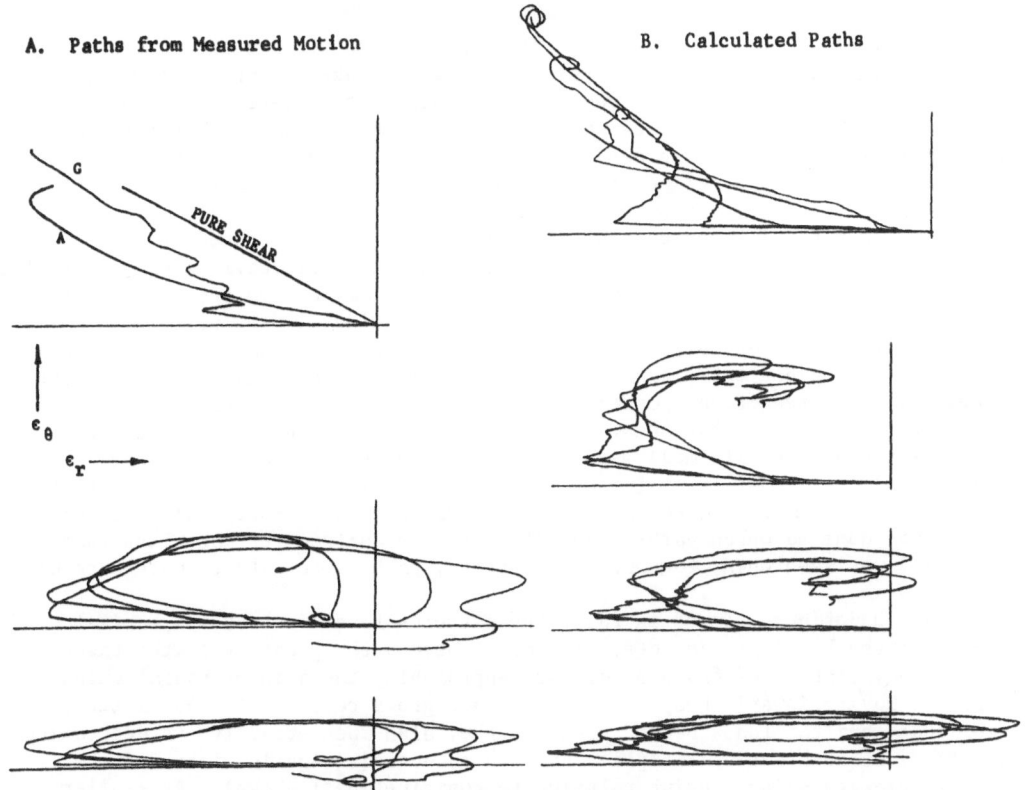

FIG. 3.   Strain Paths Deduced from Measured Motions in Various Media (left),
          and Calculated Strain Paths (right), Assuming Spherical Symmetry.
          Hoop strain, ε_h, is plotted vs. radial strain, ε_r. On the left,
          the topmost curves refer to granite (G) and alluvium (A), and the
          others to tuff and dome-salt.

medium-independence. For instance, strain-path fields are exactly known around
suddenly-pressurized spherical holes in linear elastic isotropic media, and they
vary with medium (loc. cit.). Hence, to quantify the similarity of strain-path
fields, the fields themselves had to be examined. In so doing, the shift from
stress to strain had critical practical effect: Strain paths in situ can be deduced from measured ground motions, whereas gauges for measuring stress in situ
are still under development. In particular, the slow change in waveform with
radius seen on the left of Fig. 2, and the smooth (if rapid) fall of amplitude
with range, make accurate interpolation possible among pulses differing in
radius by a factor of two or more; from discrete pulses, material velocity is
thus calculable as a function of range, and with it displacement-, and hence
strain-fields. At least two interpolation procedures are in use (Seaman, 1974;
Workman et al., 1981, p. 9), but the set of strain paths obtained from gauge output is nonetheless small because not many events have yielded pulses adequate to
define their velocity fields. Fig. 3(A) contains paths from those events, which
took place in granite, dome-salt, a desert alluvium (Perret et al., 1963), and
an ash-flow tuff. Given spherical symmetry, strain paths can be defined in full
by plotting hoop strain (ordinate) vs. radial strain (abscissa). However, for
the reason cited above, only path patterns (shape and orientation) are so displayed in Fig. 3; in any one plot on its left, strain amplitudes vary as much as
a factor of 5 (peak velocities on the right of Fig. 2 span a factor of ~40).

The events contributing to Fig. 3(A) include nuclear and chemical bursts, each in both tamped and cavity settings. Care was taken to obtain (by eye) matching patterns for the two lower plots, though closer matches are possible. The largest pattern-differences occur toward the ends of paths, but those are the paths' shakiest parts; ground motion gauges are prone to register non-zero final velocities or "baseline shifts," whose correction (as in the pulses of Fig. 2) always entails guesswork. In the upper plot of Fig. 3(A), pattern matching was minimal because granite's velocity field was defined by just the two best-certified wholespace pulses for that medium (Piledriver event); further, the use of those pulses to define spherical motion may be improper, since large nonradial motions were measured for Piledriver (Trulio, 1981b, p.12-12). Calculated spherical fields, of course, have perfect symmetry, and discretization error has negligible effect on strain paths, even at their ends. Also, though unreliable for prediction, constitutive models do describe materials whose behavior tracks that of real ones. Hence, in testing the thesis that strain-path patterns are only weakly medium-dependent, the paths from calculated fields (Fig. 3(B)) are almost as useful as those of Fig. 3(A). The patterns in Fig. 3(B) (none was calculated for granite) scatter more than in Fig. 3(A) because little was done to match paths from different fields; scatter more like that of Fig. 3(A) was seen when, for one of the plots, paths were matched more carefully.

Fig. 3 defines, almost in full, the patterns found in nearly spherical fields driven by single charges. At ranges beyond the greatest in the figures, the already flat paths flatten further, approaching the axis of radial uniaxial strain. Moving inward, hoop-strain-amplitude grows relative to radial-strain-amplitude, and the loops become rounder; they also open steadily, as decreasing fractions of peak outward displacement are recovered (negative lobes of the radial velocity pulses shrink relative to positive ones; above). At smaller radii than in Fig. 3, there is little further change in path pattern: Large outward displacement of a spherical shell gives it both tensile hoop strain and compressive radial strain that exceed even its initial compressive radial strain (uniaxial) due to shock loading; later, radial and hoop strains dwarf cubical dilatation, whence strain paths become curves of almost pure shear (a straight line of slope $-\frac{1}{2}$ in Fig. 3, if natural strains are plotted).

The empirical evidence that strain-path patterns are similar regardless of medium and source type (Fig. 3) is telling, though not profuse. Beyond that, the case for strain-field similarity has been strengthened greatly by showing that burst geometry underlies it. Summarizing the key points (above): Mainly for geometric reasons, a) the radius-vector is a strain axis and the other two principal strains are equal, b) point-like sources quickly radiate energy away into a surrounding wholespace, forcing simplicity on velocity wave-forms and hence on strain-path patterns, c) only uniaxial strain is seen at large radii, d) at small radii, strain is at first nearly uniaxial and then changes to nearly pure shear, and e) recovery from peak outward displacement becomes more complete with increasing radius, whence (plotting hoop strain vs. radial strain) open paths [d)] gradually become closed flat loops [c)]. Within these constraints, path patterns are most medium-specific during initial straining, and on open paths. For stiffer, more elastic materials, uniaxial radial compression is then partly relieved before geometry forces further radial compression. However, the effect is plain in Fig. 3(B), muted in Fig. 3(A), and hence may demand a degree of elasticity scarce in real geo-materials.

That the paths traced in different bursts and media are similar is a result of signal importance. To the extent that their patterns are the same, the

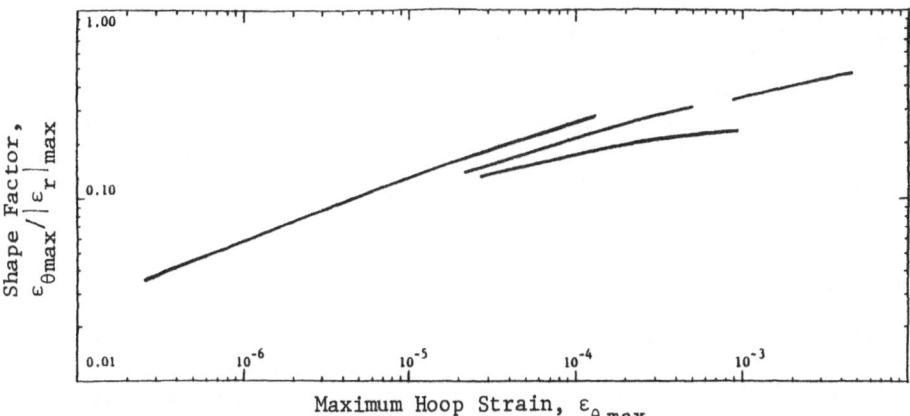

FIG. 4.   Shapes of Strain Paths as Functions of Maximum Hoop Strain, Deduced
from Motion Measured in Fields Considered Spherically Symmetric.
Shape is defined as the ratio of maximum tensile hoop strain $\varepsilon_{\theta max}$,
to maximum compressive radial strain, ε_{rmax}, excluding paths like
those in the top row of Fig. 3.

deformations they represent become independent of source and medium. So, therefore, do the kinds of material properties tests needed to supply stress-strain
curves sufficient for reliable *prediction*. Also, since the path-patterns in
those tests form a simple set, interpolation among them may be accurate even if
only a few are traversed. True, amplitude can be uncertain by as much as an
order of magnitude for a given path-pattern, but mainly because path-shape (like
velocity waveform) is insensitive to amplitude: Using peak-strain-ratio
(hoop ÷ radial) to measure the shapes of paths with radial-strain recovery of
50% or more (as in all but the top plots of Fig. 3), shape was found to vary by
a factor < 2 at fixed amplitude (Fig. 4; Workman et al., 1981, p. 33); variability of shape for open paths (Fig. 3; top plots), though less well studied, appears no greater than for closed ones. Hence, the present matrix of strain-path
tests calls for stress measurement on two or three different paths for each amplitude of a set that covers a range of interest. In practice, the number of
tests can probably be cut even now by forming separate test-matrices for soil,
soft rock and hard rock. Moreover, while tests have yet to be made over a full
matrix, stress has been measured at a few laboratories on typical paths, both
open and closed (Ko and Scavuzzo, 1981; Bogart and Schatz, 1982; Akers, 1983;
Lade, 1983). Thus, with present hardware and knowledge of strain paths, it now
appears feasible to measure all that needs to be known about material properties
for accurate prediction (not just postdiction) of spherical motion around buried
bursts — *if* the resulting stress-strain curves apply to geo-materials in situ.

This last issue cannot be decided short of in-situ measurement of stress-
strain curves. Strain is accessible in nearly spherical fields (above), but in-
situ stress measurement is frustrated by the conflicting requirements that, at
gauge faces, both tangential straining of the medium and its shear stress be unaffected by the gauge. After years of development, gauges are now becoming operational that can, in principle, record a principal free-field stress without
serious distortion, if the allied stress-axis is fixed and known (Keough,
DeCarli and Rosenberg, 1976). Those gauges were exercised very recently, along
with ground-motion gauges, in the field of a buried spherical charge. Earlier,
using samples of the medium near the charge, stress had been measured in the
laboratory along expected strain paths; similar measurements will be made on

strain-paths that the ground-motion records imply at stress-gauge locations. The
test plan also calls for calculation of the impulse delivered to material elements (using stress-gauge output) and the momenta of those elements (using
motion-gauge output); the fractional difference between impulse and momentum can
presumably be ascribed to errors in stress-gauge output. To that accuracy, differences between lab and in situ behavior will then be known, and sample-and-lab
test procedures can be modified in order to reduce them.

4.   STRAIN PATHS IN AXISYMMETRIC FIELDS: SURFACE BURSTS

     At the next level of complexity — but a long step up — lay axisymmetric
fields calculated for surface bursts on layered and homogeneous media. Yet,
study of those fields soon showed (Workman et al., 1978) that they were much
simpler, in one respect, than for general torsionless axisymmetric motion (i.e.,
with no rotation about the symmetry axis): Near the ground surface, at distances
$\gtrsim$ 1.5 crater radii from the burst, hoop strain is the smallest principal strain
by a factor almost always > 4 and usually > 8; for the few pertinent strain paths

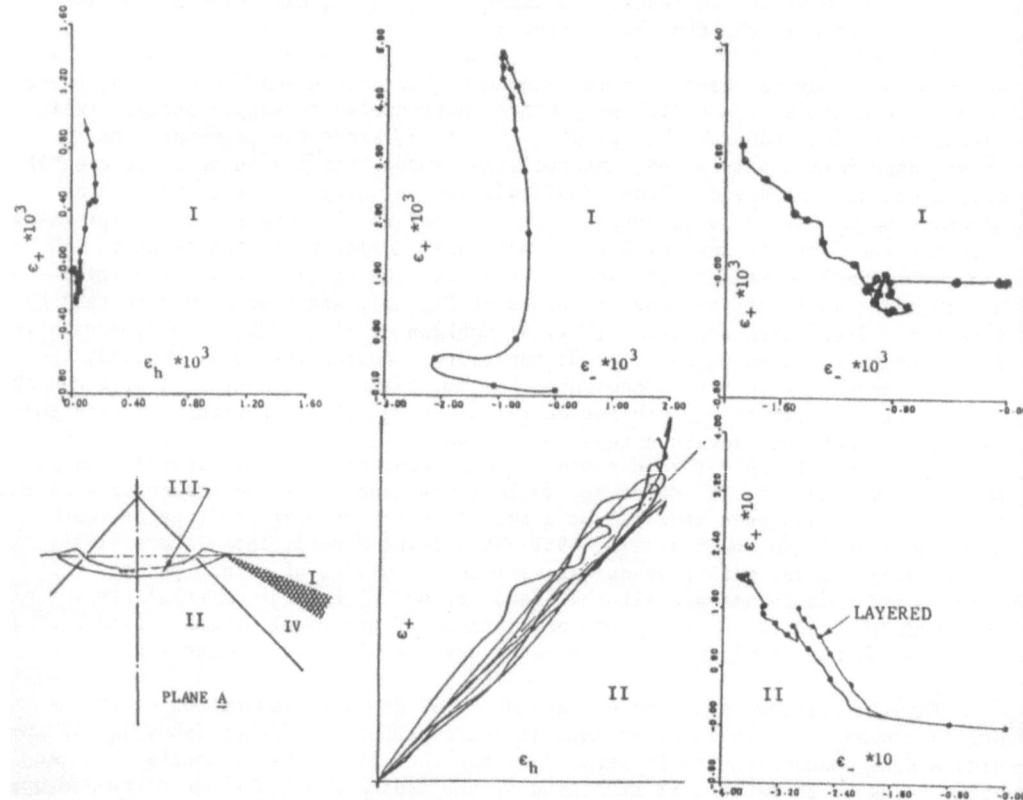

FIG. 5.   Surface Bursts: Principal Regions of Deformation (lower left)
          Assuming Axial Symmetry about the Vertical through Ground Zero.
          Typical paths in principal-strain space are shown for Regions I
          and II. Hoop strain $\equiv \varepsilon_h$, while ε_+ is the more tensile of the
          other two principal strains.

that have been measured, the factor is larger still. Hence, in the region noted, deformation can be approximated as plane strain (no hoop strain). Strain paths then do not wind and twist through principal-strain space, but occupy just a single plane; for any material element, the hoop-strain axis is normal to that plane, which, in physical space, therefore runs parallel to the plane $\underline{A}$ that contains both the element and the axis of symmetry. As in the case of strain paths for deep bursts, this simplification too has geometric roots: When a wavetrain passes through matter, torsionless axisymmetric motion becomes planar as horizontal range becomes large relative to wavetrain-length (and to displacement).

In plane $\underline{A}$, the more compressive principal strain and the more tensile are denoted ε_- and ε_+, respectively; hoop strain $\equiv \varepsilon_h$. The upper left-hand plot of Fig. 5 shows, typically, that ε_h is small near the surface in calculated fields, relative to ε_+ or ε_-. However, with ε_- as abscissa and ε_+ as ordinate, near-surface strain paths present patterns much more diverse than those of deep bursts. The kind of pattern seen most often (over 3/4 of the time) — also represented in Fig. 5 (top, center and right) — starts with compressive, nearly-uniaxial strain (along the ε_- axis); then, slight-to-full uniaxial recovery is cut short by a cross between pure shear and uniaxial stretch (along the ε_+ axis). Such overall patterns are often the net result of many brief excursions about them, in which discretization error plays no small part. On the few paths measured, uniaxial stretch prevails over pure shear, and there are relatively few jogs about the overall path-direction (Fig. 5, upper right); strain axes rotate through a large angle during the switch to uniaxial stretch (which thus can even amount to a final stage of initial uniaxial strain, though we see no reason to expect paths that simple).

The strain-path qualities noted are especially in evidence up to the time, t_D, when peak downward vertical displacement occurs. That period has been one of prime concern over free fields. Until time t_D, simple strain paths are found over most of the field and not just on its near-surface region. In particular, let the field be divided into two parts by a right circular cone, C, that opens downward and has a vertical axis, a half-angle of 45°, and a vertex above the burst point at a height of one crater radius (Fig. 5, lower left; Trulio and Port, 1982). The near-surface region, denoted "I," lies entirely outside C. Inside C and down to about ten crater depths (region II), the paths followed up to time t_D are open ones like those generated by deep bursts (section 3 above) except that compressive uniaxial strain is followed not by pure shear, but by a combination of shear and volume expansion; for both layered and monolithic sites, Fig. 5 shows that, on region II, patterns in the $(\varepsilon_-, \varepsilon_+)$-plane (lower right) do resemble those at the top of Fig. 3, and that $\varepsilon_h \approx \varepsilon_+$ (bottom, center). We note also that inside C (except for a small region where ejecta form), peak displacement and peak vertical displacement both occur at time t_D. While these results hold for all calculated fields known to us, pertinent measurements on region II are being attempted for the first time in an explosive event, as this is written.

Strain paths are depicted in Fig. 6 over a key part of a field calculated for a chemical burst on monolithic desert "alluvium." Paths like those just deemed typical of region I, and likewise II, are evident in the figure; not shown, but verified, is near-equality of ε_h and ε_+ for the plots that apply to region II (e.g., one path of Fig. 5 (bottom, center) belongs with those plots). On a given path, "v" marks the point of peak downward vertical displacement due to initial loading and relief; larger downward displacements occur much later in

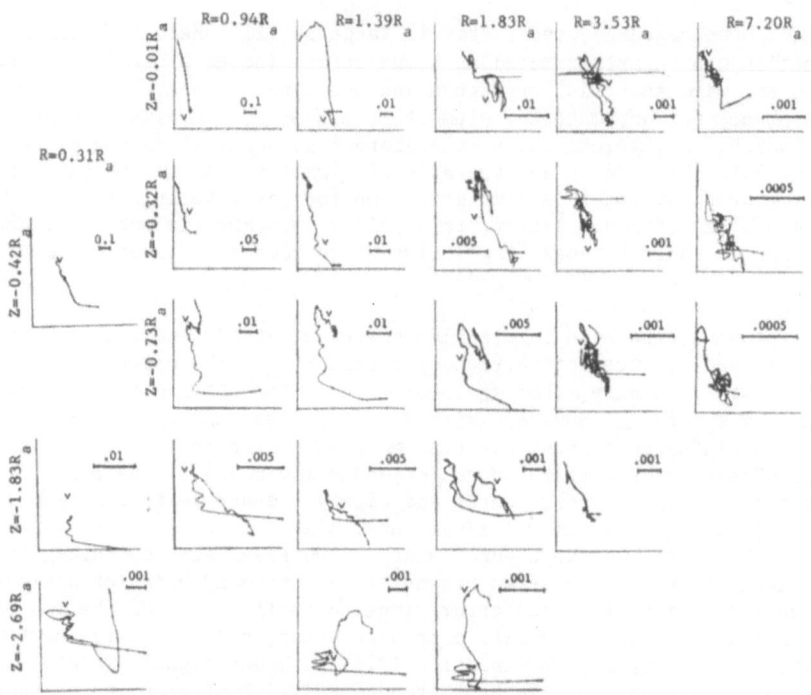

FIG. 6.   Calculated Principal-Strain Paths for a Chemical Explosion on Mono-
          lithic Alluvium. The strain axes lie in Plane $\underline{A}$ of Fig. 5; ε_+ is
          the more tensile of the two principal strains in that plane. Peak
          vertical displacement occurs on each path at point "v." Horizontal
          range, R, and altitude, Z, are given in units of the apparent crater
          radius, R_a. Strain scales are defined on the horizontal line-segment
          in each plot (.0x ≡ x%); dots on any path mark equal time increments.

a few cases and point v was not reached in a few others. With rare exceptions
motion grows more complex as it unfolds, and discretization error probably accounts more and more for the paths' detailed wriggling. However, it is not known
how exact paths for the problem posed differ from these, because a) exact solutions to such axisymmetric problems are lacking and b) time-marching methods give
numerical fields whose slow, non-uniform convergence makes it very expensive to
put close bounds on their errors by mesh refinement. Larger errors than in Fig.
6 are likely to enter the paths of Fig. 7, which were calculated for a layered
alluvial halfspace. While most patterns in Fig. 7 still show typical region I
features beyond time t_D, they tend to be more elaborate than in Fig. 6 even to
that time; layering no doubt has such an effect, in which case the effect itself
is an added source of numerical error.

     Measured motion is also subject to error, and to the further realities of
local inhomogeneity and global asymmetry. Still, in the main, we obtained strain
paths like those connected above with region I, when the first (and only) attempt
was made to measure them in situ (Trulio, 1982). For that purpose, a cluster of
gauges was deployed around each point M of strain-path measurement: Horizontal
and vertical velocities at three points of the plane $\underline{A}$ (above) define a spatially
linear field over the triangle formed by those points — and hence an instantaneous strain rate at M. Since redundancy is needed both as a hedge against gauge
failure and for error assessment, each cluster actually contained four gauge-

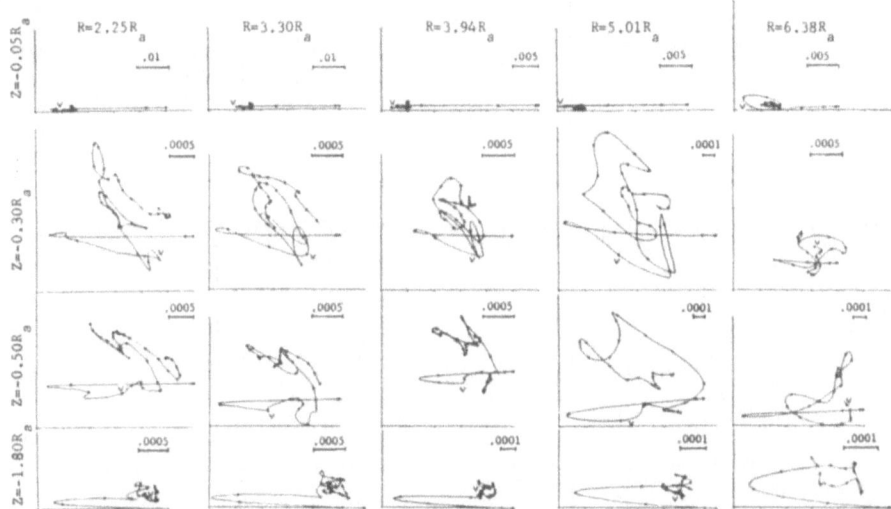

FIG. 7. Calculated Principal-Strain Paths in a Layered Half-space Loaded by
Nuclear Airblast. The notation and conventions of Fig. 6 are used
here. Layer materials include clay, sand, shale, sandstone and
siltstone.

pairs, giving it a maximum output of sixteen ideally redundant paths. Six clusters were emplaced at the same small depth along two horizontal lines, $2\frac{1}{2}$, $3\frac{3}{4}$
and 5 crater radii from the burst. Over the 44 paths actually obtained after
gauge attrition, initial uniaxial strain was clearly followed on 20 by uniaxial
stretch, on 6 by shear, and on 5 by a mixture of shear and stretch; as many as
13 had unexpected shapes (of those, all but 2 were computed using one suspect
gauge-record or more; most of the gauge canisters proved to be incompletely
grouted to the medium, but no relation was seen between bonding and gauge-record-
quality; also, as measured by one extra gauge in each cluster, hoop motion was
not significant).

A sizable and unvarnished sample of the measured paths is shown in Fig. 8.
The top and middle rows of the figure, respectively, contain ideally redundant
paths from clusters on the same line, $2\frac{1}{2}$ and 5 radii from the shot; the two left-
most paths on the bottom row came from the central cluster on that line, while
the rest are from different clusters on the other line. The one $(\varepsilon_+, \varepsilon_h)$-path
shown for each path location of the figure is typical of the clusters in that row
(the curve at the bottom right and the one next to it are orthogonal projections
of the same path). The similarity between most $(\varepsilon_+, \varepsilon_-)$-paths of this set, and
those of Figs. 6 and 7, is plain, and one of their more notable differences is
pleasing: the measured paths are less spastic than those calculated.

Ejecta come from the upper reaches of a "crater region," denoted III in
Fig. 5, of the same shape as the apparent crater but having 1.5 times its radius
and depth. Up to time t_D, region III bears such high overpressures in nuclear
bursts that shear stress can probably be ignored there for many media; if so,
then plate-slap and hydrostatic experiments in the laboratory can supply the
region's essential constitutive properties. Moreover, the methods used to measure stress-strain curves for deep bursts (section 3) should serve equally well
in region II. On region I, however, it is much harder to measure strain (as just

266   J. G. Trulio

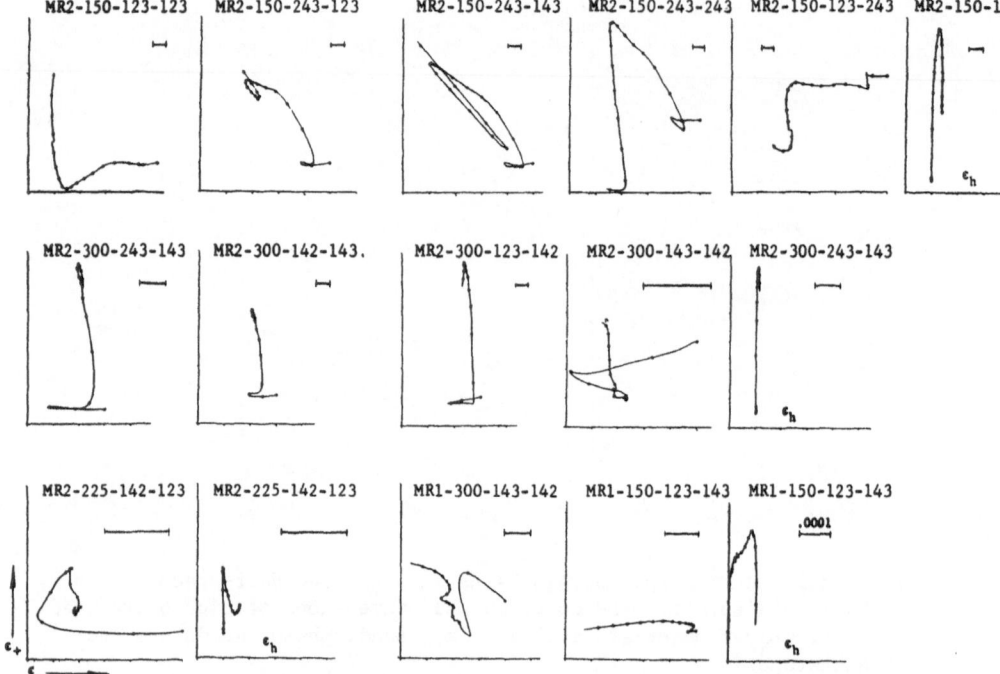

FIG. 8.   Measured Principal Strain Paths for a Chemical Explosion on Dry
          Desert Alluvium (Mill Race Event). Hoop strain $\equiv \epsilon_h$; ϵ_+ is the
          more tensile of the other two principal strains (ϵ_-,ϵ_+). As or-
          dinate, ϵ_+ appears in every plot; plots with ϵ_h as abscissa are
          so labeled, and the abscissa is ϵ_- in all other cases. Field loca-
          tions are designated by the first four digits of the path label;
          e.g., MR2-300-243-143 applies to a point on gauge-line 2, 300 feet
          (4.8 apparent crater radii) from ground zero. Lines 1 and 2 are
          89° apart and all points of strain-path measurement are 6 meters
          deep. Except for the lower right-hand plot, the horizontal line-
          segment in each plot defines a strain of .001 (.1%).

seen), while laboratory measurement of stress is essential because (like strain)
it has fixed, known axes only in the hoop direction, and very close to the symmetry axis.

     Observed axis-rotations on region I, which are plotted in Fig. 9 for each
(ϵ_-,ϵ_+)-path in the two top rows of Fig. 8, again follow a general pattern. That
pattern covers a host of detailed shapes, and is not fit by some rotations, but
is nonetheless widespread and distinctive enough to restrict sharply the kinds of
stress-strain curves needed for surface-burst prediction. Specifically, the
angle θ from the upward vertical in plane $\underline{A}$ to the ϵ_- axis quickly assumes a
value θ_0 dictated by the normal direction to the wavefront (only $\theta - \theta_0$ matters;
the vertical is an arbitrary reference line). Then θ decreases to a minimum θ_{min}
and increases to a value $\theta_{max} > \theta_0$ that persists until well after time t_D. The
angles $\theta_0 - \theta_{min}$ and $\theta_{max} - \theta_0$ vary from path to path, as do the (ϵ_-,ϵ_+)
changes subtended by θ_0, θ_{min} and θ_{max}. In fact, the basic pattern in the
(ϵ_-,ϵ_+) plane (Figs. 6-8) itself contains variables —principally amplitude,
degree of recovery from initial uniaxial compression, and the proportions of pure
shear and uniaxial stretch in subsequent deformation. Thus, on balance, path
variations in region I appear too diverse to be covered fully by interpolation

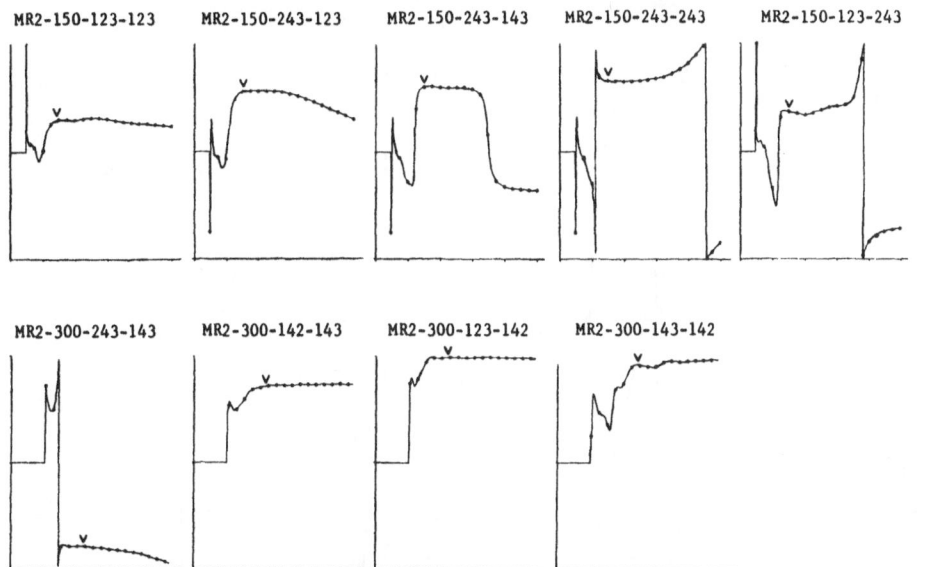

FIG. 9. Measured Rotation of Strain Axes in the Vertical Plane through both
         Ground Zero and any Given Point of Strain Path Measurement (Mill
         Race Event). Axial symmetry is assumed about the vertical, V,
         through ground zero. The rotation angle θ is measured from V to the
         axis of most compressive strain. A plot is shown for each strain
         path in the top and middle row of Fig. 8; peak vertical displacement
         occurs at about the point "v."

among measured stress-strain curves. Moreover, to force material around given
plane-strain paths when axes rotate, is beyond the state of the art. Even in the
laboratory, stress can be measured only on prescribed principal-strain paths
(though of quite general shape). Indeed, up to now, models of geo-materials have
all been based on stress-strain tests with fixed stress and/or strain axes; the
models' ability to deal with axis-rotation has not yet been put to any direct
test —a large gap in model validation.

    Even before laboratory equipment is developed to effect prescribed rotations
of plane-strain axes, stress can be measured in situ on principal-strain paths
like those of Figs. 6-8, and compared to corresponding laboratory data. In particular, Cylindrical In Situ Test (CIST) events generate such paths when they
drive cylindrical radial motion. In each of the many CIST events to date, chemical explosive filling a borehole has been detonated all at once (ideally) along
its vertical centerline. On any plane bounded by the centerline, a roughly triangular region T is formed by that line and two others, namely, the lines on
which a first-arriving wave from the cavity meets first-arriving waves from i)
the ground surface, or ii) the bottom of the charge-hole. In a homogeneous
medium of suitable isotropy, radial cylindrical motion lasts at each point of T
for a time proportional to the length of explosive. Such motion proceeds in
stages much like those outlined above for spherical fields near charges (section
3), but with one basic difference: No two principal strains are equal; instead,
one of them (vertical) is zero, and only plane strain occurs. Thus, the initial
stage of compressive radial strain (ε_-) and possible uniaxial recovery, gives
way to hoop stretch (ε_+), in qualitative agreement with the typical region I
pattern.

    To obtain paths at points of T, we turned to two computed fields (few pulses

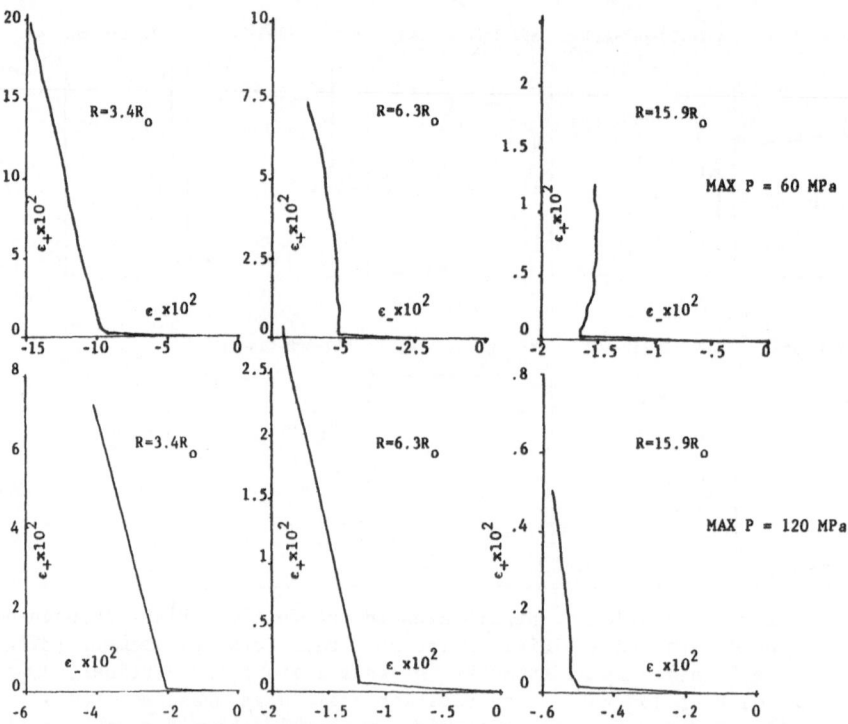

FIG. 10.   Strain Paths in Cylindrical Radial Fields Driven by Chemical
           Explosions. Motion is the same in all planes normal to a fixed
           vertical line, and is radially directed from that line. In any
           such plane, ε_+ and ε_- are hoop and radial strains, respectively.
           The paths on the first and second lines refer to fields driven
           by explosives with peak pressures of 60 and 120 MPa, respectively.

have been measured on T in CIST events). The two fields differ in the pressures
reached by their assumed explosives (noted in Fig. 10). Figures 8 and 10 make
quite plausible the use of CIST events to generate $(\varepsilon_-, \varepsilon_+)$-paths like those of
region I; further, since stress axes are fixed and known where such CIST paths
occur, stress can be measured there. Hence, while new hardware is needed to
deal with strain-axis rotation, only paths on region IV (a zone of transition
between spherical and plane strain) appear unsuited to stress-strain measurement.

5.   AGENDA FOR STRAIN-PATH MODELING

     For surface bursts, the near-term goals of strain-path work are: i) to bet-
ter define the bounds of regions I-III and the paths found on those regions; ii)
to pare the set of path-shape parameters that now apply to each region (e.g., by
finding relations among them); and iii) to divide geo-materials into classes
(e.g., soil, soft rock and hard rock) within which strain paths form relatively
simple sets. For these purposes, calculations can be almost as effective as
measurements, and offer timely access to fields in highly diverse media. Still,
strain paths must also be measured in situ, if only to confirm more fully that
burst geometry is the main determinant of their patterns. That task, and also
the day of reliable models, would be speeded if strain paths were measured in
situ by a simpler, cheaper scheme than the cluster method — but one no less clear
in principle.

For near-surface paths (region I), measuring all stress components is not
feasible outside the laboratory. Even there, controlled strain-axis rotation
must first be achieved. Meanwhile, hoop stress can be measured in situ; so can
cylindrical radial motion and its principal stresses, in CIST-like events. To
obtain the resulting stress-strain curves in the laboratory as well, is both the
immediate reason for seeking improved sample-and-test procedures, and the proof
that they work. Still, such proof, for which the curves in question are key
(from CIST and surface-burst events), will not include rotation of axes unless
a new way is found to measure stress in situ.

For region II, where all calculated paths are much simpler than on region I
(through peak vertical displacement), it may be possible to limit in-situ stress-
strain measurement to one event in each major type of geo-material. The first
suitably-instrumented shot was fired in dry desert alluvium; its data (not yet
available) extend from region II into region III. On strain paths in region IV,
only hoop stress seems measurable [$(\varepsilon_+, \varepsilon_-)$-axes rotate, and strain is not plane],
but even a subset of stress-strain curves puts a needed rein on models. Further,
all stress-strain curves measured for in-situ material, either directly or by
proven laboratory methods, become part of the strain-path model for that material. In addition, some of them form the data-base for mechanistic models — and
predicting the rest is both the measure of such a model's value and a guide to
its improvement. Laboratory stress-strain curves meet these needs about as well
as curves measured on site; for, even if the two sets of curves differ appreciably, both apply to real geo-substances (measurement error aside).

When a mechanistic model truly predicts the stresses measured along key
strain paths, it then remains to measure its accuracy in predicting actual
ground motion and stress in a surface-burst event. Thus, each such event is
both a source of in-situ stress-strain curves and a gauge of model-accuracy.
However, for the first one (in dry alluvium; above) the prior steps of laboratory
measurement of stress on key strain paths, and evaluation/adjustment of models,
were not taken. They should be carried out for future events, of which a surface
burst now appears most cost-effective: Besides its twin uses (just noted) in
developing models of such events, it can also help to validate models for contained, nearly spherical bursts, whose strain paths share basic properties with
the paths of region II.

Much of this agenda is designed to boost knowledge of strain paths in surface bursts to a level already reached for contained, nearly spherical ones. At
that level, generating new strain-path fields is secondary: For contained bursts
the main task is to fix the accuracy of ground-motion predictions made with
models based on radial and hoop stresses measured in situ (and perhaps in the
laboratory) along paths typical of Fig. 3. Clearly, such models bypass the questions of mechanism that dominate conventional modeling. Yet, success with
strain-path methods will not obviate those questions. At a minimum, more insight into response mechanisms will permit more accurate interpolation among
stress-strain curves, thereby reducing the number and cost of the curves needed
for strain-path models. Also, though the set of curves required for free-field
prediction is surprisingly small, it is clear (for surface-bursts at least;
above) that models must have some power to extrapolate from the curves that are
measurable. By design, strain-path modeling reduces the need for such power;
critical stress-strain curves are furnished by experiment. As a result, full
advantage can be taken of mechanistic models, without letting their faults limit
the reliability of ground-motion predictions.

Progress in modeling would be more rapid if complexity implied understand-

ing. As it is, no two sites are identical, and the mechanistic models used to
predict general ground motion are quilts whose every patch is intricate. Hence,
while geo-materials *are* mechanically complex, we forgo knowledge of their dynamics when we try to apply all-purpose models to explosions in the earth, without
showing that each mechanism in a given model really does affect key stress-
strain curves as the model says. However difficult, such validation of each
major model-element is the sine qua non of mechanistic modeling. Yet, for nearly
spherical fields —and hence the source problem in nuclear monitoring — that task
is of lower priority than strain-path modeling because:

   i) The pertinent spherical strain-path fields are known and simple.
   ii) It is practical — now — to measure stresses on those paths, both
       in situ and in the laboratory.

   There is little reason and no need to press on with unvalidated models and
postdiction. Headway will be made more surely, steadily and quickly by using
already-adequate resources to create strain-path models — i.e., to a) measure
stress in the laboratory on key strain paths, b) evaluate mechanistic models by
comparing their stresses to those measured, c) modify a convenient mechanistic
model, arbitrarily, so that it fits the measured stress-strain curves, d) measure
stress and motion in situ, e) deduce in-situ stress-strain curves, and f) adjust
sample-and-test procedures so that in-situ stresses are obtained in the laboratory. The accuracy of the strain-path model [c)] is then found by comparing predicted and measured [d)] stresses and motions; so is the accuracy of the stress-
strain curves given by current mechanistic models —a true overall test of their
validity.

   A pass or two through steps a)–f) will tell us, for any given medium, how
accurately buried bursts can be defined as seismic sources, and how to predict
them that well (three media predominate in nuclear monitoring). The process has
begun for dry alluvium (a fourth medium), in which a buried 20-ton charge was
fired. Step e) is now being taken with data from that shot, but sustained support is needed to see the cycle through step f) and final model-evaluation.
Then, along with nuclear monitoring needs, the goals of several organizations
can be served by repeating steps d) and e) for a near-surface burst. In addition, since a pregnant set of stress-strain curves is measured in steps a) and
e), strain-path modeling is likely to prove an efficient way to learn about the
mechanics of geo-materials — and whatever is learned will take a form of direct
use to those concerned with explosively driven ground motion.

ACKNOWLEDGMENT

   The work summarized herein was performed over the past decade for the Air
Force Systems Command (Norton AFB), the Defense Nuclear Agency, the Defense Advanced Research Projects Agency, and the Air Force Weapons Laboratory. Among
the many government personnel whose support, advice and encouragement have been
vital to its performance, thanks are due especially to then-Capt. John Kaiser,
Dr. George Ullrich, Col. Donald Gage, Col. George Bulin and Dr. Carl Romney. It
is a pleasure as well to acknowledge the assistance of Dr. James Workman, whose
comments and draftsmanship were essential in preparing the manuscript, and who
has contributed notably to the work itself almost since it began. Also in order
are sincere thanks to Mr. Neil Perl for innumerable bits of help along the way,
and to Mrs. Barbara Glaser who manages to convert my microscrawl to print.

REFERENCES

Akers, S. (1983), 2nd and 3rd Quarterly Progress Reports for DNA Task Y99QAXSB, Work Unit 00019, U.S. Army Engineer Waterways Experiment Station, Vicksburg, MS.

ATI file: R. Hoy, H. Belchic, R. Miller, N. Short, R. Bendenelli and A. Mathews have independently confirmed that there was almost no water in Cowboy salt (1980). Their signed statements on the subject are on file at Applied Theory, Inc.

Bjork, R. (1972), "Computed Response of the Hudson Moon H.E. Experiment," Systems, Science and Software Topical Report No. 3SIR-976 (Contract DASA01-69-C-0165).

Bogart, J. and J. Schatz (1983), "Specified Strain Path Testing of Geologic Materials," *Rock Mechanics, Theory-Experiment-Practice: Proceedings of the 4th U.S. Symposium on Rock Mechanics Held at Texas A and M University June 20-23, 1983*, C. Mathewson (ed.), Texas A and M University, p. 473.

Cherry, T. and N. Rimer (1982), "Verification of the Effective Stress and Air Void Porosity Constitutive Models," VELA Seismological Center Topical Report No. VSC-TR-83-1, pp. 6-15.

Heard, H., A. Abey, B. Bonner and A. Duba (1975), "Stress-Strain Behavior of Polycrystalline NaCl to 3.2 GPa," Lawrence Livermore Laboratory Report No. UCRL-51743.

Hoffman, H. and F. Sauer (1969), "Operation Flint Lock, Shot Pile Driver, Project Officers Report — Project 1.1, Free Field and Surface Motions," Defense Atomic Support Agency Report No. POR-4000.

Ialongo, G. (1973), "Prediction Calculations for the Mixed Company Event III," Defense Nuclear Agency Topical Report No. DNA 30206T, pp. 116-128 (Contract No. DNA001-72-C-0009).

Keogh, D., P. DeCarli and J. Rosenberg (1982), "Development of a High Modulus, Piezoresistive Gauge for Dynamic In Situ Soil Stress Measurements," Defense Nuclear Agency Report No. DNA-TR-82-17-V1.

Kitchens, C. (1972), "Numerical Experiments with the Compressible Navier-Stokes Equations," *Proceedings of the International Conference on Numerical Methods in Fluid Mechanics*, Vol. I, H. Cabannes and R. Temam (eds.), Springer-Verlag, p. 120.

Ko, H.-Y. (1981), "Cubical Test Data on Ralston Valley Soil," Systems Science and Software Report No. SSS-R-81-4824.

Lade, P. (1983), "Strain-Path Tests on Yuma Soil" (final report; Subcontract ATS-55-1; Contract No. DNA001-80-C-0232).

Murphey, B. (1960), "Particle Motions Near Explosions in Halite," Sandia Corporation Report No. SC-4440(RR).

Perret, W., A. Chabai, J. Reed and L. Vortman (1963), "Project Scooter," Sandia Laboratory Report No. SC-4602(RR), Chapter 4.

Perret, W. (1967), "Free-Field Particle Motion from a Nuclear Explosion in Salt,

Part I," Report No. VUF-3012 (Vela Uniform Program).

Rimer, N. and J. Cherry (1982), "Ground Motion Predictions for the Grand Saline Experiment," VELA Seismological Center Topical Report No. VSC-TR-82-25 (Contract No. F08606-79-C-0008), pp. 23-25.

Sauer, F. and J. Kochly (1971), "Operation Diamond Dust, Project 3.1, Ground Motion Measurements," Defense Nuclear Agency Report No. POR 6437 (Contract No. DASA 01-69-C-0165).

Sauer, F. and J. Kochly (1972), "Operation Diamond Mine, Ground Motion Measurements," Defense Nuclear Agency Report No. POR 6573 (final report; Contract No. DASA 01-71-C-0014).

Seaman, L. (1974), "Lagrangian Analysis for Multiple Stress or Velocity Gauges in Attenuating Waves," *J. Appl. Phys.*, p. 4303.

Terhune, R. and H. Glenn (1977), "Estimate of Earth Media Shear Strength at the Nevada Test Site," Lawrence Livermore Laboratory Report No. UCRL-52358, p. 24.

Thomas, J. (1979), "Misers Bluff Negative Phase Measurements and a Pore-Air Model for Ground Motion Simulations," *Proceedings of the Misers Bluff Phase II Results Symposium 27-29 March 1979*, Vol. I, Defense Nuclear Agency Report No. POR 7013-1, pp. 3-181,182,183.

Tittman, B. (1983), "Studies of Absorption in Salt," Rockwell International Report No. SC5320.5FR (final report; Contract No. F49620-82-C-0015).

Trulio, J. and N. Perl (1974), "Calculations in Support of the Diamond Dust and Diamond Mine Events," Defense Nuclear Agency Report DNA 3268F (Contract No. DASA 01-69-C-0138).

Trulio, J., G. Ialongo, J. McDonald and D. Srinivasa (1975), "Overdrive Calculations Related to Nuclear Explosions," Defense Nuclear Agency Report No. DNA 3542F (final report; Contract No. DASA 01-70-C-0074), Sections 4.2 and 8.

Trulio, J. (1975), "Stress Trajectory Analysis, Part I," Applied Theory, Inc. Report No. ATR-75-32-19(I) (Contract No. F04701-71-C-0016), Sections 1.2 and 3.1.

Trulio, J. (1977), "Ground Shock Deformation Fields," Report No. ATR-77-48-6 (final report; Contract No. F04704-76-C-0031), Section 2.

Trulio, J. (1978), "Simple Scaling and Nuclear Monitoring," Applied Theory, Inc. Report No. ATR-77-45-2 (final report on Phase IV of Contract No. DNA001-75-C-0304), p. 22.

Trulio, J. (1981a), "Utility of Calculations" and "Strain Path Modeling," *Proc. of the Review of Free Field Ground Shock from Contained Nuclear Events, 8-9 Dec. 1981*, R. Port (ed.), R & D Associates, Marina del Rey, CA.

Trulio, J. (1981b), "Seismic-Wave Generation: Planning of In-Situ Experiments," Applied Theory, Inc. Report No. ATR-81-57-1 (Contract No. DNA001-80-C-0360), p. 20.

Trulio, J. (1982), "Strain-Path Analysis and Testing, *Proceedings of the Strategic Structures Review Conference, 4-6 May 1982*, Vol. 1, Defense Nuclear Agency Internal Report No. DNA-IR-82-23-VI, p. 267.

Trulio, J. and R. Port (1982), "Material Properties for MX Land Basing," Applied Theory, Inc. Report No. ATR-55-82-1 (Contract No. DNA001-80-C-0232), Fig. 1.

Workman, J., J. Trulio and E. Stokes (1978), "Strain Fields Calculated from Measured Velocities: Axisymmetric Fields of Motion," Applied Theory, Inc. Report No. ATR-78-49-3 (final report; Contract No. F04704-75-C-0025), p. 44.

Workman, J., J. Trulio and E. Stokes (1981), "Modeling the Behavior of Geologic Materials in Explosive Field Events," Air Force Weapons Laboratory Technical Report No. AFWL-TR-80-66 (final report; Contract No. F-29601-76-C-0015), pp. 8-23.

Wright, J., I. Sandler and M. Baron (1973), "Ground Motion Calculations for Events II and III of the Middle Gust Series," *Proceedings of the Mixed Company/ Middle Gust Results Meeting 13-15 March 1973*, Vol. II, Defense Nuclear Agency Report No. DNA 3151P2, pp. 638-639.

APPENDIX A

At first, equations for elastic-plastic behavior were added to those previously developed for hydrodynamic motion. The constitutive models then featured i) general (nonlinear) elastic hydrostats and a linear elastic description of shear, giving way to ii) plastic flow at prescribed limits of shear strength, in accord with one flow rule or another. Also, iii) transitions between condensed and gaseous states were made hydrostatically; to deal with porous solids, iv) the hydrostats were further generalized to encompass inelastic volume changes during compression ("compaction").

Elements i)-iv) still form the backbone of computational models of explosively driven earth motion. However, a major conceptual step was taken by requiring that the stress-strain relations for compaction, shear failure and elastic deformation conform to Drucker's postulates. Equations for those three modes of deformation, which had been stitched together largely on an ad hoc basis, then form a single, internally consistent set. Yield surfaces expressing the limits of material strength in shear and compression occupy a central place in that set. By allowing yield surfaces to vary during shear failure, inelastic strains can be held to realistic levels for frictional materials —a major reason for moving to Drucker's postulates. Though a part of the apparatus for describing inelastic deformation (of which flow rules are another part), the growth and shrinkage of yield surfaces ("hardening" and "softening") make possible such wide variations in material behavior as to form a fifth basic element of the models (element v)).

Both yield surfaces and hydrostats are affected by shock heating. Those effects are usually small (shocks from bursts decay rapidly as they travel), but are included in the models (element vi)). With regard to hydrostats, the laws of thermodynamics provide a basis for so doing, whereas estimates of thermal effects on strength have largely been guesses.

Other processes are now described by the models in ways that lie outside the framework of Drucker's postulates. Prominent among them is tensile failure, as distinct from cracking in shear (element vii)). Present models of tensile failure include equations for crack nucleation and spreading, and also for the changes in strength and moduli that attend crack damage. Those equations are ancillary to the rest of the model (elements i)-vi)). So are equations for multiphase flow, of which there are at least three major sets: viii.a) pore-air equations, describing the flow of air into or out of the ground, where it interacts thermomechanically with particles of earth; viii.b) equations for the dif-

fusion of water through networks of cracks in rocks, with attendant thermomechanical effects; ix) equations for the sweep-up of dust and debris from the
ground surface by burst-induced winds.

The development and use of models incorporating all these elements was well
under way ten years ago. It has since been made clear that soils, at least, can
depart from the normality condition to which Drucker's postulates lead (when
taken with certain other assumptions). Hence, modelers are now free not only to
employ yield surfaces that vary widely in any given material element, but to relate increments of stress and inelastic strain in ways that are independent of
their variable yield surfaces. In short, the flow rule has become a tenth model-
element (element x)). More recent is element xi) —"subsidence"—a relative of
the shock-damage idea; in this case a shock breaks down a lightly cemented
matrix of saturated granular material, and the resulting particles settle (under
gravity) to a relatively well-packed state — leaving a relatively large crater.
Many current models also describe rate effects (element xii)), for which we note
here only the vast range of modeling possibilities they open: a) stress increments become linear combinations of both the incremental strain and strain-rate
tensors (though the usual constitutive tensor may "simply" become a function of
strain rate if causality is strictly enforced), and b) both the flow rule itself
(element x)) and the hardening/softening yield surface become functions of strain
rate. Another fairly recent part of models is tectonic energy release (element
xiii)); an older part provides for orthotropy in layered media (element xiv)).
Moreover, when media are heterogeneous, site structure allows some freedom of
model-variation; that structure, never exactly known, has to be idealized.

Rarely are all these elements used in one calculation, but most of them
often are — and the complexity of the individual elements has hardly been
hinted at. As a small example, nonlinear elasticity calls for a) definition of
strain energy as a function of at least the three strain invariants, b) knowledge
of how yielding (Bjork, 1972, pp. 5,6) and cracking affect that function, and c)
its integration into models that include heating and rate effects. Each of these
items is dealt with in full by all models, but mainly by default: The many
material-specific functions they entail are almost unknown for geo-materials —
as is the cumulative effect of myriad decisions (and non-decisions) modelers
make about such items when a specific geo-material is modeled.

GENERAL DISCUSSION

NEMAT-NASSER: Just to understand some of the terminology, could you please explain, number one, what it is that you calculate and how do you calculate it?
Number two, what it is that you measure and how you measure it?

TRULIO: Let me take those questions in reverse order, because it is easier. I
will refer first to nearly spherical fields from buried bursts, and to the wholespace portion of what's measured.
     The quantity measured is either velocity or acceleration, depending on what
kind of gauge has been used. In the better-designed experiments, radial motion
is measured and the other two components are measured, so that you learn about
the symmetry of the field. The other two components, you hope, just tell you
that the field is nearly spherical. In the case of the salt shots, that's pretty
much what they did say.
     Now, given measured radial velocity pulses at different radii, whether they
are found directly or by integrating accelerometer records, you have to construct
the radial-velocity field —a continuous function of position —because the

radial derivative of radial velocity is the rate of radial strain. The best known method for doing that is called LASS. It was developed at SRI at the same time we were developing one of our own. They're both just interpolation procedures. Ours was used to get the curves I showed. Its accuracy has been verified in a number of ways; for instance, we used experimental data directly by omitting a measured pulse and seeing what the interpolation procedure gave at that position; it has also been tested using calculated pulses just as if they had come from gauges.

Now, assuming that the measurements don't show the field to be strongly non-spherical, hoop strain and radial strain are computed from measured radial motion; that's what was plotted to form the curves you saw. The paths can be shown as a function of time, since rate effects might be important in some media; often [but not on the plots I showed] we put small circles along a path at equal increments in time; then you can see what strain rates are involved, and if you can control that variable, move along a path at the indicated rates.

For the calculated paths, we simply used models not much more complicated than the early ones, especially for deep bursts. They contain yield surfaces of Mohr-Coulomb/von Mises type, depending on laboratory measurements of strength. The choice of associated or Prandtl-Reuss flow rules is optional in our codes. We don't put much faith in the calculations, and in fact, in prediction tests, they haven't shown up well.

The point I was making is that their prediction accuracy really didn't matter in studying strain-path fields. The strain paths — and their shapes in particular were determined mainly by burst geometry, not by material properties. You can be wrong in your formulation of material properties and still get the right strain-path shapes; that's what's happened here.

BALADI: Our ability to predict ground shock has been hampered in the past, because you know very well that available time and funds generally limited the assessment of pretest material properties to those obtained from laboratory uniaxial strain and triaxial compression tests. However, many of the existing constitutive models are not able to simulate this data even qualitatively, much less quantitatively, e.g. in my opinion, some of the calculators use very simple prehistoric models. In order to improve our ground shock predictions, soil engineers should base their material property estimates on results from a variety of stress and/or strain paths (both lab and field) and the theoreticians should come up with models that can simulate the observed different behaviors.

TRULIO: That's the hope I am expressing. The paths I showed are the ones that are actually followed. What's needed in laboratory tests is to measure the stress around those paths. Those are the data needed for prediction of explosively driven motion. After all, the paths came from just that kind of motion.

BALADI: But do you believe that the soil engineer will ever have the time and funds to conduct these tests and come up with properties?

TRULIO: I hope the job can be done in the laboratory. Strain path measurements are starting to be made. I understand what you mean by the eleventh hour. I have just gotten results of strain path measurements for a shot that will be fired in three weeks. There is no way that people making predictions for the shot can include that information in their models. But this won't be the last opportunity to do so — or maybe I'm the real optimist here.

READ: I have two questions. You mentioned at the beginning of your talk that you'd been led to accept the strain path as less property sensitive than the stress path. I wonder if you can elaborate on the analysis that led to that

conclusion? And the second question is: The ultimate event for which your predictive tools are designed to treat will probably involve very high strain
rates. However, the analysis you described is strain rate independent. How can
these strain path techniques then account for the large range of strain rates in
the big event?

TRULIO: The stress paths are material dependent; that's why strain instead of
stress. Let me give you a very simple illustration: an explosion in a wholespace of isotropic linear elastic material. At far ranges only uniaxial strain
will occur. In strain space that's a single line on the strain axis that corresponds to the radial direction; regardless of medium, it's always that axis.
Since the medium is linear and elastic, that line maps into a straight line in
the stress plane [hoop stress vs. radial strain]. However, the line's orientation in that plane will vary with Poisson's ratio — the only material property
for the strain path. That was the thing we had to see a long time ago. It took
a while to sink in because we were so preoccupied with stress, and trying to
describe shear failure and define yield surfaces. The stress paths are material
dependent. The strain paths are almost material independent, and I guess you
would expect that, if you think about the role of geometry here. Also, strain
is a geometric property; it's kinematic.
   The second question had to do with strain rates. Let's talk about deep
bursts first. What I hope is that servo-mechanisms will be developed that allow
us to make laboratory measurements over the range of rates actually observed for
yields of interest. As I said, it is easy to put time markers on the strain
paths themselves; you know material velocity vs. time, so you can tell where you
are on path as a function of time. But nobody can use that information at the
moment. What I hope is that as the apparatus is developed further, it will get
that capability more and more. People at the Waterways Experiment Station have
some control over strain rate that doesn't seem to be available anywhere else;
but that control seems limited to uniaxial strain, not strain paths like the
ones I've shown.

VALANIS: If I propagated axial wave in a very thin rod, I would have a uniaxial
stress state. Strain then depends, of course, on the materials. In this particular case, it is the strain that's material dependent, not the stress.

TRULIO: I mean to emphasize this point: When you get done with this kind of a
model, it will give you (if it all works) what you need to know to predict motion *from an explosion.* If you are interested in the stresses that develop in
soil under a footing, it may not help you at all.

VALANIS: No, but it is a question of geometry, rather than material behavior.
Here, you have a plane strain, basically, a large object, basically.

TRULIO: In this case — the one I had to deal with — geometry dominates.

VALANIS: So it is not the material.

TRULIO: Exactly right, and there are other cases where it wouldn't dominate;
that's correct.
   [Note added in proof: Mr. Valanis raises an incisive point. In his example, geometry determines *stress*-path shape and orientation, *irrespective of material properties* — and again we know what kind of material properties testing
to do *for all media*, to get reliable models. Conclusion: For a system of any
given sort, fields of motion should be examined for both strain-path *and* stress-
path invariance (as in fact explosively driven fields were). If stress-path

shape and amplitude prove to be set mostly by geometry, then, for the main types of path, the dependence of stress amplitude on medium has to be learned, etc.]

PATCH: Jack, I have a comment and a question. First, the comment. One of the points you made was that we are hoping to avoid having to extrapolate into untested regions of stress and strain spaces. I think it is fair to point out that there is a shell of material near the burst that's strained 100, 200, even 5,000 per cent. Clearly you are not going to be running a material property test in that strain regime. Would you disagree?

TRULIO: At 5,000 per cent strain, we are probably all right again. There is an intermediate regime that's very tough.

PATCH: My second point is really a question. I would assume that your approach would be to look at strain paths like those you have shown us and to then pick out characteristic paths; that is, uniaxial strain followed by so much biaxial strain, etc. You are clearly not going to be chasing around spaghetti-shaped strain paths, are you?

TRULIO: Absolutely not. Characteristic paths will be extracted from —well, I hope more and more from measured paths in situ. A lot of the spaghetti, I believe, is discretization error.

HEGEMIER: Jack, how do you propose to do this in the laboratory?

TRULIO: Well, for deep shots (nearly spherical fields) the question is really more how people *are* making their measurements in the laboratory. Of course, they use samples taken from the field — usually cylindrical samples. At one laboratory, two lateral stresses are controlled; ultimately, only loads on the sample are controllable. More often the two lateral stresses come from the same fluid reservoir and are equal. The ends of the cylinder are then subjected to a different fluid pressure.

HEGEMIER: So, as I understand it, the best situation for you would be a closed-loop servo-controlled system in the laboratory?

TRULIO: Right.

HEGEMIER: This concerns me then. Because typical servo-controlled systems exhibit an upper frequency response of about 25 Hz. On the other hand, you are talking about wave propagation, so the strain rates in the field are extremely high.

TRULIO: They tend to be very high in stiff materials for small charges. For the bursts that concern us all the most, strain rates are not nearly as high because the yields are so large. Then it may well be that 25 Hz is in the right range.

PREPARED DISCUSSION

by Ted Belytschko

In discussing approaches such as that of Trulio, it is important to bear in mind the ultimate objectives of this class of computations, which are often called "free-field" computations: to predict the strains, stresses and velocities rising in a medium from a high-energy event, so that the performance of an embedded structure and its contents can be evaluated. Thus, in addition to the free-field computation, subsequent computations must be made with a structure embedded in the medium. The coupled structure - medium response can result in complex interaction and wave phenomena such as reflection and refraction. Furthermore, scenarios such as multiple events and reloading are often of interest.

Trulio's approach of emphasizing certain strain paths associated with burst geometry in the acquisition of the material properties is motivated by the large discrepancies which have sometimes arisen between predicted free-field motions and experimental observations. His assertion that laboratory data acquired by using strain-paths similar to that found in the field will yield better results is difficult to refute, but it does raise some disturbing questions:

(1) Does a better understanding of the strain and stress paths in the spherical burst event enhance the prediction capabilities for the structure-medium problem, which is the ultimate goal of these computations?

(2) Are the strain paths which are found in the spherical burst sufficient to characterize even a simple model which would have to underlie these calculations?

(3) How would the effects of strain-rate be accounted for in an approach of this type?

A thorough understanding of the physical phenomena in the free-field motion such as evidenced by Trulio's work is undoubtedly useful in obtaining better predictions and in determining the shortcomings of various methods. Since the material properties and constitutive equations are the major sources questionable of error in a free-field computation, it would seem that a more important objective of such studies would perhaps be to explain the source of discrepancies in previous predictions. To a large extent, the development of geomaterial models seems to consist of a sequential addition of physical phenomena and modification of models without any specific analysis of what the precise effects of the shortcomings of previous models actually were. Although such studies are perhaps carried out on an informal level, they are seldom published, and consequently the rationale underlying the development of constitutive models for geomaterials is not at all well documented.

In any case, to meet the ultimate objective of such calculations, a predictive model which works for almost any conceivable strain path is almost unavoidable. Therefore, while Trulio's work may serve an interim purpose, the task of developing geomaterial models of general applicability cannot be evaded if the ultimate goals of these computations, structure-medium computations, are to be achieved.

A subsidiary topic I would like to mention is the role of damping in these calculations. Damping and physical dispersion are the major determinants in the rise time of the stress waves in these events. Experimental measurements exhibit rise times which are long, compared to the frequencies of interest in the structure and equipment in the structure, so an accurate determination of the rise time is often important. Nevertheless, the rise times which are currently predicted by computer codes are largely a matter of chance and are influenced by a variety of numerical phenomena:

(1) semidiscretization solutions of wave propagation problems, whether they are finite element or finite difference, exhibit considerable dispersion, which reduces the slope of the wave front when the mass is lumped;

(2) in the application of the computer methods to geomaterials, substantial
   linear and quadratic artificial viscosity is usually employed in order to
   eliminate the shocks which are a consequence of the increase in the slope
   of the pressure-dilatation curve with increasing pressure.

Thus two artifices, numerical dispersion and artificial viscosity, govern
the rise time of the stress waves, and perhaps other characteristics of free-
field calculations. The viscous or damping properties of geomaterials are not
included, perhaps because they are not well understood and little has been done
to measure these properties or to incorporate them in material models which are
used in these computations. Although this is not perhaps the central issue for
this class of computations, it is well to bear in mind that this aspect of the
computed response in present day methods arises from numerical characteristics
and that their implications are not well understood.

GENERAL DISCUSSION

GOUDREAU: Ted, with respect to this frequency content and the ability to propagate waves in two-dimensional analysis, do you see this as the difficulty of the
code propagating the ground motion to the structure or the early time code that's
handing off a time history to the ground motion code? Basically, there are two
codes in the sequence, right?

BELYTSCHKO: Right, but the phenomenon is in the propagation code. The calculation I showed here was actually a one-dimensional calculation, because it was a
deep burst; for a surface burst, you have to do a two-dimensional calculation.
Then, if you calculate the response of a buried structure, you do get the severe
dispersion effect.

GOUDREAU: This is sort of at the distance where a structure would be. What I'm
asking is which generation of codes are we talking about; the whole scenario from
burst to structure; or is it the late time propagation out to the structure, or
is it the early time characterization of the wave that produces this?

BELYTSCHKO: I can't say exactly what caused it in this code but, generally, it
arises from the fact that artificial viscosity is added to eliminate the shocking
that comes about due to the stiffening behavior.

GOUDREAU: I would think that normally one-dimensional calculations could be quite
precise. I thought you were referring to a two-dimensional result; you are now
telling me it is one-dimensional.

SANDLER: The objective of this one-dimensional calculation was to predict, as
well as we thought practicable, the signals at the range of the structure. Now,
it is possible within the context of a rate independent model to "twiddle the
knob" of artificial viscosity in order to get any rise time you want, given the
choice of grid size. In this case, a particular choice was made that would produce a rise time that was consistent with what had been seen in previous experiments. As is well known, for a rate independent material which has a hardening
stress-strain curve, the material is going to shock up. There is no way to get
a finite rise time in wave propagation over long distance in such a material,
unless you introduce viscous effects.
   For this material, there were no experimental data (other than rise time
observations in just such experiments) with which to tune viscosity in the model.
Therefore, the calculation was tuned that way; it was a practical artifice, just

as artificial viscosity is, and the objective was to get a rise time which would be more or less reasonable for this case. Indeed, the shock could have been represented with greater fidelity to the underlying mathematical formulation of the problem within the context of the rate independent model; this would have led to a sharp shock at the range of interest.

BELYTSCHKO: Which would not be consistent with the experiment.

SANDLER: Which would not be consistent with previous experiments of this type.

GOUDREAU: Does that say the physical process has dissipation?

SANDLER: Absolutely.

KELLER: I would like to point out very briefly, that some of Jack's pessimism was because of the poor comparison of measurements with a calculation. I would also like to point out that the measurements he showed are nothing like the measurements we have made under similar circumstances within the last few months.
     The other thing I would like to point out is that the last paper was also dependent upon opinions of how well the calculations compared with measurements. I do not believe that the quality of many of the measurements deserves that kind of effort to match them. I also do not believe that the method proposed by Jack will match the two-dimensional or three-dimensional characteristics of the wave that was shown last. We have seen enormous vertical motions near nuclear explosions that would probably give you strain paths that are very different than you would calculate in the ideal circumstance of 1-D spherical motion. So, the measurements can tell you about heterogeneities, the medium, if you believe them; but we really need to drop back to circumstances we can understand and measure well, before we try very hard to match the measurements.

BACHE: Now, there is the most pessimistic viewpoint. Whenever you talk about field data, there is always somebody who stands up and says that the gauge records are probably faulty, so what is it we are trying to match?

CHERRY: But I think he had a good point.

BACHE: It is a good point. People say it because they think it's true, and it probably is.

CHERRY: The advancement of measurement techniques in well-defined materials has come a long way. Jack, I would like to ask you a question. Is the strain path and its magnitude both dependent on geometry only or only the shape of the strain path?

TRULIO: The shape of the strain path, which of course determines the kind of test to be made.

BALADI: But not the magnitude.

TRULIO: I showed a plot in which shape was correlated with amplitude, and there is perhaps a factor of two spread. I therefore visualize making two or three tests at different amplitudes for a single shape, or vice versa. Now, that's a fairly large data base, so I also think the strain path approach will help in developing mechanistic models. The data base itself, being dense, may give you a good deal of insight into the mechanisms at work. I don't want to create a misimpression here: The ultimate object is to get mechanistic models. Mean-

while, people with system responsibilities, and so on, need predictions they
can rely on and they don't get them. This is a way of making more reliable predictions, and, I hope, at the same time speeding the process of getting mechanistic models that work. Mechanistic models have much more demanding constraints
on them than the models I was talking about. All we ask of a strain-path model
is that it give correct predictions. We are not asking that it have equations
in it to represent the mechanisms responsible for a material's stress-strain
curves, i.e., that the predictions be right for the right mechanistic reasons.
In the extreme, there is no explicit account of mechanisms in such a model.
There are just measured stress-strain curves and formulas for interpolating
among them, but the curves are for the right strain paths; that's the idea.

HERRMANN: I would like to be sure that I know what I'm talking about before I
make this following comment. It is my understanding, although nobody has stated
so today, that these are plasticity models with yield surface. In general, one
of the features of a yield surface is that it's fitted to failure envelopes and
elastic moduli obtained from triaxial tests together with compressibility which
is fitted to compression test data and so on.

   I'd really like to come back to the comment I made yesterday. While I understand the expediency in pressing plasticity models into service because they
exist, and I would accept your arguments that they are useful, I get very unsettled when I am directed to work in progress, after 15 years worth of work in
this area, to justify the models. It is my impression that they usually need a
good deal of adjustment after the test to make sure everything comes out right.

   Now, I would suggest that while it is useful to use models of this sort in
the interim, we really ought to be looking for models that seem to have a little
more realism to them. My argument yesterday was that one of the features fitting
plasticity yield surfaces to failure envelopes is kind of curious. First of all,
this failure envelope represents a global instability of a machine specimen configuration, and in fact, you can easily move the failure point around by changing
the machine (the boundary value problem), and that failure isn't a material property at all. In addition to that feature and some other things that one does to
obtain the yield envelopes, you have a lot of latitude in choosing nonassociated
flow rules, and that's done without any attempt to think about the consequences
for the stability of the material. This is all very well as an interim step. I
wouldn't normally try to say, "look, this is the wrong thing," in the absence of
having a better suggestion, but my feeling is that the preoccupation with these
models and the limited success which is being obtained is inhibiting people from
thinking about what is really going on in these materials, thinking about some
realistic mechanics that might be incorporated into models. I would urge that
that thinking be reinstituted, and in fact, funding be made available — there
are, in fact, ideas of how brittle materials might behave — and that there be a
vigorous search for something else to supplant some of these plasticity models.

CURRAN: I want to switch my comments back to the question of making valid stress
measurements. The comment was made that we can always get out of our calculational difficulties by claiming that the stress measurements were invalid or
poor. It is true that it is a very difficult problem to measure stress. What I
would like to point out is although there are many problems involved, the key
one is trying to analyze a plastic inclusion in a plastic matrix, because that's
what a stress gauge basically is. It is in a package, and, at the high stress
level we sometimes operate at, it is in the plastic region itself. There is no
analytic solution we are aware of for that problem. So, we are in the situation
of having to understand the constitutive relations in order to measure the
stress, but be able to measure the stress to understand the constitutive relations. So, it's clearly a bootstrap operation.

TRULIO: I did say how that bootstrap operation is going to be handled in an up-coming event. There is an impulse-momentum check that's rigorous — although it would not be if just a single stress component were measured.

CURRAN: I don't mean to say it should not be done. But it is a difficult problem.

BELTYSCHKO: I would like to make a comment on Walter Herrmann's statement. I think you are being a little harsh on the area, too critical, considering the successes that have been achieved. The plasticity is used, in a sense, as a failure surface; and there have been many tests where the comparisons are quite good. The difficulty really arises more out of the fact that many of these tests exhibit lack of, for example, symmetry. In spite of what Carl says, I think a lot of these gauges are good and when you are dealing with a real site, there are problems in characterizing the material over large domains. It is not like dealing with a very small, well-controlled specimen.

   Now, if you take that into consideration, then a lot of the predictions are quite good. There are some subtle features which are important which are not predicted very well. But overall, I think there has been a considerable amount of progress in the last 15 years; it has not just been done in the name of ex-pediency. Some good thought has gone into it. The cap model has come out of this community, and I think it is a very fundamental model which applies quite well to these materials, particularly granular materials.

DIENES: I want to make a comment about the rise time associated with these shock waves; perhaps it has nothing to do with viscosity at all and may be just a scattering process that's going on. It seems to me, if you take out the finite differences and look at viscosity, the rise time associated with vis-cosity would be exceedingly short. Physically, what's going on is a lot of reflections and inhomogeneities in the rock which are responsible for the ob-served, rather long, rise time.

BELYTSCHKO: Your point is good. As in a composite material, it might be dis-persion due to inhomogeneities, but we don't know how to characterize this; and in many cases, it does make a difference in structural response. I don't know if it is damping.

SANDLER: I want to reiterate the point that Ted made about inhomogeneity and variation from point to point in the site. A number of measurements have been made in connection with the magnitude of tangential motions, compared to radial motions, from a surface burst. It is often the case that the tangential velocity can be as high as 50 per cent or more of the radial velocity and/or vertical ve-locity, indicating that there is a definite nonhomogeneity from point to point. There is asymmetry in the events. It is the kind of variation that we find very difficult to include in a deterministic approach, and that appears to be a fun-damental reason as to why many of the comparisons between calculation and exper-iments are poor.

READ: I wanted to go back to an issue that Ted Belytschko raised which I don't think has been answered. And it is this: All of this strain path work is really directed, as Ted said, at developing predictive tools for determining the motion of buried structures, that is, the structure medium interaction problem. The work that Jack Trulio is doing is, however, restricted to strain paths for free field motions and constitutive relations for that particular class of deformations; in the vicinity of the structure, however, the stress wave system is very complex, the strain fields are complex and not like the strain paths in the free field. I would like to know how your approach, which focuses on a free field, rate-independent response relates to this ultimate objective of calculating the response of buried structures.

TRULIO: You get free field stresses and motions. Those are input for the calculations of structure-medium interaction. You may not have a powerful enough model at this time to tell you what stresses will arise in the medium from that interaction, but there is a point connected with that that's very important here. It's much harder to measure stress in the free field than to measure the stress on a structure interacting with the medium. The reason is that in measuring free field stress, you want the gauge to look like the medium, but gauge materials simply aren't like the medium. To measure stress at a point on a structure, it suffices to have the gauge look either like the medium or like the structure. Gauge materials are a lot more like structural ones than like geo-materials, so it's easier to measure the stress on the structure than in the free field.

CHAPTER XI

STRAIN-SOFTENING

Ivan S. Sandler and Joseph P. Wright

Weidlinger Associates
333 Seventh Avenue, New York, New York 10001

ABSTRACT

Rate independent and rate dependent constitutive models of strain-softening in a dynamic continuum are analyzed by means of standard stability analyses, numerical examples and qualitative comparisons with physical observation. It is demonstrated that any treatment of strain-softening must be rate dependent if it is to represent the phenomenon in a physically and mathematically meaningful manner.

1. INTRODUCTION

With the advent of more accurate analysis tools, in particular improved numerical schemes and larger and more powerful computers, structural designs into the plastic range have become a reality. Designers are now considering structures that may survive under severe dynamic loadings (such as those resulting from seismic or explosive sources) which exercise the material so far into the plastic range that strain-softening occurs. In addition, strain-softening may be significant in the analysis of progressive failure of conventional structures. The rational and accurate modelling of strain-softening behavior is therefore becoming an important area of structural analysis.

Strain-softening may be loosely defined as a region of the load-deformation curve of a structural element for which increasing deformation is associated with decreasing load. One common example occurs in the compressive loading of concrete cylinders in a stiff testing machine. As the length of the specimen gradually decreases and compressive strains accumulate, the test specimen does not fail when maximum load is reached, but a subsequent decrease in the load can be observed under increasing deformation.

A self-consistent treatment of strain-softening in the analysis of structural behavior is clearly needed by structural analysts, but such a treatment is difficult to obtain for a number of different but deeply related reasons which will be discussed below. The question will be addressed from physical, mathematical and numerical viewpoints, although no specific models are proposed here. Instead, the general requirements for dynamic continuum models of strain-softening are presented.

2. THE STANDARD APPROACH

In order to highlight the difficulties associated with the modelling of strain-softening, we consider a particularly simple situation and follow through the standard approach to constitutive modelling.

Let us assume that a typical load-deflection curve for a cylinder of concrete has been obtained from some laboratory experiments. The standard approach is to interpret this curve as "stress" versus "strain" even though it is common knowledge that specimens in the strain-softening range of behavior do not deform uniformly (Bazant, 1976). In fact localized "fracture bands" occur, leading to highly nonuniform strain fields with much of the deformation concentrated in a very small portion of the volume of the specimen (Wawersick and Brace, 1971). The stress-strain relation obtained this way is then used to define the "constitutive behavior" of the material.

In the ordinary strain-hardening regime this standard approach leads to useful characterizations of material behavior. In particular it is possible for the constitutive model to reproduce a laboratory experiment when it is represented as a dynamic initial/boundary value problem with the specimen considered as a continuum (note that in the hardening regime pronounced strain localizations tend not to occur in reasonably homogeneous specimens).

If we try to represent the strain-softening situation as the dynamic loading of a continuum, serious problems arise if we simply interpret load-deflection as stress-strain; in fact, a constitutive model constructed in such a way cannot even be made to exhibit the behavior it was intended to represent! A particularly simple way to demonstrate this is to consider the dynamic end-loading of a rate independent strain-softening semi-infinite bar, Fig. 1. The stress-strain relation which characterizes the bar is shown in Fig. 2. The end of the bar, at x = 0, is subjected to the prescribed velocity-time history shown in Fig. 3.

The governing equations for this problem are

$$\rho \ddot{u} = -\sigma' \tag{1}$$

$$\varepsilon = - u' \tag{2}$$

$$\sigma = \textstyle\sum(\varepsilon) \tag{3}$$

in which $\cdot \equiv \partial/\partial t$ and $' \equiv \partial/\partial x$ where x is the (Lagrangian) spatial coordinate, t is time and ρ is the density, u the displacement, σ the stress and ε the strain of the material (σ, ε assumed positive in compression). Here $\sum(\varepsilon)$ is the functional of $\varepsilon(t)$ illustrated in Fig. 2 and defined as follows:

FIG. 1. Example problem to demonstrate effect of strain-softening

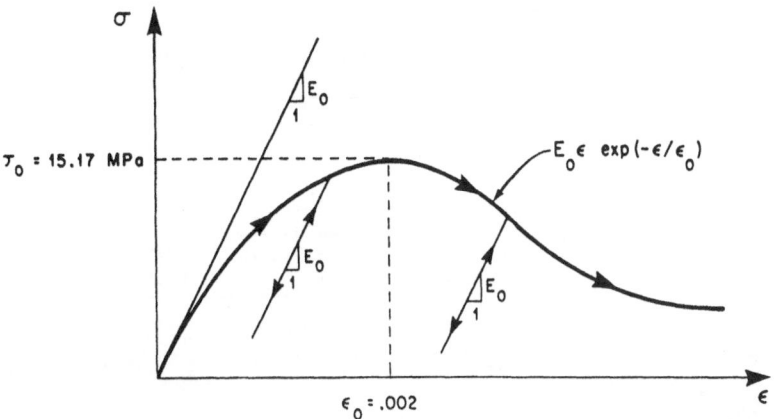

FIG. 2. Rate independent model for strain-softening

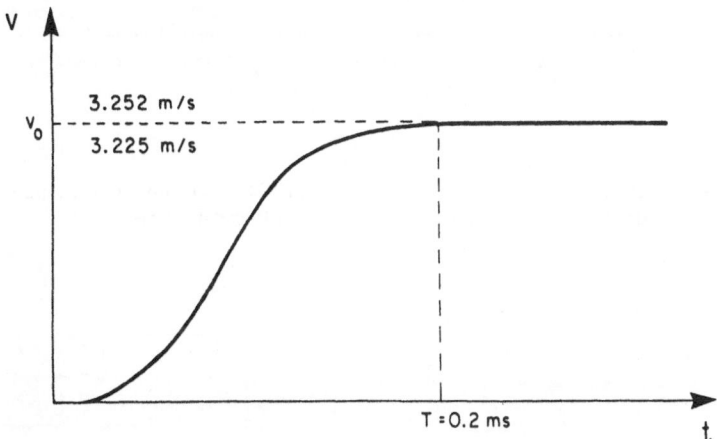

FIG. 3. Input boundary velocity for example problem

In loading, the limiting stress-strain curve is defined by

$$\sigma(\varepsilon) = E_0 \, \varepsilon \, \exp(-\varepsilon/\varepsilon_0). \qquad (4)$$

Unloading and reloading are completely reversible and linear (with Young's modulus E_0), except that the material is assumed to support no tension (so that $\sigma \geq 0$). The material constants are E_0 = 20,700 MPa, ε_0 = 0.002, ρ = 2400 kg/m^3.

The prescribed surface motion is defined by

$$v(t) = \begin{cases} v_0[1 - \cos(\pi t/T_0)]/2, & \text{for } t \leq T_0 \\ v_0, & \text{for } t > T_0 \end{cases} \qquad (5)$$

where T_0 = 0.2 ms.

Numerical results are shown for two nearly equal, but different, values of v_0 in Figs. 4 and 5. (The numerical method is based on von Neumann's finite difference method which involves a grid of equally spaced displacement values placed midway between equally spaced strain values, with time integration based on cen-

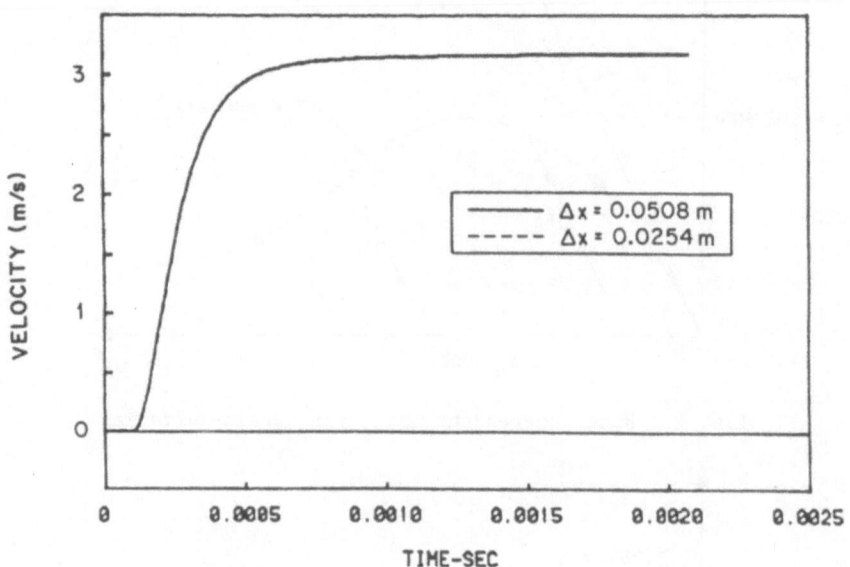

FIG. 4. Comparison of velocities at 0.305 m depth in example problem
for v_0 = 3.225 m/s for different grid sizes

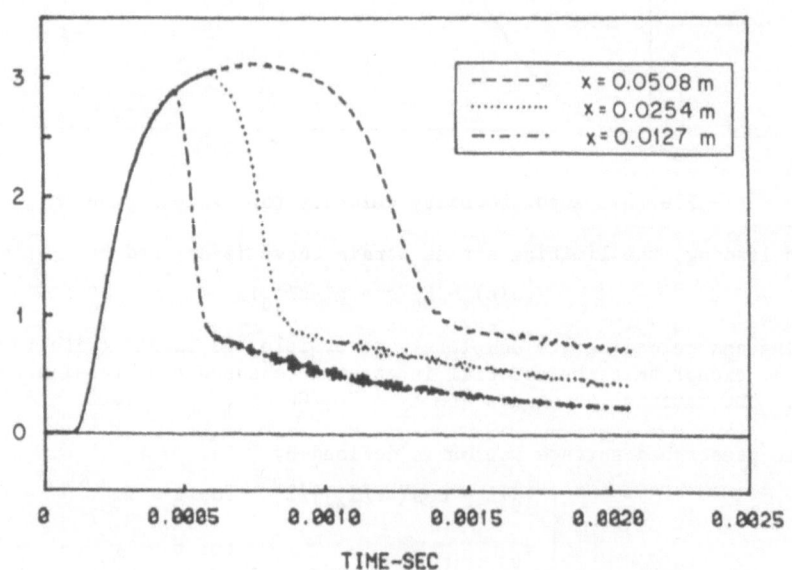

FIG. 5. Comparison of velocities at 0.305 m depth in example problem
for v_0 = 3.252 m/s for different grid sizes

tral differences. No artificial viscosity is used.)

Case A: v_0 = 3.225 m/s

In this case, the strain in the material never reaches the critical value ε_0. Consequently the modulus is always positive and the material model is exercised only in a stable regime. The numerical results are not sensitive to the grid spacing, Δx, as shown in Fig. 4, where the velocity-time history at a distance of 0.305 m from the surface is plotted for Δx = 0.0508 m and Δx = 0.0254 m. These two curves are indistinguishable from one another, illustrating that the mesh sizes are quite adequate for the solution of this problem.

Case B: v_0 = 3.252 m/s

The motion of the surface is only slightly larger (by less than one percent) than in Case A, but now the strain nearest the surface exceeds the critical value ε_0. Consequently the modulus becomes negative, which leads to velocity-time histories which are quite sensitive to the grid spacing Δx, as shown in Fig. 5. In fact the displacement at 0.305 m from the surface differs by more than a factor of two for Δx = 0.0508 m as compared to Δx = 0.0127 m.

Aside from the sensitivity to grid size, a more detailed comparison of the numerical results shows that in the first case the material follows the prescribed stress-strain curve by up to the proper loading level, while in the second case the mesh point adjacent to x = 0 is the only point for which strain-softening actually occurs. All of the other points in the grid unload elastically before the peak stress, σ_0, is reached. As the mesh size is shrunk to zero, essentially none of the material actually softens. Therefore the overall behavior of any finite length of the bar does not approximate the stress-strain relation to which the model was fit! Finally, the displacements at all points of the bar away from the surface are actually smaller in the more severely loaded case.

It can be concluded from this exercise that a rate-independent-dynamic continuum representation of strain-softening is incapable of reproducing softening behavior in a dynamic simulation of experiments.

3. STABILITY ANALYSIS

Why does the standard approach fail when it comes up against the problem of modelling strain-softening? Eighty years ago Hadamard (1903) answered this question in terms of the mathematical concept of proper posedness. If a deterministic description of a physical system is to be meaningful it must exhibit a certain degree of insensitivity to slight changes in initial and/or boundary conditions. As a physical analogue, if a laboratory experiment is to be reproducible, it must yield the same results in cases of small perturbations because it is never possible to repeat an experiment with precisely identical laboratory environments.

Following the reasoning of Hadamard, we consider the result of an infinitesimal, short wavelength, perturbation in the condition of a rate-independent strain-softening material at some instant of time. In order to do this in the simplest possible fashion, consider the uniform bar of length L, shown in Fig. 6, which is in equilibrium at σ_E, ε_E on the softening stress-strain curve shown. An infinitesimal initial disturbance or perturbation of the form $u(x,0) = \delta(x)$ is assumed at t = 0, and the effect of this disturbance on the subsequent behavior of the bar is required. If $u(x,t)$ is the subsequent displacement of point (x,t) then

$$\rho \ddot{u} = -\sigma' \tag{1}$$

$$\varepsilon = \varepsilon_E - u' \tag{6}$$

$$\sigma = \sigma_E - d(\varepsilon - \varepsilon_E) \tag{7}$$

where $d > 0$. In equilibrium $u(x,t) = \sigma(x,t) - \sigma_E = \varepsilon(x,t) - \varepsilon_E = 0$.

For simplicity, we assume that the initial disturbance $\delta(x)$ satisfies $\delta(0) = \delta(L) = 0$ as well as

$$\dot{u}(x,0) = 0 \tag{8}$$

and can be expressed in terms of its Fourier series over $0 \le x \le L$ as

$$u(x,0) = \delta(x) = \sum_{n=1}^{\infty} \delta_n \sin \frac{n\pi x}{L} . \tag{9}$$

In order to obtain the subsequent motion $u(x,t)$ of the bar, we express it, too, in terms of its time-dependent Fourier series:

$$u(x,t) = \sum_{n=1}^{\infty} q_n(t) \sin \frac{n\pi x}{L} . \tag{10}$$

Substitution of this equation into conditions (1), (6), (7), (8), (9) gives

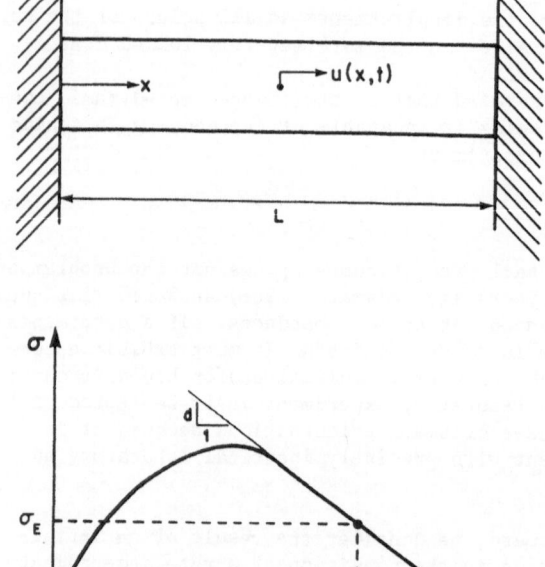

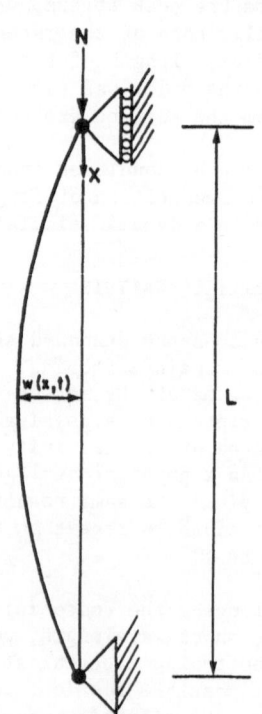

FIG. 7. Buckling of an Euler column

FIG. 6. Strain-softening bar of finite length

$$\varepsilon(x,t) - \varepsilon_E = -\frac{\pi}{L} \sum_{n=1}^{\infty} n q_n(t) \cos \frac{n\pi x}{L} \tag{11}$$

$$\sigma(x,t) - \sigma_E = \frac{\pi d}{L} \sum_{n=1}^{\infty} n q_n(t) \cos \frac{n\pi x}{L} \tag{12}$$

$$\rho \sum_{n=1}^{\infty} \left[\ddot{q}_n - \frac{\pi^2 d}{\rho L^2} n^2 q_n \right] \sin \frac{n\pi x}{L} = 0 \tag{13}$$

$$\sum_{n=1}^{\infty} \dot{q}_n \sin \frac{n\pi x}{L} = 0 \tag{14}$$

$$\sum_{n=1}^{\infty} q_n \sin \frac{n\pi x}{L} = \delta(x). \tag{15}$$

For these equations to be valid over $0 < x < L$ for all n it is necessary that the functions $q_n(t)$ satisfy the following ordinary differential equation

$$\ddot{q}_n - \frac{\pi^2 d}{\rho L^2} n^2 q_n = 0, \qquad n = 1,2,\ldots \tag{16}$$

subject to

$$\dot{q}_n(0) = 0 \tag{17}$$

$$q_n(0) = \frac{2}{L} \int_0^L \delta(x) \sin \frac{n\pi \xi}{L} d\xi \equiv \delta_n. \tag{18}$$

The solution for $q_n(t)$ becomes

$$q_n(t) = \delta_n \cosh(n \frac{\pi}{L} \sqrt{\frac{d}{\rho}} t) \tag{19}$$

and the perturbation $u(x,t)$ is

$$u(x,t) = \sum_{n=1}^{\infty} \delta_n \sin \frac{n\pi x}{L} \cosh(n \frac{\pi}{L} \sqrt{\frac{d}{\rho}} t). \tag{20}$$

If the perturbation $\delta(x)$ is such that $\delta_n = 0$ for all $n > m$, then the rate of growth of the disturbance is given by $e^{m \frac{\pi}{L} \sqrt{\frac{d}{\rho}} t}$, i.e.,

$$|u(x,t)| < A e^{m \frac{\pi}{L} \sqrt{\frac{d}{\rho}} t}. \tag{21}$$

However, an arbitrary perturbation excites all modes of the bar and the response is immediately unbounded (and no solution can be found) for $t \to 0$.

The fact that the future behavior of the bar (the effect of the disturbance) cannot be constructed in general even for a short time is due to the fact that the governing equations (1), (6), (7) are elliptic instead of hyperbolic in the softening regime, i.e., when $d > 0$. It is well known that for such a system construction of a time-marching solution is not possible when arbitrary initial conditions are prescribed.

In order to clarify the situation still further, consider the similar stability analysis of the classic Euler buckling of a column. Viewed as a dynamic problem, Fig. 7, the governing equation is

$$EIw'''' + Nw'' + \rho\ddot{w} = 0. \tag{22}$$

Proceeding as before, w can be expressed as

$$w = \sum_{n=1}^{\infty} q_n(t) \sin \frac{n\pi x}{L} \tag{23}$$

which leads to

$$\ddot{q}_n + \frac{n^2\pi^2}{\rho L^2} \left(EI \frac{n^2\pi^2}{L^2} - N \right) q_n = 0. \tag{24}$$

In the solution of this equation three cases arise as follows:

a) If $n^2 > \frac{NL^2}{\pi^2 EI}$, then

$$q_n(t) = A_n \sin \omega_n t + B_n \cos \omega_n t \tag{25}$$

where $\omega_n = \frac{n\pi}{L} \sqrt{(EI \frac{n^2\pi^2}{L^2} - N)/\rho}$. $\tag{26}$

b) If $n^2 = \frac{NL^2}{\pi^2 EI}$, then

$$q_n(t) = A_n t + B_n \tag{27}$$

or

c) If $n^2 < \frac{NL^2}{\pi^2 EI}$, then

$$q_n(t) = A_n e^{P_n t} + B_n e^{-P_n t} \tag{28}$$

where

$$P_n = \frac{n\pi}{L} \sqrt{(N - EI\, n^2\pi^2/L^2)/\rho} < \frac{N}{\sqrt{\rho EI}} . \tag{29}$$

In a), b), c) above, A_n and B_n are chosen to satisfy initial displacement and velocity conditions. No matter which modes are excited by the initial disturbance w(x,0), the solution satisfies

$$|w(x,t)| < Ae^{\frac{N}{\sqrt{\rho EI}} t} . \tag{30}$$

Therefore the series expansion for w(x,t) provides the solution for all t > 0, so that the dynamics problem is properly posed.

Note that for

$$N < N_B = \frac{\pi^2 EI}{L^2} \tag{31}$$

the beam does not buckle because case c) above does not arise. For $N \geq N_B$ at least one mode will grow large as $t \to \infty$. This produces the *static* instability known as Euler buckling. In this case it is therefore demonstrated that even though a physical "instability" occurs, namely buckling, the formulation remains dynamically well-posed and the model is physically reasonable.

These examples highlight the fundamental difference between material and geometric instabilities. They indicate that the representation of the geometric instability of strain localization by the material instability of rate independent strain-softening may not be satisfactory for problems where the material is treated as a dynamic continuum.

The mathematical question of stability in the sense of Hadamard, i.e., proper posedness, appears to be fundamentally related to the physically based stability postulates of Drucker (1956). According to the latter, the work done by any "external agency" required to disturb a stable system must be nonnegative; this implies that the energy required for the growth of a disturbance should not come from the system itself. In an unstable rate independent system an infinitesimal disturbance can instantaneously receive a finite amount of energy from the system, thus completely (and haphazardly) changing the state of the system.

4.   RATE DEPENDENT MODELS OF STRAIN-SOFTENING

Since Hadamard's example demonstrates the futility of constructing rate independent dynamic continuum models of strain-softening, the question naturally arises as to whether the introduction of rate dependence improves the situation. As will be shown below the answer to this question is "yes."

Let us begin by performing yet another stability analysis of the bar in Fig. 6, this time replacing the rate-independent constitutive equation (7) with the rate-dependent viscoelastic behavior

$$\sigma = \sigma_E - d(\epsilon - \epsilon_E) + g\, \dot{\epsilon} \tag{32}$$

in which $d > 0$ and $g > 0$. Once again we seek solutions of the form (10). In this case, however, Eq. (16) is replaced by

$$\ddot{q}_n + \frac{n^2\pi^2}{L^2}\frac{g}{\rho}\dot{q}_n - \frac{n^2\pi^2}{L^2}\frac{d}{\rho}q_n = 0 \tag{33}$$

subject to (17) and (18). The solution to Eq. (33) is

$$q_n(t) = \frac{\delta}{\alpha_n + \beta_n}\left[\beta_n e^{\alpha_n t} + \alpha_n e^{-\beta_n t}\right] \tag{34}$$

where

$$\alpha_n = \frac{n^2\pi^2}{2L^2}\frac{g}{\rho}\left[\sqrt{1 + \frac{4L^2\rho d}{n^2\pi^2 g^2}} - 1\right] > 0 \tag{35}$$

and

$$\beta_n = \frac{n^2\pi^2}{2L^2}\frac{g}{\rho}\left[\sqrt{1 + \frac{4L^2\rho d}{n^2\pi^2 g^2}} + 1\right] > 0. \tag{36}$$

From the inequality

$$1 + \frac{4L^2\rho d}{n^2\pi^2 g^2} < 1 + \frac{4L^2\rho d}{n^2\pi^2 g^2} + \frac{4L^4\rho^2 d^2}{n^4\pi^4 g^4} \tag{37}$$

it is clear that

$$\sqrt{1 + \frac{4L^2\rho d}{n^2\pi^2 g^2}} < 1 + \frac{2L^2\rho d}{n^2\pi^2 g^2} \tag{38}$$

so that $0 < \alpha_n < d/g$ for all n. Therefore, $|q_n(t)| < Ae^{d/g\ t}$ for all n and the series for $u(x,t)$ converges. Hence

$$|u(x,t)| < Ce^{d/g\ t} \tag{39}$$

and a bounded solution exists for any t > 0 in spite of the fact that, because $\alpha_n > 0$, the disturbance will always grow large as $t \to \infty$. This indicates that strain-softening, although *statically* unstable, nevertheless leads to dynamically well-posed problems if rate-dependence is introduced.

This state of affairs is very much like that of the buckling column, in that the rate dependent formulation allows perturbations to grow in time, but these remain bounded at any finite time and a time-marching solution can be constructed. Rate dependent formulations of strain-softening are therefore physically and mathematically acceptable even though the strain localizations may be "smeared out" in the constitutive behavior.

In order to indicate the potential of rate dependent models to represent strain-softening, a particular model was constructed and exercised. In this model the constitutive behavior is taken as

$$\dot{\sigma} = E_0\dot{\varepsilon} - G(\sigma,\varepsilon) \tag{40}$$

where

$$G(\sigma,\varepsilon) = \begin{cases} D[\sigma - F(\varepsilon)]^2 & \text{for } \sigma > F(\varepsilon) \\ 0 & \text{for } \sigma \le F(\varepsilon) \end{cases} \tag{41}$$

in which $F(\varepsilon)$ represents the quasistatic stress-strain curve shown in Fig. 2 and defined in Eq. (4). The value of D was arbitrarily chosen as 200/MPa-s.

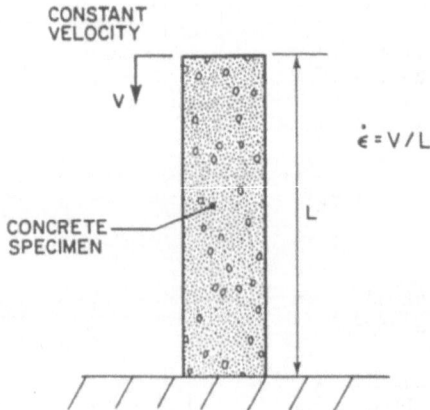

CONSTANT VELOCITY

v

$\dot{\varepsilon} = V/L$

L

CONCRETE SPECIMEN

FIG. 8.   Idealized constant strain rate test

In order to exercise this model, the problem depicted in Fig. 8 was solved for L = 0.1524 m at a strain rate of 0.2/s. The behavior of the model, treated as a dynamic continuum, is compared to the quasistatic curve $\sigma = F(\varepsilon)$ in Fig. 9. It can be seen that this "viscoplastic" model behaves in a manner which is qualitatively similar to much of the strain-softening data (Bazant, 1976).

Because the stability analyses indicate that exponential growth in time is possible (indeed virtually certain) with this model, a study of the sensitivity of the response to grid size was made. In order to highlight the numerical sensitivities which can arise in such a case, the bar of Fig. 8 is assumed to be loaded at a low strain rate of approximately 0.01/s from the inflection point, $\varepsilon_E = 2\varepsilon_0$, of the quasistatic stress-strain curve. The results for three different grid sizes are shown in Fig. 10, in which the "specimen average" values of stress (load/area) and strain (end displacement/length) are plotted. The figure clearly shows a definite sensitivity of the results to grid size although all of the curves approximate the appropriate softening behavior. (This sensitivity of the results is qualitatively similar to sen-

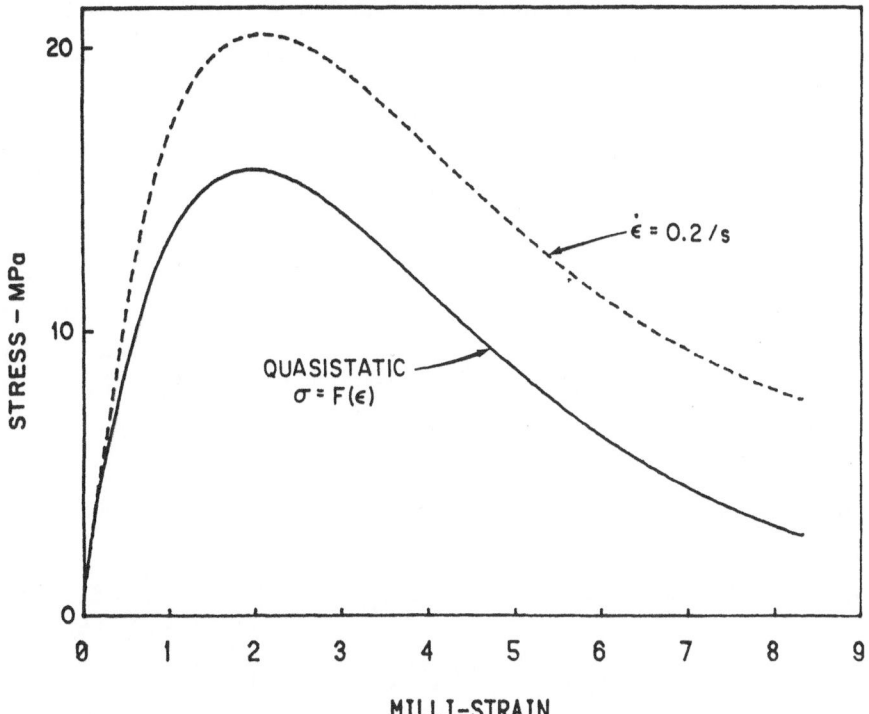

MILLI-STRAIN

FIG. 9. Behavior of rate dependent model

sitivity of actual experiments in the strain-softening range, where strain
localizations and stick-slip phenomena do not lead to precisely repeatable details in the load-deformation histories from specimen to specimen, even though
the gross features of strain-softening may indeed be reproducible. Furthermore,
rate dependence in the form of relaxation and creep is observed (Wawersik and
Brace, 1971) in this regime.)

It is clear from the foregoing that numerical solutions for rate dependent
models such as the preceding one can present practical problems because of the
exponential growth possible in the response. In order to represent such growth
properly, computer resources might be stretched beyond practical limits. New
numerical techniques might be required to approximate the underlying continuum
solution. We hope to develop such techniques in the future.

5. CONCLUSIONS

The self-inconsistency of rate independent models of strain-softening for
dynamic continuum problems is demonstrated by means of numerical examples as
well as through stability analyses. Further analysis is presented to show that
the inclusion of rate dependence allows properly posed descriptions of strain-
softening in dynamic continuum mechanics.

Even though rate dependent models of strain-softening are stable in the
sense of Hadamard (and so are physically reasonable) they still manifest the
physical instabilities always observed when strain-softening occurs. Numerical
treatment of such models may present some practical challenges because the solu-

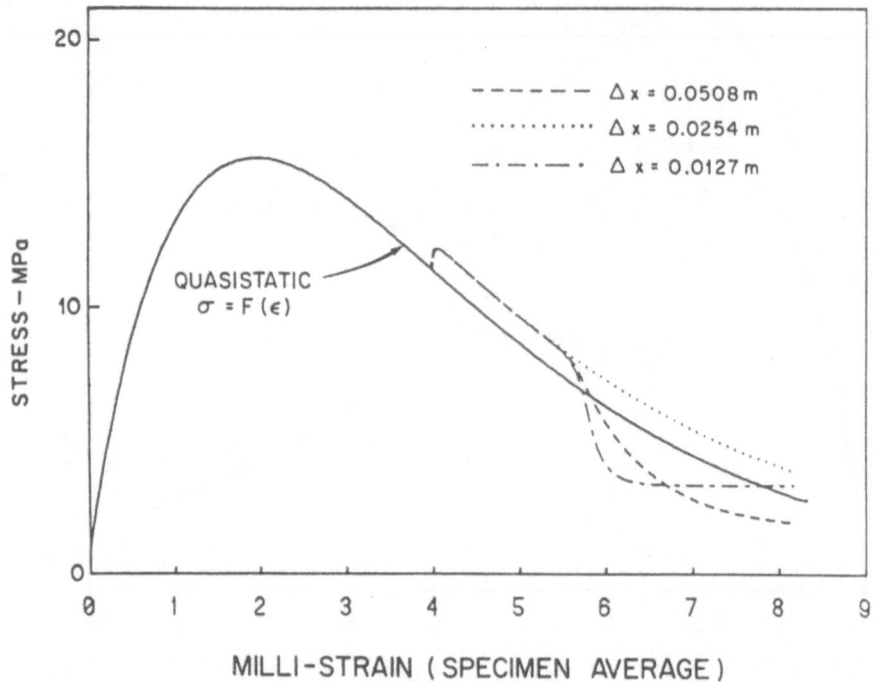

FIG. 10.   Effect of grid size on computations using the rate dependent
           model for $\dot{\epsilon} \approx .01/s$

tions grow exponentially with time and may be difficult to approximate within
computational meshes of limited size.

ACKNOWLEDGMENT

     The authors would like to thank their colleagues at Weidlinger Associates
and Columbia University who contributed to this work through numerous helpful
and stimulating discussions (to put it mildly).

REFERENCES

Bazant, Z.P. (1976), "Instability, Ductility, and Size Effect in Strain-Softening Concrete," *J. Engrg. Mech.*, ASCE, *102*, 331.

Drucker, D.C. (1956), "On Uniqueness in the Theory of Plasticity," *Quarterly of
Appl. Math.*, *14*, 35.

Hadamard, J. (1903), *Leçons sur La Propagation des Ondes*, Hermann, Paris, France.

Wawersik, W.R. and Brace, W.F. (1971), "Post-Failure Behavior of a Granite and
Diabase," *Intl. J. Rock Mechanics and Mining Science*, *3*, 61.

GENERAL DISCUSSION

NEEDLEMAN: I think there are three concepts you mentioned that are very important ones but distinct. One is the concept of sensitivity to changes. The other is the concept of boundary layers, and the third is the concept of material stability.

Regarding the concept of sensitivity to small changes, a material like rubber, at small strains with Poisson's ratio near one-half, is nearly incompressible. Then, if you were to consider a shear plus an oscillation that gave a small volume change, you'd find large differences in the stress response. But I think one would agree one had stability.

SANDLER: Well, you can measure the bulk modulus of rubber. The incompressibility may be a convenient simplification —

NEEDLEMAN: But I'm saying with a finite bulk modulus you get large changes in stress, small changes in strain.

SANDLER: You don't get finite changes in stress from infinitesimal changes in strain.

NEEDLEMAN: Yes, but the change can be large.

SANDLER: Fixing the shear modulus and making the Poisson ratio equal to a half is equivalent to not defining the compressibility.

NEEDLEMAN: What I'm saying is you get a sensitivity to small changes. Anyway, that's one point. Another point is —

SANDLER: But you know how to measure compressibility. You know how to measure it and you know how to describe it. That's the important thing.

NEEDLEMAN: That's the very important point. I agree. The second thing is that boundary layers arise in elliptic problems. For instance, $\varepsilon^2 y'' + y = 1$ gives a boundary layer effect.

And finally there's material instability which has its own set of problems. But I think sensitivity to relatively small changes and boundary layers can arise in circumstances where you don't have material instability. They are not necessarily coupled.

SANDLER: I agree. In fact, that's my point. Generally, what we observe as instability is not material instability. It tends to be geometric instability in the most common situation. Buckling of a beam, for example, is not a material instability.

NEEDLEMAN: Well, I think there is material instability, as well, but one has to be careful about that.

SANDLER: Yes, if we wish to study sensitive physical systems, we must be careful about how we try to quantify them.

ODEN: Let me ask, before I recognize Sia, how do you define material instability, the same way that Ivan is doing it?

NEEDLEMAN: I would define it in terms of the Hadamard condition.

NEMAT-NASSER: I want to emphasize the last part of the comment that Alan was making. I think once the physics tells you that something else is happening, although it is nice to make the equations well posed, one should stop and take a look at the physics. Now, that one-dimensional picture that you have proceeded to analyze and make its corresponding differential equation well posed, really ceases to exist, as I indicated with one example in the case of rock, and it becomes indeed a three-dimensional problem.

The stress-strain curve for mild steel shows upper and lower yield limits. But with proper techniques one obtains a smooth unstable response curve. And what emerge in a simple torsion test are wedges of elastic and plastic regions. Please note that this instability occurs at very, very small strains. One obtains either shearing off instability, or instability consisting of alternating plastically deformed wedges and elastic wedges. So one no longer has a one-dimensional state of stress. It's very complicated. At this point one has to stop and take a look at the physics. This problem has been completely solved, and it is in the literature, so one can relate to it.

SANDLER: I agree with everything you've said. The point of my comments concerns the word "stop." You used the word "stop," and I agree with you. You have to stop and take another look. The trouble is that computer programs are not intelligent; they do not stop. They continue to calculate. And if they do so with models which do not include all of the relevant physics, you have a serious difficulty on your hands. That really, I think, is the main practical thrust of what I'm trying to say. The concepts of stability are not new. I think they're at least eighty years old. It seems that Hadamard understood all of this, but the conditions that he formulated are sometimes overlooked when people approach things numerically today.

HUTCHINSON: I'm just rising to take issue with Sia in his claim that this problem is completely solved. I wouldn't bother to mention this normally except that I think it is important in connection with the concerns that have arisen here to mention that that problem hasn't been solved at all, as I understand it. There have been crude attempts to explain how the bands go in, but such basic questions as the number of bands, for example, have not been solved at all. There is an approximate solution method proposed, but it's not what I would call a "complete solution." I say this because I think, and this is a very nice example, that this is a very hard problem which cannot yet be solved by the tools we now have.

NEMAT-NASSER: I hate to see John's comments are misunderstood! I'm sure he didn't intend to take away from what I was trying to point out. What I was trying to point out was not whether this particular paper anticipates precisely for a given set of experiments what will be the exact final response.

What I am trying to emphasize is that once material shows this type of response, instability, then the simple torsion test that you all rely on, at strains less than one percent, ceases to be a simple problem, and that we have a very complicated three-dimensional stress pattern. So in that sense, I want to say that John was supporting this statement. He is saying that the problem becomes even more complicated than what I was suggesting. Is that correct, John?

HUTCHINSON: Yes.

HUGHES: I would like to come back to your problem involving oscillating strain. I was trying to follow as you went through it, but I seem to come up with something very different. You drew a very strong conclusion, and one should scrutinize it. It seems to me what happens in your picture is as follows: Let's assume that in the first "bump" on your strain path you go to the right so there is some yielding. At that point there, of course, you can't get outside the yield surface. All along you're walking along the yield surface. So that point, P, let's say, is the point of stress that corresponds to the peak. Now, when you reverse the strain direction from P, you go through a *double* excursion. Now, when you come back you're just going to hit P. That's an entirely elastic process. You never reload plastically. Consequently, you "ring" elastically thereafter.

SANDLER: Well, that would be true if in fact you have hardening.

HUGHES: It is true independent of hardening. It also could be true for the specific case of an equilibrium path, but if I superimpose upon this a general loading, then your ω·a (frequency times amplitude), formerly constant, is now growing. And, of course, you expect to accumulate and drift when you're putting in that type of an input.

SANDLER: You can always envision an underlying path which keeps you on the yield surface and then have the perturbations always continue from a state on the yield surface. Your point is well taken. The problem requires a further definition of detail. If the nominal strain path is exactly constant, you're right. Only your first loading will be plastic. But I think you can always envision a slowly loading nominal path which would keep you on the yield surface so that this would always be true.

HUGHES: I think it would have to be shifted off of zero.

KRIEG: Tom covered the point I was going to make. There is a further point, that as ω·a goes to infinity, P approaches zero.

SANDLER: No, I don't think so.

KRIEG: ω·a goes to infinity, a goes to zero, P approaches zero.

SCHUSTER: Ivan, as you know, we've looked at this problem also, and I think that in some ways we may be doing a disservice by picking to examine what may be pathological cases! I think the solution to your problem, as Dick Nelson of UCLA and Marv Ito of our shop pointed out, is that you've chosen a velocity which just puts you over the softening condition. Consequently, you get strain-softening only in the very first zone; and therefore, the calculation is very sensitive to the zone size.
      We know you can't resolve anything less than the zone size and if you choose a problem in which everything happens within the first zone, if you decrease the zone size and the pulse still remains within the zone, you can generate all kinds of erroneous results.

ODEN: What's probably happening here, however, is that there is no solution whatsoever beyond this point, and your finite element or finite difference approximations are giving you numbers; they're giving you an erroneous solution.

PREVOST: But how did you construct the solution?

SANDLER: Let me say how we did construct the solution. We used a straightforward marching scheme. The code integrates the equations of motion forward. It calculates stresses for whatever the strains happen to be at any particular time.

And you can always march forward. If I just let the code run, it will produce numbers. That's all I will say about that calculation, though.

TRULIO: Your point about having no solution just reminded me of something that came up many years ago. For design purposes, it would often help to be able to run problems backwards from an answer that you like. That was actually tried. But on looking into the idea carefully it turned out that with shocks in the system, there was more than one solution from any given time onward. When the codes were run backwards in time, errors in numerical solutions determined the end answers.

ODEN: The most elementary heat conduction problem in one dimension: Suppose you want to specify the temperature at time T. What initial conditions are required to produce that temperature? That's an ill-posed problem and in general does not have a solution. There are a lot of them.

SANDLER: Well, I think what we most often classify as physical instabilities fall into that category. If it is known that a column buckled under a load and that it buckled by bending in some particular direction, well, then you can say that it must have had some eccentricity in that direction. But to go the other way and predict a priori in which direction the buckling is actually going to occur when an axial load is applied is a much different problem.

PREPARED DISCUSSION*

by G. A. Hegemier and H. E. Read

The purpose of this paper is to discuss some recent theoretical work by Wu and Freund (1983) as it relates to the present paper by Sandler and Wright on strain-softening, and to briefly review several experimental studies conducted on rock and soil which provide important insight into the nature of strain-softening. We wish to explore the question of whether or not strain-softening is a real material property or merely the result of inhomogeneous deformation caused by the experimental technique.

1. MATHEMATICAL OBSERVATIONS AND INTERPRETATION

The wave propagation problem posed by Sandler and Wright falls into the category of "deformation trapping." A number of authors have investigated this problem from different standpoints. The recent work by Wu and Freund (1983), however, is more relevant to the present stability problem and it is instructive at this point to review this work.

The problem considered by Wu and Freund is mathematically identical to that given by Sandler and Wright. The physical interpretation, however, is different. It will be convenient here to adopt Wu and Freund's problem which may be stated as follows: Let the half space x > 0 (Fig. 1) be occupied by a material described by the constitutive relation

$$\tau = F(\gamma) \tag{1}$$

*Research sponsored by the Air Force Office of Scientific Research (AFSC), under Contract F49620-81-C-0033. The United States Government is authorized to reproduce and distribute reprints for governmental purposes notwithstanding any copyright notation hereon.

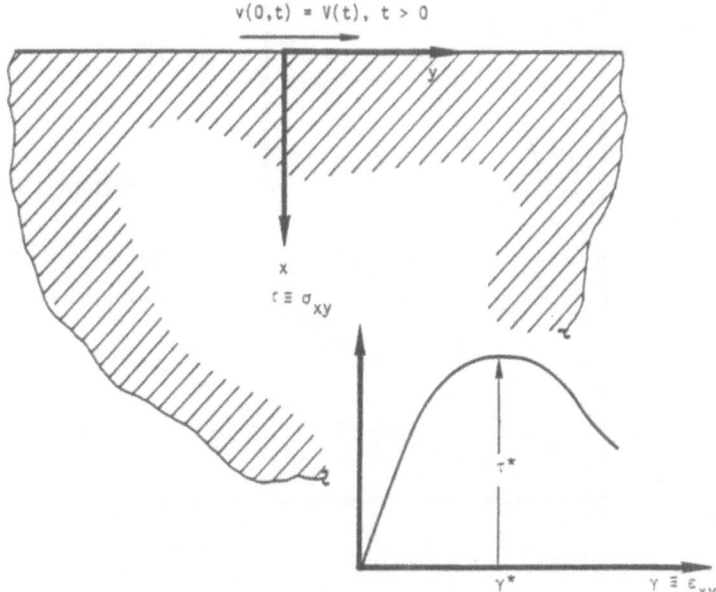

SHEAR OF HALF SPACE
(Strain-Softening Material)

$v(0,t) = V(t), t > 0$

FIG. 1.   Problem Configuration

where τ, γ denote shear stress, shear strain, respectively, and F has a continuous derivative. The material is stress free and at rest for time up to t = 0, and thereafter the material is subject to the boundary velocity

$$v(0,t) = V(t), \quad t < 0 \qquad (2)$$

where $V(t)$ is a nondecreasing function of time. The equations of momentum balance and kinematic compatibility are (small deformations assumed)

$$F'(\gamma)\gamma_x - \rho v_t = 0, \quad v_x - \gamma_t = 0 \qquad (3)$$

where ρ denotes density, subscripts denote partial derivatives and a prime denotes an ordinary derivative. For $F' > 0$ the system of partial differential equations (3) is hyperbolic, and the solution subject to quiescent initial conditions and the boundary condition (2) is unique and has the form of a simple wave. Accordingly, the particle velocity propagates away from the boundary x = 0 along a straight line characteristic in the x,t-plane at a speed c and slope dt/dx of

$$dt/dx = c^{-1}, \quad c = \sqrt{F'(\gamma)/\rho}. \qquad (4)$$

Now, let the boundary velocity be sufficiently large that $F'(\gamma) = 0$ at $\gamma = \gamma^*$ and $t = t^*$. Then (see Fig. 2), the speed of propagation of this strain level is zero and the associated characteristic is aligned with the t-axis in the x,t-plane. Thus, the boundary itself becomes characteristic and it is no longer possible to communicate boundary data to interior points in the half space. As V increases for $t > t^*$, a discontinuity in the velocity field develops at x = 0. This discontinuity, however, can only exist if the above characteristic remains aligned with the t-axis. Thus, for $t > t^*$ and points in the neighborhood of the

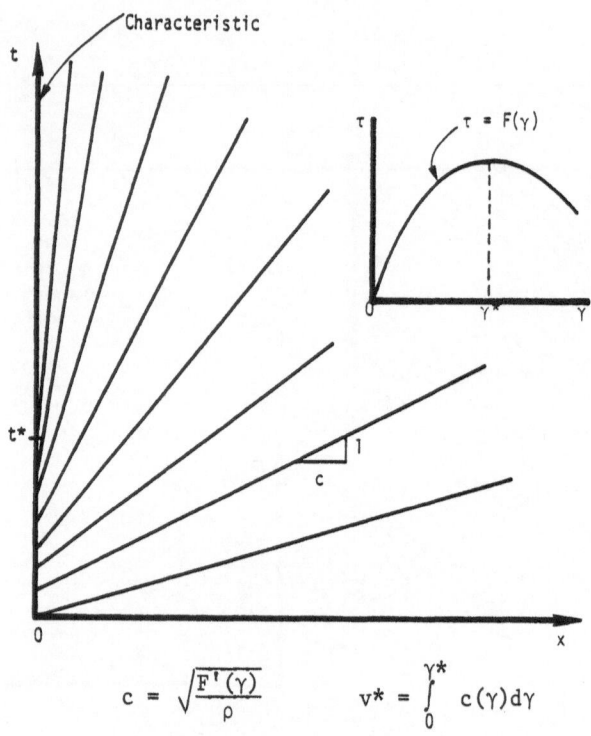

$$c = \sqrt{\frac{F'(\gamma)}{\rho}} \qquad v^* = \int_0^{\gamma^*} c(\gamma)d\gamma$$

FIG. 2.   Evolution of Characteristics

boundary, the fields v, τ, γ become uniform at levels v^*, τ^*, γ^*, respectively, and there is a discontinuity of magnitude $V(t) - v^*$ across $x = 0$, where

$$v^* = \int_0^{\gamma^*} c(\gamma)d\gamma. \tag{5}$$

The evolution of the foregoing discontinuity is depicted in Fig. 3.

The vanishing of the tangent modulus $F'(\gamma)$ at $\gamma = \gamma^*$ leads to the concept of "deformation trapping" in general and "shear banding" for the problem as posed. Here one argues that, since the speed of propagation of increasing strain levels vanishes at a critical strain, strains greater than the critical level are trapped at the boundary and are accumulated into a deformation (shear) band. The infinite strain rate associated with the discontinuity at $x = 0$ indicates that rate effects become important as $F'(\gamma) \to 0$ and that a rate dependent model is necessary to determine the fields within this band.

Wu and Freund have examined the development of the shear band noted above by adopting a rate dependent model and conducting a boundary layer analysis. One model considered has the linear rate form

$$\tau/\tau_0 = f(\gamma/\gamma_0) + m\gamma_t, \qquad \gamma_t \ge 0,$$

$$f(z) = z^n/(1 + az^{n+1}), \tag{6}$$

where a, n, m are material constants and τ_0, γ_0 are reference states. A solution was sought for $t > t^*$ which satisfied the initial conditions $v = v^*$, $\tau = \tau^*$, $\gamma = \gamma^*$ at $t = t^*$, the boundary data $v(0,t) = V(t)$, and $v, \tau, \gamma \to v^*, \tau^*, \gamma^*$ as $x \to X$, where

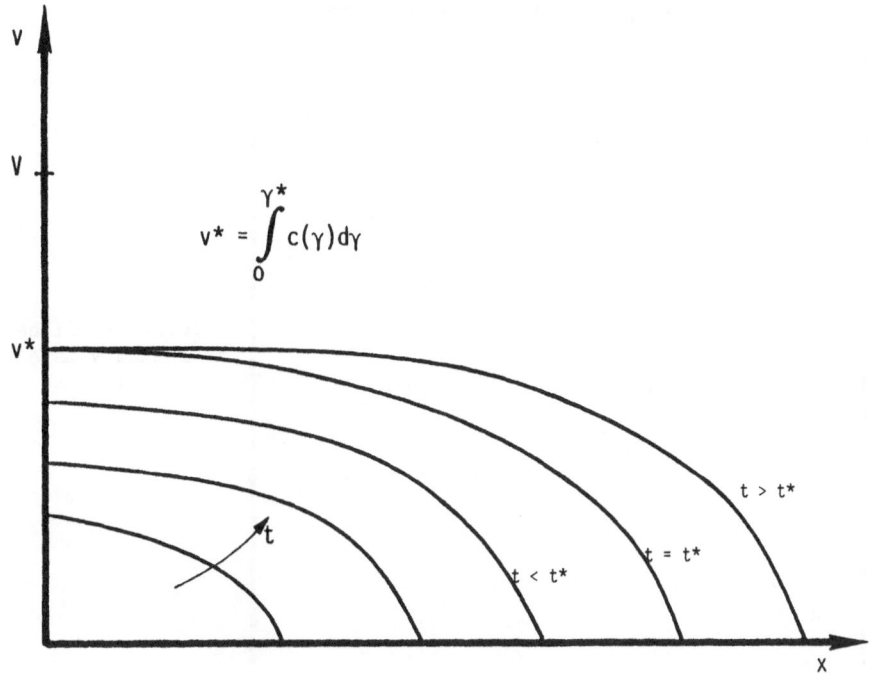

FIG. 3. Velocity Profiles, Inviscid Case

$X \gg m\sqrt{\tau_0/\rho}$. Typical boundary layer solutions for $(v- v^*)/c_0 = 0.1$ at $x = 0$ are illustrated in Figs. 4a,b. As can be observed, a well developed band of intense shear deformation develops adjacent to the boundary. Wu and Freund show that the boundary layer thickness is sensitive to the particular form of the rate-dependence in the constitutive model.

The implications of the example treated above are considerable and may be stated as follows: (1) information concerning the descending branch of the stress-strain curve cannot be communicated to points interior to the boundary in the case of a rate-independent model, (2) in view of the singularity at the boundary, a conventional numerical (finite difference) treatment of the strain-softening problem is not feasible for a rate-independent model, (3) the actual deformation field associated with strain-softening is highly nonhomogeneous, and (4) numerical treatment of the rate-dependent case may be difficult due to boundary layer effects.

The formation of localized deformation bands implies that the usual assumption concerning homogeneous deformation and stress fields in the laboratory is not valid if strain-softening occurs. Thus, the conventional laboratory tests to determine material constitutive parameters are not appropriate in the presence of strain-softening.

The above discussion presupposes that strain-softening is a material characteristic. While shear bands are observed in the case of soils and metals under certain conditions, strain-softening in brittle materials such as rock and concrete can be attributed to geometric effects that occur during laboratory testing. This point is amplified in Section 2.

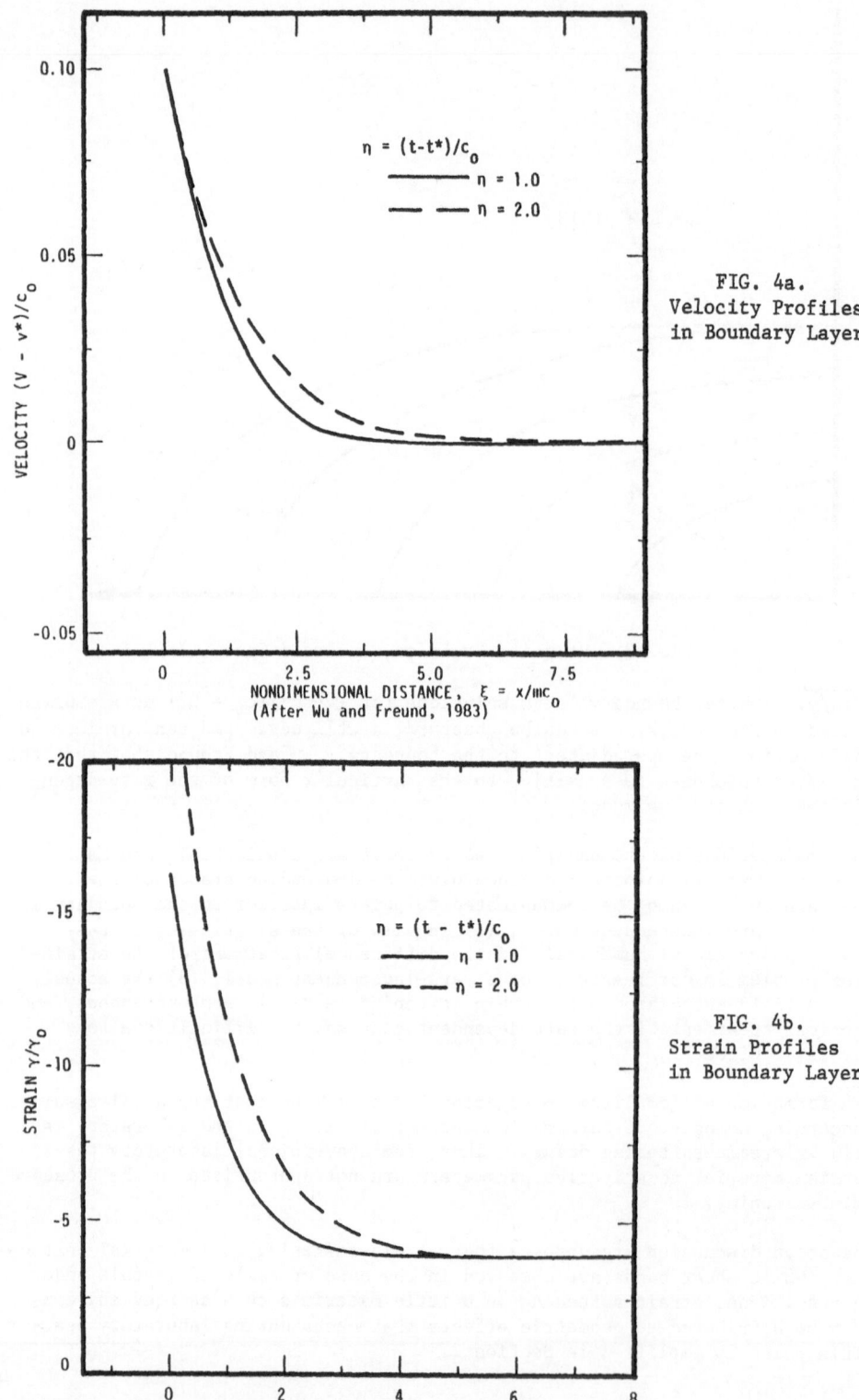

FIG. 4a.
Velocity Profiles
in Boundary Layer

FIG. 4b.
Strain Profiles
in Boundary Layer

2.   EXPERIMENTAL OBSERVATIONS AND INTERPRETATION

Numerous results from laboratory tests conducted with displacement control on materials such as dense sand, rock and concrete reveal that, under uniaxial compression and triaxial compression up to some critical confining pressure, the stress-strain relation first hardens to a peak and then softens. The softening branch of the curve occurs concurrently with increasing inhomogeneous deformation of a specimen. In soils, the inhomogeneous deformation usually results from either shear banding or the development of rigid cones near the specimen's ends caused by friction between the soil and the end plates. In concrete and rock, the inhomogeneous deformation results from the development of large cracks and slabbing, and also the formation of rigid cones near the specimen ends.

Deman (1975) used X-ray techniques to investigate the stress fields produced in cylindrical samples of dry sand by triaxial compression. The tests were conducted with and without lubrication of the end plates. Typical results from these tests are depicted in Fig. 5a. Although both curves exhibit a peak, the softening is much more pronounced when the end plates are not lubricated. Fig. 5b shows the measured deformation fields for the lubricated and non-lubricated cases. The non-lubricated case produces substantial nonhomogeneous deformation, with the deformation being essentially confined to a wedge-shaped ring surrounding rigid cones adjacent to the end plates.

Hettler (1981) performed triaxial compression tests on flat specimens of dense dry sand (same as that used by Deman) with lubricated end plates. Typical results for several confining pressures are shown in Fig. 6. From this, it is seen that no significant softening occurred. Based on these results, Dresher and Vardoulakis (1982) recently concluded: "These results mean that the true material softening is very slow and can be neglected for relatively large strains after the limiting state has been reached."

Hudson et al. (1971) conducted a systematic series of uniaxial compression tests on a rock* to determine the influence of specimen geometry and specimen size on the observed strain-softening characteristics. All of the experiments were done at the same strain rate ($\varepsilon = 1.5 \times 10^{-6}$ sec^{-1}), using a closed loop, servo-controlled testing machine. The end plates were not lubricated. The results from this series of tests are depicted in Fig. 7, which illustrates the effects of both specimen size and specimen geometry on strain-softening. Each curve shown represents the average of three tests. The $\sigma - \varepsilon$ curves were obtained by scaling the force by the original cross-sectional area, rather than the continuously decreasing cross-sectional area. As the figure reveals, specimen geometry has a very pronounced effect on the measured strain-softening relation. For short specimens, no softening is evident; as the L/D ratio increases, the softening becomes increasingly prominent.

Figures 8 and 9 show cross-sectional and end views of some of the test specimens of Hudson et al. in the advanced states of failure. From these, it is seen that gross slabbing of material has occurred, resulting in a decrease in the effective cross-sectional area.

It appears that the significant influence which the L/D ratio exerts on the specimen response can be explained by reference to Fig. 10. For large L/D, the slabbing and shear failure lead to large reductions in the effective cross-sectional area, while for small L/D ratios, the reduction in cross-sectional is

---

*Georgia Cherokee marble.

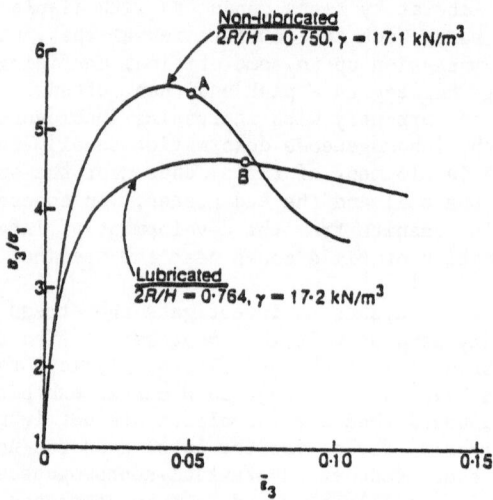

Stress ratio versus axial strain for dense sand -- triaxial
compression (after Deman, 1975)

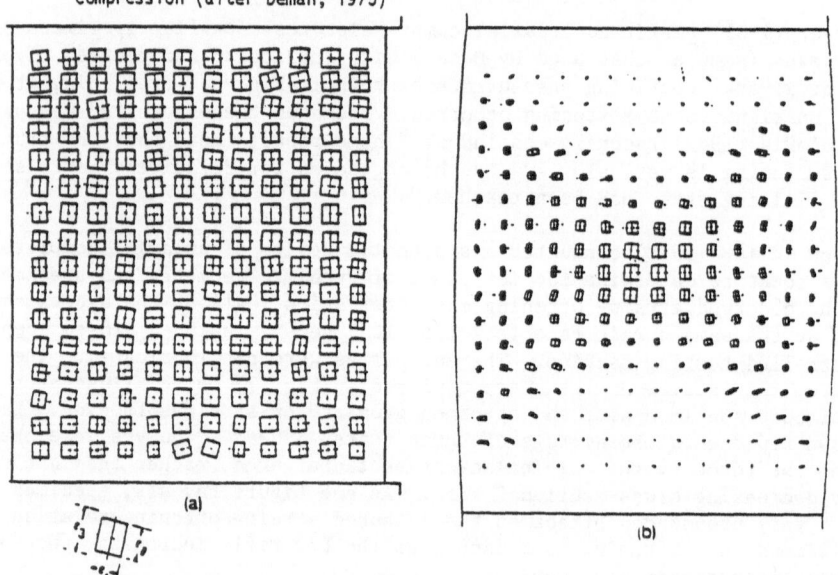

FIG. 5.   Internal deformation field for *dense sand*; (a) *lubricated*
          end platens; (b) *non-lubricated* end platens (after Deman, 1975)

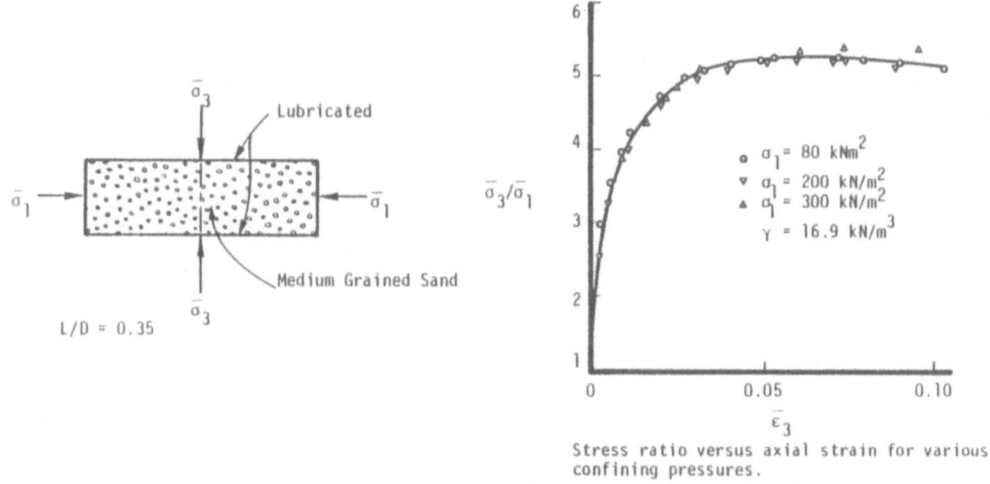

Stress ratio versus axial strain for various confining pressures.

FIG. 6.   Triaxial Compression Tests on Flat, Lubricated Specimens of
          Medium Grained Sand (Hettler, 1981)

very small. These observations support the conclusions that strain-softening is
not a material property, but is essentially due to scaling the applied force by
the original cross-sectional area, rather than the actual cross-section, which
diminishes as failure progresses. Unfortunately, it is extremely difficult — if
not impossible — to experimentally track the effective cross-sectional area at
each stage of the failure process for specimens with large L/D ratios. Experimental data from such specimens would appear to reflect the true stress-strain
curve.

4. CONCLUSIONS

Treatment of the strain softening problem posed by Sandler and Wright leads,
for a rate-independent model, to a boundary discontinuity and to the conclusions
that the material interior to the boundary never experiences the descending
branch of the stress-strain curve. The boundary singularity renders conventional
numerical treatment infeasible. Inclusion of rate-dependence leads to deformation trapping, i.e., a boundary layer region wherein the fields are highly nonhomogeneous. Numerical treatment of such boundary layer problems may be difficult.

Examination of laboratory data reveals that "observed" strain softening in
the laboratory most probably is the result of geometrical effects and not of
material behavior. The highly nonhomogeneous stress and strain fields observed
in the neighborhood of specimen "failure" renders the assumption of homogeneous
fields invalid and, consequently, data from such specimens cannot be utilized in
the usual manner to determine material model parameters.

REFERENCES

Deman, F. (1975), "Achsensymetrische Spannungs-und Verformungsfelder in trockenem
Sand," Dissertation, University of Karlsruhe.

Dresher, A. and I. Vardoulakis (1982), "Geometric Softening in Triaxial Tests on

Granular Material," *Geotechnique, 32,* No. 4, 291.

Hettler, A. (1981), "Verschiebungen starrer und elastischer Grundungskörper in
Sand bei monotoner und zyklischer Belastung," Dissertation, University of Karlsruhe.

Hudson, J.A., E.T. Brown and C. Fairhurst (1971), "Shape of the Complete Stress-
Strain Curve for Rock," *Proc. 13th Symp. on Rock Mechanics,* University of Illinois, Urbana, Ill.

Wu, F.H. and L.B. Freund (1983), "Deformation Trapping due to Thermoplastic Instability in One-Dimensional Wave Propagation," Brown University, Providence,
R.I., Report MRL-E-145, April.

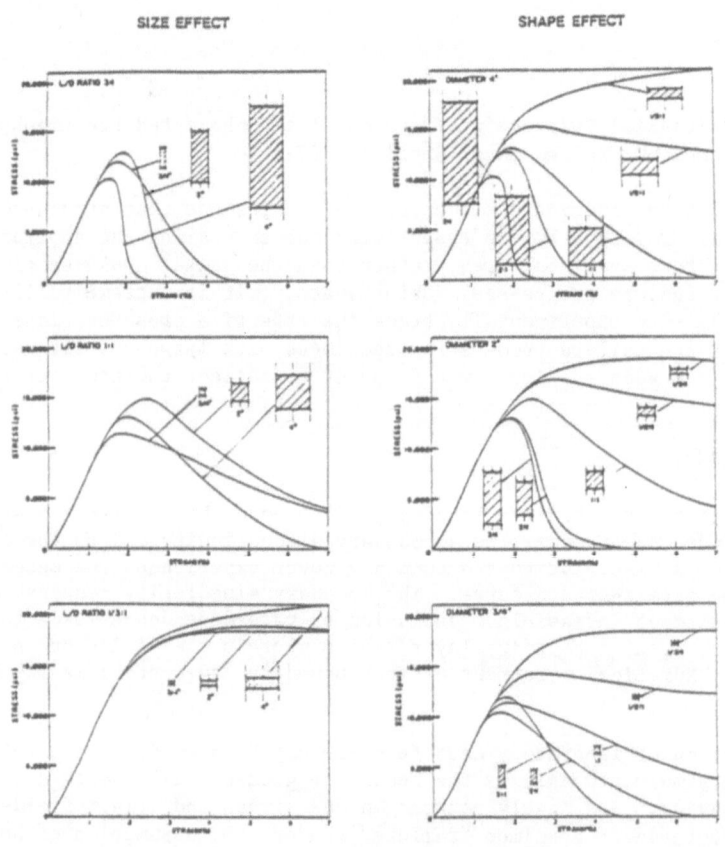

FIG. 7.   Influence of Specimen Size and Shape on the Complete
          Stress-Strain Curve for Marble Loaded in Uniaxial
          Compression (Hudson et al., 1971)

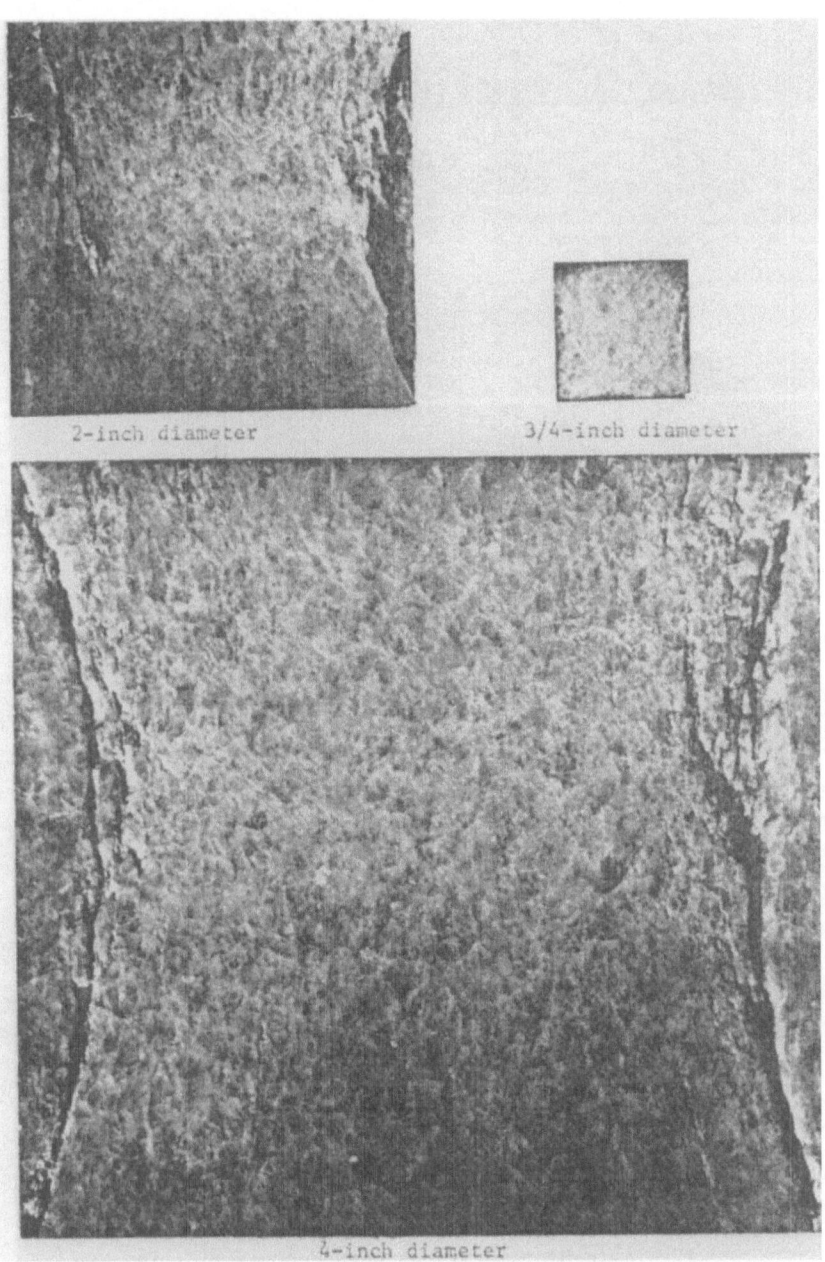

2-inch diameter   3/4-inch diameter

4-inch diameter

FIG. 8.   Cross-section of 1:1 specimens at an advanced
          state of failure. (Georgia Cherokee marble)

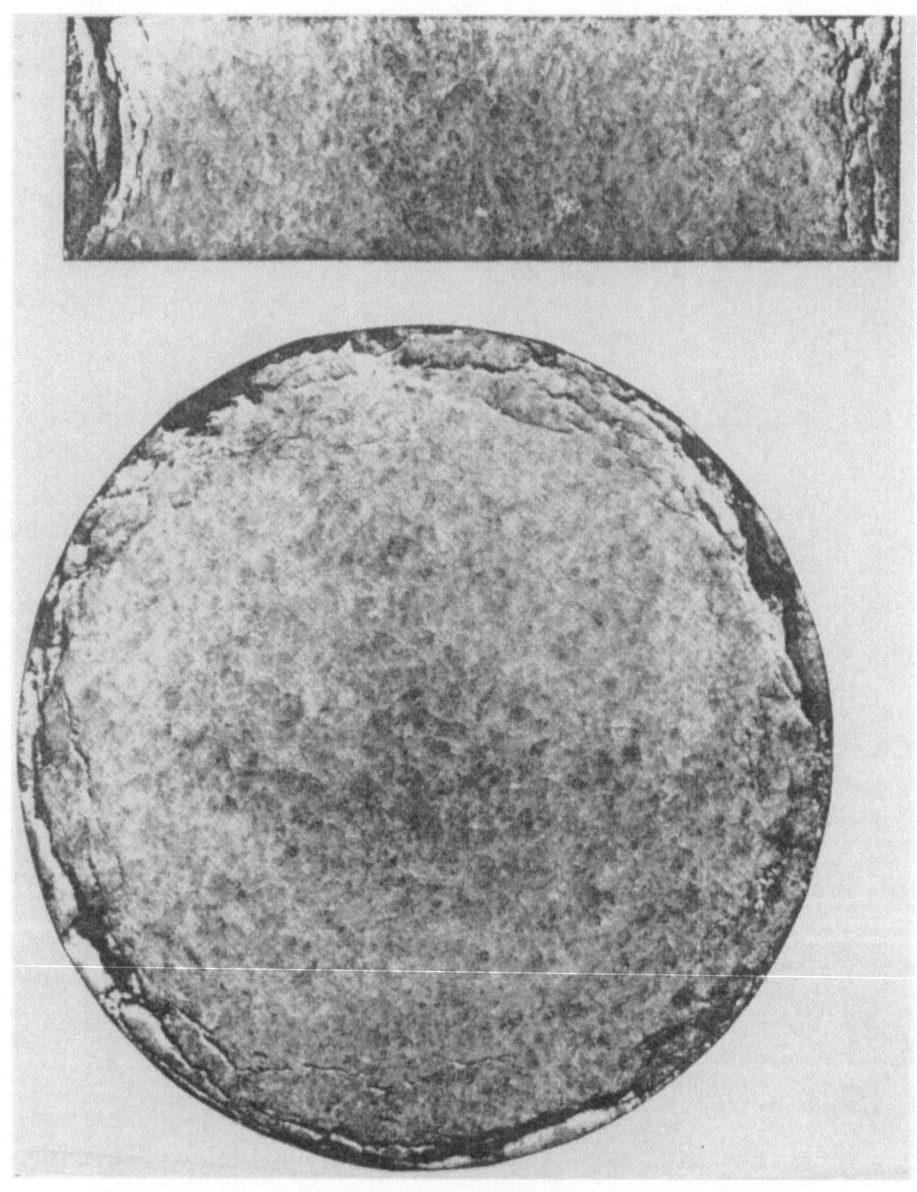

FIG. 9.   Cross-section and end view of 1/3:1 specimen
          (4-inch diameter) at advanced state of failure.
          (Georgia Cherokee marble)

DEFINITIONS

$$\sigma_{TRUE} = \frac{F}{A(\varepsilon)}$$

$$\sigma_0 = \frac{F}{A(0)}$$

LARGE L/D RATIO

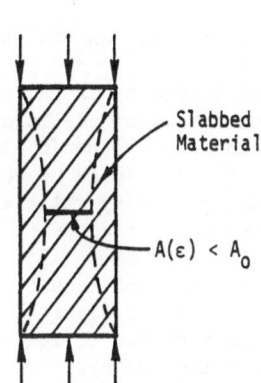

Slabbed Material

$A(\varepsilon) < A_0$

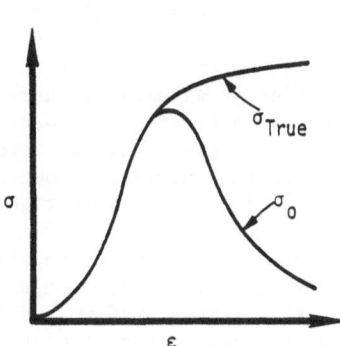

SMALL L/D RATIO

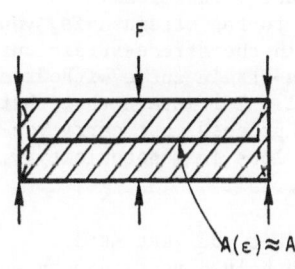

$A(\varepsilon) \approx A_0$

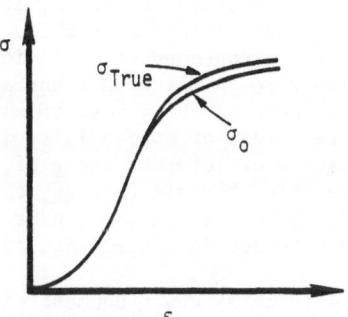

FIG. 10.   Effect of Stress Definition on Shape
of the Stress-Strain Curve

GENERAL DISCUSSION

HUGHES: I'd like to make a few comments, and admittedly these thoughts just came into my mind as you were putting those characteristic diagrams on the board. I think perhaps my thoughts are imprecise, but I sense a relationship between a well-studied area of a transonic flow problem and the present problem. There you have the same type of equation. You have elliptic zones and hyperbolic zones, exactly the same thing. Here you have it in space-time, but there you have it exclusively in space. In transonic flows, as perhaps you know, these regions are sometimes separated by shock waves. It seems to me that one can solve these problems. There are boundary conditions and initial conditions, whatever, that you can set to make well-posed problems.

I'm just wondering if adding all of this viscosity isn't just spreading a discontinuity layer that is somewhat fundamental to the problem?

HEGEMIER: The transonic problem is a spatial problem, as you have indicated. This is not a spatial problem. In the case of the transonic problem there are regions with different character which are separated by characteristics. This is a very different situation. Here you cannot communicate information from the boundary to the interior. That doesn't happen in a transonic flow problem. The addition of viscosity allows information to once again be communicated from the boundary to the interior. However, adding viscosity will lead to a boundary layer. Numerically, you may therefore go from the frying pan to the fire.

ODEN: Well, I will take my prerogative as chairman and insert a quick comment. My reaction is that there are not enough axes on these figures. What is shown is stress versus strain. If we had another axis or two to indicate other features of the response, then it might be more enlightening.

One other point is that nowhere in this discussion have you heard plasticity or unloading; this is strictly a feature of the geometric property of the stress-strain curve. Why can't we think of a situation, therefore, like this: Suppose you plot the free energy in a thermodynamic process and this free energy can be a function of strain, but it could be a function of other parameters, such as moments of dislocation densities; it could be functions of strain rates and so forth. And there is one axis where I will put strain. Now, it's perfectly possible to have a non-conventional yield — free-energy functions.

If you were to slice this surface parallel to the strain axis, you may perceive that there can be a loss of monotonicity in the stress-strain curve. There can then be a loss of monotonicity in the stress-strain curve without a change in temperature or internal state of the material. I maintain that that's probably impossible, that in fact any thermodynamic process must carry you on a path along this surface where you cannot suffer that loss in monotonicity without a concomitant change in other features of the material.

CURRAN: I think my short comment will support what you just said.

But this general problem has, I think, been solved years ago by Whittam where he described just this series of problems.

If the hyperbolic nature of equations breaks down, it's because the original constitutive relations weren't sophisticated enough. You put in some more terms which restore the hyperbolic nature. It introduces a new fast wave speed. If then the thing still shocks up or does something, traps strain, you bring in another term.

Whittam was actually considering hydrodynamics, but he showed how you could keep adding terms every time you got into trouble in your force law or in your constitutive relation. Eventually, if you keep going, you run into the speed of light as your leading characteristic, and everything else will diffuse behind it.

So I think what Gil showed was, if you add a term, in this case a loosely called
viscosity, you can get a characteristic which carries information out away from
the boundary. You certainly cannot have a characteristic parallel with the
boundary because then the compatibility relations on that characteristic will
compete with the imposed boundary conditions. I think this is a very important
point, that when we run into these kinds of troubles, it is the constitutive relation community that's challenged, not the numerics.

HERRMANN: You've got your one-dimensional example right. It's exactly what
Von Karman looked at during the Second World War, and the critical velocity is
the Von Karman critical velocity. What they were doing was measuring dynamic
properties by dropping weights on the end of wires and essentially putting a
constant velocity on the end of the wire in tension; at the critical velocity
the wire breaks right at the end. If you do this in compression, then you get
the Taylor bar experiment. Two-dimensional effects take over in "uniaxial
stress." If you do it in uniaxial strain, of course, the compressibility of the
material makes the stress-strain part concave up again, so that other effects
take over.
    I think there is a clue here, to my way of thinking, when you have this
kind of a situation it leads to an instability, and the one-dimensional continuous solution you expect ceases to exist, at least in this example.

BELYTSCHKO: I just want to address the question of whether solutions exist for
finite element models of this class of problems. The problem I considered is
similar to the problem that Wu and Freund looked at, and it's the same problem,
incidentally, that Von Karman and Taylor looked at — Fig. A. The problem does

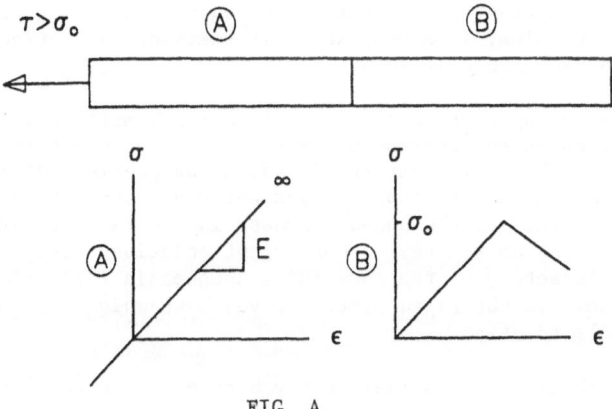

FIG. A

have difficulties. I think Ivan Sandler made a very good point in calling our
attention to some very serious shortcomings in the finite element models that
are used.
    On the other hand, you may take a very simple problem, consisting of two
materials, one of which is elastic, the other with a strain softening part. The
solution to this problem can be constructed using the d'Alembert method, but you
have to introduce a delta function into the strain (displacement) at this point
(the interface between the materials). In other words, at the time that the wave
arrives there, you reach the critical stress; physically there is rupture at this
point. Consequently, this problem is very similar to having two bodies glued together. There are mathematical difficulties in this problem in that you have to
use a delta function. However, if you look at the finite element solution and
compare it to the d'Alembert solution you find that you do have convergence in

the H^1 norm except at the interface where the strain does not converge. The
smaller you make your element, the higher the strain. But outside of that one
element, you have convergence to an analytic solution.
     So mathematically, this is not a complete explanation; many difficulties
remain. But to me, the fact that you are able to converge to a solution does
indicate there is some hope in using this type of model for computational work.

ODEN: Are you saying that you prescribe a Dirac delta for the solution or for
its derivative?

BELYTSCHKO: You have to include it in the solution.

ODEN: You have to throw it in. So what you've effectively done is to incorporate a failure criteria in the model, right?

HUGHES: That's not right. You want a jump in displacement, not a delta function in displacement.

BELYTSCHKO: Oh, yes, I said it wrong; the displacement is discontinuous.

ODEN: Then you will get a delta function for the displacement gradient.

SANDLER: This exercise highlights a key issue. It's a very intelligent way to
solve the problem. But you don't have a single continuum there. You have two
continua. You are giving up continuity of displacement. If you use a code that
enforces the equation of continuum mechanics for the bar as a whole, you won't
be able to get that nice solution.
     There is an important point there. Although this is a good (intelligent)
way to solve the problem, it's not, strictly speaking, a continuum approach,
and it has to be recognized as such.

BELYTSCHKO: I'm saying there are difficulties here, particularly in the mathematics. I don't think the current mathematics of finite elements can deal with
this. On the other hand, in most strain-softening problems of interest, you
have failure in a confined region, and you get your strength out of the remaining
part of the structure. So the non-uniqueness in the failing part is not of great
concern. And as long as you have a convergent solution outside that domain, the
situation is satisfactory — the same thing happens in rigid plasticity. You
have non-uniqueness in the rigid part, but you have uniqueness over the remainder
and in the overall solution.

SANDLER: The problem can be solved, but you have to use the right tools.

CLIFTON: I think that the point that needs to be made with regard to these dynamic problems and the possibility of strain-softening is that one should really
view these problems within the framework of rate dependent plasticity.

CLIFTON: Then the stability problem is not one of looking for zero wave speeds.
In rate dependent plasticity, the wave speeds are the same as the wave speeds
for the instantaneous elastic response of the materials. There can still be instabilities such as localized regions in which very high shear strain rates develop — as, for example, in the deformation trapping problem. But, it's a different kind of instability to analyze. Whenever you are analyzing a problem
with a high frequency disturbance or a jump in a derivative of particle velocity
you should look at it within the framework of a rate dependent theory. The
principal part of that system of equations determines the response to the high

frequency disturbance. And that part is totally hyperbolic. There is no problem associated with zero wave speeds.

Another point that hasn't been emphasized is that all these analyses assume failure can be described in terms of the constitutive response for homogeneous deformation. That may often be a very good assumption, but it's clear that there are cases where it's not. An obvious one is the Griffith crack problem. A more subtle one is a microscale instability that leads to behavior different from that assumed for homogeneous response.

CHAPTER XII

SOIL PLASTICITY

George Y. Baladi and Behzad Rohani

U.S. Army Engineer Waterways Experiment Station
Corps of Engineers, Vicksburg, Mississippi

ABSTRACT

The essential features of the stress-strain-strength response of soils subjected to both hydrostatic and deviatoric states of stress are briefly discussed. The fundamental equations of the work-hardening elastic-plastic constitutive models are summarized for both the associated and the nonassociated flow rules, and the ramifications of using these flow rules for modeling the stress-strain response of frictional materials are pointed out. A procedure for fitting elastic-plastic models (based on the associated flow rule) to a given set of laboratory test data is outlined, and the application of such models for characterization of the stress-strain response of soil is demonstrated. Finally, some results from large-scale numerical calculations of wave propagation in earth materials using elastic-plastic soil models are presented.

INTRODUCTION

The development of mechanical constitutive equations for geologic materials has received considerable attention in the past decades primarily for the following reasons: (1) with the advent of computer technology and the development of new methods of numerical analysis, a number of complex geotechnical problems may be solved provided appropriate constitutive relationships for the materials of interest are available; and (2) constitutive equations provide a rational means for interpretation and organization of laboratory test data for various states of stress and deformation. Therefore, a number of constitutive relationships have been developed based on the continuum representation of the actual geologic materials. Broadly speaking, these constitutive relationships fall in one of the following classes of material models: (1) nonlinear elasticity, (2) incremental elasticity, (3) endochronic models, and (4) incremental elastic-plastic models. A review of the literature indicates that currently in the field of geotechnical engineering the use of elastic-plastic models is more prevalent than the other three classes of models. The reason for this preference is the ability of the elastic-plastic theories to model most of the essential features of the stress-strain-strength properties of the geologic materials and to provide stable and unique solutions to realistically posed boundary-value problems.

This paper presents a review of the application of work-hardening elastic-

plastic models in geotechnical engineering. In particular, the application of a
class of elastic-plastic models, commonly referred to as "cap" models, is presented in detail. The paper is divided into five parts. In Part I some of the
essential features of the mechanical response of soil are discussed and summarized. The fundamental equations of the work-hardening elastic-plastic models
based on associated and nonassociated flow rules are given in Part II. Part II
also contains a procedure for fitting work-hardening models based on the associated flow rule (such as cap-type models) to a given set of material properties. Application of cap models for characterization of material response for
various test conditions is demonstrated in Part III. Part IV contains some
typical results from large-scale numerical calculations of wave propagation in
earth materials using cap models. Finally, a summary and recommendations are
given in Part V.

PART I: ESSENTIAL FEATURES OF THE MECHANICAL BEHAVIOR OF SOIL

     The stress-strain response of soils subjected to externally applied loads
is quite complicated and, unlike the properties of most engineering materials,
is greatly affected by such factors as soil structure, degree of voids saturation, drainage conditions, loading rate, loading history, and current stress
state. It is beyond the scope of this presentation to discuss in detail the
abundant experimental results concerning the deformational properties of various
soils. The discussions in this presentation are limited to some of the aspects
of the behavior of soils which are essential for the formulation of mechanical
constitutive relationships.

     Under hydrostatic states of stress, soils generally exhibit a nonlinear be-
havior and undergo permanent volumetric strain during a load-unload cycle of
deformation (Fig. 1). The permanent volumetric strains that take place are dependent on the initial state of compaction of the material and the degree of
voids saturation. For a very dense soil, the volumetric strains are usually
very small. On the other hand, large volumetric strains can result from compression of a highly flocculated soil. Also, as portrayed in Fig. 1, during an
unloading-reloading cycle most soils display a small amount of permanent volumetric strain. The shearing stress-strain response of soil is far more complex
than the hydrostatic response. Typical shearing stress-strain relationships for
soils are shown qualitatively in Fig. 2, indicating the dependency of the shearing strength on confining pressure or superimposed hydrostatic stress, P_c. Most
soils also exhibit changes in volume when subjected to a shearing (or deviatoric)
state of stress or deformation. Depending on the initial state of compaction,
the volume change can be either contractive or dilative. Under a deviatoric
state of stress, dense sand or overconsolidated clay tends to dilate while loose
sand or normally consolidated clay contracts (Fig. 3). The stress-strain curves
associated with dilative materials (curve 1, Fig. 3) show a well-defined peak
stress at relatively low strains. With increasing strain the strength of the
material decreases, resulting in a brittle-type failure. This type of response
is referred to as "strain-softening" behavior. In the case of contractive materials (curve 2, Fig. 3), there is a relatively slower rate of increase of
stress with strain. With increasing strain, stress increases steadily or remains more or less constant. This type of response is characteristic of
"strain-hardening" materials.

     The behavior of the stress-strain curves for soil is strongly dependent on
the applied confining pressure and the time rate of loading. This dependence,
however, is not unique for all soils and varies greatly with respect to the co-

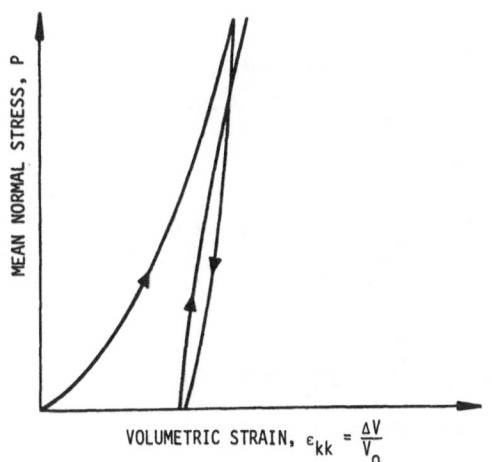

FIG. 1. Typical Behavior of Soil
Under Isotropic Consolidation Test

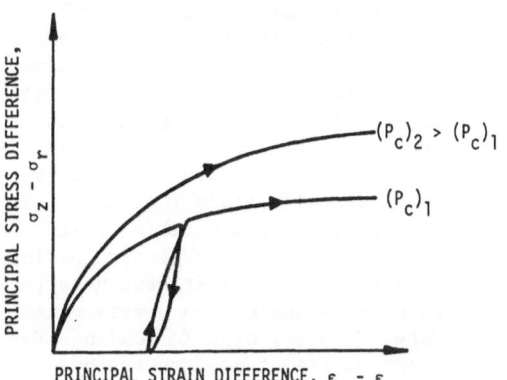

FIG. 2. Typical Triaxial Stress-
Strain Curves for Soil

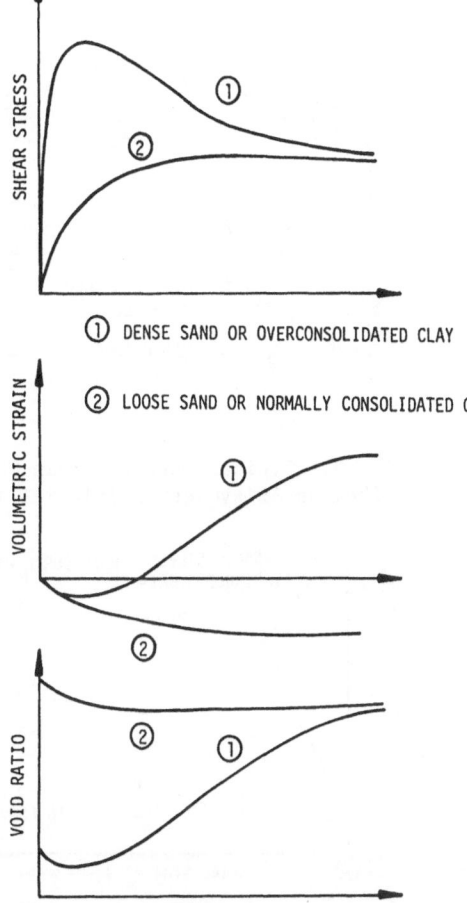

① DENSE SAND OR OVERCONSOLIDATED CLAY

② LOOSE SAND OR NORMALLY CONSOLIDATED CLAY

FIG. 3. Typical Behavior of Soils
Tested Under Drained Triaxial Test
Conditions

hesive and frictional properties of the materials. In general, in the case of
purely cohesive soils, the shear strength of the material is found to be independent of the confining pressure but appreciably affected by the rate of strain
or rate of shear. On the other hand, the shear strength of granular materials
is strongly dependent on confining pressure and generally is not affected by the
rate of shear. The shear strength of mixed soils (soils exhibiting shearing
resistance due to the frictional and cohesive components) depends on both confining pressure and rate of shear.

Fig. 4 depicts qualitatively typical stress-strain relations under axisymmetric triaxial test conditions for various rates of strain. As indicated in
Fig. 4, the soil specimens are isotropically consolidated to the same confining
pressure (point 1). The samples are then sheared at different rates of strain
by increasing the vertical stress, σ_z, while the radial stress, σ_r, is held
constant. The important behavior to be observed from Fig. 4 is that, as the
rate of strain, $d\epsilon_z/dt$, increases, the strength of the material also increases.

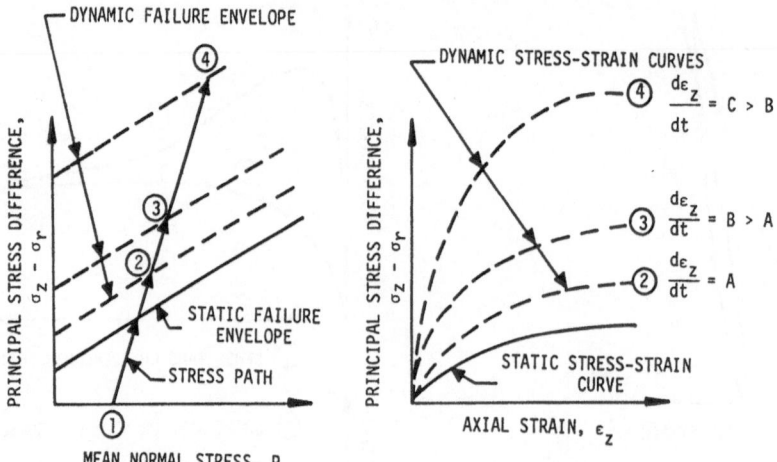

FIG. 4. Typical Stress-Strain Curves for Different Strain Rates
from an Axisymmetric Triaxial Test

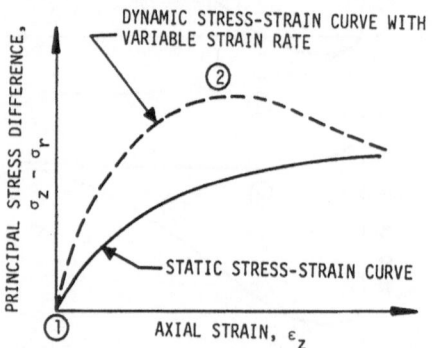

FIG. 5. Dynamic Stress-Strain Curve
for a Variable Strain Rate from an
Axisymmetric Triaxial Test

Therefore, associated with each strain rate, the material possesses a unique failure envelope which may be referred to as the "dynamic failure envelope." The implications of the dynamic failure envelope can best be realized from the results of a variable strain rate test. It is possible for a material that strain-hardens under static loading conditions to exhibit strain-softening behavior due to strain rate effects during dynamic loading. For example, Fig. 5 depicts the hypothetical result of such a test superimposed on the corresponding result from a static test. Similar to Fig. 4, the two stress-strain relations in Fig. 5 are associated with axisymmetric triaxial tests and are isotropically consolidated to the same confining pressure. In the case of the dynamic test, the strain rate during the initial part of the test (point 1 to point 2 in Fig. 5) is increasing. Beyond point 2, the strain rate decreases. During the initial part of the test, the strength of the material continuously increases because of the increasing strain rate. Beyond point 2, the strength of the material decreases because of the decreasing strain rate, resulting in a "falling" or softening stress-strain curve.

The behavior of saturated soils under undrained conditions differs considerably from the corresponding response in a drained test. Stress-strain pore pressure response curves for saturated soils tested in undrained shear conditions are shown in Fig. 6. The three specimens were first isotropically consolidated to the same effective mean normal stress level (point 2) and then sheared. The curves marked "2→3" show the typical response of a normally consolidated clay or a very loose sand. The curves marked "2→5" show behavior typical of an overconsolidated clay or a very dense sand. Within the extreme

— — — — EFFECTIVE-STRESS PATH

—·—·—— TOTAL-STRESS PATH

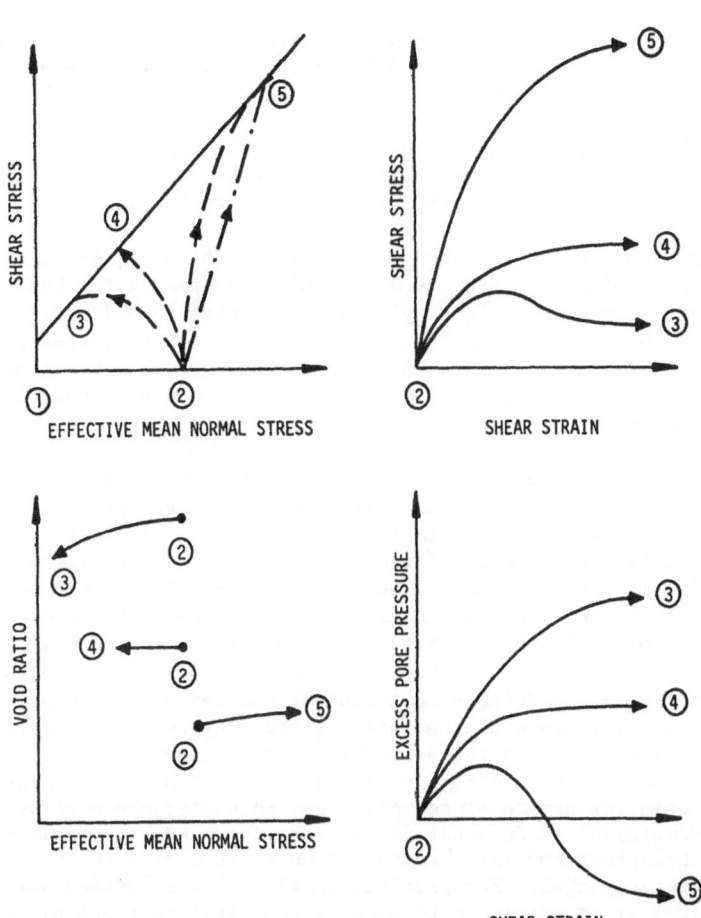

FIG. 6. Typical Behavior of Saturated Soil Tested Under Undrained Triaxial
   Test Conditions

limits of these loose and dense soil responses, there is a graduated response,
typified herein by the curves marked "2→4." The latter response depends on the
state of compaction (consolidation) of the material. It is clear from this figure that the effective stress is the only part of the total stress that affects
soil shear strength. A comparison of Fig. 3 with Fig. 6 indicates that the
shearing stress-strain response for very dense sand or overconsolidated clay and
very loose sand or normally consolidated clay in the drained condition contrasts
markedly with the corresponding responses in the undrained condition. That is,
a softening stress-strain response corresponds to a dilative behavior in a
drained test (Fig. 3) but a contractive behavior in an undrained test (Fig. 6).
Conversely, a work-hardening response corresponds to a contractive behavior in a
drained test and a dilative behavior in an undrained test.

Another useful laboratory test for investigating the deformational characteristics of soil is the uniaxial strain test. Typical qualitative results from

such a test are presented in Figs. 7 and 8. Fig. 7 portrays the stress-strain
response of the material and Fig. 8 shows the corresponding stress path. The
loading portion of the stress-strain curve is usually S-shaped, indicating some
initial resistance to natural cementation. As the applied load increases, the
material densifies and the response becomes similar to the hydrostatic curve
shown in Fig. 1. The slope of the unloading curve is initially very stiff, indicating a "locking" response, but it gradually decreases as the unloading continues. The uniaxial strain stress path (Fig. 8) is generally continuously concave to the pressure axis throughout the loading cycle regardless of the shape
of the stress-strain curve. Most existing experimental data indicate that the
unloading stress path falls below the loading, as shown in Fig. 8.

Another important response observed for some earth materials is anisotropic
behavior. This phenomenon can best be described by comparing the behavior of an
isotropic soil and a transverse-isotropic soil subjected to hydrostatic stress.
For the isotropic soil (Fig. 9), all strains are equal under hydrostatic states
of stress. However, as indicated in Fig. 10, in the case of a transverse-isotropic soil, the strains in the plane of isotropy, ε_r, are different from those
in the axial direction, ε_z.

Developing a single constitutive relationship that would model all the mechanical stress-strain response of the soil discussed in the preceding para-
graphs is a formidable task. Furthermore, it is neither necessary (from practical points of view) nor desirable (from mathematical considerations) to develop such a constitutive relationship. The constitutive relationship should contain only the basic features of the material response which are relevant to the
solution of the problem at hand. For example, for ground motion calculations
from a surface burst under superseismic conditions, the uniaxial strain response
of the soil is the dominant feature that must be modeled by the constitutive relationship. The initial slope of the uniaxial stress-strain curve determines
the arrival time of the wave to the point of interest and must be preserved by
the constitutive model. The radial stresses during wave propagation are greatly
affected by the uniaxial strain stress path, and this response must be correctly
modeled. If the material of interest is rate-sensitive and the loading rate
varies greatly throughout the calculations, then it is desirable to adopt a time-
dependent constitutive model. For problems dealing with saturated soil and
short-time loading, it is necessary to adopt a constitutive model based on effective stress concepts so that pore pressure generation can be modeled in the calculations.

Different types of elastic-plastic models have been used successfully to
model most of the basic features of the stress-strain properties of soil discussed here; see References. The fundamental equations of work-hardening elastic-plastic models are summarized in Part II of this paper.

PART II: FUNDAMENTAL EQUATIONS OF WORK-HARDENING ELASTIC-PLASTIC MODELS

The basic premise of the elastic-plastic constitutive models is the assumption that certain materials are capable of undergoing small plastic (permanent)
as well as elastic (recoverable) deformation during each loading increment.
Mathematically, the total strain increment is assumed to be the sum of the elastic and plastic strain increments

$$d\varepsilon_{ij} = d\varepsilon_{ij}^E + d\varepsilon_{ij}^P . \tag{1}$$

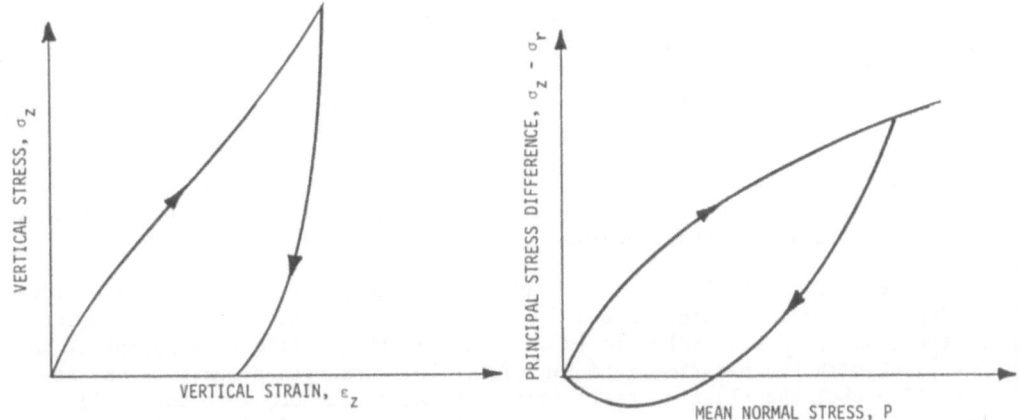

FIG. 7. Typical Behavior of Soil   FIG. 8. Typical Uniaxial Strain
   Under Uniaxial Strain Test   Stress Path for Soil

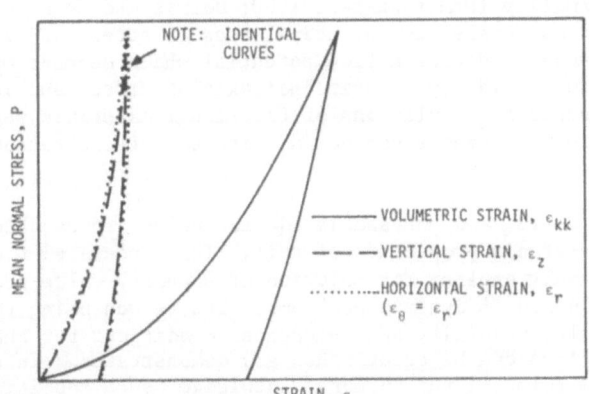

FIG. 9. Typical Behavior of Isotropic Soil Under Hydrostatic
   Loading and Unloading

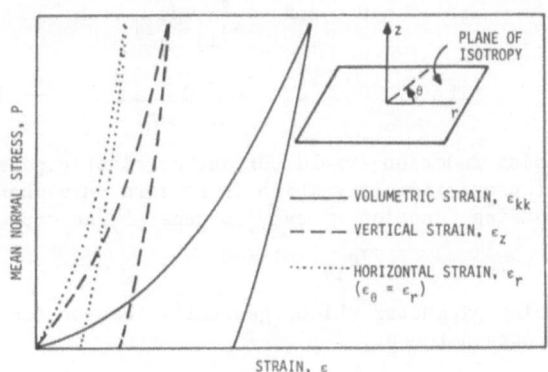

FIG. 10. Typical Behavior of Transverse-Isotropic Soil Under
   Hydrostatic Loading and Unloading

Within the elastic range, the behavior of the material can be described by an elastic constitutive relation of the type

$$d\varepsilon_{ij}^{E} = \frac{dJ_1}{9K(\sigma_{mn})}\,\delta_{ij} + \frac{1}{2G(\sigma_{mn})}\,dS_{ij} \qquad (2)$$

where $J_1 = \sigma_{kk}$ = first invariant of the stress tensor, $S_{ij} = \sigma_{ij} - (J_1/3)\delta_{ij}$ = stress deviation tensor, δ_{ij} = Kronecker delta, $K(\sigma_{mn})$ = elastic bulk modulus, and $G(\sigma_{mn})$ = elastic shear modulus.

There are basically two approaches in the theory of plasticity for calculating the plastic strain increment tensor. In the first approach, referred to as the associated flow rule, it is assumed that the plastic flow potential is identical with the plastic yield function. This assumption leads to the restriction that the plastic strain increment vector is normal to the yield surface. This normality condition, however, cannot be verified experimentally for frictional materials such as soil (Lade et al., 1975,1978). In the second approach, referred to as the nonassociated flow rule, the plastic flow potential is different than the plastic yield function and the theory is not bound by the restriction of normality (Nemat-Nasser, 1980; Dorris and Nemat-Nasser, 1982). Also, for granular materials, the principal axes of stress and increment of strain do not coincide. Use of a flow potential which depends on the stress invariants leads to coaxiality of principal axes of stress and increment of strain. Micromechanical formulations of frictional materials seem to yield results with strain rate tensor noncoaxial with the stress tensor (Nemat-Nasser, 1983).

Both the associated and nonassociated flow rules are capable of modeling the *overall* stress-strain properties of soil. The associated flow rule, however, is more commonly applied for solution of boundary-value problems, especially for the solution of dynamic problems. The reason being that using the associated flow rule, stability and uniqueness conditions for the solution of this class of problems can be established and demonstrated. In the case of the nonassociated flow rule, to the authors' knowledge, such conditions have not been fully established.

Adopting the associated flow rule, the plastic strain increment tensor can be calculated to be

$$d\varepsilon_{ij}^{p} = \begin{cases} d\lambda\,\dfrac{\partial f}{\partial \sigma_{ij}} & \text{if } f = 0 \quad \text{and} \quad \dfrac{\partial f}{\partial \sigma_{ij}}\,d\sigma_{ij} > 0 \\[2ex] 0 & \text{if } f < 0 \quad \text{or} \quad f = 0 \quad \text{and} \quad \dfrac{\partial f}{\partial \sigma_{ij}}\,d\sigma_{ij} \le 0 \end{cases} \qquad (3)$$

where f is the loading function (yield surface or plastic potential) and $d\lambda$ is a positive factor of proportionality which is nonzero only when plastic deformation occurs. The loading function f can, in general, be expressed as

$$f(\sigma_{ij}, \kappa) = 0 \qquad (4)$$

where κ is a hardening parameter which, generally, can be taken to be a function of the plastic strain tensor.

For the cap models, the loading function f is assumed to be isotropic and to consist of two parts: an ultimate failure envelope (or surface) which serves to limit the maximum shear stresses that the material can sustain, and a strain-

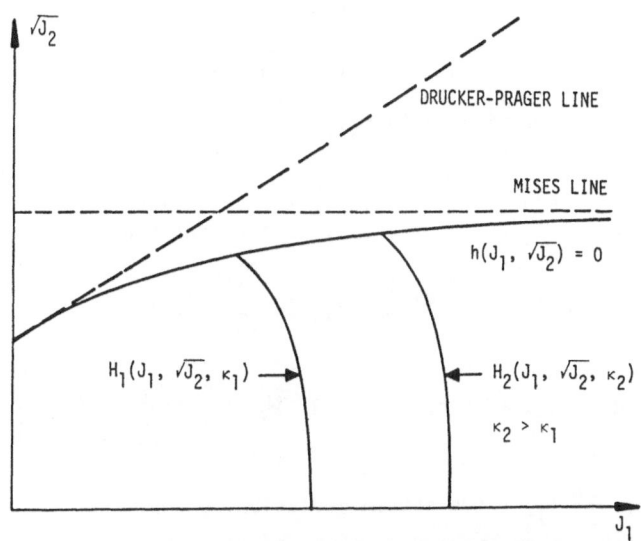

FIG. 11.   Loading Function for Cap Model

hardening surface or "cap" (Fig. 11).   The failure envelope portion of the loading function is denoted by

$$f = h(J_1, \sqrt{J_2}) = \sqrt{J_2} - f(J_1) = 0 \qquad (5)$$

and the strain-hardening surface by

$$f = H(J_1, \sqrt{J_2}, \kappa) = \sqrt{J_2} - F(J_1, \kappa) = 0 \qquad (6)$$

where J_2 = the second invariant of the stress deviation tensor.   The hardening parameter κ can be taken to be a function of plastic volumetric strain or maximum plastic volumetric strain,

$$\kappa = g\left(\varepsilon_{kk}^P\right) \qquad (7)$$

$$\kappa = g\left(\varepsilon_{kk}^P\big|_{max}\right). \qquad (8)$$

Eq. (7) must be used if the material compacts under applied shearing stresses. On the other hand, Eq. (8) must be used if the material dilates or first compacts and then dilates.

The elastic moduli K and G must be restricted to

$$K = K(J_1, \varepsilon_{kk}^P) \qquad (9)$$

$$G = G(\sqrt{J_2}, \varepsilon_{kk}^P) \qquad (10)$$

in order to ensure that the response of the material in the elastic range is path-independent.

The proportionality factor $d\lambda$ can be obtained from the consistency criterion $df = 0$ and takes the following form:

$$d\lambda = \frac{3K \frac{\partial f}{\partial J_1} d\varepsilon_{kk} + \frac{G}{\sqrt{J_2}} \frac{\partial f}{\partial \sqrt{J_2}} S_{ij} de_{ij}}{9K\left(\frac{\partial f}{\partial J_1}\right)^2 + G\left(\frac{\partial f}{\partial \sqrt{J_2}}\right)^2 - 3 \frac{\partial f}{\partial J_1} \frac{\partial f}{\partial \kappa} \frac{\partial \kappa}{\partial \varepsilon^p_{kk}}} \cdot \tag{11}$$

Combining the previous equations results in the following expressions for the total strain increment tensor and the stress increment tensor

$$d\varepsilon_{ij} = \frac{dJ_1}{9K} \delta_{ij} + \frac{dS_{ij}}{2G} + d\lambda \left(\frac{\partial f}{\partial J_1} \delta_{ij} + \frac{S_{ij}}{2\sqrt{J_2}} \frac{\partial f}{\partial \sqrt{J_2}}\right) \tag{12}$$

$$d\sigma_{ij} = K d\varepsilon_{kk} \delta_{ij} + 2G de_{ij} - d\lambda\left(3K \frac{\partial f}{\partial J_1} \delta_{ij} + \frac{G}{\sqrt{J_2}} \frac{f}{\partial \sqrt{J_2}} S_{ij}\right) \tag{13}$$

where $e_{ij} = \varepsilon_{ij} - \frac{\varepsilon_{kk}}{3} \delta_{ij}$ = the strain deviation tensor. Eq. (12), or (13), is the general constitutive equation for an elastic work-hardening plastic iso-tropic material. To use either of these equations it is necessary only to spe-cify the functional forms of the elastic moduli K and G, the hardening param-eter κ, and the loading function f and, of course, to determine experimentally the numerical values of the coefficients in these functions.

For application in the finite-element analysis it is necessary to express Eq. (13) in terms of a stiffness matrix, $C_{ijk\ell}$, containing both the elastic and plastic behavior responses of the material; that is,

$$d\sigma_{ij} = C_{ijk\ell} d\varepsilon_{k\ell}. \tag{14}$$

The elements of the elastic-plastic stiffness matrix can readily be computed using the preceding equations. In indicial notation the stiffness matrix takes the following form

$$C_{ijk\ell} = D_{ijk\ell} - (1-r) \frac{D_{ijuv} \frac{\partial f}{\partial \sigma_{uv}} Q_{k\ell}}{\frac{\partial f}{\partial \sigma_{mn}} D_{mnsq} \frac{\partial f}{\partial \sigma_{sq}} - \frac{\partial f}{\partial \kappa} \frac{\partial \kappa}{\partial \varepsilon^P_{mn}} \frac{\partial f}{\partial \sigma_{mn}}} \tag{15}$$

where

$$Q_{k\ell} = \frac{\partial f}{\partial \sigma_{mn}} D_{mnk\ell} \tag{16}$$

$$\frac{\partial f}{\partial \sigma_{ij}} = \frac{\partial f}{\partial J_1} \delta_{ij} + \frac{1}{2\sqrt{J_2}} \frac{\partial f}{\partial \sqrt{J_2}} S_{ij} \tag{17}$$

$$D_{ijk\ell} = \left(K - \frac{2}{3} G\right) \delta_{ij} \delta_{k\ell} + G(\delta_{ik} \delta_{j\ell} + \delta_{i\ell} \delta_{jk}). \tag{18}$$

The parameter r in Eq. (15) is the ratio of the strain increment at the end of which the plastic behavior is first encountered to the total strain increment.

The plastic strain increment tensor can also be calculated based on the nonassociated flow rule. A general nonassociative flow rule is developed by

Nemat-Nasser and Shokooh (1980) for finite plastic deformation of porous metals
and geologic materials. Following their terminology, the deformation rate tensor can be expressed as

$$D_{ij} = L_{ijk\ell} \overset{*}{\sigma}_{k\ell} \tag{19}$$

in which

$$L_{ijk\ell} = \frac{1}{4G}(\delta_{ik}\delta_{j\ell} + \delta_{jk}\delta_{i\ell}) + \left(\frac{1}{9K} - \frac{1}{6G}\right)\delta_{ij}\delta_{k\ell}$$

$$+ \frac{1}{H}\left(\frac{S_{ij}}{2\sqrt{J_2}} + \frac{\partial \tilde{G}}{\partial J_1}\delta_{ij}\right)\left(\frac{S_{k\ell}}{2\sqrt{J_2}} - \frac{\partial \tilde{F}}{\partial J_1}\delta_{k\ell}\right). \tag{20}$$

In Eq. (19), $\overset{*}{\sigma}_{k\ell}$ corresponds to the Jaumann rate of change of stress tensor. The
functions $\tilde{G}$ and $\tilde{F}$ in Eq. (20) describe, respectively, the plastic flow potential $\tilde{g}$ and the plastic yield function $\tilde{f}$ via the following expressions:

$$\tilde{g} = \sqrt{J_2} + \tilde{G} \tag{21}$$

$$\tilde{f} = \sqrt{J_2} - \tilde{F} . \tag{22}$$

Both $\tilde{G}$ and $\tilde{F}$ are functions of the first stress invariant, the total plastic
volumetric strain Δ, and total distortional plastic work ξ. The parameter H in
Eq. (20) is related to $\tilde{G}$ and $\tilde{F}$ through the following relation

$$H = \left(3\frac{\partial \tilde{G}}{\partial J_1}\frac{\partial \tilde{F}}{\partial \Delta} + \sqrt{J_2}\frac{\partial \tilde{F}}{\partial \xi}\right) . \tag{23}$$

It is noted that the above formulation includes, as a special case, the small
strain plasticity theory with normality and the associative flow rule. If $\tilde{f}$ and
$\tilde{g}$ are identical, the theory reduces to the associative flow rule.

Plasticity models based on the associative flow rule have been used extensively by the authors for the solutions of geotechnical problems. A general
procedure for the numerical implementation of this type of model is presented
by Baladi (1977) and Sandler and Rubin (1979). The remainder of the discussions
in this paper will be centered around cap models based on the associated flow
rule.

Procedure for Fitting Elastic-Plastic Models to a Given Set of Laboratory Data

The procedure for obtaining the functional forms and parameters used in
elastic-plastic models constructed for use in boundary-value/initial-value problems is based on the quantities and types of data obtained from laboratory tests
on representative samples of the material of interest.

The data generally consist of stress-strain and strength properties from
uniaxial strain and triaxial compression tests. Sometimes hydrostatic, proportional loading, direct shear, or other laboratory test data are also available.
The first step in the fitting procedure is to use the unloading portions of the
available stress-strain data to determine the appropriate elastic behavior of
the model. For example, as long as the model behaves elastically, unloading
behavior in hydrostatic tests indicates the form of the bulk modulus K while
the form of the shear modulus G can be ascertained from triaxial stress test
unloading data. Moreover, the combination $K + 4G/3$ can be deduced from uni-

axial strain test unloading data. Other tests, if available, may also be used
to check or adjust the overall fit of the elastic portion of the model.

The next step in the fitting procedure is to establish the failure envelope,
i.e., that portion of the yield surface which limits the shearing stresses that
materials can withstand. The failure envelope is generally obtained by fitting
failure data from triaxial stress and proportional loading tests. These data
are fit by a function of the stresses and are usually assumed to involve only
the first stress invariant and the second invariant of the stress deviator tensor.

The remaining step in the fitting procedure is the most difficult. The
hardening surface portion of the model is obtained by a trial-and-error procedure in which the shape of the hardening surface and a hardening rule are assumed
and the loading stress-strain behavior of this assumed model is computed and
compared to the representative material property data. If the fit requires improvements, a new set of parameters is tried and the procedure is repeated.

Obviously, the success of such a trial-and-error procedure and the rapidity
with which it converges are strongly dependent on the experience of the modeler.
Knowledge of the effects on the model behavior of changes in its parameters is
important for rapidly obtaining satisfactory fits. For example, the fitting
procedure is greatly simplified by the knowledge that the hardening rule strongly influences the stress-strain behavior for triaxial stress situations and the
stress path for the uniaxial strain condition. In addition, an adequate hardening rule has often been obtained in the past from analyzing only the plastic
volumetric strain occurring during cyclic hydrostatic tests. Figure 12 presents
the fundamental equations of the cap model and the type of data needed to determine each function.

Comparisons of laboratory test data with model behavior for several types
of cap models are presented in Part III of this paper.

PART III: COMPARISON OF MODEL BEHAVIOR WITH LABORATORY TEST DATA

In the past ten years the authors have developed several types of cap models
for describing some of the features of the stress-strain response of the earth
materials discussed in Part I. Furthermore, these models have been successfully
implemented in large-scale computer codes for numerical analysis. In the following sections we will compare the behaviors of five of these models with the appropriate laboratory test data in order to demonstrate the versatility of the
elastic-plastic cap models.

General Total-Stress Model. — The total-stress model is one of the simpler
forms of cap models and has been used extensively for ground motion calculations
(Baladi, 1973, and Sandler et al., 1976). The model is restricted to time-independent strain-hardening responses but is capable of reproducing the nonlinear
hysteretic behavior of most soils. Figures 13-15 portray the behavior of the
model and the corresponding experimental data for a normally consolidated sand
under uniaxial states of strain (UX) and triaxial test conditions (TX). It is
observed from these figures that the model is capable of reproducing the basic
features of the experimental data. In the case of the UX test both the experimental data and the model prediction indicate that the stress-strain curve con-
caves to the stress axis during loading (Fig. 13). In the case of the TX test,
both model behavior and test data show that during loading the stress-strain

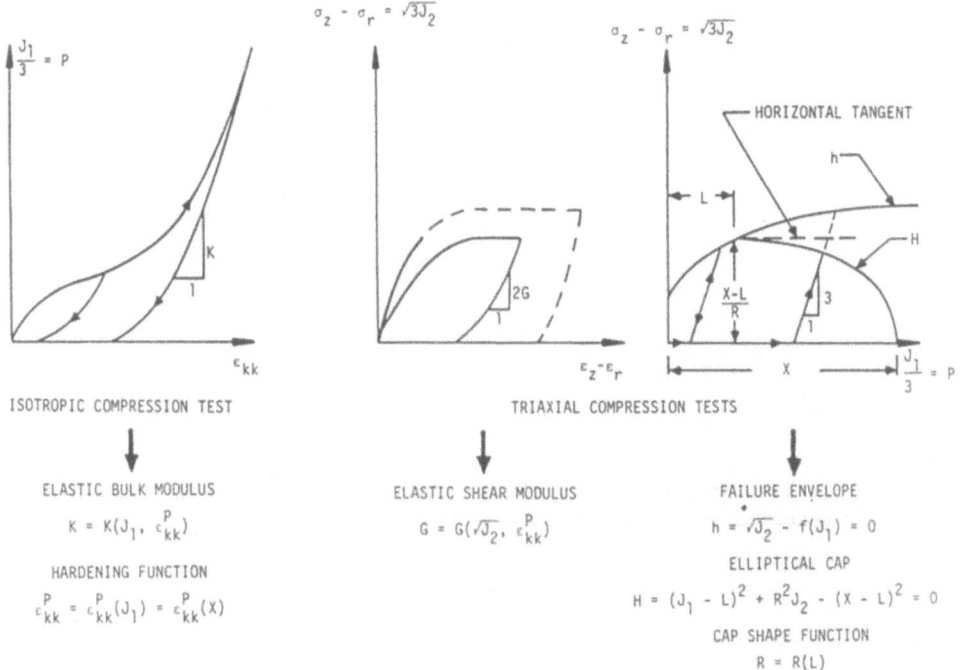

FIG. 12.   Fundamental Equations of the Cap Model and the Type of Data It
Requires for Fitting

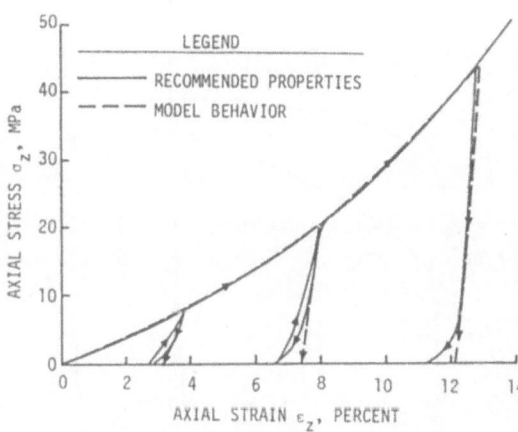

FIG. 13.   Axial Stress-Axial Strain
Relation for UX Test; Recommended
Properties Versus Model Behavior

curves concave to the strain axis for
all confining stress (σ_r) levels
(Fig. 15).   The curvature of the UX
test stress path (Fig. 14) is also
correctly simulated by the model.

Strain-Softening Model. — In
order to describe the stress-strain
response of overconsolidated and very
sensitive clays, the authors developed a rate-independent strain-softening cap model by allowing the
failure envelope to expand and contract isotropically as a function of
plastic work (Baladi and Rohani,
1979a).   This model is capable of
reproducing the stress-strain and the
corresponding volumetric response of
overconsolidated clays subjected to a
deviatoric state of stress.   The response of the model under TX test
conditions is correlated with corresponding experimental data for a heavily
overconsolidated clay in Fig. 16.   Fig. 16 contains the results of both the
stress-strain and the volumetric strain responses, and it indicates that the
general features of the data are reproduced by the model.

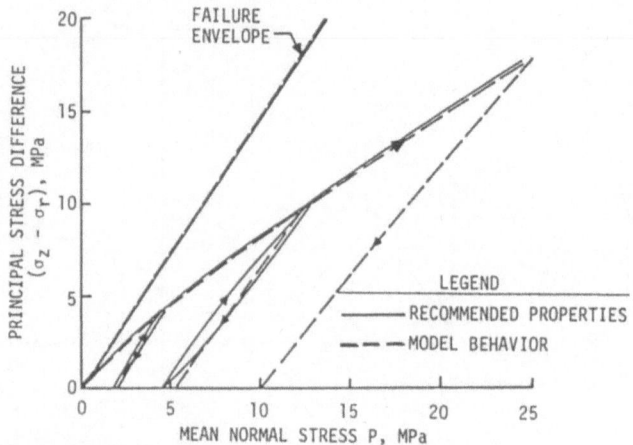

FIG. 14.   Failure Envelope and Stress Path for UX Test;
Recommended Properties Versus Model Behavior

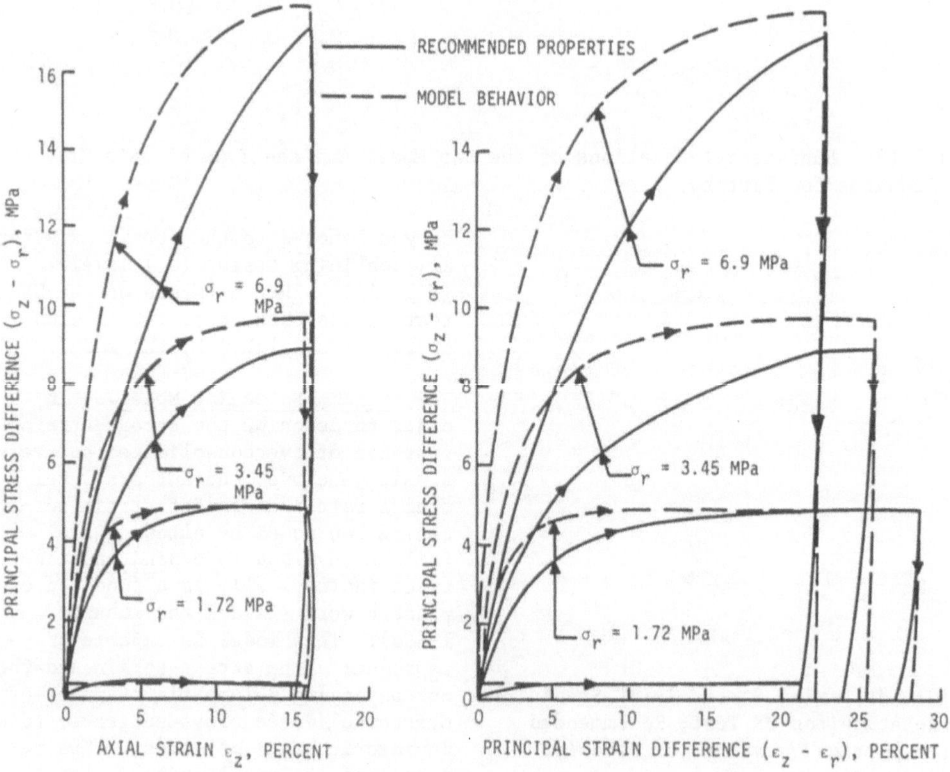

FIG. 15.   Stress-Strain Relations for TX Test; Recommended Properties
Versus Model Behavior

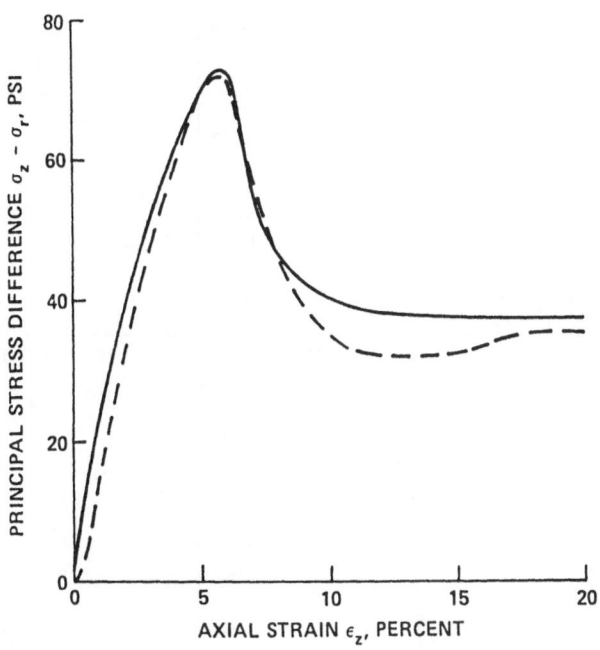

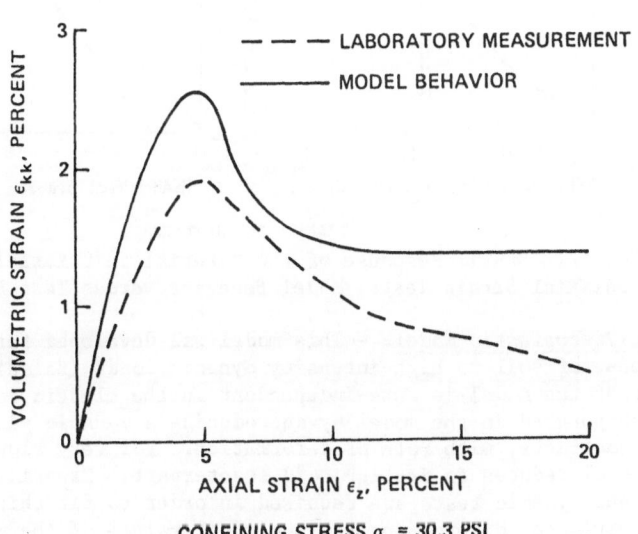

FIG. 16. Principal Stress Difference and Volumetric
Strain Versus Axial Strain Response from Consolidated-
Drained Triaxial Test of Heavily Overconsolidated Oxford
Clay; Laboratory Measurements Versus Model Behavior

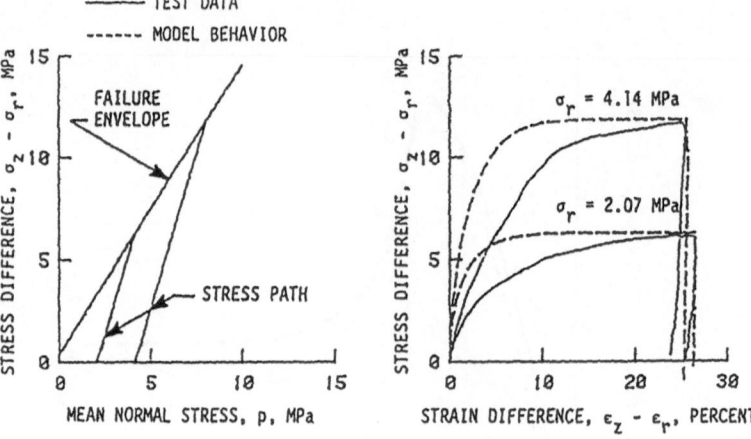

a. Triaxial shear response

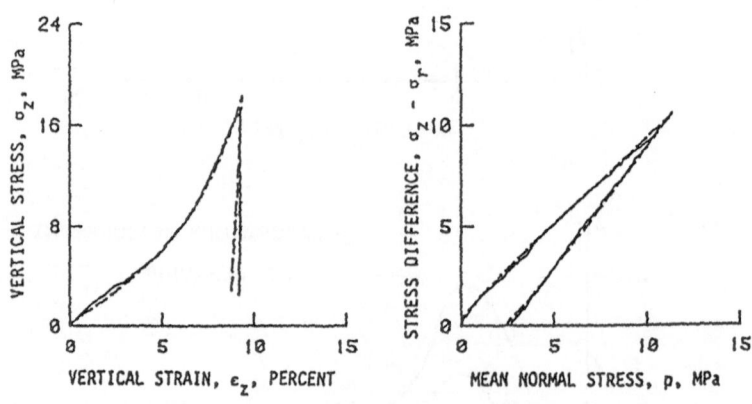

b. Uniaxial strain response

FIG. 17.   Static Response of the Material in Triaxial Shear and
Uniaxial Strain Tests; Model Behavior Versus Test Results

Elastic-Viscoplastic Model. — This model was developed for characterization
of the response of soil to high-intensity dynamic loads (Baladi and Rohani,1982).
The behavior of the model is rate-independent in the elastic range. Rate dependency is incorporated in the model by introducing a dynamic yield surface which
expands, or contracts, with rate of deformation. For very slow rates of deformation the model reduces to its inviscid counterpart. Experimental data from
both static and dynamic tests are required in order to fit this model. Static
data are required for determining the numerical values of the material constants
which govern the static response of the model. Dynamic properties are needed to
determine the parameters of the dynamic yield surface. Comparisons of model behavior with experimental data for a clayey sand are shown in Figs. 17 and 18.
Fig. 17 contains the static response of the material and indicates that the inviscid part of the model describes the response of the material accurately. The
dynamic response is shown in Fig. 18 and consists of two stress-strain curves
from a UX test and the corresponding strain-time histories. The dynamic data
were obtained for loading rise times on the order of a few tenths of a milli-
second. Also, as noted from Fig. 18, the entire load-unload cycles for these

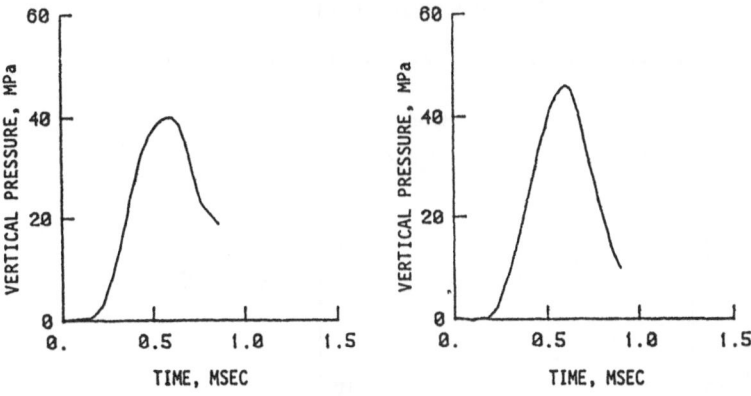

a. Input pressure-time histories

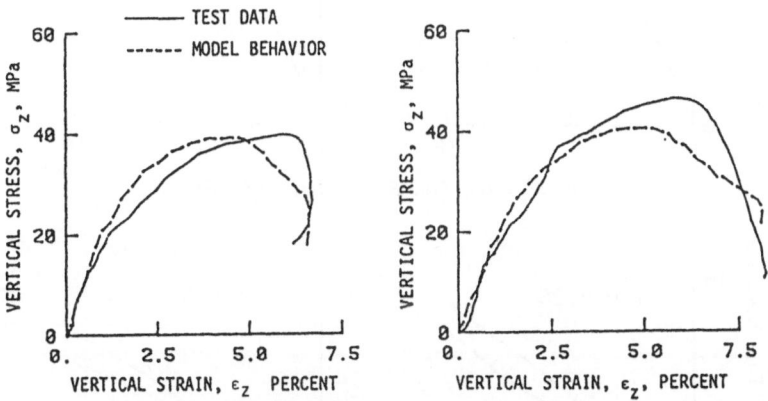

b. Resulting dynamic stress-strain relations

FIG. 18. Dynamic Response of the Material in Uniaxial Strain
    Configuration; Model Behavior Versus Test Results

tests were completed in slightly less than 1 msec. Using the strain-time histories as input, the viscoplastic model was used to predict the dynamic UX
stress-strain response of the material. The experimental stress-strain curves
and the corresponding model behavior for the two dynamic tests are shown in
Fig. 18. The agreement between the dynamic test data and the model behavior is
very good.

    Anisotropic Model. —The general total-stress cap model was converted to a
transverse-isotropic constitutive model by introducing the concept of a "pseudo-
stress invariant" which replaces the classical J_1, J_2 parameters (Baladi, 1977,
and 1980). The model would automatically degenerate to an isotropic cap model
when the coefficients in the expressions for the pseudo invariants are set to
unity. Correlation of the transverse-isotropic model with test data is shown in
Fig. 19 for a natural clay under a deviatoric state of stress. Fig. 19 includes
both the stress-strain and volumetric responses of the material for stress
ratios (m) of 0.25 and 0.5, where $m = (\sigma_2 - \sigma_3)/(\sigma_1 - \sigma_3)$. It is noted that the
stress-strain results for the two stress ratios are different, indicating direc-

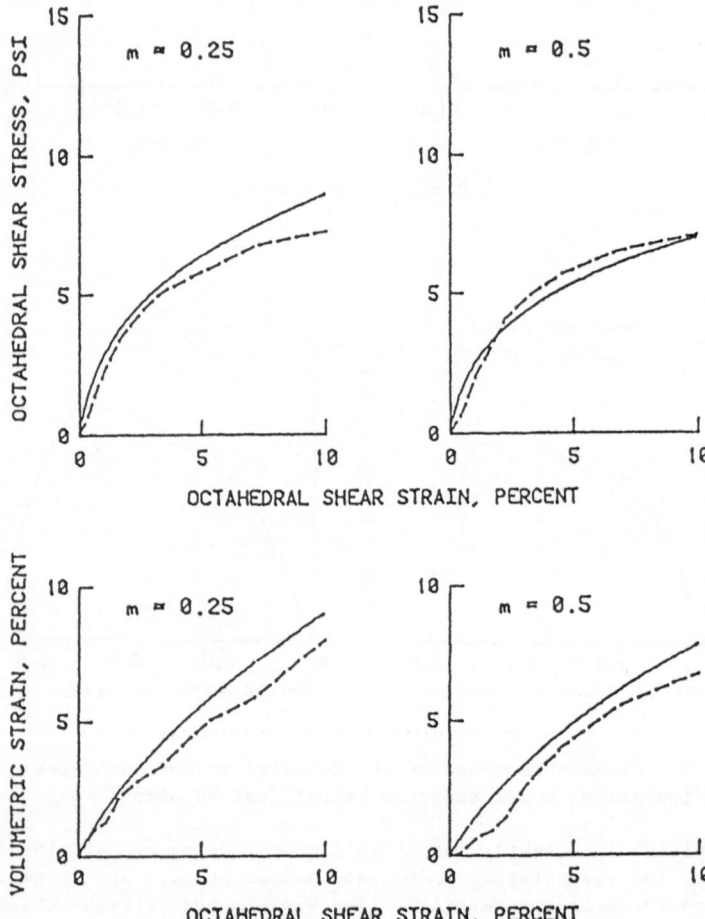

NATURAL SOIL, CLAY X

Constant Mean Stress Test (Pc) With Pc = 20.0 PSI

——————— MODEL PREDICTION

– – – – –· ACTUAL DATA

FIG. 19. Octahedral Shear Stress and Volumetric Strain Versus
Octahedral Shear Strain for a Natural Clay; Model Prediction
Versus Laboratory

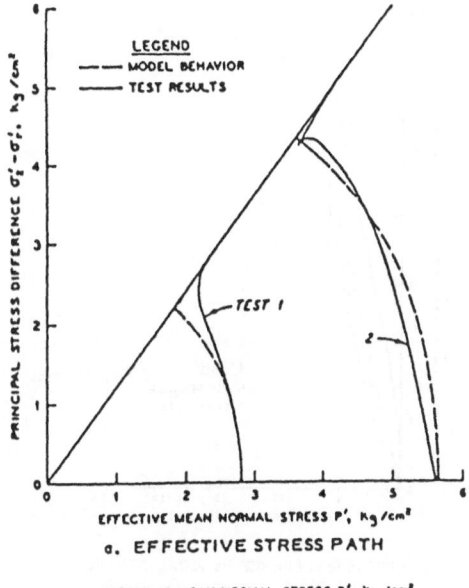

a. EFFECTIVE STRESS PATH

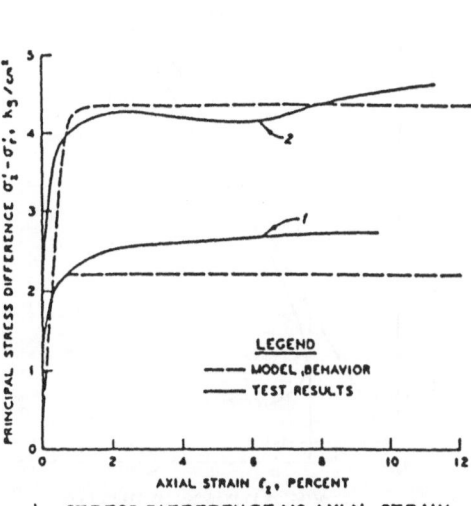

b. STRESS DIFFERENCE VS AXIAL STRAIN

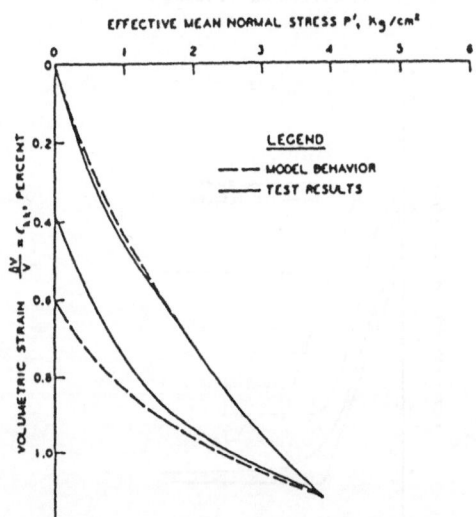

c. ISOTROPIC CONSOLIDATION RESPONSE

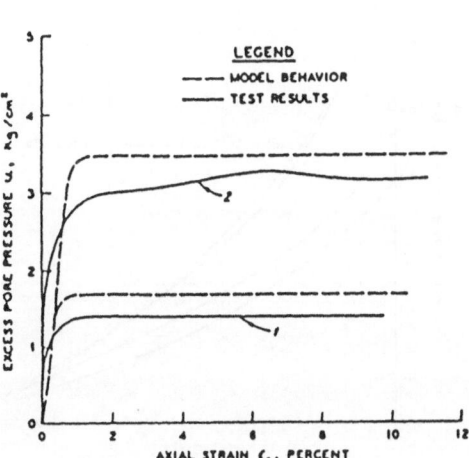

d. EXCESS PORE PRESSURE VS AXIAL STRAIN

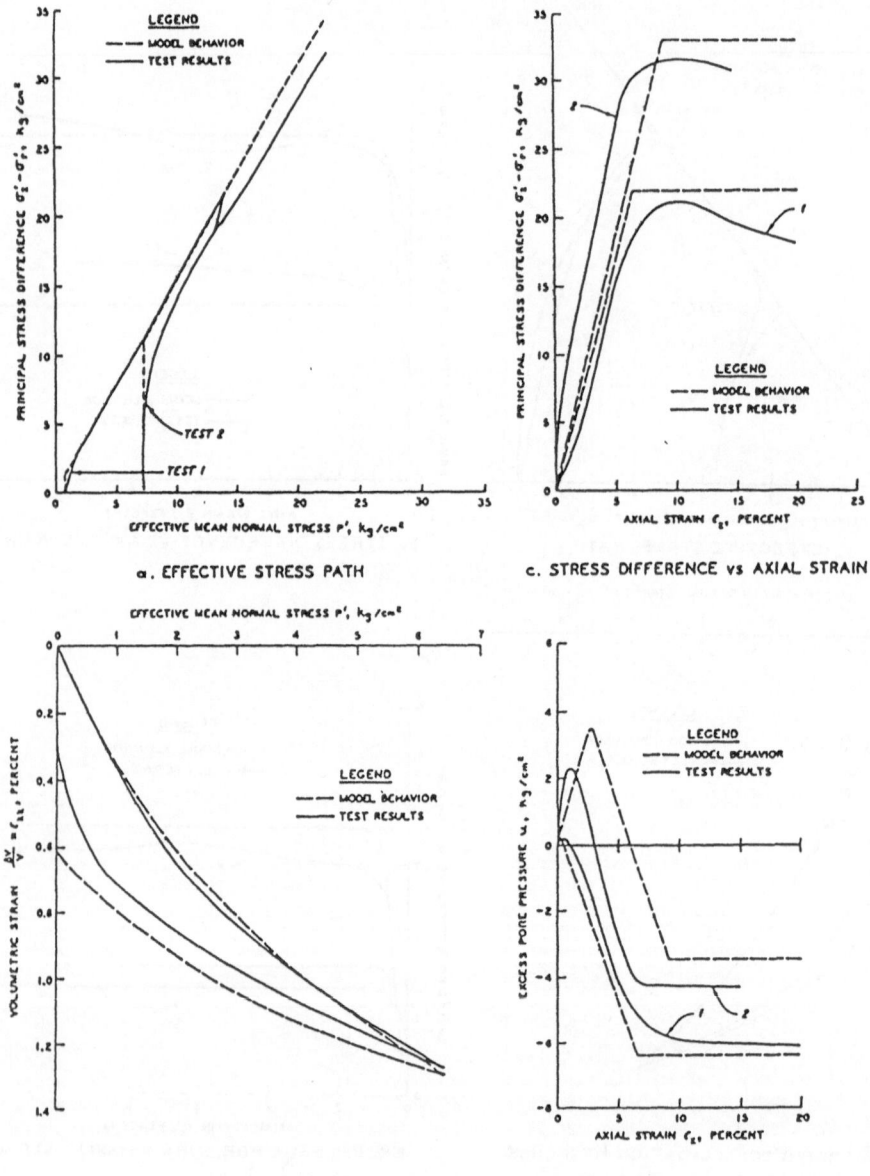

a. EFFECTIVE STRESS PATH

c. STRESS DIFFERENCE vs AXIAL STRAIN

b. ISOTROPIC CONSOLIDATION RESPONSE

d. EXCESS PORE PRESSURE vs AXIAL STRAIN

FIG. 21.   Comparison of Model Behavior with Experimental Data
(Test Series 1, Relative Density = 76 Percent)

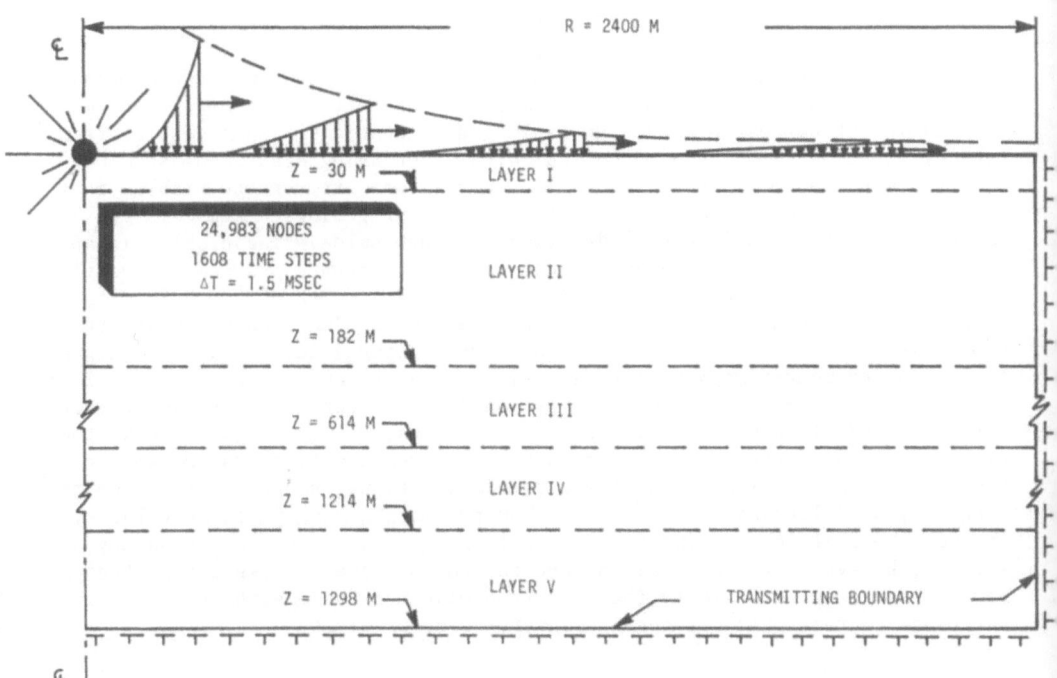

FIG. 22.   Geometry of the Problem

tional dependency of the response of the material. This behavior is accurately
reproduced by the model for both the stress-strain and volumetric responses.

Effective-Stress Model. — An effective-stress cap model was developed for
analysis of the response of saturated sand in an undrained condition. The basic
idea for developing this model was the fact that during an undrained condition
the increment of volumetric strain is zero (or can be calculated from the volumetric response of the mixture). For low stress levels the assumption of incompressibility of the mixture is reasonable and leads to accurate predictions. The
effective-stress model (Baladi and Rohani, 1979b and 1979c) has been used for
predicting the generation of pore water pressure under undrained conditions for
saturated sand at different relative densities. Fig. 20 shows the behavior of
the model and the corresponding experimental data for a relative density of 20
percent. At this relative density the pore water pressure increases during the
test and reaches its peak value when the effective-stress path reaches the failure envelope. The results for a relative density of 76 percent are portrayed in
Fig. 21. It is observed that at this relative density the pore water pressure
initially increases and peaks when the effective-stress path reaches the failure
envelope. The pore pressure then drops and actually becomes negative as the
test is continued. The pore pressure response in Fig. 20 is a direct consequence of the tendency of the material to compact under shearing stresses. On
the other hand, the behavior in Fig. 21 is indicative of the dilative behavior
of the material under shear. As noted in Figs. 20 and 21, the effective-stress
model accurately reproduces both behaviors.

PART IV: TYPICAL RESULTS FROM LARGE-SCALE NUMERICAL CALCULATIONS OF WAVE
           PROPAGATION IN EARTH MATERIALS

Various versions of cap models have been incorporated into two-dimensional
wave propagation codes for the prediction of states of stress and ground motions
induced in natural earth masses by explosive detonations. Some results from a
typical two-dimensional ground motion calculation using elastic-plastic cap
models are presented in this part. The general geometry of the problem is shown
in Fig. 22. The medium consists of five layers of geologic materials and is
loaded at its free surface by a decaying pressure pulse. The problem is analyzed as an axisymmetric two-dimensional finite-difference problem and consists
of 24,983 nodes. As indicated in Fig. 22, beyond a depth of $Z = 1298$ M and a
range of $R = 2400$ M from ground zero, the medium is simulated by transmitting
boundaries. Problem output consists of stress-, particle velocity-, and particle displacement-time histories at selected ranges and depths. The typical
output for a range of $R = 468$ M, corresponding to a peak surface overpressure of
6.96 MPa, and various depths is presented in Figs. 23-25. Fig. 23 shows the
time histories of vertical and horizontal stresses and Figs. 24 and 25 portray
the time histories of horizontal and vertical motions, respectively. The high-
frequency oscillations observed in the output (stress and particle velocity)
are not physical and are due to numerical integration. The low-frequency oscillations, however, are physical and are due to the layering aspects of the problem. These results indicate that cap-type models provide stable and unique
solutions to realistically posed boundary-value problems.

PART V: SUMMARY AND RECOMMENDATIONS

It has been demonstrated that the elastic-plastic cap models are capable of
reproducing many of the essential features of the mechanical response of geologic materials. Furthermore, this type of constitutive model provides stable
and unique solutions to realistically posed boundary-value problems and is currently being used for the numerical solution of many geotechnical problems.
There are, however, several aspects of material characterizations that are not
currently simulated by this type of model that require further investigation and
refinement. Probably the most important aspect is the proper characterization
of the tensile behavior of the material in a general three-dimensional problem.
This problem has received considerable attention in the past several years, but
its solution is still elusive. Alongside the tension problem is the treatment
of shear fracture and Bauschinger effect in dynamic problems involving cyclic
loadings. Several attempts have been made in the past to simulate the response
of the soil under cyclic loading using cap-type models. These attempts have not
been completely successful. Therefore, future efforts in the area of soil plasticity must address the treatment of tensile strength, shear fracture, and
Bauschinger effect. Furthermore, it is desirable to examine the behavior of
cap-type models under a broader range of stress and/or strain paths for the
specific material of interest.

Another important issue which must be resolved with regard to application
of plasticity models in geotechnical engineering is the question of the associative flow rule versus the nonassociative flow rule. As indicated before, the
assumption of normality inherent in the associative flow rule cannot be verified
experimentally. This notwithstanding, cap models based on the associative flow
rule are capable of simulating many of the mechanical responses of soils and at
the same time provide stable and unique solutions to dynamic boundary-value

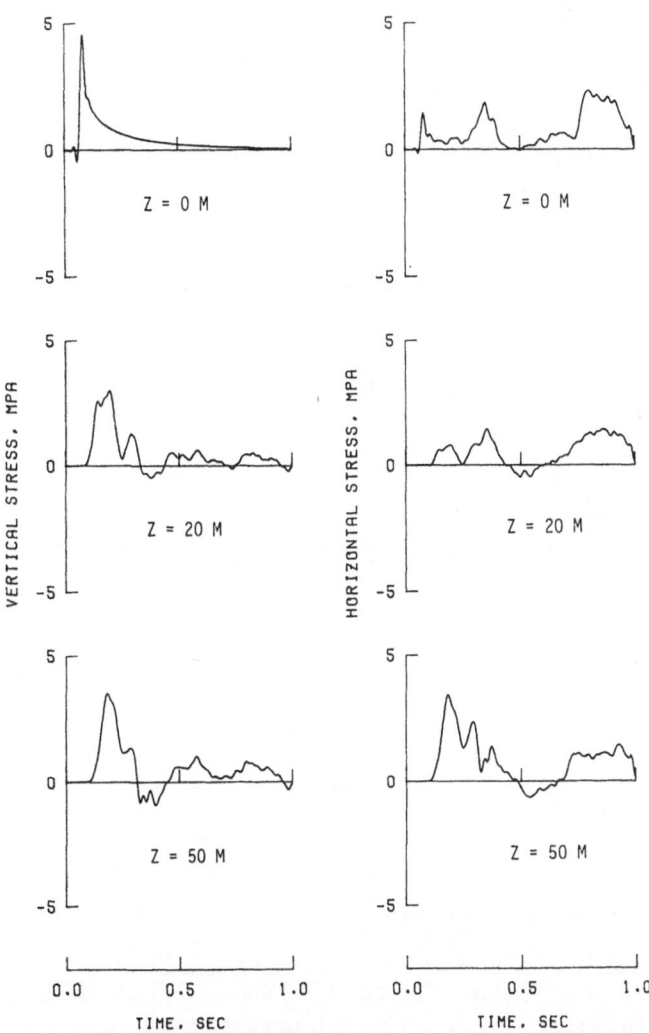

FIG. 23. Vertical and Horizontal Stress-Time Histories;
468-M Range, 6.96-MPa Contour

problems. On the other hand, the nonassociative models are not bound by normality conditions and are also capable of simulating the response of the geologic
materials. It appears that the application of the nonassociative models for the
solutions of boundary-value problems (especially dynamic problems) has not been
fully demonstrated. Future efforts in soil plasticity must address this issue.

REFERENCES

Baladi, G. Y. (1973), "The Latest Development in the Nonlinear Elastic-Nonideally
Plastic Work Hardening Cap Model," *Symposium on Plasticity and Soil Mechanics*,
Cambridge, England, September 1973, 51-55.

Baladi, G. Y. and I. Nelson (1974), "Ground Shock Calculation Parameter Study;
Influence of Type of Constitutive Model on Ground Motion Calculations," Techni-
cal Report S-71-4, Report 3, U.S. Army Engineer Waterways Experiment Station,
CE, Vicksburg, Miss.

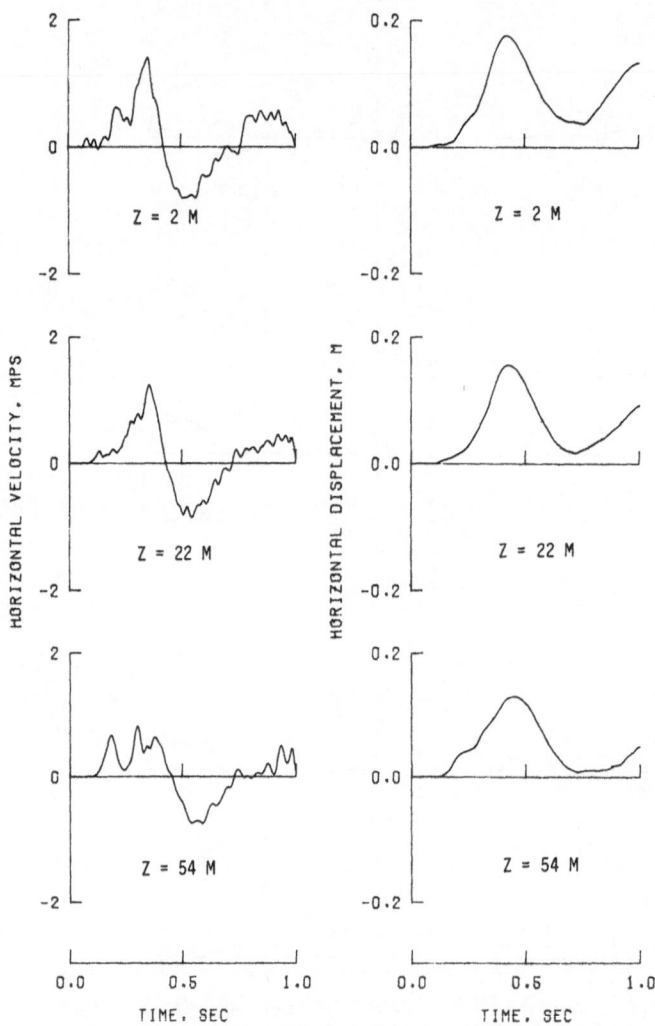

FIG. 24. Horizontal Velocity- and Displacement-Time Histories;
468-M Range, 6.96-MPa Contour

Baladi, G.Y. (1977), "Numerical Implementation of a Transverse-Isotropic Inelas-
tic, Work-Hardening Constitutive Model," *Transactions of the 4th International
Conference on Structural Mechanics in Reactor Technology*, Vol. M, Methods for
Structural Analysis, San Francisco, Calif., August 1977.

Baladi, G.Y. and B. Rohani (1979a), "A Work-Softening Model for Soil," *Proceedings of the Third Engineering Mechanics Division Specialty Conference*, ASCE,
pp. 530-534.

Baladi, G.Y. and B. Rohani (1979b), "Elastic-Plastic Model for Saturated Sand,"
Journal of the Geotechnical Engineering Division, ASCE, Vol. 105, No. GT4, Proc.
Paper 14510, pp. 465-480.

Baladi, G.Y. and B. Rohani (1979c), "An Elastic-Plastic Constitutive Model for
Saturated Sand Subjected to Monotonic and/or Cyclic Loading," *Proceedings of*

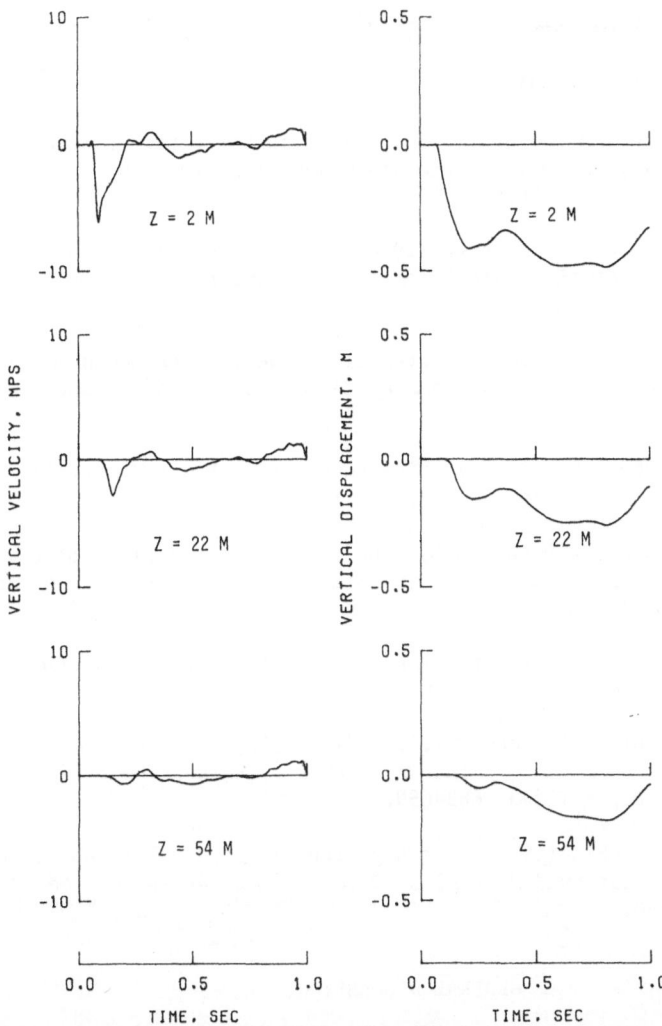

FIG. 25. Vertical Velocity- and Displacement-Time Histories;
468-M Range, 6.96 MPa Contour

the *Third International Conference on Numerical Methods in Geomechanics*, Aachen,
Germany, pp. 389-404.

Baladi, G.Y. (1980), "An Elastic-Plastic Transverse-Isotropic Constitutive Model
for Clay," *Proceedings of the Workshop on Limit Equilibrium, Plasticity and Generalized Stress-Strain in Geotechnical Engineering*, McGill University, May 28-30,
pp. 650-681.

Baladi, G.Y. and B. Rohani (1982), "An Elastic-Viscoplastic Constitutive Model
for Earth Materials," Technical Report SL-82-10, U.S. Army Engineer Waterways
Experiment Station, CE, Vicksburg, Miss.

DiMaggio, F.L. and I.S. Sandler (1971), "Material Models for Granular Soils,"
Journal of the Engineering Mechanics Division, ASCE, Vol. 97, No. EM3, Proc.
Paper 1212, pp. 935-950.

Dorris, J.F. and S. Nemat-Nasser (1982), "A Plasticity Model for Flow of Granular Materials under Triaxial Stress States," *International Journal of Solids and
Structures, 18,* No. 6, 497-531.

Lade, Paul V. and James M. Cuncan (1975), "Elastoplastic Stress-Strain Theory
for Cohesionless Soil," *Journal of the Geotechnical Engineering Division,* ASCE,
101, No. GT10, Proc. Paper 11670, 1037-1053.

Lade, Paul V. and Horacio M. Musante (1978), "Three-Dimensional Behavior of Re-
molded Clay," *Journal of the Geotechnical Engineering Division,* ASCE, *104,* No.
GT2, Proc. Paper 13551, 193-209.

Nelson, I. and G.Y. Baladi (1977), "Outrunning Ground Shock Computed with Different Models," *Journal of the Engineering Mechanics Division,* ASCE, *103,* No.
EM3, 377-393.

Nemat-Nasser, S. (1980), "On Constitutive Behavior of Fault Materials," *Solid
Earth Geophysics and Geotechnology AMD, 42,* 31-37.

Nemat-Nasser, S. and A. Shokooh (1980), "On Finite Plastic Flows of Compressible Materials with Internal Friction," *International Journal of Solids and
Structures, 16,* 495-514.

Nemat-Nasser, S. (1983), "On Finite Plastic Flow of Crystalline Solids and Geo-
materials," *J. Appl. Mech.* (50th Anniv. Issue), *50,* 1114-1126.

Sandler, I.S., F.L. DiMaggio, and G.Y. Baladi (1976), "Generalized Cap Model for
Geological Materials," *Journal of the Geotechnical Engineering Division,* ASCE,
102, No. GT7, Proc. Paper 12243, 683-699.

Sandler, I.S. and D. Rubin (1979), "An Algorithm and a Modular Subroutine for
the Cap Model," *International Journal of Numerical and Analytical Methods in Geo-
mechanics, 3,* 173-186.

GENERAL DISCUSSION

CHERRY: Can you tell me the dimensions of the samples tested to produce those
curves?

BALADI: Yes, it's 2.5 inches high, three inches in diameter.

CHERRY: And have you tried to do eight inches high and ...

BALADI: No, but we have tried samples four inches high.
     Here are some more data (Fig. A). To tell you the truth, we believe these
data. Look at the unloading curves; they are all parallel and not much scatter
in them. The loadings, however, are different.

BELYTSCHKO: I think the last slide you showed perhaps pertains to the same
point I made yesterday afternoon. I think someone pointed out to me that this
damping we see, the ramp time, the fact that we do not see shocking in the soil
which has this hardening behavior, is probably due to essentially dispersion
which arises due to the presence of inhomogeneities. Now, I don't think you
want to include all of this in the probabilistic sense. What we need is some
understanding of what is the distribution of inhomogeneities so that we can add

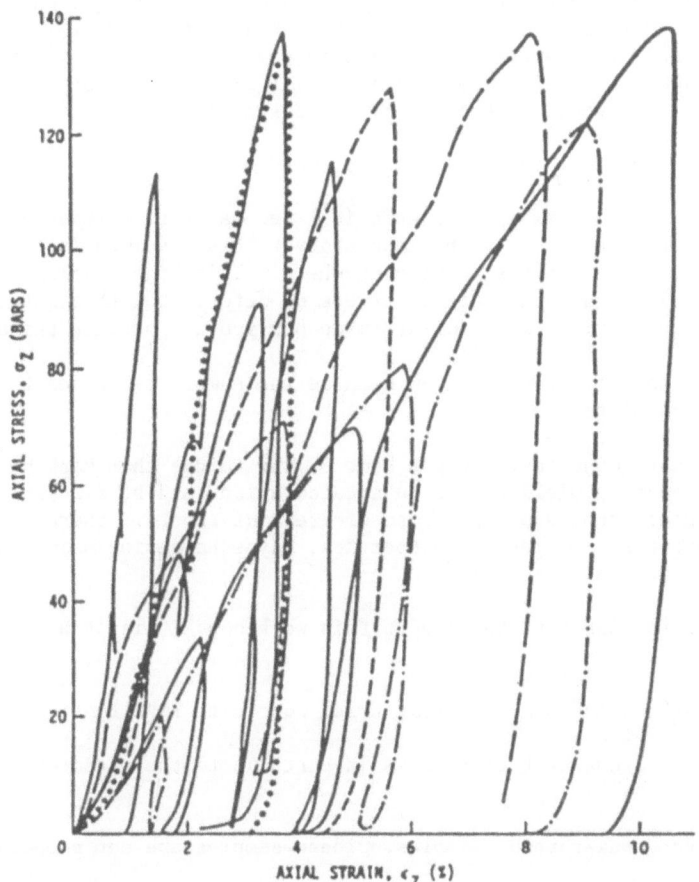

FIG. A. LBGR undisturbed sands — typical WES US test data

damping in a realistic fashion in the codes.

NEMAT-NASSER: This is uniaxial strain. It's confined. It doesn't deform laterally. It's a uniaxial deformation. The hardening is just due to confinement.

BALADI: These are uniaxial strain tests.

BELYTSCHKO: Well, hardening is always found in soils.

BALADI: I agree. That's why I mention that we have to try probabilistic approaches, micromechanical approaches, etc.

SANDLER: I think what Ted is getting at is that wave propagation in statistical media exhibits dispersion in its own right. Even if any particular point in the medium doesn't represent a material in which dispersion would occur! Some of the long rise times that are observed in the field most likely are due to that kind of dispersive effects. Now, I think what you're saying, Ted, is that if you took a statistical model of all these curves and every single one of them had the concave upward behavior, you would not produce the dispersion unless you used some statistical spatial variation in several dimensions. If, on the other hand, you put in a multidimensional spatial distribution of properties and varied these, you would probably get some dispersive effect.

BALADI: I think you'd get the same answer, Ivan.

HEGEMIER: This is a good example of when it's necessary to go back and take a look at what you're observing in the laboratory. This specimen is totally confined. There's load platen restraint everywhere. It's highly unlikely, therefore, that there is a uniform state of stress or deformation throughout the specimen. Consequently, it's not clear to me how you can utilize the data.

BALADI: Well, I do not agree with you because the sample is loaded through a fluid and not load platen.

HUGHES: I was just going to reiterate some of the things that Mark Wilkins said. He did randomize yield points in a calculation, and he did very much change the character from shock waves to smeared out fronts. There is a possibility to do that for a variety of properties, maybe hardening moduli or other things, yield points, etc.

BALADI: Yes. Unfortunately, throughout this workshop, I didn't hear anybody orking in this area.

IGHES: I'm just quoting somebody that tried something like that.

CLIFTON: Is there a chance that this is in part due to the preparation of the sample?

BALADI: They are "undisturbed" samples. These samples are not prepared in the laboratory.

CLIFTON: It's a question of whether or not the sample is "undisturbed."

BALADI: Let me elaborate on that. The samples came from the field in a steel tube. The steel tube and the sample are sliced in order to get individual samples to be tested in the uniaxial strain device. This is the type of data you have to deal with all the time.

A VOICE: Just to throw in something, in metals the safety factor is two on stress and twenty in life because we have similar things like that. I think your suggestion to use probability is a very good one.

VALANIS: I was going to comment on Prof. Clifton's comment. I think he's absolutely right. A method of preparation being the pre-history of the specimen, that is what has happened to the specimen in the past in terms of its strain life, for instance, or stress life. The fact that it lies at different levels may cause different pre-histories.

TRULIO: On the matter of sample disturbance, material properties can be measured in situ for uniaxial strain; such experiments are now being done. The effort is very worthwhile, I think, because of the real possibility of sample disturbance, and because you can find stress from the equations of motion if uniaxial motion has been measured.

BALADI: I agree with you, but there is one point here we are overlooking. When you measure the uniaxial property in the field, you're probably measuring the mean property.

TRULIO: That is what we're interested in.

BALADI: No, because the mean property when used in a computer code does not lead to a mean output waveform. Based on probability theory, using mean input of the data does not lead to the mean output unless you know the distribution of the data and use it.

CHERRY: Presumably, the initial slope of those loading curves is proportional to the compressional wave velocity in the material. Have you tried to correlate those initial slopes with a velocity log in the hole where the cores have been taken?

BALADI: The measured compressional wave velocity in the field is a mean value unless you perform several measurements.

CHERRY: You see, the initial slope of all those curves — of a single curve there is proportional to the P wave velocity.

BALADI: I was going to say it. You cannot use a mean slope (a single slope) to all these curves.

CHERRY: Yes, but you can measure that velocity in situ. Those are laboratory measurements of a similar property. You can decide by trying to correlate the two.

BALADI: The velocity in situ is also a random variable. Therefore, it is not a unique number. If you obtain a mean in situ velocity and use it as input to a computer code, this mean velocity input does not lead you to a mean answer.

CHERRY: That's not true. The velocity is a point function in situ, and you can measure that. Well, I mean, over an interval.

BALADI: It then becomes a mean over that interval!

CHERRY: But you can measure it over a two-inch interval, the same length that you've measured that core.

TRULIO: You can do "redundant" measurements and look at the scatter.

KRIEG: In the WIPP project, where the concept of nuclear waste isolation in a deep salt repository is being studied, we looked at the creep of rock salt. In laboratory tests we found that the creep rate from test to test varied by an order of magnitude. We were curious as to what effect this variability had on the creep closure of rooms and drifts. So we took the mean of these laboratory creep tests and used it in a deterministic calculation. Then we took the statistics for the laboratory data and applied that randomly in a spatial manner to our computational grid of our practical problem. The results were tabulated. We took the statistics and again randomly laid it on the grid, of course with different distribution from the first. The answers were of course different on the second run. We did this fifteen times and took the mean of those results. As you say, George, the mean of the answers from these fifteen runs is different than the single run using the mean from the laboratory tests. Now, the conclusion for all this is that yes, you can do statistical evaluations. It costs fifteen times what it costs to do the one deterministic analysis. If you had taken any one of those analyses, you would have come up with erroneous results in the sense that there's a lot of spread from analysis to analysis.

BALADI: Let me answer that. You need the spread (distribution) in your answer so you can determine the coefficient of variation in your answer. The mechanical properties of soil are random variables. These properties change from one point to the next. A single perfect field measurement, I think, leads to a mean value. If you use this mean value in any calculations, whether code calculations or closed form solutions, this mean value does not lead you to the mean value of the answer.

KRIEG: That's right. The purpose of this study was to find whether the spread in experimental results on the WIPP project will lie within the same error bar as the theoretical answer that we got.

VARDOULAKIS: One method to improve the quality of experimental results from undisturbed specimens is to use large specimens. And there are methods of sampling undisturbed large specimens of soft rocks. For example, I have in mind methods for specimens with diameters up to one meter. This means that one can have several features of soft rock or natural soil, like fissures and faults, and this will improve the confidence and perhaps reduce the total cost.

SCOTT: I think I'd like to demolish the notion of an undisturbed sample. We've heard a great deal during this meeting about the terrible properties of soils. They are nonlinear, irreversible, hysteretic, frictional, etc. Consequently, any method of obtaining a soil or rock core involves unloading it. When you reload it again in the laboratory, you can't possibly get it back to its virgin condition again. We have a problem with the uncertainty principle in general, that you can't measure any in-place soil property without disturbing the property that you're trying to measure. In addition, you can't put any instrument down in the soil to measure displacements or stresses without boring a hole into the material in order to emplace your gauges. When you bore a hole into the material, you establish a stress field around the hole, and thus around the gauge, which has other problems itself — stiffness relative to soil, density, etc. So you can never measure *anything* properly in soils; the same is true to some extent in rocks, although the problems diminish as the material gets stiffer or harder, of course.

PREPARED DISCUSSION

by Jean H. Prevost

1. INTRODUCTION

The report presented by Baladi and Rohani (1983) addresses the very difficult problem of defining an adequate constitutive model for the inelastic response of soils subjected to complex loading paths. The authors propose relatively simple models arguing that "it is neither necessary (from practical point of view) nor desirable (from mathematical consideration)" to develop a complex (and more accurate) constitutive model. All the models presented are Cap models (Drucker et al., 1957) based on classical isotropic plasticity theory with associated flow, and are variations and refinements of the basic Cap model pioneered by DiMaggio and Sandler (1971). First and foremost, it must be commented that in the hierarchy of the many elastoplastic constitutive soil models presented in the literature, these are truly *by far* the simplest models. However, they are also somewhat outdated as further discussed hereafter. The most obvious limitations of these models are: (1) they do not adequately model soil stress-induced anisotropy; (2) they are not applicable to cyclic loading conditions. It may be argued that plastic models based on isotropic hardening rules are adequate for situations in which only loading (and moderate unloading) occurs, however it is unlikely that such restrictions can be met at every point in general boundary value problems, especially in wave scattering problems. Therefore, without supporting field data, one must be suspicious of predicted responses because of the material model limitations. Further, the models are not free either of some subtle (physical and/or theoretical) difficulties as discussed in Section 2. Some of the authors' comments about uniqueness and stability of solutions raise some interesting questions which are discussed in Section 3. Finally, Section 4 contains a brief description of a modern type of plasticity theory, free of the above mentioned limitations, and used by several researchers in attempts to more accurately model the behavior of soil media.

2. GENERAL COMMENTS

2.1 Elastic Moduli

In order to ensure that the elastic response is path-independent, Baladi and Rohani (1983) postulate that (their Eqs. 9 and 10)

$$K = \Lambda + 2G/3 = K(J_1, \varepsilon_{kk}^P), \qquad G = G(\sqrt{J_2}, \varepsilon_{kk}^P) \tag{1}$$

where J_1 = trace $\underset{\sim}{\sigma}$; $2J_2 = \underset{\sim}{s}:\underset{\sim}{s}$ = trace $(\underset{\sim}{s} \cdot \underset{\sim}{s})$ with $\underset{\sim}{s} = \underset{\sim}{\sigma} - J_1/3\underset{\sim}{\delta}$. However, it has been shown experimentally (see, e.g., Richard et al., 1970) that

$$G = G_1^* \left(\frac{J_1}{J_1^*}\right)^n \tag{2}$$

with $n = 1$ for cohesive soils, $n = 0.5$ for cohesionless soils; G_1^* = shear modulus at confinement J_1^*. That G is also a function of $\sqrt{J_2}$ is rather suspicious since, for instance, it would imply that G is a function of the lateral stress coefficient at rest $(K_0 = \sigma_H'/\sigma_V')$.

2.2 Corners in Yield

In the cap models presented by Baladi and Rohani (1983) (see their Figs. 11 and 12), two yield (failure) surfaces are used. There is a corner at the intersection of the two surfaces where two infinitely closed stress states must lead to two radically different flow conditions. This obviously is a rather unsatisfactory feature of the models within the context of a continuum theory.

3. ASSOCIATIVE vs. NONASSOCIATIVE FLOW — UNIQUENESS AND STABILITY

Baladi and Rohani (1983) advocate the use of an associative plastic flow rule to ensure stability and uniqueness of solutions, claiming that such conditions have not been fully established for nonassociative flow rules. This last statement is not entirely correct since it is well known that uniqueness and stability (of the incremental solution) require that

$$\int_\Omega \underset{\sim}{e}:\underset{\approx}{C}:\underset{\sim}{e} \ d\Omega > 0 \tag{3}$$

for every non-zero second-order tensor $\underset{\sim}{e}$; Ω = domain occupied by the solid body; $\underset{\approx}{C} = \underset{\approx}{C}(\underset{\sim}{x})$ = fourth-order tensor of elastoplastic moduli, viz. (for small deformations)

$$\underset{\approx}{C} = \underset{\approx}{E} - \frac{(\underset{\approx}{E}:\underset{\sim}{P})(\underset{\sim}{Q}:\underset{\approx}{E})}{H' + \underset{\sim}{Q}:\underset{\approx}{E}:\underset{\sim}{P}} \tag{4}$$

with $\underset{\approx}{E}$ = tensor of incremental elastic moduli; $\underset{\sim}{P}$ and $\underset{\sim}{Q}$ = symmetric second-order tensors (normalized in any convenient way) such that $\underset{\sim}{P}$ gives the "direction" of plastic deformations and $\underset{\sim}{Q}$ is the outer "normal" to the yield surface in stress space; H' = plastic modulus. If $\underset{\sim}{P} \neq \underset{\sim}{Q}$ a nonassociative flow rule is assumed and $\underset{\approx}{C}$ does not exhibit the major symmetry (viz., $C_{ijke} \neq C_{keij}$). *Sufficient conditions* for Eq. (2) to hold can easily be established by, for instance, restricting the plastic parameters to be such that pointwise stability is ensured, viz., by restricting $\underset{\approx}{C}(\underset{\sim}{x})$ to be such that for every $\underset{\sim}{x} \in \Omega \subset \mathbb{R}^n$, and for every nonzero second-order tensor $\underset{\sim}{e}$, there is an $\varepsilon > 0$ such that

$$\underset{\sim}{e}:\underset{\approx}{C}:\underset{\sim}{e} > \varepsilon \ \|\underset{\sim}{e}\|^2 \qquad \forall \ \underset{\sim}{x} \in \Omega. \tag{5}$$

Existence and uniqueness proofs are then classical and deduced from Korn's inequalities; see, e.g., Duvaut and Lions (1972), and Fichera (1972). Application of Eq. (5) to Eq. (4) yields (for symmetric $\underset{\sim}{e}$):

$$H' > \frac{1}{2} \ [\sqrt{(\underset{\sim}{Q}:\underset{\approx}{E}:\underset{\sim}{Q})(\underset{\sim}{P}:\underset{\approx}{E}:\underset{\sim}{P})} - \underset{\sim}{Q}:\underset{\approx}{E}:\underset{\sim}{P}]. \tag{6}$$

Note that if $\underset{\sim}{e} = \underset{\sim}{\alpha} \otimes \underset{\sim}{\beta}$, then Eq. (5) is the strong ellipticity condition.

Therefore, there are no theoretical reasons for restricting the flow rule to be associative. Further, it has been shown (see, e.g., Rice, 1976, Rudnicki and Rice, 1975) that nonassociative flow rules are very versatile since they allow loss of ellipticity for positive hardening ($H' > 0$) and an accurate description of observed instabilities and emergence of shear bands in geological media.

4.   ISOTROPIC/KINEMATIC SOIL MODELS

Much effort has been devoted in the past few years to the development of a plasticity theory which accurately describes the behavior of soil media. There is not yet firm agreement among researchers. However, the general trend has been to adopt a combination of isotropic and kinematic plasticity (see, e.g., Mroz et al., 1978, 1979, 1983; Dafalias et al., 1980, 1982; Prevost, 1977, 1978), because it offers extreme versatility in describing observed nonlinear behavior, stress-induced-type anisotropy, Bauschinger effects and softening behavior of real materials under cyclic loading conditions. Some researchers prefer a two-surface plasticity while others prefer a multi-surface plasticity. Both theories suffer inherent limitations, namely: storage requirements for the multi-surface theory, *a priori* selection of an evolution law for the two-surface theory. However, neither limitation is more degrading than the other. This is further discussed in Prevost (1982). These theories have been shown to offer great versatility, accuracy and flexibility in describing the behavior of real soil materials, and are very *significant* improvements over the models presented by Baladi and Rohani (1983).

REFERENCES

Baladi, G.Y. and B. Rohani (1983), "Soil Plasticity," *Proc. Workshop on the Theoretical Foundation for Large-Scale Computations of Nonlinear Material Behavior*, Northwestern Univ., Evanston, Illinois.

Dafalias, Y.F. and L.R. Herrmann (1980), "A Generalized Bounding Surface Constitutive Model for Clays," in *Application of Plasticity and Generalized Stress-Strain in Geotechnical Engineering*, ASCE, Eds. R.N. Yong and E.T. Selig, 78-95.

Dafalias, Y.F. and L.R. Herrmann (1982), "Bounding Surface Formulation of Soil Plasticity," in *Soil Mechanics: Transient and Cyclic Loads*, John Wiley; Eds. G.H. Pande and O.C. Zienkiewicz, 253-282.

DiMaggio, F.L. and I.S. Sandler (1971), "Material Models for Granular Soils," *J. Eng. Mech. Div.*, ASCE, *97*, No. EM3, 935-950.

Drucker, D.C., R.E. Gibson and D.J. Henkel (1957), "Soil Mechanics and Work-Hardening Theories of Plasticity," *Transactions ASCE*, *122*, 338-346.

Duvaut, G. and J.L. Lions (1973), *Les Inequations en Mecanique et en Physique*, Dunod, Paris.

Fichera, G. (1972), "Existence Theorems in Elasticity," *Handbuch der Physics*, Springer-Verlag, Vol. VI a/2, 347-390.

Mroz, Z., V.A. Norris and O.C. Zienkiewicz (1978), "An Anisotropic Hardening Model for Soils and its Application to Cyclic Loading," *Int. J. Num. Anal. Methods Geomech.*, *2*, 203-221.

Mroz, Z., V.A. Norris and O.C. Zienkiewicz (1979), "Application of an Anisotropic Hardening Model in the Analysis of Elastoplastic Deformation of Soils," *Geotechnique*, *29*, 1-34.

Mroz, Z. and St. Pietruzxzak (1983), "Constitutive Model for Sand with Anisotropic Hardening Rule," *Int. J. Num. Anal. Methods Geomech.*, *7*, 305-320.

Prevost, J.H. (1978), "Plasticity Theory for Soil Stress-Strain Behavior," *J. Eng. Mech. Div.*, ASCE, *104*, 1177-1194.

Prevost, J.H. (1977), "Mathematical Modeling of Monotonic and Cyclic Undrained Clay Behavior," *Int. J. Num. Anal. Methods Geomech.*, *1*, 195-216.

Prevost, J.H. (1983), "Two-Surface versus Multi-Surface Plasticity Theories: A Critical Assessment," *Int. J. Num. Anal. Methods Geomech.*, *6*, 323-338.

Rice, J.R. (1976), "The Localization of Plastic Deformation," *Proc. 14th Int. Cong. Theoretical and Applied Mechanics*, Delft, Ed. W.T. Koiter, *1*, 207-220.

Richart, F.E., J.R. Hall and R.D. Woods (1970), *Vibrations of Soils and Foundations*, Prentice-Hall.

Rudnicki, J.W. and J.R. Rice (1975), "Conditions for the Localization of Deformation in Pressure-Sensitive Dilatant Materials," *J. Mech. Phys. Solids*, *23*, 371-394.

GENERAL DISCUSSION

TRULIO: Let me try to state more clearly the point I was trying to make before; I think Ted was on the same point. It's not a question whether a soil property like bulk modulus is distributed or not. I think we all agree it is. It's a random variable with a distribution. There is, though, a real question as to whether the distribution in the laboratory is the same as the distribution in the field, and that really needs to be answered.

When measurements of motion are made, I don't think we see as much scatter in supposedly redundant motions as there is in the kind of stress-strain data you showed — although there are certainly examples to the contrary.

PREVOST: I can answer that because I don't do any experiments myself, so I'm quite relaxed talking about people doing experiments — well, I do some kind of experiments but not the kind of experiment that we are talking about now. I think that Dr. Baladi's slide was exaggerated because my experience is that you can have much better data than what he has shown here. Obviously, this is not true for every laboratory in the world, just for a handful of laboratories, which can really do high-quality testing with high reproducibility. So it is not as bad as he mentioned, but still there is a problem in sampling and testing with soils.

However, you can get good properties if you reconstitute your specimens by some magical procedure, which obviously can be criticized because you may still disturb some properties. But what I want to emphasize is that you can get good soil data.

DIENES: I was interested in the aspect of the curves that were shown before for the soils, that they aren't repeatable. I'd just like to mention another class of processes that isn't repeatable that I think is very interesting, and it has to do with propellants. When you fire propellants into a target, the experiments are not repeatable. Sometimes they blow up and sometimes they don't, though we have presumably identical samples. It's really very important to understand that, and I don't think you can do it with the methods of continuum mechanics. I've had some success in trying to understand what's going on using statistical models, what I call statistical crack mechanics and what Curran calls Nag Frag.

There are a lot of processes that I think you cannot use continuum mechanics for, because what's important is what's happening on the microscale. Maybe those problems in soil mechanics can be attacked by micromechanics. I think they might be.

HERRMANN: I think anybody associated with a rock mechanics or soil mechanics lab sees predominantly data such as Baladi showed, a very ubiquitous feature in testing earth materials because they're variable. However, I think anybody that looks at real geologies does not conclude that those properties are randomly distributed. Usually we don't look in enough detail to see the distribution. If we did, we could incorporate the variation and reduce some of that scatter. We've done a little bit of that at the WIPP site, for example, although there is a lot of controversy.

Now, the real problem with the statistical approach is that you're interested in the mean response. You're always interested in sure safe. If you want a building to not fall down, you're interested in sure safe. You're always interested in the limits at some level of confidence. I would argue that the attempt to give mean values is academic because nobody really cares. You've got to look at the confidence limits, and it's the uncertainty band that is crucial more than the mean value.

SANDLER: Jean, you pointed out two differences between the cap models and the newer multiplasticity models. You pointed out the use of multiple yield surfaces and the use of kinematic hardening. I agree that the use of multiple yield surface is probably a fundamental difference and may very well contribute a better modeling of soil behavior. But the kinematic hardening has been incorporated into many of the cap models, and the original cap model can be argued to have kinematic hardening volumetrically, although there was no kinematic hardening in shear in that particular one. Other, subsequent versions have used kinematic hardening over the years, so I would say the single difference is in the use of the multi-surface plasticity.

PREVOST: No, I don't believe so. I'm aware of what you are referring to. I would say that this is at best an ad hoc procedure that you are using.

SANDLER: Well, they're all ad hoc.

BALADI: I'd like to answer Prof. Prevost about the exaggeration in the data. At the Waterways we don't deal with "university material." Instead we deal with real materials. Whenever you deal with real material you always get scatter in the data.

KELLER: I just wanted to ask a question. I hadn't heard anyone say anything about composite materials. Obviously, geologic materials are made up of strong and weak materials, clay layers and very solidified layers, and soils are made up of cobbles and so forth. The measurements indicate that some parts of the medium are strong, some are weak. Is there any work that has been done on what's really controlling the behavior? One would expect that the weak materials would do the most yielding. In fact, when you're averaging, it seems to me you're violating an intuitive assumption that the weak material is controlling the behavior. I'm just wondering, has anything been done on that?

BALADI: You're right. If you have composite materials made up of weak and strong materials, the weak materials control the overall behavior. But if you apply probabilistic analysis properly, the results will show that the overall behavior is controlled by the weak material.

SCHUSTER: I agree with the probabilistic analysis. However, the problem that we run into is that frequently we verify these models or calculations by going out and performing a single experiment. That test is not probabilistic since it produces a single result which we have to match.

Our recent experience, the one-ton test that Carl Keller described briefly today and some of the recent tests show substantial correlations between many of the calculations and the experiments, so I don't think the problem of correlating calculations and field data is quite as hopeless as your scatter-gram showed. In addition, we are currently interested in much higher stress levels than you've showed today. At these stresses the unloading paths, which most of the time is where your material resides, all coalesced. I think we have to remember that the tests we compare calculations with are really unique. It is those experiments or calculations that determine whether we build a silo or declare it is safe or not. It's not the probabilistic approach that's going to do that.

BALADI: I agree with you. However, if you have enough redundant field measurements which lead to a scatter in the data, then what do you do? If you perform a probabilistic analysis and you calculation-bound the measurements, you will have a better handle for explaining your analysis. You will have much better understanding of your analysis versus the field measurements than trying to make a one-to-one comparison.

ROESSET: Of course, there have been all these cases where they have done probabilistic analyses and then they did the experiment and found out the experiments were entirely out of the bounds.

NEMAT-NASSER: I would like to make a comment and raise a question which seems to me to be relevant to the geological materials. These relate to the length scale over which we can reasonably expect any kind of reproducibility.

I would like to take you back to some of the slides that I have showed on polycrystals. In the case of polycrystals, at the micro-level you can see all kinds of complicated events. If you wish to predict these events in a sample, you will find that it would be awfully difficult. The scale that we are looking at, a piece of metal, is large enough to kind of cover these micro-events.

Now, in the case of geological materials it seems to me that somebody has to at least look at the possibility of a length scale over which we would expect some kind of deterministic certainty.

SANDLER: With respect to the comment about reproducibility of some of the wave-forms that were shown previously, I think it's a function of where you look. The near-surface materials tend to be less uniform and hence lead to less reproducibility than deeper materials, especially if you're away from major faults or away from where water table variations might be important. The variations in azimuth from a surface burst are probably stronger and lead to a broader distribution of observables than you would get from a buried burst where the material is more confined by being nearly one-dimensional (spherically symmetric).

The length scales near the surface are much shorter because you tend to have very highly layered materials with rapid variations in properties from the surface to just a few meters below, whereas at depths, that tends not to be true. So reproducibility is a complicated question or, rather, a complicated set of issues. There is probably no simple single answer that really addresses questions of reproducibility or length scale.

CHAPTER XIII

SHORT PRESENTATIONS, DISCUSSIONS, AND SHORT CONTRIBUTIONS

Chairman: *S. Nemat-Nasser*

NEMAT-NASSER: The Organizing Committee decided to devote this session to short presentations. Except for Ron Scott, who was invited to make a twenty-minute presentation, we have managed to provide ten minutes each for brief reviews of recent advances in constitutive modeling, including two presentations on "micro-mechanics."

Short Presentation by R. F. Scott

I must say, coming here from the field that I represent and considering things at the level at which I usually perform them, I feel something like a tortoise in a convention of hares. There isn't really anything I can say that hasn't already been said except, perhaps, that some reiteration or emphasis is worthwhile.

As far as the flow rule is concerned, it seemed to me worth referring again to this business of uniqueness. In the case of a non-associated flow rule, which I think covers all of Dr. Wilkins' material yesterday, uniqueness may in fact exist, but we can't show it. And that means that his solutions to all of these problems may be all right or they may not. I don't think this has been clearly enough emphasized. Since solutions of this kind have been obtained, the question must be asked: what are the ways in which one can demonstrate that a particular solution of that class would be correct?

The only method I can think of (apart from a mathematical proof) is a lab-oratory or field test. The drawback is that the larger the scale at which the field test is conducted, the less information is available from it to check the response of a computer model. Compared to the field test, the computer model presents results in excessive detail. There is no way we can check phenomena in the field to this level of detail.

That brings me to something that Dr. Trulio was talking about yesterday, which is the question of the geometrical constraints on some of these problems. I think there's a fairly old and well-known example of this. Many years ago, at the beginning of World War II, the Corps of Engineers Waterways Experiment Station was seeking ways to design air fields for bigger and heavier bombers. They were interested in seeing whether the models of those days, linearly elas-tic or rigid plastic, would describe the stress conditions in the soil under-neath the pavement, in order to work out a design method.

They dug up a stretch of ground and placed pressure meters horizontally in the ground in order to measure vertical stresses. Then these were covered over and a vertical load placed on the ground surface. The idea was to compare the measured vertical stress distribution with that calculated from theory. However, what happens is predictable with a little thought. When you put a uniform load of finite area on the surface of material, there results a more-or-less Gaussian distribution of vertical stress underneath it, regardless of the material properties. The vertical stress distribution is governed by the equilibrium condition.

The measured stresses are not sensitive to the material properties. They just record the equilibrium distribution, with minor variations.

The overburden could be clay, sand, gravel, concrete, rock, or anything else you like and you'd still get the same distribution. A comparison with a computed result would not be a decisive test of a model.

If the pressure cells had been imbedded vertically and recorded the distribution of horizontal stress it would have been possible to distinguish much better between the classes of material and material response, as indicated by computer calculation. So I think one wants to look very carefully at comparisons in terms of the sensitivity of the properties measured and computed.

There are two more subjects I'd like to touch on: the formation of slip lines, and micromodels.

As far as shear banding or bifurcations are concerned, I think that Gil Hegemier presented a very nice discussion of that as far as tests are concerned and what you can get out of them. I only want to add one thing more to Gil's statements. That is that the test behavior of a soil sample also depends on the rigidity of the confining chamber or the confining plates that you put around it. There have been some three-dimensional soil experiments done with extremely ingenious apparatuses, all of which are called in the literature "true" triaxial tests. To its progenitor the apparatus is always true. Some of them involve compressing the sample in three dimensions with a rigid plate system, in which you can therefore observe no instability, and no bifurcations occur whatsoever. They're inhibited by the rigid boundaries.

Whereas, on the other hand, if you have a cubical sample with flexible membranes pressurized with air or water around it, then the soil is free to develop slip at shear zones and bifurcate. As soon as it does so, in fact somewhat before slip is actually visible, you no longer have a single-element test. The results of such a test are therefore not applicable to any kind of continuum mechanics analysis or the establishment of continuum relations, involving stress-strain, or constitutive laws, etc. The shear banding question is going to be tackled by Dr. Vardoulakis in more detail, after me, so I'm not going to say anything more about that.

On the other subject, microscopic studies, as applied to soils, I personally don't feel terribly optimistic about the micromodel as the solution in the future. This is illustrated by Fig. 1. This is a picture of a two-dimensional model of clay. Clay is formed of little platelets of clay minerals, with diameter-to-thickness ratios of 50 or more. The solid lines in this diagram represent a random model of little plates of clay in contact. One would say that it would be very nice if we can do a deterministic analysis of some random collection of clay like this, under load. Then the results might give some idea of

PHYSICAL MODEL–MODEL 1

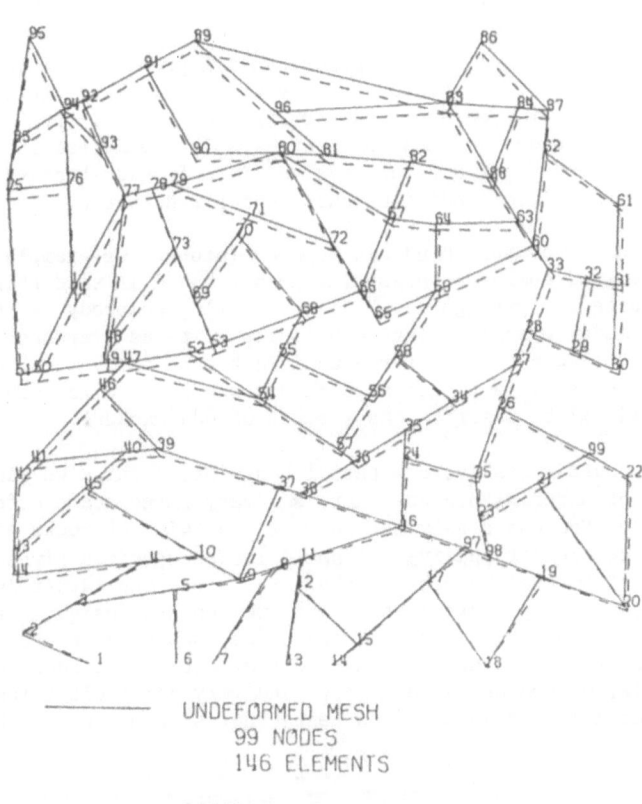

UNDEFORMED MESH
99 NODES
146 ELEMENTS

DEFORMED MESH, ELASTIC RESPONSE

FIG. 1

what a macroscopic constitutive law might contain. The analysis could ultimately
be extended to three-dimensional models.

The nice thing about computers is that you can put a pure load on the
boundaries. We can put uniform stresses (rubber membrane in a laboratory test)
or pure displacements (rigid plate in laboratory test) on all of the boundaries
to conduct a test on this material. Its performance indicates what the constitutive relationships on the macroscopic level are. But the first problem that
you have to face when you put this model into the computer is, what are the
microscopic relations that you're going to give to the particle contacts? So
you start reading books on physical chemistry to find out what happens at the
junction between clay platelets. Since, so far as I can tell, at the mechanical
level of forces and displacements, nobody knows, before very long, you're totally lost. That is to say: you have at least as big, and probably bigger, a problem knowing what goes on at the micro-level as you have knowing what happens
macroscopically.

Ideally, you would like to put some relationship at the particle level and see what it does to the macro-constitutive relations. Then change the relation, quantitatively or qualitatively, change the cohesion, the friction, if in fact you can talk about cohesion and friction on a molecular level, and examine the effect on the behavior in the large.

It seems to me that almost any geomechanical problem that you look at in this way, for example if the picture represents a solid and the clay particles are cracks as in a rock, you have just as big a microscopic problem in saying what happens at the sides and ends of a crack. Is the material really plastic there, or time-dependent or are you involved in molecular bonding? In other words, what are the micro-conditions under which fracture occurs?

I think studies of this kind can form an interesting example of the very old and now outdated computer phrase (I haven't heard it said this week, for instance) "garbage in, garbage out." Incidentally, somebody at the beginning of the meeting made an interesting verbal slip. It was a spoonerism; he meant to say "ad hoc." I don't know if you noticed but he said, "od hac." Since among teenagers nowadays a hacker is somebody who plays with computers, I thought we could all be described as a bunch of odd hackers.

In sum, to obtain realistic material behavior, I think we want to test bigger samples of material because there's a very-large scale effect in geo-mechanics tests. You can easily see that if you think of rocks with faulting or cracking at particular spacing. A small intact specimen gives a behavior quite different from what happens in the large, since the latter is dictated by the fracture and crack geometry, etc. When we want to design a very large con-crete or earth dam that rests on rock and we're looking at problems of engineer-ing geology, an attempt is usually made to test a block of rock several meters in size. That's, of course, an expensive and very difficult thing to do. It's also hard to accumulate statistics on tests that size by doing a hundred of them, for instance.

***

BELYTSCHKO: This is not really a question, but perhaps some of the people from the micromechanics community could give us an idea of what the accomplishments have been in metal plasticity. If I remember, in the workshop on thermal visco-plasticity that Nemat-Nasser organized several years ago, there was considerable pessimism as to whether anything had been learned, for example, from dislocation models of plasticity as to the behavior of metals. Perhaps you could update us on that field.

HUTCHINSON: I think within the context of this meeting, the main thing one ob-tains from the micromechanics is the structure of constitutive laws. That is, we obtain guidelines for generalizing phenomenological relationships. That's, I think, been the primary success of micromechanics.

KELLER: I just wanted to comment, since you say you test very large samples. I've read a little bit about the advantage of testing large samples, and it's been sort of discouraging. The cost is very discouraging. I was wondering if you feel that it's hopeless to try to obtain a reasonable estimate of what a large-sample response might be from measuring small samples?

SCOTT: I think that it depends on the economics of the situation. You can work at any level you care. You can go out to a test site, as somebody mentioned the other day, and if you're given only a few days' notice, you can just look at a geological map and compare the materials of that site with what you know of the mechanics of such materials in the literature and construct a site model. That doesn't cost very much. It involves a graduate student for a couple of days. And then with more warning or budget, you can go out and drill a few holes, which don't cost very much, take a few samples, which, as I have said, are always disturbed, test them in some laboratory, get some properties; and you have a slightly better idea of the site conditions and model. You can just go on from there. Clearly, when one is looking at nuclear weapons tests, etc., one would ideally like to test as large samples as possible — samples that are perhaps of the size of the grid that you're going to use in a finite element analysis.

ASARO: I don't know that these comments are necessarily about micromechanics and metal plasticity. But maybe I'll just comment briefly about the progress in using micromechanics in that area. For ten years or more there's been the ability to model all the way from the single-crystal level to the polycrystal level and —

SCOTT: I'm talking about geomaterials.

ASARO: Yes, I suspected that and that's why I prefaced my remarks by saying they may not be necessary. But because Ted asked for the progress review, I'll make them brief. As John said, it has been possible for the last ten years to go from the dislocation, at least single-crystal level which is actually the most expedient place to start, to the polycrystal level and obtain strong suggestions for the form and essential structure of the constitutive laws. This sort of modelling has been used with tremendous advantage over the years. Now, in fact, it's possible to go even further and actually do extremely detailed computations on the single-crystal and polycrystal level and actually look at lattice rotational effects and microcrystalline texturing effects. It's now possible to take still another step, and that is toward fully rate dependent polycrystalline models, which will include the very important effects of material anisotropy. But I think in the metal plasticity area, we have an advantage. The fact that these materials are crystalline makes our job of defining the essential features of the micromechanics a lot easier than it does in what you were talking about. So I realize that what you meant by micromechanics pertained to that class of materials which indeed are more complex to deal with.

NEMAT-NASSER: Thank you very much. I think in this connection there are also possibilities in the area of geomaterials. And for that reason, we have the next presentation, where Dr. Horii will indicate to you how a micromechanics of rock failure can be formulated; we are not talking about crystal levels, or the dislocation level, but at the level which can be defined as "micro" in rocks.

Short Presentation by H. Horii: "Micromechanics of Rock Failure in Compression"

The main objective of this presentation is to reveal the mechanisms of failure in compression of brittle solids containing micro-flaws. Rocks are typical examples of the material of this kind. A rock sample fails by axial splitting under uniaxial compression and by faulting under triaxial compression. The strength of the specimen and the overall failure angle depend on the confining pressure.

Under farfield compression, preexisting flaws undergo deformations resulting in local tension at their boundaries. Then tension cracks nucleate and propagate

curving into the maximum compressive direction. This has been considered to be
one of the mechanisms of the microcracking in compression. As a mathematical
model, we consider an infinite plate containing a preexisting flaw PP' and out-
of-plane cracks PQ and P'Q' as shown in Fig. 2. Along the preexisting flaw, the
normal displacement is continuous, and we assume a simple slip condition, having
both frictional and cohesive resistance. Along the out-of-plane cracks, stress-
free conditions are satisfied since these are tension cracks.

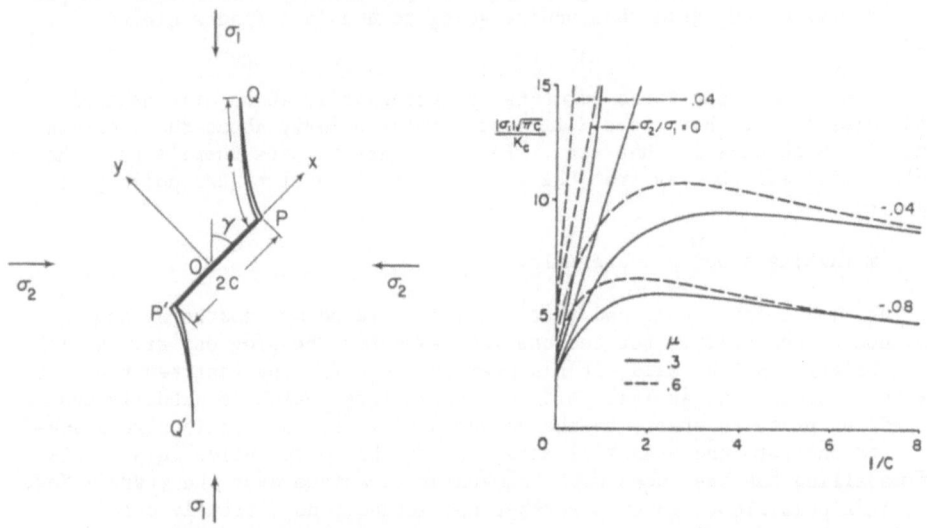

FIG. 2. A preexisting flaw PP' and   FIG. 3. Compressive force required to
out-of-plane cracks PQ and P'Q'.   attain the associated crack length.

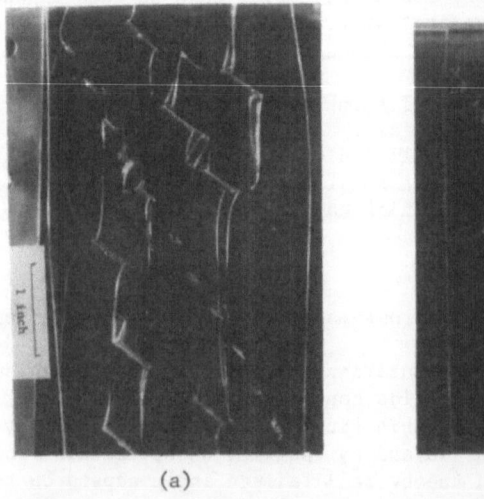 

(a)   (b)

FIG. 4. (a) Axial splitting of a specimen containing a row of small flaws and
several larger flaws under uniaxial compression; and (b) Formation of a high
crack density zone under triaxial compression.

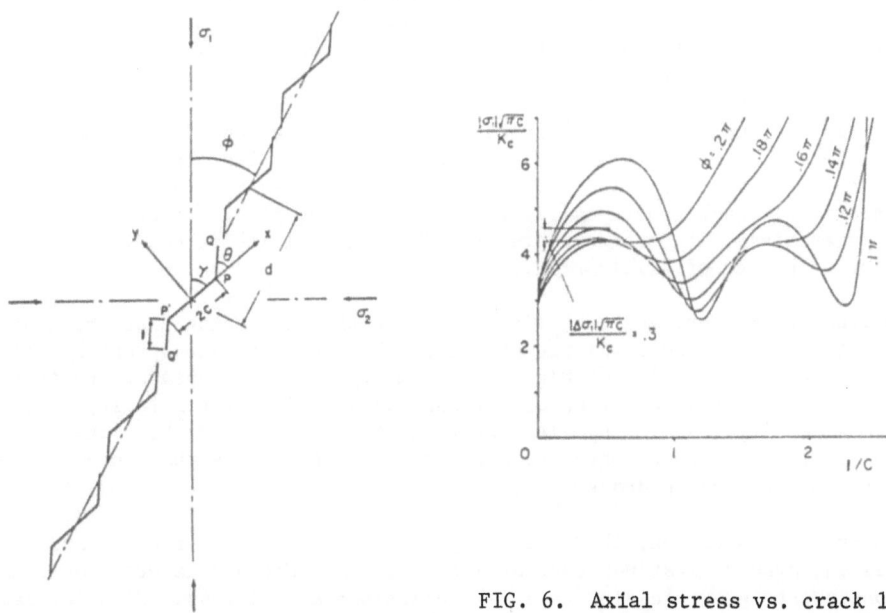

FIG. 6. Axial stress vs. crack length
for indicated overall orientation.

FIG. 5. A row of preexisting flaws
  PP' and tension cracks PQ and P'Q.

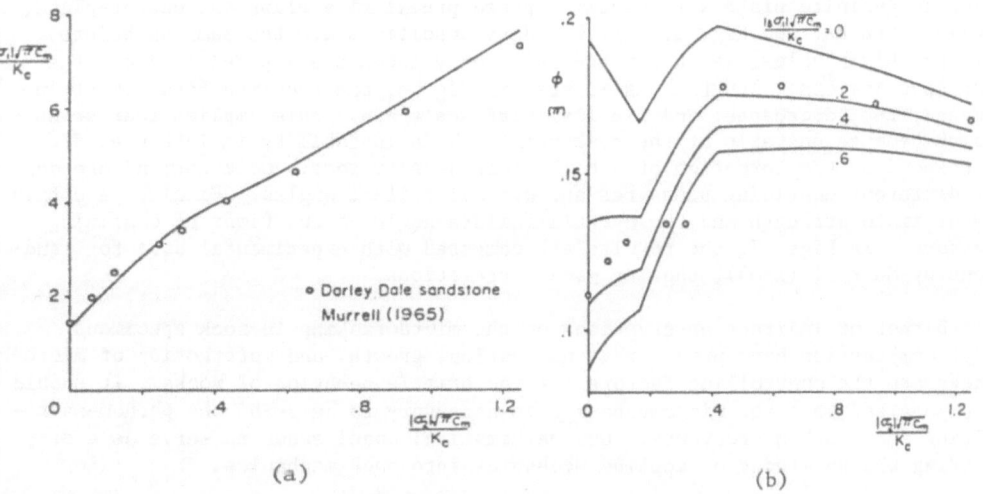

(a)   (b)

FIG. 7. Variation of (a) the ultimate strength and (b) overall failure angle
  with confining pressure.

To solve this problem, we consider a pair of antisymmetric dislocations
near the preexisting flaw. Replacing out-of-plane cracks by distributed dislocations, the problem is reduced to a singular integral equation for the dislocation density. Solving the singular integral equation numerically, we obtain
the stress intensity factors at the crack tips. Using a fracture criterion that

cracks propagate in the direction of the maximum tension, we obtain the crack profile incrementally. Setting the opening mode stress intensity factor, K_I, equal to the fracture toughness, K_c, we obtain the relation between the applied load and the crack length.

In Fig. 3, the required axial load is plotted as a function of the corresponding crack length. With lateral compression acting, the crack growth is stable and it ceases at a finite length. If, on the other hand, small lateral tension exists, crack growth is first stable and then after a certain crack length is attained, it becomes unstable. This instability is considered to be the major mechanism of axial splitting.

Hallbauer, Wagner, and Cook (1976) observe that the crack distributions in rock samples are uniform until the stress reaches the ultimate strength. At the ultimate strength, a region of high crack density suddenly appears. Further straining leads to the formation of a shear failure plane (or a fault) in this high crack density zone. They also report that each crack in this zone is parallel to the maximum compressive direction. This implies that they are not shear cracks but tension cracks.

In the rock specimen, there are many flaws of different sizes. Under axial compression, cracks first nucleate at larger flaws. Without lateral confinement they continue to grow, leading to axial splitting; see Fig. 4a. With lateral confinement, on the other hand, they cease to grow. At certain stress level, tension cracks suddenly develop at suitably arranged smaller flaws because of their interaction, and a region of high crack density forms; see Fig. 4b.

To study the interaction between suitably arranged, smaller flaws, we consider an infinite plate with equally spaced preexisting flaws and out-of-plane, straight cracks; see Fig. 5. The boundary conditions are the same as before. Solving this problem, we obtain the relation between the applied load and the corresponding crack length. As is seen in Fig. 6, the required force first increases, then decreases, and finally increases again. This implies that crack growth becomes unstable at the peak point. This instability is interpreted to correspond to the formation of a high crack density zone. Calculations are done for different confining pressures and overall failure angles. Finally we obtain the ultimate strength and the overall failure angle as functions of confining pressure. In Figs. 7, the results are compared with experimental data for sandstone by Murrell (1965), showing good correlations.

Direct or indirect observations on the microcracking in rock specimens under compression have proved that nucleation, growth, and interaction of microcracks are the controlling factors for the brittle behavior of rocks. It should be emphasized that the micromechanics is necessary to describe the phenomena resulting from such microevents. Our mathematical model seems to serve as a step to bring the knowledge of applied mechanics into rock mechanics.

REFERENCES

Hallbauer, D.K., H. Wagner, and G.W. Cook (1973), "Some observations concerning the microscopic and mechanical behavior of quartzite specimens in stiff, triaxial compression tests," *Int. J. Rock Mech. Min. Sci. & Geomech. Abstr., 10,* 713-726.

Murrell, S.A.F. (1965), "The effect of triaxial stress systems on the strength of rocks at atmospheric temperatures," *Geophys. J. Roy. Astr. Soc., 10,* 231-281.

CURRAN: Willie Moss and Professor Gupta from Washington State have run a model very similar to yours. I think you're probably aware of it. I think this is extremely good work and strikes a blow for micromechanical modeling of the type which I mentioned previously. This gives us a lot of insight as to what form the continuum constitutive relations should take.

DIENES: You said the crack extended in the direction of maximum compression. I'm just wondering what would have happened if you had chosen a different criterion for crack extension, like energy density.

HORII: I believe same results are obtained by different criteria.

NEMAT-NASSER: Except if you use the criterion of maximum energy density, all other commonly used criteria seem to give essentially the same results. The differences are so small that you cannot tell for this kind of calculations. You may as well use the simplest fracture criterion.
  That's it. Thank you for your attention.

Short Presentation by D.R. Curran

  My talk is also along the same lines, that is, micromodeling. In order to keep it short, I want to restrict it to one particular micromodel we've been playing with lately. The model is based on an idea that goes back to Batdorf and Budiansky in 1949, the idea of forcing all of the slip to occur in a few discrete planes. Alan Needleman and Bob Asaro also referred to this type of approach earlier this week.

  The reason we took this approach was not primarily to describe plasticity, but was because we wanted to build a model for shear banding, and shear bands are noted to occur on discrete planes. We thought we would put in a few discrete planes and allow all the shear bands to occur in those planes, nucleate, grow and coalesce and so on to make fragments. In this way we could handle the shear band — induced anisotropy. Since we were doing that, my colleague Lynn Seaman thought it would be simple if we also made all the plastic flow occur on those same planes. Lynn chose nine planes which are sufficient to accommodate any kind of deformation. All the slip is forced to occur in those planes. For a two-dimensional problem you can get by with six planes, which are drawn in Fig. 8. Sheary is just a name for the subroutine that we were developing to do this.

  The way a computer code works is to first impose an incremental deformation. Next, we have to partition that deformation among all the planes. A nice feature of computer codes is that large distortions and rotations can be handled. So we let the computer keep the books on the rotations and the distortions and concentrate on what we're going to use as a yield condition on each plane. What we've done on each plane is to follow Schmid's Law, just as was done by the earlier work. Actually, we modified this by including damage; we also can create shear bands and start making surface there. The shear band surface has friction. So we modified the yield to account for friction by using a Mohr-Coulomb model for the percentage of the plane which is shear banded. However, I want to confine my remarks to plasticity now, so the main point is that each plane can be hardened or softened separately by any algorithm you'd like. The next step is to partition the strain and get the stresses on the slip planes.

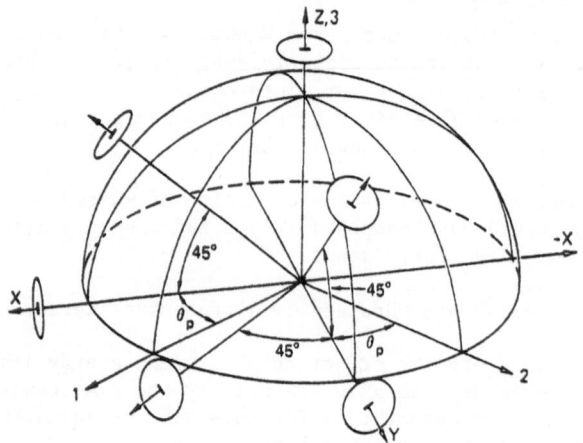

FIG. 8. Relative locations of the X-Y directions,
         principal directions (not necessarily in order),
         and initial orientation of the shear band and
         damage groups.

Now, of course we can't use global normality because we don't have a continuum theory in the large. But we have to have something that takes the place
of normality, because we have to partition the slip somehow among all those different planes. The way Lynn Seaman has done this is by stress relaxation. In
other words, we introduce a viscosity, and we have a trial stress on each plane.
If the trial stress exceeds the yield condition on that plane, it's allowed to
relax to the equilibrium yield stress according to the viscous relaxation time.
We presently use the same relaxation time for each plane. Although the viscosity
could be considered to be real, in this case it's only a scheme to partition the
slip on each plane during a quasi-static deformation.

We have used this model to run a single cell through the different deformation histories. The interesting thing is that what comes out is almost impossible in global terms, and we're still working on how to describe the results. If
the material started out isotropic, it rapidly can become very anisotropic as
some planes harden or soften and others don't. You can see that you will probably
generate in the large some sort of star-shaped yield surface as some planes are
hardening and the ones in between are not hardening. At any rate, it's very difficult to describe in any sort of continuum terms. Nonetheless, it's giving us
answers, and a rather simple model is producing extremely complicated behavior.

Let me now summarize a few advantages and disadvantages of this kind of approach. The global yield surface is an output, not an input. You don't worry
about normality, but of course you do have to worry about something else, how
you partition the strains. The model is naturally anisotropic, and one advantage
of any microscopic model is that if you have microscopic data, you can test the
model directly. A disadvantage is that such models take more computer time than
simpler continuum models. Our normal experience is that any one of our micromodels costs about two or three times as much time in each cycle than a simpler
model. On the other hand, in a typical problem, most cells are not being exercised in that mode. That is, in most cells you can use a simpler model, and you
only turn on the complicated models in regions where all the action is, like at
the tip of a crack. Another disadvantage is that we have decreased accuracy.

What we have done is to take this particular plasticity description and run it through some deformation cycles that we can compare against analytic solutions. We find that we end up in stress space as much as 10% away from the right answer.

My final comment about micromodeling is that I do think it can produce insight and guidance for continuum plasticity and for failure constitutive relations.

***

BUDIANSKY: Well, I certainly agree with your conclusion; micromodeling can do good things. I just want to question some assertions you made, kind of casually. Your yield surface is output, not input. Very good, however, it is not star-shaped: it is convex.

CURRAN: Yes, excuse me, you're right. It is convex in the small, although the envelope of yielded states appears strangely star-shaped in the large.*

BUDIANSKY: I don't want to see star-shaped yield surfaces coming out of a model that I feel very good about. Second, you said you don't have to worry about normality. Well, you don't have to worry about it explicitly, that's true; but you got normality.

CURRAN: On each plane, normality is obeyed, yes.

BUDIANSKY: And hence, you have extended normality where planes intersect in stress space.

CURRAN: Yes.

BUDIANSKY: So you got normality.

CURRAN: Exactly, yes, but it didn't help us partition the strains on the slip planes.

BUDIANSKY: Number three — this is not a criticism; this is a statement. That was a cute way you figured out how to partition the plastic strain among the various mechanisms. I just want to mention that there is a theorem that says the partitioning is unique and there are programming algorithms that can let you determine it without the artifice of going through your viscosity; although for all I know your technique may be the most efficient one there is.

CURRAN: I was unaware of that theorem; I certainly want to get the reference.

NEMAT-NASSER: Bernie, are you meaning the number of active slip systems?

BUDIANSKY: Not only which are active but how much plastic strain they contribute.

NEMAT-NASSER: Once you decide which are active and when their number is less than, say, five, then I suppose there would be no problem. But if you have larger number of active slip systems, I don't know. Are you referring to this or are you saying something else?

HUTCHINSON: Bernie, you're invoking an additional assumption.

---

*Note added in proof: The apparent star shape in the large may be due to failure to account for material rotation.

BUDIANSKY: Wait a minute. The stress —it's the stress rate that is unique. And there may be non-uniqueness, that's true, in the ideally-plastic case.

NEMAT-NASSER: Let me add one more thing to that because, you know, if you use the very simple approach of the Taylor-type model and get the yield surface, which of course does not satisfy the self-consistency or any kind of local equilibrium conditions, you always get a convex yield surface if you assume Schmid Law, and that you do get corners. There may be many of them, but you do get corners.

VALANIS: You say that your yield surface is an output, but Schmid's Law that you are putting in at the micro-level is a yield condition.

CURRAN: I think Bernie has already taken me to task on that.

VALANIS: Comment number two, there is a paper that is going to appear in the proceedings of the conference at San Antonio. In common slip systems, there is a way of partitioning the total plastic strain uniquely. It has been worked out for more than five slip systems. There is a proof of uniqueness.

NEMAT-NASSER: Without additional conditions?

VALANIS: Without additional conditions. Schmid's Law.

Short Presentation by K.C. Valanis

Well, I stand here with trepidation because I'm going to talk about the endochronic theory of plasticity. Psychologically, I feel like a Methodist minister preaching in St. Peter's Cathedral in Rome. I did see the stake outside in the yard. I will offer myself for burning after the meeting is over. Anyway, this is a case study that I'm presenting here, a comparison between theory and experiment in a case where the stress distribution is not homogeneous. The study was conducted by myself and Dr. J. Fan in the course of his doctoral dissertation. It involves a rectangular plate with two edge notches. The load is applied in a direction normal to the notches and in a cyclic manner. Strain gauges, 0.2 mm, were placed between and in the line of the notches. Strain gauges were also placed along a line emanating at the root of the notches and normal to their plane; see Figs. 9 and 10. Measurements of strain were made throughout the history of the loading.

Now let me talk briefly about the constitutive equation which is used in this particular case. It is very simple. It is a linear functional of the plastic strain history with respect to the kernel $\rho(z)$ insofar as the deviatoric response is concerned:

$$\underset{\sim}{s} = \int_0^z \rho(z-z') \frac{\partial \underset{\sim}{e}^P}{\partial z'} dz', \quad d\zeta = \| d\underset{\sim}{e}^P \|, \quad dz = d\zeta/f(\zeta), \quad \sigma_{kk} = 3K\varepsilon_{kk},$$

$$d\underset{\sim}{e}^P = d\underset{\sim}{e} - \frac{1}{2\mu} d\underset{\sim}{s}. \tag{1}$$

The hydrostatic response is represented by an elastic law. The entity dz is an increment of the path in plastic strain space and is given by the norm of the increment of plastic strain times a constant k_1, which is trivial, is in fact arbitrary; and it can be assigned any positive value; just like employing a second

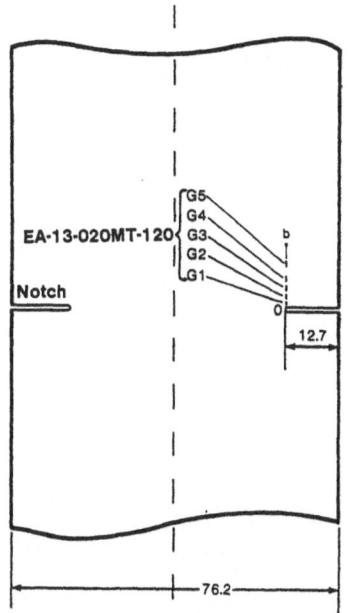

FIG. 9. Arrangement of strain
gauges for determining the distribution of strain ε_y along the
vertical line ob at the top of
notch tip.

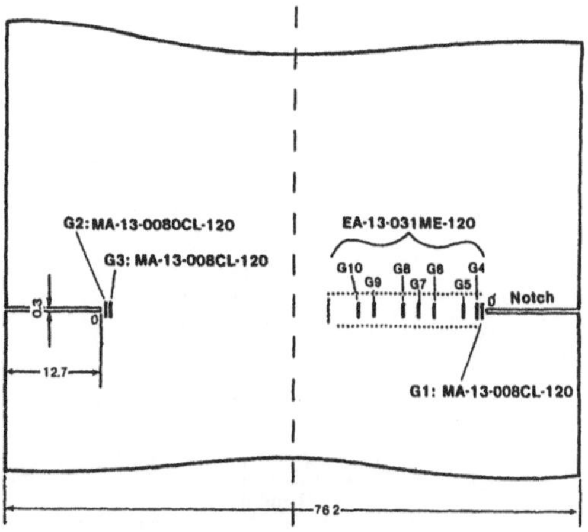

FIG. 10. Arrangement of strain
gauges for determining the distribution of strain ε_y along the
notch line 00'.

of time as the unit of measurement in the Newtonian time scale.

On the other hand, dz reflects the hardening behavior of the material. So
you can see that here we have basically the following material functions. We
have kernel $\rho(z)$ that determines the deviatoric stress as a function of the history of plastic strain. We have the function $f(z)$ that determines the hardening
behavior of the material. We have the bulk modulus K, and we have the shear
modulus which is the slope of the shear stress-strain curve near zero. So we
have two functions and two constants that determine the behavior of the material.

Now, in the material that we used, it transpired that the function $\rho(z)$ is
the slope of the initial stress-strain curve, and the hardening function f is
determined simply by the ratio of the peak stresses at the ends of the cycles in
a cyclic test performed at constant strain amplitude. So one cyclic test actually determines both functions ρ and f. Now, $\rho(z)$ has a weak singularity at the
origin. But to make the computation a little bit easier, we approximated $\rho(z)$ by
three exponentials, one of which has a very, very high multiplier in front and
also a very high exponent. We have used a circular rod, and under axial cyclic
loading have determined the material functions $f(z)$ and $\rho(z)$. Then we have used
a plate made of the same material, very high-purity copper, in which we made two
notches, put it under cyclic loading conditions, and measured experimentally the
strain distribution, as I pointed out before, Figs. 9 and 10.

Then we made a computer calculation using a finite element technique using
precisely these two functions $f(z)$ and $\rho(z)$ that we found by using the circular rod. So even the geometry of the specimens is different.

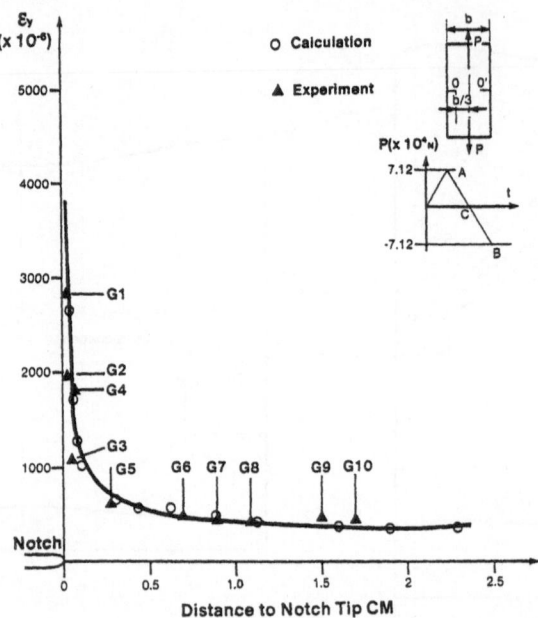

FIG. 11.   Comparison between experimental and calculated
distribution of strain ε_y along notch line 00' at positive
peak A.

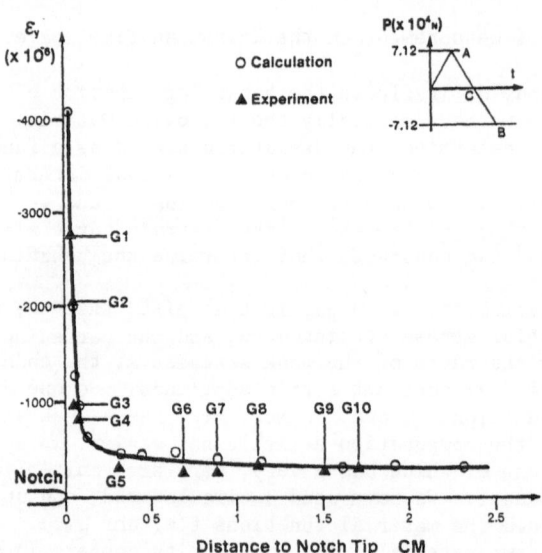

FIG. 12.   Comparison between experimental and calculated
strain distribution ε_y along notch line 00' at negative
peak B.

The comparison between the theory and the experiment is illustrated in Figs. 11 and 12. Similar results are obtained at C and after additional 1/4 cycle of loading. I'm comparing here the distribution of strain.

In Fig. 12 you see calculation and experiment being compared at the point B after we've gone into compression, with a very satisfactory outcome. So you see how the celebrated Bauschinger effect is actually being closely predicted in a highly heterogeneous strain field.

***

GOUDREAU: A brief question. How many coefficients were used in your constitutive functions?

VALANIS: We represented it by three exponentials; that is, three numbers in front of the exponent and three exponential coefficients.

GOUDREAU: I'm interested in the computational efficiency of your model. I would say that I think this certainly would do a much better job than, say, the bilinear isotropic kinematic hardening model that Sam Key had indicated is in most of the codes today. I would also think that the multiple yield surface kinematic models can probably do as well. And unless someone can come up with a theoretical argument against this approach to plasticity, I'm very interested in exploring the computational economy of implementing this kind of a model in contrast to some of the other isotropic/kinematic hardening models we have. So I'm interested in the theoretical critique, if there is any.

NEMAT-NASSER: Let me make one comment here. The way I understand it, the number of constants is in fact irrelevant in Kirk's theory. The question is how many tests he needs in order to fix the constants.

VALANIS: One test. You put a specimen in a uniaxial machine. You carry out a cyclic tensile-compressive test under fixed strain amplitude conditions, and that's how you find $\rho(z)$ and $f(z)$.

GOUDREAU: I can appreciate that and, of course, the efficiency of the model as to how well it can extrapolate to other conditions. Here we saw one example. Other tests might have to be exercised. But it seems to me to have computational potential in terms of the economy of representation.

PREVOST: I'm curious as to how you get the moduli tensor out of the theory, the tensor of moduli that you have to use in the multi-dimensional experiment? The equations that you had were one-dimensional equations. I didn't see any matrix of moduli.

NEMAT-NASSER: The question is, how do you go from one-dimensional stress-strain curve to a three-dimensional stress-strain relation?

VALANIS: This is a three-dimensional stress-strain relation. I mean, I have here a tensorial relation: my relations are in tensor form.

KEY: I believe anybody reasonably skilled in tensor analysis can convert that from an isotropic behavior to an anisotropic behavior. The real issue is the endochronic part. That's the interesting part.

GOUDREAU: Just a final point—basically you end up with a tensor history variable for each term of your exponential series. And maybe Jean (Prevost) could have another comment, but it seems to me that you have a fitting capability that may be more economic than, say, piece-wise linear representation of that initial spline curve.

VALANIS: Well, somebody mentioned the question of finding a closed form solution for a problem and comparing it with experiment. Now, I have just one whole file of slides which I do not have time to show. But you can obtain the cyclic response in terms of a closed form solution. That is, it doesn't matter what the strain history is; you can get a closed form solution for the stress response and you can then compare directly with experiment. We have done that for the work of Jhansale and Topper, for instance, that they used on steel. The agreement between the closed form solution and the experiment is absolutely unbelievable. I mean, it is in terms of just two constants. We have set $f(z) = 1$ and $\rho(z) = \rho_0/z^\alpha$. Well, α is between 0 and 1, and in this case is .862.

We get an absolute coincidence between the experimental data and the theory.

HUGHES: I have some questions about what you said. Is this closed form solution for the constitutive equations themselves?

VALANIS: Yes. For the homogeneous field.

HUGHES: That would seem to indicate that you potentially have fast algorithm because you've already got a closed form solution for the constitutive equation. Even if it was expensive to evaluate, because of exponentials, you might be able to employ asymptotics, or something like that, that might make it economical. Do you have a computer routine? That would be something that people could put into big codes and try some things with.

VALANIS Yes, I do. Actually, my student did it on the computer, and I did it at home on the calculator — by the time he got back his program, etc., I think we ended up at the same point. Using a calculator I did it as fast as he did.

HUGHES: Everything looks fast when you're dealing with a stress point.

GOUDREAU: One last thing. Maybe Sam can concur or disagree. But the recursion relations we use with an exponential visco-elastic kernel seem to me to be directly applicable here. And the hardening function is just like a temperature shift function of some sort. It seems to me that the computing arsenal we've already got can apply directly.

Short Presentation by I.G. Vardoulakis

I want to discuss briefly strain softening in granular media. I will talk about triaxial testing in a large triaxial apparatus as it is presented in a recent paper by Hettler and myself, and if I have time, I will present some recent results on shear band imperfection sensitivity.

Figure 13 shows a very large triaxial cell developed at the Institute of Soil Mechanics in Karlsruhe, where we could test large flat specimens of sand with a diameter of 78 cm and a height of 28 cm. The apparatus is guided by four guides so that there is no tilting of the upper plate, has enlarged, lubricated and smooth-end platens to avoid friction. Of course, I will present to you only a small piece of the whole experimental program.

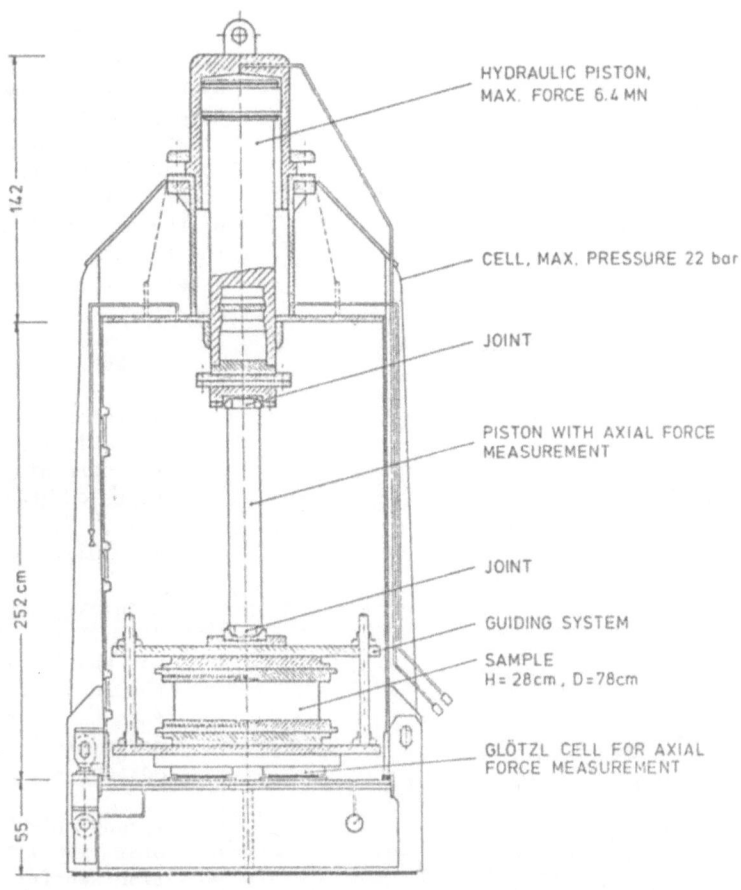

FIG. 13

In Fig. 14 the stress ratio strain curves from a series of experiments with varying density are shown. Because bulging can be suppressed strain softening is not much pronounced. The formation of rigid cones is also presented, which would also produce geometric softening. So, strain softening exists, but it's small. The volumetric strain curves show again the property of diminishing dilatancy with increasing porosity, and the results of carefully prepared specimens in the lab are repeatable. The bifurcation modes that will occur in a triaxial test with enlarged and lubricated end platens are barelling, bulging and shear banding.

One can suppress the diffuse bifurcation modes by using flat and lubricated specimens. One cannot avoid shear banding if it has to occur.

How dramatic barreling is, is presented here (Figs. 15a and b) for a dense sand specimen where optical records of various diameters have been taken. One can easily see how unreliable the results are, especially for the dilatancy characteristics of the material. Less catastrophic is a situation with respect to the friction angle.

If, again, stress is evaluated at the top and at the bottom of the specimen,

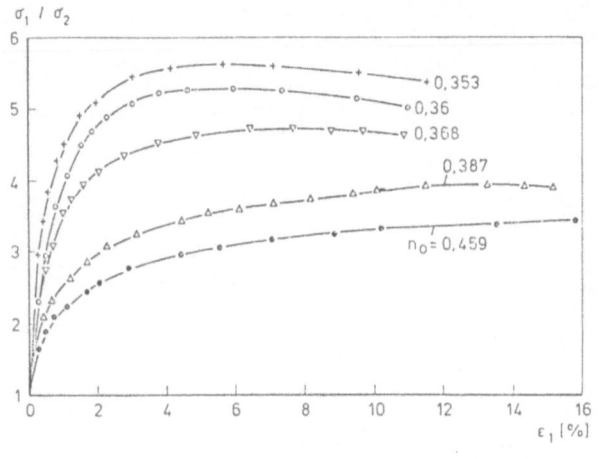

FIG. 14a

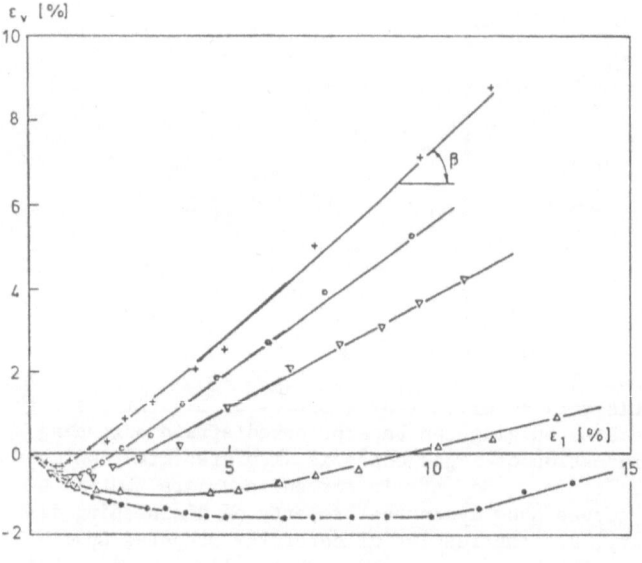

FIG. 14b

softening is appreciable.

In order to have shear banding in triaxial compression tests as predicted by theory, some critical value of strain softening must be reached. So shear bands do occur in the softening regime of the stress-strain curve and the triaxial test, on dense samples and not on loose samples.

REFERENCES

Drescher, A. and I. Vardoulakis (1982), "Geometric softening in triaxial tests on granular material," *Géotechnique, 32,* 291–303.

Rudnicki, J.W. and J. R. Rice (1975), "Conditions for the localization of the deformation in pressure-sensitive dilatant materials," *J. Mech. Phys. Solids, 23,* 371–399.

Vardoulakis, I. and G. Graf (1982), "Imperfection sensitivity of the biaxial test on dry sand," *IUTAM Conf. Deformation and Failure of Granular Materials,* Delft, 31 Aug. – 3 Sept., 485–491, A. A. Balkema.

Vardoulakis, I. (1983), "Rigid granular plasticity model and bifurcation in the triaxial test," *Acta Mechanica, 49,* 57–79.

***

BALADI: The softening behavior does not show up in your figures because they are plotted in terms of the stress ratio. Because, in an actual case, some of the data points show hardening behavior and some show softening behavior. If you plot these data points in terms of stress ratio, then the softening behavior will disappear.

VARDOULAKIS: No, this is not the reason because the lateral stress is constant. So essentially the plot is for a normalized axial stress.

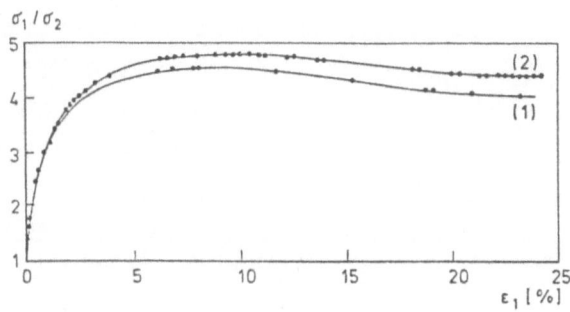

FIG. 15a

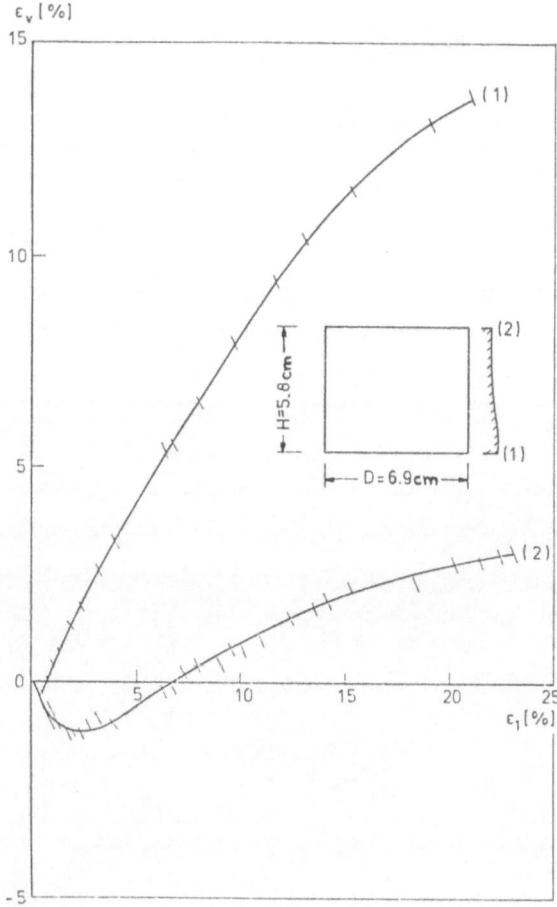

FIG. 15b

A Challenge to Numerical Analysts by T. C. Bache

During these last two and a half days, I've heard theoreticians and experimentalists explaining how complex the problems of material characterization are. On the other hand, I've heard calculators use very simple models which most of the time are based on theories that are ten or fifteen years old, and they are able to match the features that they want to match, at least in the presentation we've heard. There's a pretty big gap here, isn't there? Complicated models being developed theoretically, and complicated behavior being observed experimentally; yet simple models being used in the calculations, and the calculators are asserting that these calculations meet our needs. Something's wrong: Either the calculators have not yet told us what important features there are that they're unable to deal with, or the theoreticians ought to look for a new line of work because all these things are really irrelevant and what's necessary to be done has already been done. I think the former is really the case, and I would hope that the people involved in calculations —during this break we're about to have —would search their souls a bit, and would be able to come back and tell the rest of the community what fundamental issues there are that they are unable to address at this time, and tell the theoretical community what it is that they can do to help.

Short Contribution by D. Krajcinovic:

"Continuous Damage Mechanics; A Macro Theory Inspired by Micromodels"

As pointed out in the introductory lecture of this Workshop delivered by Prof. Nemat-Nasser, inelastic response of a material is a macroscopic reflection of the structural rearrangement on the microscale. It is important to note that different modes of microstructural kinetics affect the macro-response in a different manner. The theory of plasticity in its many reincarnations can model most of the phenomena in loading (at a possible loss of simplicity and physical insight) to a satisfactory degree. The situation is, however, less satisfactory in unloading since one of the basic tenets of plasticity requires that the unloading path parallels the initial portion of the loading path in the stress-strain space. For most of geological materials and concrete this appears to be a rather questionable assumption.

While it is at least in principle possible to amend the plasticity theory to model different unloading patterns as well, it appears more reasonable, more effective, and more productive to acknowledge the difference between various modes of microstructural kinetics in formulating a rational analytical model. In other words, in developing a continuum theory a reasonable strategy would be to introduce a separate kinematic (internal) variable for every mode of microstructural rearrangement which has a significant effect on the response.

A theory of this type, characterized by an additional internal variable being a measure of the microdefect density locally, is known as Continuous Damage Mechanics. Naturally, for the modelling to be successful, the introduced variable should, in an appropriate (smoothed) sense, reflect the actual geometry of the microdefects. A geometrically consistent formulation, neglecting the details of the actual shape of the defects as second order effects, would call for: a scalar for the spheroidal microvoids (Davison et al., 1977), and a set of axial vectors for flat and planar (penny-shaped) microcracks (Davison and Stevens, 1973, and Krajcinovic, 1983a).

The original Kachanov's idea according to which the damage is best measured by the void area density in the observed plane is still unsurpassed in its physical clarity and appeal. This idea in a general case must be considered in conjunction with a characteristic length (depending on the stress diffusion angle).

It is important to keep in mind that in case of planar microcracks a single internal variable will not be a correct representation. This conclusion readily follows from the experimental evidence according to which only some of the cracks with preferred orientation are active while the others are dormant.

Once the internal variables are selected the subsequent procedure follows conventional algorithms (Krajcinovic, 1983a, b). The selection of particular expressions for the Helmholtz free energy and the dissipation rate density function (invariant for the full group transformations) are the two most important aspects of the procedure.

Despite a relative paucity of specific experimental data a host of interesting analytical results have already appeared in the literature. The theory was successfully used for modelling of creep rupture, behavior of concrete in unconfined tests, etc. These results, even though far from being complete, clearly emphasize the benefits derived from the development of macromodels based on a rational representation of the microstructural kinetics.

REFERENCES

Davison, L. and A.L. Stevens (1973), "Thermomechanical Contribution of Spalling Elastic Bodies," *J. Appl. Phys., 44,* 667-674.

Davison, L., A.L. Stevens and M.E. Kipp (1977), "Theory of Spall Damage Accumulation in Ductile Metals," *J. Mech. Phys. Solids, 25,* 11-28.

Krajcinovic, D. (1983a), "Constitutive Equations for Damaging Materials," *J. Appl. Mech., 50,* 355-360.

Krajcinovic, D. (1983b), "Creep of Structures — A Continuous Damage Mechanics Approach," *J. Struct. Mech., 11,* 1-11.

Short Contribution by J-H. Prevost, J.M. Roesset, I. Sandler, R.F. Scott (Ed.) and I.G. Vardoulakis: "Geotechnical Issues"

The application of the theory of plasticity to model the behavior of geotechnical materials under arbitrary states of stresses is relatively recent when compared to the applications in the study of metals. It is, however, an area which is expanding rapidly, as illustrated by the number of sessions devoted to this topic at professional meetings and the ever-increasing number of specialty conferences. It should be noted, on the other hand, that nonlinear behavior of soils was accounted for, even if crudely, in the simplest and earliest solutions developed in soil mechanics since a substantial fraction of the practical problems in Geotechnical Engineering deal in fact with limiting or failure conditions. Study of the mechanical behavior of intact and jointed rocks has developed more recently.

All geomaterials are highly nonlinear and time-dependent. The definition of a yield surface is in fact rather arbitrary and more of a mathematical convenience than a physical reality, the purely elastic range being extremely small or non-existent. Even for small strains elastic moduli are functions of the stresses and their values in situ are hard to determine from laboratory tests. Some of the basic characteristics of soil behavior are discussed by Baladi; see Chapter XII. Baladi describes also one of the families of models used in practice today, the cap model, which is an outgrowth of the original Cam-clay model developed by Roscoe and coworkers in Cambridge. A number of other models have been developed in the past few years; the current number proposed is more than a dozen. They range from nonlinearity elastic, with no accounting for plastic flow strains, through hypoelastic with a yield surface, to incremental plastic with multiple yield surfaces. Yield, failure, elasticity, and effects of cyclic loading are accounted for in a variety of ways.

Here, we address briefly the principal topics of current concern. They are: (1) the nature of the constitutive relations proposed, in particular, in the area of incremental plasticity, the flow rule; (2) the characterization and treatment of the frequently observed phenomenon of slipline generation, or bifurcation; (3) the urgent need for the validation of the requirements of various models through adequate, repeatable and accepted experimental work.

By reasons of space, we are compelled to omit consideration of other models, and other effects, such as time, temperature and strain-rate dependence. It is also not clear if pore pressure effects are being accounted for correctly across the entire range of geological materials. For example, at what porosity or fracture size and frequency can the effective stress hypothesis be eliminated from the constitutive formulation?

Flow Rules in Incremental Plasticity.—The question of the use of associated or non-associated flow rules remains an important issue in soil and rock modelling. Although a number of studies have been performed in which the investigators claimed to have demonstrated that soils do not follow an associated flow rule, these conclusions are generally based on an assumed yield condition. In all of the studies in which clear violations of normality (by clear we mean that deviations from normality are well in excess of the uncertainties or variations in the data due to lack of reproducibility) have been documented, the direction of measured "plastic flow" was compared to a "normality" direction obtained by taking the gradient of an assumed yield function in stress space. Because it is not possible to directly measure yield surfaces, but, instead, they must be inferred from a large group of independent experiments, these studies may simply have indicated that inappropriate yield surfaces had been assumed, rather than that non-normality of flow was demonstrated. Other choices of yield surfaces could have led to the observed behavior; even with the associated flow rule.

One way to demonstrate the normality of "flow" is to measure tangent moduli and test for symmetry of the resulting material stiffness tensor. It can be shown from plasticity theory that, independent of the choice of yield surface or hardening behavior, normality leads to symmetry of the tangent modulus tensor of the stress-strain curve. Experiments to test for this symmetry in actual soil behavior could lead to a definite conclusion with respect to the adequacy of normality in soil models without reference to any specific yield surface assumptions. It would be extremely useful if experimenters could find a way to test for symmetry.

The question of normality is also important from the theoretical standpoint of uniqueness and stability. It is well known that the use of associated flow rules leads to properly posed mathematical boundary value continuum problems for which robust numerical schemes can be constructed. If non-associated flow rules are used in rate-independent plasticity, the proper posing condition for a dynamic initial/boundary value problem may be lost. Specifically, these dynamic problems may no longer be hyperbolic, but, instead, become what mathematicians refer to as ultra-hyperbolic. For such problems wave propagation speeds become complex and forward time-marching schemes cannot be constructed for arbitrary initial conditions. It is possible that such models lead to properly posed quasistatic or equilibrium problems if inertia effects are completely eliminated from the analysis. Such models may therefore find utility in non-dynamic situations.

Uniqueness may exist for solutions developed from non-associated flow rule constitutive relations; it just cannot be demonstrated. At present, therefore, any solution obtained must be viewed with suspicion. The only current way of demonstrating the reasonableness, but not necessarily the correctness, of such a solution is to compare it with the results of a mechanical test performed under the same boundary conditions. Very large dynamic problems involving geomaterials are not easily testable in this way, to say the least. It is possible that the performance of tests in a centrifuge would alleviate this difficulty. In a centrifuge not only is much of the scaling correct, but in addition, the

boundaries are well-defined, and thus the domain of numerical analysis is clearly limited. This avoids many of the difficulties associated in nature with
boundaries at infinity.

Test Requirements. — There is an enormous range of materials developed as
a result of geological processes. The range is represented by spectra of both
material constitutive relations and the numerical coefficients referred to these
relations. In addition, for many of these materials, the moduli are of the same
order of magnitude as the stresses; which are also developed in many cases by
the gravitational field. Finally, the variability occurs on a very small scale,
so that the material properties may vary very substantially over short distances.
This means inherently that more tests are required for geotechnical materials
than for metals [if an ABC 999 steel is to be characterized; it could be done
(but hasn't) once and for all]. It probably also means that a wider range of
constitutive relations must be available. Other differences have been noted
due to friction, dilatation, and pore pressure effects in geological materials.

Only a few, generally idealized (standard sands, prepared kaolin clays,
Westerly granite) geotechnical materials have been reasonably well-studied in
laboratory tests. Other materials (e.g. tuffs) remain poorly defined. Even for
the much tested substances gaps exist, mostly because of the deficiencies in the
testing apparatus. Only a few "true" triaxial pieces of equipment are available
and there are still arguments about how "true" they are (rigid versus flexible
loading surfaces). Even there, no rotation of principal stresses can occur.
Other equipment with different boundary conditions (hollow cylinder tests) must
be employed, which involves questions about properties obtained for one material
from a variety of apparatus.

There is also the problem of scale. Most laboratory tests on geological
materials use specimens with dimensions of a few inches (tens of centimeters)
and the properties of these do not represent the properties of samples a few
meters to tens of meters in dimension. The reason for this may lie in grain
size, size distribution, or crack size and spacing. The majority of geomechanical tests is carried out at slow strain rates, and relatively few tests, usually uni-axial, at high strain rates. We do not know if laboratory tests have
ever been conducted at extreme strain rates. If combined stress tests are to
be performed at high strain rates, new apparatus designs are required. Inclusion of elevated temperatures further complicates the technical problems of
testing.

The parameters of a plasticity model must be determined from these laboratory tests. Yet the properties of these samples may be very different from
those of the soil in situ due to a number of factors (sample disturbance, time
effects, etc.). While correction factors are typically applied to laboratory
measurements in order to predict the values of the elastic moduli at very low
levels of strain in the field, the procedure is at best an approximation. It
is even more difficult to extrapolate the nonlinear behavior of the samples to
derive constitutive equations for the soil in the field. Thus, while it is important to conduct laboratory tests not only to determine the parameters of a
plasticity model, but also to verify its adequacy (by using the model to predict the results of other tests), there is also a need to conduct in situ experiments to assess the accuracy of the models and their numerical implementation.
As has been repeatedly pointed out, the results of some nonlinear problems are
relatively insensitive to the detailed characteristics of the constitutive model.
In such cases, and when this insensitivity can be guaranteed, it will be possi-

ble to use relatively simple or crude models to predict satisfactorily the basic features of the response. In other cases, however, this will not occur and the selection of an appropriate set of constitutive equations may be more critical.

Field tests are, unfortunately, expensive and sometimes hard to justify just for the sake of verifying a particular constitutive model. Modelling of some types of dynamic loads, such as those caused by an earthquake excitation is also hard to accomplish. It is possible, however, to make use in some instances of existing facilities to perform some testing and to instrument other facilities to get adequate data if an earthquake occurs. The problem in the first case is that the time available may be short and that it may be difficult to obtain research funds for the testing with such short notice. It would appear that some funds should be kept in reserve to cover these situations. The problem in the second case is that the instrumentation must be carefully planned and that it must be maintained probably for years before any results are obtained.

The difficulties of measuring stress, and, to a lesser extent, velocity and displacement in geomaterials are not often appreciated. Very commonly the stress gauge has a density different from that of the surrounding material, and its stiffness is not the same as an equivalent volume of the material. In general, of course, its constitutive response is enormously different from that of soil or rock, since to be interpretable, the gauge has to be at least elastic, and normally is linearly elastic. In consequence, the gauge develops its own local stress field and the stress it indicates from preliminary, usually hydrostatic calibrations, is not that of the material in its absence. This is particularly true for the first unloading of the gauge, and all subsequent cycles of stress in one test.

Shear Band Formation and Bifurcation. — Although essentially all constitutive relations proposed for geomaterial substance assume the behavior is homogeneous and continuous, a very large number of real-life cases of stressing of these materials demonstrates the development of sliplines, faults, or shear bands, a phenomenon now generally referred to as bifurcation.

(a) Numerical Implementation

The ability to model accurately bifurcation phenomena and associated emergence of shear bands in geological materials plays an essential role in the solution of many boundary value problems in geotechnical engineering. The modelling task is formidable. It involves modelling effects both at the constitutive and numerical simulation levels. Non-associative plastic flow rules seem to offer promising tools (see, e.g., Rudnicki and Rice, 1975) to tackle the phenomenon at the constitutive level. However, the ideas and concepts are new, and have not been thoroughly investigated yet.

Numerical strategies for capturing bifurcated solutions are in their infancy. It is only recently that such numerical solutions within the control of classical plasticity have been obtained (see Needleman, 1979; Prevost and Hughes, 1981). However, there remain questions as to how to further advance the solution in the post-bifurcation mode. Also, the matter of which constitutive theory is to be used within the bifurcated band, is still a subject of great controversy [e.g., Palmer and Rice (1973), who advocate the use of a stress versus displacement relation; Mehrabadi and Cowin (1980) double sliding/rotating model].

(b) Experimental Considerations

In selecting a yield criterion for soils, K.H. Roscoe, et al. (1963) presented the results of "special" triaxial compression tests for which very precise optical records of the failure patterns had been made. This series of experiments has shown that it is difficult to interpret the experimental data of such tests due to the appreciable bulging or necking of the samples. Kirkpatrick and Belshaw (1968) and Deman (1975) used the x-ray technique to investigate the strain field inside cylindrical samples of dry sand in triaxial compression tests performed with or without lubrication of the end platens. Those experiments and past experience have shown that rough end platens support the development of rigid cones at the ends which are mainly responsible for a global geometrical softening, as shown recently by Drescher and Vardoulakis (1982). Lubrication prevents the formation of these cones; the deformation is uniform for moderate strains, although bulging occurs at larger strains (Deman, 1975). Bishop and Green (1965) extensively studied the influence of slenderness and of end restraint, and arrived at similar conclusions. In triaxial extension tests, the dominant failure mode is localized necking that cannot be suppressed by any refinements in the boundary conditions or changes in the sample geometry (Reades and Green, 1976).

The existing experimental data has given rise to theoretical analyses of bifurcation (Vardoulakis, 1983) and imperfection sensitivity (Drescher and Vardoulakis, 1982) in the triaxial test. The main results of the bifurcation analysis of ideal triaxial tests on dry sand can be summarized as follows: (1) bulging is not possible if the slenderness of the sample is smaller than a critical value. If bulging occurs, then it takes place close to the limiting state and in the hardening regime of the stress-ratio strain curve of the tested sand; (2) shear band formation in the compression test is only possible in the softening regime of the stress-ratio strain curve; (3) diffuse and localized necking in the extension test will inevitably take place close to the limiting state and in the hardening regime of the stress-ratio strain curve. On the other hand, imperfection sensitivity analysis by Drescher and Vardoulakis (1982) provided an estimate of the apparent increase of the measured friction angle of the material caused by the end restraint. All these theoretical results have been recently tested and verified in an extensive experimental study by Hettler and Vardoulakis (1983).

Bifurcation analyses of the plane-strain biaxial test (Vardoulakis, 1980, 1981) yielded the result that the dominant failure mode in this test is shear-band formation. In both the triaxial and the biaxial test the shear band orientation, θ, (measured from the σ_1-direction) is given approximately by the formula

$$\theta \approx \theta_B = \arctan \sqrt{\lambda\delta}, \quad \lambda = \sqrt{\left(\frac{\sigma_2}{\sigma_1}\right)_p}, \quad \delta = \sqrt{\left|\frac{\varepsilon_2}{\varepsilon_1}\right|_p}, \tag{1}$$

where p indicates the values of these ratios at the maximum (peak) stress-ratio state. For plane strain conditions Eq. (1) approximately coincides with the experimental formula proposed by Arthur, et al. (1977) and is confirmed by the experimental results by Vardoulakis (1980):

$$\theta_B \approx 45° + \tfrac{1}{4}(\phi_p + \nu_p) \tag{2}$$

where ϕ_p and ν_p are the peak values of the friction and dilatancy angles.

Micromechanical Considerations. —Various studies (Cundall and Strack, 1979; Rowe, 1962; Horne, 1965; Scott and Craig, 1980; and Christoffersen, Mehrabadi, and Nemat-Nasser, 1981) have been performed on the relation between individual grain interactions and the macroscopic response of granular materials. Although the investigations have shed light on the mechanism of dilatancy and its contribution to overall material behavior, they have not yet led to guidelines as to constitutive relations in the large. One reason may be the limited scale of the studies applied to a medium whose response is statistical in nature. Larger-scale computations may be more productive.

REFERENCES

Arthur, J.R.F., T. Dunstan, Q.A.J.L. Al-Ani, and A. Assadi (1977), "Plastic Deformation and Failure in Granular Media," *Géotechnique, 27*, 53-74.

Bishop, A.W. and G.E. Green (1965), "The Influence of End Restraint on the Compression Strength of a Cohesionless Soil," *Géotechnique, 15*, 243-266.

Christoffersen, J., M.M. Mehrabadi and S. Nemat-Nasser (1981), "A Micromechanical Description of Granular Material Behavior," *J. Appl. Mech., 48*, 333-344.

Cundall, P.A. and O.D.L. Strack (1979), "A Discrete Numerical Model for Granular Assemblies," *Géotechnique, 29*, No. 1, 47-65.

Deman, F. (1975), "Achsensymmetrische Spannungs- und Verformungsfelder in trockenem Sand," Veröffenlichungen des Instituts für Bodenmechanik und Felsmechanik der Universität Karlsruhe, Heft 62.

Drescher, A. and I. Vardoulakis (1982), "Geometric Softening in the Triaxial Test on Granular Material, *Géotechnique, 32*, 291-303.

Hettler, A. and I. Vardoulakis (1983), "Stress-Strain Behavior of Sand in Triaxial Tests," submitted for publication.

Horne, M.R. (1965), "The Behavior of an Assembly of Rotund, Rigid Cohesionless Particles," *Proc. Roy. Soc. Lond.*, Parts I and II, *A286*, 62-97; Part III, *A310*, (1969), 21-34.

Kirkpatrick, W.M. and D.J. Belshaw (1968), "On the Interpretation of the Triaxial Test," *Géotechnique, 18*, 336-350.

Mehrabadi, M.M. and G.C. Cowin (1980), "Prefailure and Post-Failure Soil Plasticity Models," *J. Eng. Mech. Div.*, ASCE, *106*, 991-1003.

Needleman, A. (1979), "Non-Normality and Bifurcation In-Plane Strain Tension and Compression," *J. Mech. Phys. Solids, 27*, 231-254.

Palmer, A.C. and J.R. Rice (1973), "The Growth of Slip Surfaces in the Progressive Failure of Overconsolidated Clay," *Proc. Roy. Soc. Lond.*, *A332*, 527-548.

Prevost, J.H. and T.J.R. Hughes (1981), "Finite Element Solution of Elastic-Plastic Boundary Value Problems," *J. Appl. Mech., 48*, 69-74.

Reades, D.W. and G.E. Green (1976), "Independent Stress Control and Triaxial Extension Tests on Sand," *Géotechnique, 26*, 551-576.

Roscoe, K.H., A.N. Schofield and A. Thurairajah (1963), "An Evaluation of Test Data for Selecting a Yield Criterion for Soils," in *Laboratory Shear Testing of Soils*, ASTM Special Publications, No. 361, 111-128.

Rowe, P.W. (1962), "The Stress Dilatancy Relation for Static Equilibrium of an Assembly of Particles in Contact," *Proc. Roy. Soc. Lond., A269,* 500-527.

Rudnicki, J.W. and J.R. Rice (1975), "Conditions for the Localization of Deformations in Pressure-Sensitive Dilatant Materials," *J. Mech. Phys. Solids, 23,* 371-394.

Scott, R.F. and M.J.K. Craig (1980), "Computer Modeling of Clay Structure and Mechanics," *Proceedings ASCE, 106,* GT1, 17-33.

Vardoulakis, I. (1980), "Shear Sand Inclination and Shear Modulus of Sand in Biaxial Tests," *Int. J. Num. Anal. Meth. in Geomech., 4,* 103-119.

Vardoulakis, I. (1981), "Bifurcation Analysis of the Plane Rectilinear Deformation on Dry Sand Samples," *Int. J. Solids Structures, 17,* 1085-1101.

Vardoulakis, I. (1983), "Rigid Granular Plasticity Model and Bifurcation in the Triaxial Test," *Acta Mechanica, 49,* 251-277.

CHAPTER XIV

COMMENTS BY SESSION CHAIRMEN
STATEMENTS BY ORGANIZING COMMITTEE
CLOSING REMARKS

Chairman: *G. A. Hegemier*

HEGEMIER: This is a very important session. Its purpose is to summarize the main results of all previous sessions, and to provide recommendations for improving the theoretical foundation for large-scale computations of nonlinear material behavior.

The format of the session is as follows: First, I will call upon the chairman of each previous session to furnish a ten-minute summary of that session, including recommendations for productive remedial research. Next, I will call upon each member of the organizing committee, who did not serve as a session chairman, to make a summary statement. Finally, I will call upon Dr. Snowden of DARPA to close the Workshop.

Our first speaker is Professor Budiansky.

COMMENTS BY SESSION CHAIRMEN

Comments by B. Budiansky

I'm not sure I'm about to fulfill what the job of the chairman was supposed to be. What I have decided to do, was write down issues or thoughts that I felt were the most important and not necessarily related only to the session of which I was chairman. So, here they are (in the sequel these items are referred to by their numbers, e.g., Item 1):

1. *Constitutive Relations:*

   (a) Mathematical relations vs. computational rules
   (b) Stress and strain rates vs. incremental approximations

2. *Beyond Mises Theory:*

   - Conceptual framework
   - Convexity, normality, associated flow laws
   - A platform for progress

3. *Beyond Mises Theory:*

   - Material-definition tests
   - Test standards?

4. *Beyond Mises Theory:*

   - Corners?
   - Shear band, localization predictability

5. *Beyond Mises Theory:*

   - "Open relations openly arrived at": documentation!

6. *Communication and Interaction:*

    - Mathematical and experimental test-beds?

7. *A Challenge for the 80's?*

    - Constitutive relations for material X

There is not going to be anything here that hasn't been said by one or more people, but I felt they were worth focusing attention on.

I'm going to restrict my comments to what I suppose is the central question of this meeting, the constitutive relations of plasticity. And the first point (see Item 1, above) I would like to make is that it would be useful to discriminate carefully between mathematical relations and rules for their computational implementation. I think it is not only not neat, but can be harmful, to mix up the formulation of a constitutive equation with what you do when you try to solve a problem. It would be more useful, I think, to talk about stress rates and strain rates, bearing in mind that they really want to enter into a particular problem at every instant of time, and *then* discuss the incremental approximations that are the most appropriate ones to make when you go to solve a problem. So, I feel most happy when I see a paper or a report on a particular calculation that starts out by saying, "The law that I'm using is thus and so," and then later says, "Here is how I used it."

Now, to more substantive problems: Beyond Mises Theory (Item 2 above). I think it is generally accepted that when calculations are done on the basis of von Mises Theory or J_2-flow theory, there is no argument. I think everybody does it right. Some people implement it better than others. I learned a lot from Tom Hughes' lecture, when he discussed the analysis of the radial-return rule and how good it was as an implementation of Mises Theory. But the issues come when we want to go beyond Mises Theory, because no one believes Mises Theory is really right. It might be good enough, but it is not really right.

How does one go beyond Mises Theory? Well, I would like to suggest it is important to have a conceptual framework on which to base your generalization of this most elementary theory of plasticity. I would suggest that, certainly within the context of metal plasticity (although there are analogous situations in soil plasticity about which I know much less) an essential part of this conceptual framework resides in the concepts of convexity, normality and associated flow laws. These are ideas that have been eked out painfully as unifying concepts in the field since 1928, I guess, if not earlier, when Mises invented the idea of a plastic potential.

The profession understands subliminally wherein the merit of these concepts resides. As has been mentioned many times, they guarantee uniqueness at least in small strains; perhaps more to the point, they are consistent with implications of micromechanical analyses; indeed, sometimes they stem rigorously from micromechanical analyses. Other people like them because they feel comfortable with Drucker's grand unifying hypotheses —Drucker's postulate. All of those things seem to culminate in the belief that this is an important and useful platform from which to develop further. Of course, we have heard a lot in this meeting about non-associated flow laws, and we know there are cases where flow laws want to be non-associated. There have been important experiments, for example, by Owen Richmond, in the last few years, that make it very clear that there are circumstances in which the associated flow law doesn't work.

But the point I would like to make is this: If one is going to deviate from these principles, one should do so for very good reasons and do so very consciously, and not accidentally. The reasons for going away from these principles,

I think, should be compelling.

Now, why am I saying all this? Not necessarily because answers you get when you deviate a little bit are going to be suddenly horrible; they *might* be, but I don't know that they will be. I am saying this because without this kind of intellectual framework, we really don't have a comfortable basis on which to proceed further.

Next, beyond Mises Theory, Item 3: This item is something that may not work, but I want to suggest the possibility that maybe we are ready for the beginnings of more or less standardized tests for defining the material. Everybody knows it is a good idea to get a tensile stress-strain curve for a material, and I know there may not be such a thing. I know there can be crazy slip lines going in when you least expect them. There is really no such thing as a single element specimen. I know that; nevertheless, we are not going to say we never want to talk about a tensile test anymore. So, we have a tensile test but that's not enough. We should begin to standardize, first of all, a shear test or a torsion test. I find it a little incredible that at this stage of the game, there is no standard shear stress-strain curve test. If there is, I don't know one. Of course, *that's* not enough. We need more tests involving polyaxial states of stress to begin to characterize the material.

An analogy I would make —maybe it is a rotten one —is with respect to fracture mechanics. Fracture mechanics was born fairly recently, but it's had a very, very exciting and vital development. In fracture mechanics, experimentalists and theoreticians, university types, industrial types and government types have been in constant communication. Extraordinarily, there is no communication gap there that is serious. How and why it developed I don't know exactly, but there is no serious communication gap, and indeed, there are now standardized fracture tests. We don't have the beginnings of any kinds of standardized tests that have the aim of beginning to characterize the plastic stress-strain behavior of material. Maybe we can start to make a beginning of this; perhaps ASTM, which is supposed to be an organization that does this kind of thing, can begin.

Item 4 shows the next issue I want to mention: consideration of corners, mainly with respect to the problem of the prediction of shear bands and localization, which was mentioned by quite a few people as being something that should not have to be inserted in an ad-hoc way, but should be a consequence of the constitutive laws that we use and the consequent numerical solutions of problems. So, attention should be paid to corners or pseudo-corners, because this can be, I think, a crucial generalization of Mises Theory.

And now, Item 5 sort of duplicates something I've already said, and that is that I would like people to say where their models came from. In the papers they write they might say, "The stress-strain relation I used was thus and so, and I got it from Irving down the street," or "I got it because I went in the lab and I made this measurement and that measurement, and on the basis of these two measurements, I extrapolated." As I see the literature (and I say this with some diffidence, because I haven't looked at that much literature), what I have looked at usually relegates this item a very secondary position. I think this should not be done, because we won't make progress if we persist in that kind of communication.

The next item (Item 6) which was touched on by Gil Hegemier is communication and interaction. One possible vehicle for communication and interaction is maybe setting up mathematical and experimental test-beds. I couldn't possibly give details of how this might work, but the idea would be: "Here is some material. Here is an experiment we are going to do with it. We are going to make a sphere out of it and hit it with a hammer. And now, you, you and you go to the laboratory, do some stress-strain tests, make a constitutive equation, make a numerical procedure, get me the answer. And you, you, you and you do the experiment and then let's get together and see how we did."

And finally (Item 7) I present a challenge for the 80's. Is it too much to hope that we could take material X, whatever it may be, make sure first of all that it is a reproducible material, and then get its constitutive relations? This of course doesn't exist today for any material, X, Y or Z. I don't mean get the whole thing, no, not over a huge temperature range and not over a huge strain range and not of course for every conceivable stress path, but for a certain number. What I mean is this: For this material X that's reproducible and is plastic, can we reach by 1990 or 2000 the ability, let's say, to do these (a) (σ_x, τ_{xy}) stress paths in the laboratory, measure strains (see Fig. 1) and then predict the outcome, the strains that will result from the tests in (b)? Well, I know the answer is no. So, let's do these tests in (b) as well, measuring strains. Can we then predict the outcome of the stress path (c) in the

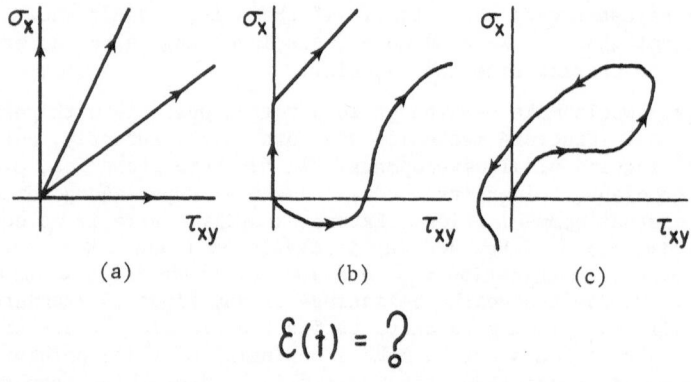

(a)   (b)   (c)

$$\mathcal{E}(t) = \underset{o}{?}$$

FIG. 1

strain space? Do as many tests as you like. Can we then write down a law, formulate it in as an efficient way as possible, so that from then on, with the provisions I mentioned concerning temperature and strain, we can predict what happens to a hunk of metal? I think this would be a fine challenge to accept. As I say, we can't even begin to do this today for any material that I know of.

Comments by K. S. Pister

One of the dangers in phonetic transcription is that ethnic misidentity takes place. Some of you saw this, but this is a classic one we can put down for history: "O'Learyian." We have a system of coordinates called "O'Learyian." These O'Learyian coordinates, along with that famous Irish mechanician, "Tim O'Shenko," can make the likes of Michael Carroll very happy when I tell him back in Berkeley that "O'Leary" had something to do besides throwing overalls in somebody's chowder, or whatever that was.

I think there is something we can learn if we think about the organization of the solution to the problem that we are all interested in here. As much as everyone in this room enjoys and is stimulated and rewarded by his particular pursuit, none of this we are talking about really is of interest ultimately to the problem solver, who says, "I have to create, conceive, design and build a system." No matter how good our constitutive equations are, no matter how good our computations are, if you put it in Air Force terms, the people at BEMO want a blue-sky launch and they don't give a damn about how you get there, as long as you achieve that, and you have to build it. So, I think we have to talk about the whole system.

The integrated system simulation, first of all, is a set of objectives that essentially characterize the activity that's needed in the simulation, the analysis and experimentation side of dealing with a model of a system; Table 1. We

TABLE 1	TABLE 2
INTEGRATED SYSTEM	INTEGRATED SYSTEM DESIGN
SIMULATION — ANALYSIS AND EXPERIMENT	SYNTHESIS AND VALIDATION
Constitutive Modeling	Load and Environment Specification
Algorithms and Software	System Performance Criteria
Controlled Laboratory Experiments	Performance Constraints
Field Tests	"Optimal" Design - Sensitivity Analysis, Trade-offs

have been talking about constitutive modeling with a great deal of emphasis. The algorithm builders and software writers have talked about the use of the models and the solution of the field problems. We have talked about controlled lab experiments and field tests. These are all activities that no one wishes to suppress. They are all essential inputs to the simulation problem.

The problem is, we have another group of activities that has to go on, and these are shown in Table 2*; and these activities have to do with the synthesis of the problem, the synthesis and ultimately the validation. We've talked about the problem of defining the loads and the environment in which the system is to operate. There is no sense in getting a wonderful stress driver if you don't know what it is on the system boundary that's exciting the system. We have to define the loads and environment that the system is going to operate in. We ought to be able to write down performance criteria in a form more specific, more quantitative, than in the case of a strategic missile at blue-sky launch.

What is it about the system that we have to sit down and evaluate through our simulation? What are the constraints on performance? Can we allow this or that level of stress or displacement or rotation? What state or condition is of importance here? We need these things in order to make an "optimal" design; I put optimal in quotes here, because I don't have any illusion that we are going to solve some sort of an optimization problem when we design these complicated systems. But in principle, we should be driven by the desire to do the best that we possibly can do in developing the engineering concept and then implementing it. A part of this, which reflects a great deal of the discussion, is the sensitivity analysis that falls out when you are trying to do an "optimal design." What happens if I change this parameter in the loading, this parameter in the structural model, this parameter in the geology? How does that affect the performance constraints, performance criteria, the costs, whatever objective function you have, or constraint functions? Finally, what are the trade-offs if I change this or that? All of these are activities that, in my view, ought to be integrated along with the other activities we are talking about.

I would hope the DARPA managers, the NSF managers, and all of us take some

---

*Editor's Note: The oral presentation showed items in Table 1 as rows and those in Table 2 as overlay columns to form an organizational matrix.

interest in seeing to it that we don't allow these activities to go on independently of one another. That's the message that's meant to be conveyed by the
entire matrix, Tables 1 and 2. There's nothing here that one hasn't seen before. But I think it needs to be reemphasized here that, if I were a guy in
Washington that had megabucks and I wanted to do something for the country, I
would insist when Tom Hughes got a half million dollars and Ted Belytschko got
so many dollars and the guys at S-CUBED are doing this or that, I would pull
them together and say, "If you guys don't work together, neither one of you is
going to get the money." Or even in an institution like Livermore, we'd make
Mark Wilkins and Jerry Goudreau talk at least weekly to one another.

A VOICE: Very weakly.

PISTER: Touché. That's a hard one to follow. In the remaining negative time,
I don't know ...

GOUDREAU: We had breakfast this morning together.

PISTER: That a boy, Jerry.

In the remaining time, I will now depart from my own gratuitous remarks and
quickly go over comments that the lecturers in my session supplied. These are
now specific details in the activities that I've just been talking about: the
constitutive modeling, numerical implementation, large scale computation activities; Table 3.

I think we are all agreed now, at least I hope so, that we need to exercise
models — not exercise them in a complicated boundary value problem, but exercise them in the sense of stress point algorithms to tell us something about the

TABLE 3

CONSTITUTIVE MODELS, NUMERICAL
IMPLEMENTATION, LARGE SCALE COMPUTATION

1. Conduct comparative analysis of response of current models, e.g., Tresca with
   normality vs. Tresca with radial return. Precisely characterize, both qualitatively and quantitatively, the differences.

2. Study response of "triangular" and "star-like" yield surfaces with radial
   flow — physical implications of results.

3. Upgrade large deformation constitutive models in codes with kinematic hardening mechanisms.

4. Homogeneous Testing
   • More emphasis on multiaxial, path dependent and cyclic tests
   • Study stress point drivers on small computers with 3-D graphics
   • Apply identification and parameter estimation techniques to constitutive
     modeling problem *(Inverse Problem)*

5. Inhomogeneous Testing
   Apply field codes and computer power to assess test geometry implications of
   models.

6. Large Scale Computing
   Improve iterative schemes for large, linear systems of equations and implicit
   time integration.

differences that are found when we make perturbations of certain classes of models—and we just suggested one particular model here out of many.

Looking at the response of so-called triangular, that is, the J_3-type and star-like yield surfaces with radial flow, what are the physical implications of playing some computational games with these models and then upgrading large deformation models and codes with kinematic hardening mechanisms? These are examples of the kinds of activities that I think need to be supported and conducted.

Then, in the area of "homogeneous" testing (I don't think you can really conduct homogeneous tests, but, abstractly, homogeneous tests), we need a great deal more emphasis on multiaxial path dependent and cyclic testing. This is something that the model builders love to ask experimenters to do, and experimenters have great difficulty in doing. We need to spend more time on this activity. Then, the stress point driver, the idea of the constitutive model builder sitting down at his work station with a stress point driver, that is, with a constitutive model, and playing games with it, playing Star Wars games with your constitutive model and having a nice graphical display, so you can see if I gave, as Bernie suggested, a certain stress history, what it looks like in strain space, and plotting that out and calling up real data and visually interacting with the stress point driver.

I've got here in Table 3 explore the application of identification and parameter estimation techniques to the modeling problem.   The modeling problem (try to understand the homogeneous test in terms of a model) is an inverse problem, and I think Kimsey is going to say a little bit about this. There is a great deal of literature available on the inverse problem that is not being tapped by the people doing constitutive modeling. The people that deal with time series analysis, the people looking at distributed inverse problems in other areas of mathematical physics are far ahead of the constitutive model builders in mechanics. Inhomogeneous testing, field codes, those are mathematical field codes and computer power to assess test geometry application of models —this is Goudreau's field; and finally, large scale computing, introduce schemes for large linear systems of equations and implicit time integration, what Jerry was talking about in his bit codes at Livermore.

In spite of the intensity and enthusiasm that I have and that others have here, I think it is important to step back for a moment and look at some recent history —and what I mean by "recent history," I mean like the last 30 years. I guess I'm finding myself more often than not one of the older people present, which is a hard thing to accept in life, but that's the way it goes. The time line is progressing. Back in the 50's, I recall that the first problem that one of my graduate students ever solved on a computer was a transcendental equation on a card punch calculator, up in an old wooden building at Berkeley, that you had to hard wire yourself. Then, we moved through the 1620, the 650, the 704, the 7040, the 6400, the changing of main frame manufacturers, and now we are back to the VACs. Think about the incredible change in computing power that's been made available to us. But at the same time, I'm kind of disheartened by the relatively small change in model building that's occurred in that period. We haven't really put the effort into model building that we have into computing power. What happened out at Livermore —I was chatting with Mark Wilkins at lunchtime —he started to compute on a Univac out there and then they went to Larks and Stars, etc. I remember when Mark started, he verified this, he didn't even put tensile or shear strength in materials; metals had absolutely no tensile strength whatsoever. Mark has changed his view, too; he's put in strength and he's done continuum mechanics. That's an example of a physicist crossing over.

I think we need a lot more of that crossover, and I hope my remarks will stimulate it.

Comments by D. R. Curran

Well, I'm happy to report our committee was able to reach at least a tentative consensus, and I'm sure we were all very happy, listening to Karl, to realize, thank God, that we all sit somewhere on his matrix. Furthermore, there should be no worry; we are all willing to do anything for money, even to talk to each other or cross over, if need be.

So, with that preamble, I'll show our summary and recommendations. I guess it will surprise nobody to see that our recommendations really do agree with everybody else's. It is a real love thesis here today.

First (see Table 4), our conclusion was that simple but respectable elastic-plastic constitutive relations have given very good results in large scale numerical calculations, but there are effects that are not incorporated in these models: vertex softening, large deviations from normality, anisotropic hardening, etc. That's not to say they aren't, in some cases, actually in the models or could easily be put in the models, but it is just to say that they have not contained those features.

TABLE 4

SESSION 3.   SUMMARY (KEY/HUTCHINSON, WILKINS/NEEDLEMAN)

Simple (but respectable) elastic-plastic constitutive relations have given good results in large scale numerical calculations, but there are effects not incorporated (vertex softening, large deviations from normality, anisotropic hardening, etc.)

What are the implications of using different constitutive relations?
Should we fear surprises from these unincorporated effects?

Possible sources for surprises:
Different elasticity models (measures of $\dot{\sigma}$, etc.)
Anisotropic hardening models
Instabilities
- Material versus structural
- Non-associative versus associative flow rules
- Vertex softening
- Effects of rate sensitivity

Experimental incompatibilities (Lab experiments exercise different paths than those exercised by code runs; the need for simple shear tests, etc.)

The question is: What are the implications of using different constitutive relations? Should we fear surprises from these unincorporated effects, despite the fact that so far, I think, there's been very little indication that we are in trouble? There are possible sources, of course, for surprises; different elasticity models, for example. We heard things from Sam Key and John Dienes on what happens if you use different measures of stress rate. There is a question that the other speakers referred to this morning, of anisotropic hardening models. There's the whole question of instabilities, and of course, we have had some discussion along the lines that we need to be able to distinguish between material versus structural instabilities. We need to worry about deviating from associated flow laws and go to various nonassociated flow laws. How far can we deviate before we are in trouble?

We had things like vertex softening that Alan Needleman talked about, and there have been some comments about the fact that rate sensitivity can often fight back against instabilities. Is the fact that we haven't seen in these large scale computations a lot of problems due to some fortunate cancellation of effects of, for example, rate sensitivity protecting us from some sort of softening? And then, something I called "experimental incompatibilities" has also been a subject of discussion in our sessions; namely, that the lab experiments often exercise different load paths in stress or strain space than those exercised by the code runs, the need for simple shear tests, and so on and so on, that's been discussed.

Well, those were our thoughts on what we believe were the issues here in the meeting. Of course, we have come up with some recommendations which we've tried to make motherhood enough to get by, but specific enough to have a little content anyway; see Table 5.

TABLE 5

RECOMMENDATIONS
(1) Implications of varying the hardening rules should be studied.

(2) Implications of features relating to instabilities should be studied.
                Deviations from normality
                Vertex softening
                Rate sensitivity

(3) Study methods of efficiently incorporating more complicated constitutive relations in numerical codes.   Calculations with the codes to study (1) and (2) above.
(4) Experiments to study and/or verify features predicted during (2) and (3) above.

First, we think that we should study the implication of varying the hardening rules. We should study the implications of features relating to the instabilities; in other words, we should find ways to study: deviations from normality, vertex softening, rate sensitivity, etc. How do we do that? That's what Items 3 and 4 are about here, and of course, our recommendations are pretty much the same as the ones of the previous two speakers, except that I think I have put them in reverse order from the last speaker.

I think one way to go is to first put these more complicated constitutive relations in the numerical code. I think that's easier and faster than trying to think of better experiments in a way; because I know, in my experience, the code people normally have very complicated constitutive relations, or can easily put them there. The reason they haven't exercised them a tremendous amount is because they didn't think they needed to. So, perhaps, first we should use the codes as they are to start more systematically looking at these sorts of things and what they do for various rather simple experiments or computational experiments; then, these should tell us something about what experiments we want to try to do.

It has been suggested that examining the simple shear of a rod would be a very good experiment to do in a much more precise way than has been done to date. But I would certainly want to do the calculation first and see if, indeed, that experiment is sensitive to the things we want to find out. Once we find out what experiments we need to do, then DARPA should come and give unlimited amounts of money to the people who want to try to do those experiments and see if the codes provide a positive correlation between experiment and theory.

Comments by R. J. Clifton

I am going to be careful not to use the word "we," or the word "us," because what I will describe probably doesn't really reflect a consensus. There was no attempt to come to a consensus in our session. In fact, the comments I would like to make really don't pertain directly or in particular to our session, but to some issues in metal plasticity that have been raised here and that I think might be clarified by my bringing them up one more time.

The first is one that Tom Hughes introduced —and I had to struggle with it a little bit, but I think I understand what's going on now. I am referring to this notion of elastic processes and plastic processes and to looking at them by means of the traditional approach and by the approach that he was describing, which I would call a finite element method approach. In both cases, as shown in Fig. 2, we have a yield surface, which is a function of stress, plastic strain and perhaps some parameters that characterize the current structure as determined by the previous deformation history. The interior of the region bounded by that surface, we call the elastic region denoted by Ω. In order to characterize these processes, we ask if a particular deformation process from a given state or stress is elastic or plastic.

Now, the way I understand what has been said is that we consider a thought experiment: In the finite element approach we make a probe of an incremental strain which is a total strain rate times a time element Δt; we compute a trial stress, which is the stress we had before, plus the increment in stress that you would get due to an elastic response to this strain increment. Then we ask the

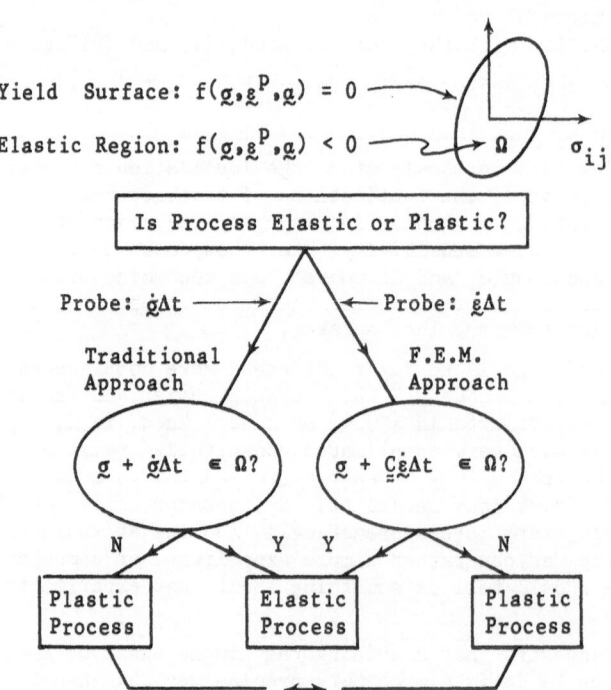

FIG. 2. Elastic Processes and Plastic Processes: Two
approaches to characterizing whether a process
is elastic or plastic are equivalent.

question: Does that stress state belong to Ω? If "yes," we say that the process is elastic.

In the traditional approach (I have just coined the word "traditional") you consider a small increment in stress equal to the stress rate times Δt. You probe to see whether or not the original stress plus this stress increment is within the region Ω; if so, you say that that process is elastic. Now, if the process is elastic according to one approach, you end up at a point that's inside Ω, for which elastic relations hold, so that the process would be deemed elastic by the other approach. So, in fact, these probes can be viewed as equivalent. Since all elastic processes are defined as "elastic" through either approach, they are the same processes. Since for either approach any process that is not elastic is plastic by definition, it follows that the plastic processes are the same. Thus, there appears to be a one-to-one correspondence between these two approaches to characterizing whether a process is elastic or plastic.

Another point raised by Tom Hughes was picked up by Alan Needleman, but I would like to bring it up one more time. This is the notion of flow rules based on "radial return." In the von Mises case where the yield surface is a circle in the space of the principal stress deviator, Fig. 3, a probe outside the yield surface followed by radial return brings you to a point on the yield surface that is approximately where you would end up if you integrated to obtain the solution to the problem for an imposed strain increment. So, radial return works very well. On the right side of Fig. 3 I show the case that Tom pointed out very nicely as one for which radial return does some things that are not very good. When the yield surface has a vertex, and you are working with a theory that incorporates normality, then a probe in the allowable sector at a vertex should bring the stress state back to this vertex; radial return would tend to return the stress state to other points. Thus, it appears to be quite clear that normal return is preferable to radial return for yield surfaces with vertices; normal return can be expected to be more satisfactory than radial return for all yield surfaces, except the von Mises surface for which the two approaches are the same. I think there is general agreement on this point, but I raised it one more time to see if anyone disagrees.

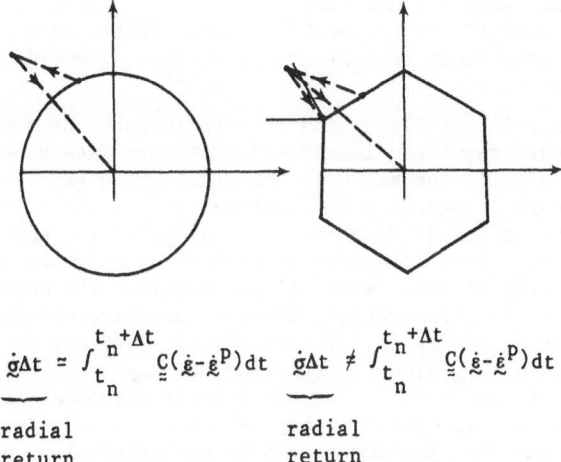

$$\dot{g}\Delta t = \int_{t_n}^{t_n+\Delta t} \underset{=}{C}(\dot{\varepsilon}-\dot{\varepsilon}^P)dt \qquad \dot{g}\Delta t \neq \int_{t_n}^{t_n+\Delta t} \underset{=}{C}(\dot{\varepsilon}-\dot{\varepsilon}^P)dt$$

radial   radial
return   return

FIG. 3. Flow Rules and Radial Return: "normal return"
         preferable to "radial return" when Mises theory
         is not used.

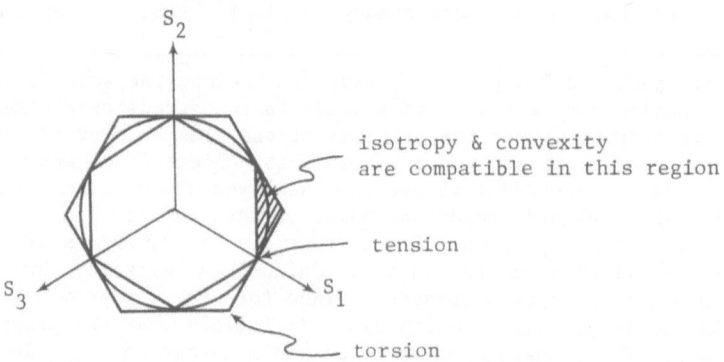

FIG. 4. Nonconvex Yield Surfaces vs. Anisotropy.

Another matter that has been raised is the notion of nonconvex yield surfaces. In discussing this matter with Mark Wilkins after his talk, I found out
why he introduced nonconvex yield surfaces. The reason is that experimental
results for tension and for torsion did not coincide when one attempted to relate
them by using, say, the J_2 invariant of the stress deviator tensor as a measure
of the yield stress and an equivalent strain obtained from the corresponding
measure of plastic strain. This conflict between tension and torsion experiments and convexity can be visualized by looking at the diagram of the principal
stress-deviator space shown in Fig. 4. The von Mises circle passing through a
data point for a torsion test is shown along with the inscribed and circumscribed
Tresca hexagons. If the result of a torsion experiment does not fall within the
cross-hatched region between the two hexagons then it is not going to be possible
to construct, within the framework of an isotropic theory, a yield surface that
is convex.

Confronted with such a conflict one needs to consider various alternatives.
It seems to me that if the torsion result falls outside the cross-hatched region,
this is an indication that the material is responding in an anisotropic way, especially at late times after considerable plastic deformation has occurred; this
can be due to the development of texture. Therefore, I would challenge the assumption of isotropy before giving up on convexity.

There has been a lot of discussion of rate independent models. I would like
to make a few comments about the idealization of rate independence for metal
plasticity. Certainly, in the kinds of low strain-rate experiments that Erhard
Krempl described there is strain-rate sensitivity. In the high strain-rate experiments that I showed Monday night there is definitely a sensitivity of flow
stress to strain rate. Thus it seemed to me that experiments indicate the rate
sensitivity of the flow stress. Also, if you consider the mechanisms of plastic
deformation due to dislocation motion, these mechanisms certainly suggest that
they are time dependent processes. The mathematical theory for elastic/visco-
plastic materials is a very nice theory. The principal part of the partial differential equations is the same as for the elastic response. All wave speeds
are real and non-zero. Viscoplastic response comes in only as the lower order
terms. As far as numerical implementation is concerned, I am suggesting that the
procedures are straightforward. For the finite element method that was described
—which involved the computation of the stress increment from the total strain
increment — one could compute the plastic strain rate from the current state of

stress, plastic strain and structure. This is a direct calculation for visco-plasticity models. From the total strain rate and the plastic strain rate one can compute directly the required stress rate. There may be some of you who want to comment about the integration of systems of the type obtained for rate dependent models, especially in the case of quasi-static problems. However, in the case of dynamic problems, it seems to me that the examples shown by Gordon Johnson indicate that there is no fundamental difficulty in carrying out calculations using rate dependent models.

It appears that all the session chairmen have some comments about single element experiments. I would say that there are a couple of experiments that I am familiar with that are reasonably good approximations of single element experiments. One is dynamic torsion, which, in some laboratories (SRI and Brown), is used for strain rates as high as 10^4 sec^{-1} and strains of over 50%. It seems to me to be a very nice experiment. I showed some results Monday night for the pressure-shear experiment. Results have been around from that experiment for a couple years now. It seems to me that this is an experiment where you can combine the effects of pressure and shear strain rate in the range of pressures and strain rates that are of interest in a lot of dynamic problems. Thus, there appear to be techniques that give very nearly single element experiments.

Finally, in our session there were negative remarks made about micro-mechanical models. I think that some negative views of micro-mechanical models are related to a mixing of size scales. If you see a calculation of polycrystalline plasticity, in which dislocation velocity is written as a function of the macroscopic stress and the shear strain rate is written as bNv, there is obviously a mixing of notions of single slip in single crystal plasticity with polycrystalline plasticity. I agree that you can infer about anything from such a model. But if you do a very detailed micro-scale study of these processes, which includes experiments on single crystals —measurements of dislocation densities before and after loading, measurements of the mobility of dislocations, and prediction of the response of well-characterized experiments as a way to understand the dislocation mechanisms — you can then build an understanding that allows you to describe the response of polycrystalline specimens. I think that is a very legitimate approach, and I would like to see more of it done.

Comments by T. C. Bache

It is interesting to note how every chairman takes a different format here, so I will do something different yet.

First of all, I had a chance to make my grand summary remarks just before lunch. I will be interested to see if, when this is over, there is any response from any of the calculators.

Having worked with Ted Cherry and Jack Trulio for close to ten years now, I knew better than to attempt any consensus; so what you will hear is my assessment about what was important about what they said in the context of this meeting and the issues, for those people who really haven't seen this sort of thing before. So, I will just quickly go over a brief summary of each of the presentations.

First of all, the Cherry presentation summarizes most briefly what Ted has done. He considers very simple models, applies them to different materials, very different materials, such as granite and porous tuffs. All that he does when he changes from one material to another is to change the elastic properties and to change the way in which the porosity is represented; there are four parameters, basically, that go into the model. The results, as you saw, are that many important data features are matched, and indeed, Ted claims some predictive power and he has made some predictions. But I put a question mark here, because that has not been done too much, and also, the experiments that are being predicted are not very different from the ones the model was normalized to; so, it may be that you are in a linear part of the variations and so you may not be predicting very much.

The material model that Ted has in his code and that he presented is a simple material model, an elegantly simple one. You might say that the primary justification is really a match to the ground motion data. It is only tenuously based on lab data, not because he doesn't want to base it on lab data, but just because the right tests aren't available. We are trying to remedy some of that now with the program that Carl Keller is running. The basic conclusion, if you accept this model and other models of a like kind that are available in the ground motion prediction community, is that to match the data, the material strength must drastically reduce during loading; some people many years ago called this "shock conditioning" of the material, having no idea of the physical mechanism, but Ted presented his ideas yesterday.

The issue really is: What is the mechanism for this strength reduction, if that's indeed what goes on? And of course, to accept that that goes on, you have to accept this paradigm that Ted set up. So the issue for people to deal with is: What is a mechanism for the strength reducing drastically when the material is loaded? Is it water in the rock or some kind of effective stress behavior, as Ted said, or something else? An alternative to this, however, is the paradigm is all wrong; maybe there is an entirely different way to approach the problem and we ought to sort of start over; maybe plasticity, and in the sense that Ted talked, isn't really the right way to approach these materials at all. This is an issue for the theoretical and experimental community, and us all.

With that, I will move into the Trulio presentation, summarizing it in the same very sketchy sense.

Jack, basically, to my way of thinking, concludes that current models have little predictive power and he can site many examples to show that —with all the caveats one has about whether the gauges worked and the measurements were right and everything else that goes into these, whether he knew the material properties, etc., etc. Now, he recommends field measurement of relative stress-strain curves. It is really presented as a way to bypass, at least with present knowledge, the need for a detailed understanding of the physics.

The problem, as it was pointed out, is you can't expect to measure all the strain paths at the various strain rates that you are interested in; so, you really must have some model in mind. There is no way to use these strain path measurements without putting in something that you would have to dignify with the name of a model, even though it may not have detailed physics in it. So, the issue is, to my way of thinking (and maybe it is something I don't understand — I don't know how other people feel about it): Is this a fundamentally different approach at all? I mean, it's just another way to measure some data that you have to put in some kind of a model; it's just a different kind of model you might end up with, and it may be almost entirely an empirical model or an ad hoc model even worse than the ones we have now, but it would have to be some model. I don't really see it as a fundamentally different approach, but rather as an approach to measure a new kind of data; strain paths with associated stresses. And these measurements, indeed, if taken would provide important new data that successful models should match, that would make it possible to make new steps forward, not only matching some velocity time history, but also strain paths, get the right strain path and make sure you have that right and the right stress that goes with it, and then you would start to be satisfied that you are doing things more or less right.

Comments by J. T. Oden

As I said earlier today, I think the most dramatic thing about this meeting is to see how a collection of models of nature, in which everyone admits they have limited faith, and numerical methods, which we now are convinced have numerous deficiencies, can lead in many cases to results that actually look like they simulate a good part of what is actually observed. One wonders secretly whether or not someone is pouring through these reams of paper and trying to find those that in fact match experimental data; see Fig. 5.

FIG. 5

My comments will be very brief, and they have to do with the following
topics: 1) Accuracy and Reliability of Codes; 2) Efficiency; 3) Inverse Problems; 4) Adaptive Methods; 5) Distinction between Model and Approximation. I
think it is quite clear that an intrinsic step in our ability to model plastic
materials is reliability in computed results. It is an intrinsic feature of
everything we have been talking about, but I'm worried, in fact, whether we
know enough about the numerics to really warrant that level of reliability.

In the particular session this morning, where we looked at strain softening
effects, there was one of many examples of cases where an attempt to model a
certain feature of a phenomenon led to a situation where numerical approximation
will just give "results"; the results may seem to be physically plausible, but,
in fact, there exists no solution to the problem at hand. There are similar examples that one can point to in much more innocent looking situations; elementary
Stokesian flow, for example, where abuse of various operators appearing in the
governing equations can lead to substantial instabilities, and these are frequently very subtle. You cannot always detect them with innocent looking numerical experiments.

I think there is a strong need for a parallel research effort into reliable
methods for checking the accuracy of these computations. To be sure, this is an
extremely difficult task. The machinery available today really only applies to
a relatively small class of principally linear problems and some nonlinear problems, and the type of models that we've looked at here go far beyond those
levels of complexity. Nonetheless, I think if we want to add more of an element
of science and less art into this field, that some serious thought into techniques for estimating the reliability of numerical solutions in these highly
nonlinear problems is a necessary ingredient in this research activity. Efficiency, of course, is always a goal of these types of computations, but I would
hope that efficiency is not sought at the expense of reliability. And I think
in some cases, we may be seeing precisely that.

Karl mentioned the inverse problem. This is an active area of research,
mostly in information theory, communication theory, control theory. It certainly
has not progressed to a stage where it could have great impact on the problems
of the complexity that we are speaking of here, but, nevertheless, it would be
good to bring into the circle of specialists looking at these issues, those who
have dealt with the mathematical apparatus connected with handling the inverse
problem; that is, given a response, given a set of initial conditions and introducing these into a "black box," what can you say about the mechanisms in the
black box that would produce this response?

Adaptive methods: The notion of adaptive methods is a hot issue now in
numerical analysis; the ability to refine the mesh and get higher resolution of
the approximation and various points in the mesh. I mention it here, not so much
as a computational device, but as an area which involves methodologies that
should be given some thought within the general area of computational plasticity.
One feature of the adaptive methods is that they attempt to construct an estimate of the error in a solution on the basis of a computed result. Reliability
and accuracy can, to a limited extent, be determined on the basis of a priori
estimations of error. One can characterize the equations governing a certain
model, determine their well-posedness, and in particular, consider as a key
issue the determination of whether or not the conditions sufficient for well-
posedness are carried over to the discrete model. Many times they aren't. When
they aren't, that normally signals the onset of numerical instability. Here, I
am talking about the reverse side of this issue: the development of techniques
which provide posteriori estimates of the quality of the solution.

As I noted earlier, this is an area of active research. Significant strides have been made in certain areas of computational fluid mechanics. I think some methods in use there which have been developed only in the last few months, as a matter of fact, might be put to good use in studying the reliability and accuracy of the results produced in large scale plasticity codes.

I want to also echo a comment made by Professor Budiansky on the importance of the distinction between the model and its approximation. Too frequently in the engineering literature, you see these two mixed up; the individual does not state explicitly what the model is, what are the differential and integral equations describing his problem, and, independently, what techniques are used to approximate it. We sometimes find ourselves caught up in a situation where our objective is to mimic what we think reality ought to be, rather than to follow a systematic sequence of steps where we identify physical phenomena, we construct what we think is an appropriate mathematical model of it, and then we attempt to approximate it; to do otherwise is just playing games with the computer. I think if we are going to merely use the computer as a device to mimic what we conceive to be nature, then a much better outfit to handle that sort of enterprise would be the Walt Disney Co.

Comments by J. M. Roesset

I am going to take five minutes only. In fact, most of what I wanted to say, Professor Pister has already said much better than I could. From that point of view, all I would like to do is express our support, my support, at least, for everything that Professor Pister said.

What I would like to do, then, is take this couple of minutes to say something in defense of geotechnical engineers. The first night I was here, someone told me civil engineers were neither civil nor engineers. We showed this morning that sometimes we may not be very civil. But I would like to say something in favor of geotechnical engineers, although I'm not one of them. You have to realize that geotechnical engineers have been recognizing that they're dealing with nonlinear problems since the early stages of soil mechanics. If you go to the crudest methods in soil mechanics, they always accounted, even if it was in a crude way, for the fact that they were dealing with a nonlinear problem. You also should realize that geotechnical engineers work in everyday life with problems for which there is not very much money, very simple problems where they have to go to a failure state, call it shear bands, whatever you want, but they have to define that state; they have to go there and make predictions, so the building doesn't collapse, or whatever. So, they are faced with a lack of time and megabucks to solve the problem, and they have to do something about it.

I would like to point out, in this respect, that I think it is important to recognize that we need two things at this moment in soil plasticity or application of plasticity to soils. On the one hand, we want to continue to develop better models, obviously, models that are as refined and as accurate as possible. On the other hand, we have to get a better idea of what kind of model we need for each particular problem. Obviously, there are many problems. Some types of results for specific problems may be rather insensitive to the details of the model. Then you can use a very rigorous and very expensive model, but it may be sufficient to use a very simple model and get something which is satisfactory, and that's an important distinction.

Unfortunately, right now, in many cases, we don't know how sophisticated our model has to be to solve the problem. So, an important thing is to be able to use the models that we have, to exercise them and to try to find out how sen-

sitive different types of results are to the details of the model. If they are
not sensitive, fine; if they are sensitive, we know we have to refine them. Unfortunately, this is the kind of thing that is not very glamorous research. If
you say, "This is what I am going to do," everyone is going to come back and
say, "That's not research, that's just engineering." Well, someone has to do
that kind of work. That's the only way we are going to get a feeling about the
problem, and I think this is an important thing to keep in mind. We also need
very badly to verify this model with some experimental work. Now, we can get
some checking from lab tests, that's good; but it is not nearly enough. Among
other things, the properties we derive from lab tests have little to do with
the soil in situ. We don't know very well what the relation is between lab and
in situ properties. While lab testing is a good step, we need more than that.
Full scale tests or tests in situ are very expensive; they are hard to come by.
Obviously, we have to take advantage of any situation where we can do them to
evaluate models.

There are other types of tests, like centrifuge tests, that are being done
now and which may be very helpful for this kind of thing. In fact, two of the
people here, Professors Scott and Prevost, do a lot of work with centrifuge.
But there is clearly a very intense need for more experimental work and further
evaluation of the different models that we have.

Thank you.

***

HEGEMIER: I'd like to thank all the chairmen at this point. I'd now like to
call Tom Hughes as part of the organizing committee to make a few statements.

***

STATEMENTS BY ORGANIZING COMMITTEE

HUGHES: I think we should be addressing the aims of this workshop and where we
stand right now. The motivation behind the workshop was that there were some
unusual plasticity models that were currently incorporated in computer codes
and, perhaps, the implications of those models were not understood by the computational community — perhaps those implications should be made known, perhaps
something should be changed. As Gil said, I think that, in fact, this is the
case. There are models that are certainly unusual. There have been people
that have made suggestions that this is not the way to do things. Nevertheless, the people that have created these models have, I would say, some good
reasons for creating them the way they have, and we are really waiting for some
specific criticisms of these models. Are they going to create phenomena that
are not what we would like to see, or aren't they?

I think you just can't cast aspersions at models and then walk away and
say, "This is not the way to do it." These models are "in place." They are
being used for critical engineering calculations; they are being used for calculations that have national security implications. This is extremely important.
The theoretical community, if it feels that there is something wrong, has to
make things explicit. It can't be in this kind of vague, qualitative way; it has
to be extremely precise: "Something is going to go wrong and here is what it is."
So I would suggest very strongly that models of this type be analyzed. They
should be contrasted with competing models or models where things are just a
little bit different. They might ostensively be a little bit different, but
there might be serious implications. I don't feel comfortable with just turning
one's back on this. I think this is a problem that potentially creates another
round of problems. There will be a new round of models like this that perhaps
will be unusual again from somebody's point of view. We've got to bring the

theoretical community and the computational community more together on this point. These are objects that are worthy of study. They cannot be dismissed, in my opinion.

I think that's what's happened in the past. I think one must realize, too, that the reason these models are in place is because alternatives, perhaps, based on what may be sounder principles, have not been put forth and have not been available when people in the computational community have created models for codes. I don't see better mousetraps available; all I see are criticisms, and that's never enough.

I would like to make another point. We've talked about constitutive models and large scale calculations; even theoretical people talk about constitutive models and solving boundary value problems very quickly. I think there is an intermediate regime that some people have mentioned and should be mentioned again and really emphasized. You'll note in papers in which constitutive models are presented that a number of tests have been used to identify the constants in the models; and you will always have the constitutive model virtually interpolating the data. Then, one seems to leap from that point to very involved boundary value problems, either theoretically or computationally.

I suggest there is an intermediate regime that is being missed and should be explored. On the constitutive level, on the stress point level, what one should be doing is looking at the implications and predictions of constitutive theories that are not a direct consequence of the identification of that model. For example, take a multiple-yield-surface theory. You might identify that model via certain one-dimensional tests; but yet, when you do computations to solve boundary value problems, that model will be exercised in regions of stress space that are a product of the model and not the one-dimensional test used in its identification. And the point is, this should be explored before you do a boundary value problem. This can be explored on the stress point level, as we have pointed out several times, by driving the algorithm with strain histories. I don't see that being done routinely. I see it done very, very rarely; and then you look at the results of an enormous calculation and you have no context or intuition for interpreting what's going on. I'm not accusing everyone uniformly, but I'm saying this is the rule rather than the exception.

If you really study stress point behavior, you start to develop an intuition about how the model behaves. You have a much better context for understanding boundary value problem phenomena. So, I think that should become part of the theorist's game and part of the numericist's game. It's such a small computation, it can hardly be described as numerical analysis. Theoreticians should have stress point routines that their students run on micro-computers; they should look at families of stress-strain histories, cyclic, going in one direction, going in another direction, etc., etc. I think that's an extremely important point in building up intuition for these very complicated models that are being proposed these days. Individuals do not seem to be building up intuition the way they used to do in the old days for simpler models. There is a tendency to be satisfied with just interpolating data points; everybody seems to do it and it's always perfect. Then you solve that big boundary value problem and start scratching your head. What's going on? What do the results mean?

ASARO: It occurred to me sitting just now and listening to the previous speakers that as one of the organizers, I should feel very good about the job we have done. This must indeed have been a successful workshop. I knew this earlier just based on the intensity of interest these past three days and by the interaction and exchange of information, all of which appear as primary goals of this workshop as mentioned in the statement of intent in the first page of the preparatory book.

But also, listening to the chairmen make such beautiful summaries and incisive
comments makes it again clear that important issues were understood and had made
an impression.

There are two points I would like to reiterate: one very close to my personal interests which involves the need and the usefulness of crossing disci-
plinary boundaries and interaction between people with different backgrounds.
It seems to me from my own perspective, starting out as a metallurgist and now
working in the area of solid mechanics and materials science on subjects such
as metal plasticity and fracture, that over the last ten years, a remarkable
amount of progress has been made in the understanding of metal plasticity and
fracture. In fact, if you trace the origins of that progress — if you ask where
did it come from and why has the last ten years been so terribly productive as
compared to the last sixty years, i.e. since the days of G.I. Taylor and his
contemporaries — the reason is that there are a number of groups throughout
the country where there is very real, very strong interaction between mechanics
people, computational people, and in more recent years, computational solid
mechanicians and metallurgists. That sort of interaction is vital, given the
scope and complexity of the problems; and it is, obviously, the kind of thing
we want to encourage here.

The other point that has been mentioned by everyone and the chairmen emphasized — it was listed on several of the transparencies they showed — and Tom
Hughes just stated very nicely again, is there's a need, before one starts to do
large computations, to understand what the constitutive theory is.

Bernie Budiansky said, first define the constitutive law for yourself and
other people to see. Tom Hughes suggested putting it on a micro-computer and
running it through with a wide range of specified deformation histories. That's
a very good idea which I totally endorse. I would go one step further and say
not only should the model be thoroughly exercised — you have got to do that to
come to grips with it since even the most innocent looking models are very complicated — but go one step further and also compare with experimental data.

The step further though that I'm referring to and, I think, that was a
major point of the Sandler-Hegemier talk and discussion, is the need to ask
about the implications of constitutive models for the character of governing
equations. It turns out that if you want to look at complex phenomena like
localized shearing, you first have to have a high degree of anticipation as to
the character — I mean that both in the qualitative and mathematical sense —
of the result you are going to find. So before you start solving boundary value
problems, the equilibrium conditions should be invoked along with the constitutive law, to make sure the problem makes sense and that you understand it, at
least to the point where you have confidence in what you are dealing with. One
example that caught a lot of attention a couple of years ago at the Stanford
workshop on finite strain plasticity was kinematic hardening models. The first
reference that I am aware of — and there are some before this — was Prager's
1961 book, in which it is very clearly pointed out that hypoelastic relations,
based on Jaumann rates, oscillate. Well, we know that. There's no mystery in
that; it's straightforward kinematics. And one of the milestones — perhaps mini-
milestones — I would say, in the first day of this workshop, was that we pinned
down these important points. Everybody I think understands them — it's all in
the evolutionary laws for internal tensorial variables. Some sensible suggestions were made as to how useful reformulations might be made, and then, some
cautions were raised, cautions which go back to this need for exercising models
thoroughly before you use them in large calculations.

Now, what are some of the future needs in constitutive laws? It seems to me that future needs include the development of constitutive laws to account for strain path dependence of hardening in metal polycrystals; these laws and models should almost assuredly account for texture and anisotropy, because that is really one of the primary origins of strain path dependence. For example, Rod Clifton mentioned a previous discussion where Mark Wilkins had pointed out to a number of us some of his interests in this very question of strain path dependence. I was showing Mark that a good deal of the observed strain path dependence of strain hardening can be accounted for by texture development. It seems to me then this subject would be a good area for some nice collaboration.

I would like to keep in touch with you, Mark, on this and show you in time what we come up with at Brown in our modelling of polycrystal large strain behavior. This also happens to be an area where micromechanics is playing a very real and extremely important role in continuum constitutive theory.

Incidentally, I would point out to anyone who is interested in developing polycrystal models from single crystal descriptions that you must start with rate dependent models if you want to describe anisotropy.

What you find, as Mark Wilkins realized, is that if you compare strain hardening in, for example, axisymmetric compression and plane strain compression — in Mark's case it was tension versus shear — using the von Mises equivalent stress and equivalent strain, there is a large disparity in the equivalent stress-strain curves; this is due to anisotropy, a large part of which has origins in texture. We are making good progress on describing textures and are now able to carve out flow potentials and pseudo-yield surfaces, things which should be of real use. We certainly want to keep in touch with all your interests in this strain path dependence.

One thing which I think might be helpful for future get-togethers, is if some of the people who do these very nice large scale dynamic calculations would answer a question that I'd find helpful in trying to assess the importance and concerns with constitutive models. We've seen some very nice comparative results, i.e. involving comparisons with experiments of some penetration problems. I remember those in particular. But what I'm wondering about is to what extent some of those phenomena are rather insensitive to constitutive description and are more dominated, say, by inertial effects. It might be nice to gain some appreciation for that, because when you consider phenomena, such as failure modes, shear bands being a prime example, we have confidence that constitutive details are really very important. In·fact, that's where the desire for constitutive models comes from. It's the importance of constitutive details in modelling localized deformation modes which motivate much of the intensive interest that you see around the country, at universities and elsewhere, in the development of large strain constitutive models.

***

HEGEMIER: Sia, do you have a short comment?

NEMAT-NASSER: I really mean it. I am going to be short, because everything that I could say has already been said.

I wish to emphasize the importance of the scale effects in material modelling. I think many of the problems that were discussed in relation to geophysics relate to the scale effects. We would have the same problem if we look at polycrystals at reduced scales. It would be almost impossible to make a prediction at local levels in grains of a polycrystal. We have to define a scale, and once a scale is defined, then I think things may fall into place. In addition, the microscale may be very large, indeed, and the macroscale may indeed be over a kilometer, depending on the considered problem. This will also affect the time scale.

Little has been said about micromechanics, although we did have a few presentations this morning on micromechanics. There are some people who claim you can do everything by micromechanics. (One person refused to participate in this meeting because we did not include micromechanics. He feels you can do everything that you do in macromechanics, by micromechanics.) At this stage, I don't want to be that ambitious. I would like to repeat what John Hutchinson said: If we capture the essential microscopic features, then micromechanics can provide guidance for developing good macroscopic models, in accord with physics.

Finally, I want to emphasize the important problem of characterizing the evolution of defects, e.g. the formation and growth of shear bands and/or microcracks. There is a transition from a more diffused distribution of micro-defects into a more localized distribution, which precipitates the final failure. This is something we don't understand and I think micromechanics can be quite useful in providing guidance.

***

HEGEMIER: There are several individuals who have indicated a desire to make a few remarks. The first is Professor Belytschko.

BELYTSCHKO: This is a short comment. I think there has been a tremendous preoccupation here with metal plasticity, which is an area where most of the time we feel quite comfortable in making calculations. About the only very difficult area that we have seen pertaining to metal plasticity is the penetration problem where we have essentially phenomena such as shear banding — and other very difficult phenomena; these calculations are today, I believe, still on rather tenuous grounds. There is a lot of basic work that needs to be done before we can really make good predictive calculations in areas such as penetration. However, there are other areas which haven't really been touched upon at this workshop, which are very important to defense problems, where there are tremendous uncertainties in calculations. One area, for example, is cratering. Cratering calculations involve a tremendous number of assumptions; much research is needed that's not available. And if I were to make a (cratering) calculation today, I would have very little confidence in the result.

Calculations concerning the failure domain of structures and of geological media are another area where we haven't even begun to touch the surface, but they are very important. Today, we would like to calculate the vulnerability or survivability of structures under both weapons environments and natural disasters. These calculations involve phenomena which are extremely complex. Yet, such computations are being made quite routinely in various laboratories and in industry. I think if we were to investigate these with the same skepticism that we have investigated other computations in this workshop, we might find a lot more room for controversy.

PATCH: I haven't had much to say, because I have been rather overwhelmed by what has been an intense learning experience. I must say, I tend to agree with one of the comments that was made; that is, that this workshop has been generally oriented, and I think perhaps rightly so, toward metals and what I would consider to be somewhat more simple materials. With that remark and the fact that I'm more interested in the behavior of earthen materials in mind, I would like to briefly respond to the question posed earlier by Tom Bache. The question was: In what areas can the theoreticians help the calculators?

While I will attempt to respond to that question as briefly as possible, I

don't think that I can take the time to give the question the complete response
that it deserves. I believe that one of the greatest needs of the earthen calculators is better material response data, rather than more models or complex
theories; and we need data in at least three areas.

One of the most serious deficiencies is in the area of intermediate and high
strain rates. Over the years we have gathered a lot of static material property
data, and there has been a lot of discussion about the validity of some of the
data from these tests. I think these concerns are well taken. Indeed, a lot of
these single element tests are almost ill-posed, and that's a serious problem.
But even more serious in my view is the complete lack of data at high strain
rates that's anything other than gas gun, uniaxial loading data. Secondly,
large strain data is badly needed. Currently we don't have anything that exceeds more than 10 or 15% strain. Finally, for certain important military systems, the multicycle, load-unload-reload response of the medium is really
critical. So, the material test area is one where I think the calculators need
a lot of help.

Next I would like to make some comments on the question of the uniqueness of
the numerical models. Some point has been made that some of the calculations,
for whatever reason, are not doing that bad a job. But in my experience, models
that are doing a relatively good job can be, in some of their fundamental aspects, quite different. Thus we can certainly use some help in identifying
those models which work over the widest range of applications.

From my parochial point of view, I think that the uniqueness question has
been a little overstated. It has been implied that there exists an infinite
number of models to fit the data; yet, this infinite manifold of models is
limited by some very real and practical constraints. So in my experience, yes,
the models are different, but when you look at their gross behavior, they all
have certain very similar aspects. That is, they may get to a final state in
slightly different ways, but if you look at stress-strain plots, they look very
similar, even though the underlying models may be touted as being quite different. So, in one sense at least, there does seem to be more fundamental unity
to some of the work that has been done than one might guess. At least that is
my personal impression.

The issue of code reliability has certainly come to the floor in this conference. What we need as calculators is a set of non-trivial check problems
that would really exercise some of the features of these models, and I would
like to see check problems that aren't pathological in the sense that they are
really different from real world problems.

We have discussed situations where the answers can be really dominated by
the geometry of the problem, essentially one-dimensional cases. For proper code
validation, we need to exercise different geometries and different strain paths
before we can separate some of these models. To find some kind of a check problem or a series of check problems to really let us exercise the codes would be
a real help.

I would really like to say that I agree wholeheartedly with what has been
called the "joy stick" approach to investigating the behavior of material models.
Numerical models *absolutely* have to be exercised through a wide range of stress-
strain paths on the computer to see that you don't have non-physical behavior
and programming bugs. At another level, the calculators have a duty to show
how well the numerical model matches the material property data that've been
provided.

CHERRY: As far as Tom's conclusion is concerned, if I can paraphrase it, the theoreticians should retire and all we need now are calculators and measurers. I think that conclusion was arbitrary, capricious and dumb, and Tom is not arbitrary, capricious or dumb. I think he said it just to stir things up a little. Is that what you said, Tom?

BACHE: No, I don't think so.

CHERRY: Maybe it was the lunch that I had.

BACHE: I said either there are some problems that you ought to tell us that need to be solved or the theoreticians ought to retire. So if you don't want the theoreticians to retire, tell us what some of these problems are.

CHERRY: Well, let me say first, if Sabodh Garg and Hal Read retire, then I'm going to follow them out the door, so don't retire. There is a real need for theoreticians, calculators and experimentalists to function in the same environment. I'm not going to presume to explain the scientific method to this group. We have strived at S-CUBED, however, to foster that method, with the theoretical aspects of the modeling and the calculational aspects of the modeling being taken care of in S-CUBED, and much of the experimental work recently being performed at SRI under the direction of Alex Florence and Carl Keller.

As far as the model I presented is concerned, we obviously need more data, and probably all the model is going to provide us with is a foundation to build on. There are many things in that model that I'm very concerned about. I made a horribly crude assumption about the variation of pore fluid pressure that is thoroughly suspect; various people gave their opinions about what it was supposed to model. That aspect of the model certainly needs to be investigated. The fact that I've done a great deal of violence with the laboratory rock mechanics' quasi-static measurements is of concern to me. I have thrown out all of the deviatoric strain paths that are measured from a quasi-static test, and that is really a worry. In other words, the model that I presented will not reproduce the strength paths that one finds in a quasi-static triaxial experiment.

I am concerned that we are not following an associated flow rule. And I am not sure what the consequences of including an associated flow rule in this model are. We need to find out. I very much agree with the people who suggested that we do that.

HEGEMIER: It is my pleasure to call upon Dr. Snowden of DARPA for closing remarks.

CLOSING REMARKS

by W. E. Snowden

This workshop was convened in response to the expressed interest of two government agencies —DARPA and the National Science Foundation — in improving our overall capabilities for solving complex problems involving computer simulation of the response of materials to dynamic loading. A specific objective — stated in my opening remarks — was to foster further communication, and, perhaps to provide some impetus for subsequent collaborative efforts among those groups whose work is critical for improving large-scale computational capabilities: the theoretical solid mechanics specialists, solid mechanics/materials experimentalists, experts in numerical methods, and, of course, the developers/users of large-scale computer simulation programs.

With respect to communication, it is quite apparent from the general intensity of the discussions that have taken place here that we have achieved our initial objective. It is important that we find some way to maintain the momentum of technical exchange that has been generated. The report of the DARPA/BDM panel that I referred to in my opening remarks contained the recommendation that a second workshop be held six months after the first for more focused discussions. It seems likely, based on the early positive assessments we make as to the success of this meeting, that we will proceed with that recommended course of action. However, it is important that the intervening period not be one of silence and inactivity.

Beyond the above comments, I think it is inappropriate for me to attempt to describe my reactions to these proceedings in any detail. It is my intention to work closely with the panel — Professors Nemat-Nasser, Hegemier, Asaro, and Hughes — to further define future objectives representing DARPA interest in the areas of constitutive modeling and large-scale computations. But I am also most interested in getting the individual inputs any of you might care to provide, both in assessing this workshop and in suggesting significant focal points for future activities.

Recall that in my opening remarks I stressed our interest in collaborative efforts directed toward this clearly interdisciplinary problem area. Your thoughts on how this can be achieved —in as specific a manner as you might want to describe -- will be gratefully received.

In closing, I want to thank all of you for your attendance and participation. I also want to thank the members of the panel for their outstanding efforts in putting this workshop together, and the supporting staff at Northwestern University for their excellent work in handling important logistics problems. I look forward to our next meeting.

PROGRAM OF THE WORKSHOP

SUNDAY, OCTOBER 23

   7:00-9:00 p.m. : Ice-Breaking Complimentary Hospitality and Registration
                          (Norris University Center, Rooms 2B & 2C)

MONDAY, OCTOBER 24

   8:15 a.m.   : Registration (Norris University Center, Room 2D)

   8:45 a.m.   : Welcome Address, D. MINTZER, Vice President for Research
                          and Dean of Science,
                          Northwestern University

   8:50 a.m.   : Introduction, S. NEMAT-NASSER, Northwestern University

Norris University Center, Room 2G

Session I: Chairman: B. BUDIANSKY, Harvard University

   8:55-9:15 a.m. : Speaker: W. E. SNOWDEN, Materials Science Division,
                          Defense Sciences Office, DARPA
                       "Comments on the Aims of the Workshop"

   9:20-10:10 a.m.; General Lecture: S. NEMAT-NASSER, Northwestern University
                       "Theoretical Foundations of Plasticity"

  10:10-10:30 a.m.: General Discussion

  10:30-11:00 a.m.: Coffee Break

Session II: Chairman: K. S. PISTER, University of California, Berkeley

  11:00-11:50 a.m.: General Lecture: T. J. R. HUGHES, Stanford University
                       "Numerical Implementation of Constitutive Models"

  11:50-12:10 p.m.: General Discussion

  12:45- 2:00 p.m.: Lunch, Allen Center Dining Room

Session II (cont.): Chairman: K. S. PISTER, University of California, Berkeley

   2:00-2:50 p.m. : General Lecture: G. L. GOUDREAU, Lawrence Livermore
                         National Laboratory
                     "Large-Scale Computations"

   2:50-3:10 p.m. : General Discussion

   3:10-3:45 p.m. : Coffee Break

Session III: Chairman: D. R. CURRAN, SRI International

   3:45-4:15 p.m. : Speaker: S. W. KEY, RE/SPEC Inc., Albuquerque, NM
                       "Finite Deformation Plasticity"

   4:15-4:30 p.m. : General Discussion

   4:30-4:45 p.m. : Discusser: J. W. HUTCHINSON, Harvard University

   4:45-5:00 p.m. : General Discussion

5:30-7:30 p.m. : Dinner, Norris University Center, Louis Room South

Special Session: Chairman and Coordinator: R. J. ASARO, Brown University
  8:00-10:00 p.m.: Open Discussion and Short Presentations

TUESDAY, OCTOBER 25
                    Norris University Center, Room 2G

Session III (cont.): Chairman: D. R. CURRAN, SRI International
  8:30-9:00 a.m. : Speaker: M. L. WILKINS, Lawrence Livermore National
                              Laboratory
                              "Material Models in Large-Scale Computations"
  9:00-9:15 a.m. : General Discussion
  9:15-9:30 a.m. : Discusser: A. NEEDLEMAN, Brown University
  9:30-9:45 a.m. : General Discussion

Session IV: Chairman: R. J. CLIFTON, Brown University
  9:45-10:15 a.m.: Speaker: K. D. KIMSEY, U.S. Army Ballistic Research
                              Laboratory
                              "Calculation of Penetration"
 10:15-10:30 a.m.: General Discussion
 10:30-10:45 a.m.: Discusser: M. E. BACKMAN, Naval Weapons Center
 10:45-11:00 a.m.: General Discussion
 11:00-11:30 a.m.: Coffee Break
 11:30-12:00 noon: Speaker: W. HERRMANN, Sandia National Laboratories
                              "Experimental Determination of Constitutive
                              Properties"
 12:00-12:15 p.m.: General Discussion
 12:15-12:30 p.m.: Discusser: E. KREMPL, Rensselaer Polytechnic Institute
 12:30-12:45 p.m.: General Discussion
 12:45- 2:30 p.m.: Lunch, Allen Center Dining Room

Session V: Chairman: T. C. BACHE, Science Applications, Inc.
                              and Ministry of Defence, Blacknest, U.K.
  2:30-3:00 p.m. : Speaker: J. T. CHERRY, S-CUBED, La Jolla, CA
                              "Ground Motion Calculations in Hard Rocks"
  3:00-3:15 p.m. : General Discussion
  3:15-3:30 p.m. : Discusser: L. W. MORLAND, University of East Anglia,,U.K.
  3:30-3:45 p.m. : General Discussion
  3:45-4:15 p.m. : Coffee Break
  4:15-4:45 p.m. : Speaker: J. TRULIO, Applied Theory, Inc., Los Angeles, CA
                              "Strain-Path Modeling for Geomaterials"
  4:45-5:00 p.m. : General Discussion
  5:00-5:15 p.m. : Discusser: T. B. BELYTSCHKO, Northwestern University

5:15-5:30 p.m. : General Discussion

6:30-9:30 p.m. : Banquet, Orrington Hotel, John Evans Room

WEDNESDAY, OCTOBER 26

Session VI: Chairman: J. T. ODEN, The University of Texas at Austin

  8:30-9:00 a.m.:   Speaker: I. SANDLER, Weidlinger Associates, New York, NY
                               "Strain Softening"

  9:00-9:15 a.m. :   General Discussion

  9:15-9:30 a.m. :   Discusser: G. A. HEGEMIER, University of California,
                               San Diego

  9:30-9:45 a.m. :   General Discussion

Session VII: Chairman: J. M. ROESSET, The University of Texas at Austin

  9:45-10:15 a.m.:   Speaker: G. Y. BALADI, U.S. Army Engineer Waterways
                               Experiment Station
                               "Soil Plasticity"

10:15-10:30 a.m.:   General Discussion

10:30-10:45 a.m.:   Discusser: J. H. PREVOST, Princeton University

10:45-11:00 a.m.:   General Discussion

11:00-11:30 a.m.:   Coffee Break

Special Session: Chairman: S. NEMAT-NASSER, Northwestern University

11:30-12:45 p.m.:   Short Presentations and Discussions

12:45- 2:00 p.m.:   Lunch, Allen Center Dining Room

Session VIII: Chairman: G. A. HEGEMIER, University of California,
                               San Diego

  2:15-4:15 p.m. :   Comments by Chairmen

              :   Open Discussions

              :   Closing Remarks

                      * * * * *

LIST OF WORKSHOP PARTICIPANTS

ARYA, Santosh K.   TRW - Defense Systems Group, Mail Station 134-9835,
                                1 Space Park, Redondo Beach, CA 90278.

ASARO, Robert J.   Division of Engineering, Brown University, Providence,
                                RI 02912.

BACHE, Thomas C.   Science Applications, Inc., P. O. Box 2351, La Jolla,
                                CA 92038; and Ministry of Defence, Blacknest, Brimpton
                                (near Reading), Berkshire RG7 4RS, U.K.

BACKMAN, Marvin E.   Engineering Sciences Division, Code 3894, Naval
                                Weapons Center, China Lake, CA 93555.

BALADI, George Y.   U.S. Army Engineer Waterways Experiment Station, P.O.
                                Box 631, Vicksburg, MS 39180.

BELYTSCHKO, Ted B.   Department of Civil Engineering, Northwestern Univer-
                                sity, Evanston, IL 60201.

BUDIANSKY, Bernard   Division of Applied Sciences, Harvard University,
                                Pierce Hall, Cambridge, MA 02138.

CHERRY, J. Theodore   S-CUBED, P.O. Box 1620, La Jolla, CA 92038; presently
                                at Science Horizons, Inc., 710 Encinitas Blvd.,
                                Encinitas, CA 92024.

CHRISTIANO, Paul   Department of Civil Engineering, Carnegie-Mellon
                                University, Schenley Park, Pittsburgh, PA 15213.

CLIFTON, Rodney J.   Division of Engineering, Brown University, Providence,
                                RI 02912.

CURRAN, Donald R.   Shock Physics and Geophysics Department, Poulter Lab-
                                oratory, SRI International, 333 Ravenswood Ave., Menlo
                                Park, CA 94025.

DIENES, John K.   Los Alamos National Laboratory, Group T-3/B216, Los
                                Alamos, NM 87545.

GARG, Sabodh K.   S-CUBED, P.O. Box 1620, La Jolla, CA 92038.

GOUDREAU, Gerald L.   Lawrence Livermore National Laboratory, Mail Stop L-122,
                                P.O. Box 808, Livermore, CA 94550.

GROSS, Michael B.   Science Applications, Inc., 2450 Washington Ave., San
                                Leandro, CA 94577.

HEGEMIER, Gilbert A.   Department of Applied Mechanics and Engineering
                                Sciences, University of California, San Diego,
                                La Jolla, CA 92037.

413

HERRMANN, Walter   Sandia National Laboratories, Organization 1500, P.O.
                            Box 5800, Albuquerque, NM 87185.

HORII, Hideyuki   Department of Civil Engineering, Northwestern Univer-
                            sity, Evanston, IL 60201.

HUGHES, Thomas J. R.   Division of Applied Mechanics, Stanford University,
                            Durand Bldg., Stanford, CA 94305.

HUTCHINSON, John W.   Division of Applied Sciences, Harvard University,
                            Pierce Hall, Cambridge, MA 02138.

JOHNSON, Gordon R.   Honeywell, Inc.; 5901 South County Road 18, Edina, MN
                            55436.

KELLER, Carl   Defense Nuclear Agency, Kirtland Air Force Base,
                            Albuquerque, NM 87115.

KEY, Samuel W.   RE/SPEC Inc., P.O. Box 14984, Albuquerque, NM 87191.

KIMSEY, Kent D.   USA AMCCOM, ARDC, Ballistic Research Laboratory, Attn:
                            DRSMC-BLT(A), Aberdeen Proving Ground, MD 21005.

KRAJCINOVIC, Dusan   Department of Civil Engineering, Mechanics and Metal-
                            lurgy, University of Illinois at Chicago, Box 4348,
                            Chicago, IL 60680.

KREMPL, Erhard   Department of Mechanical Engineering, Rensselaer Poly-
                            technic Institute, Troy, NY 12181.

KRIEG, Raymond D.   Sandia National Laboratories, Division 1521, Kirtland
                            Base East, Albuquerque, NM 87185.

McFARLAND, Clifton   Defense Nuclear Agency, 6801 Telegraph Road, Alexandria,
                            VA 22310.

MORLAND, Leslie W.   School of Mathematics and Physics, University of East
                            Anglia, Norwich NR4 7TJ, U.K.

MOSS, William C.   Lawrence Livermore National Laboratory, Mail Stop L-387,
                            P.O. Box 808, Livermore, CA 94550.

NEEDLEMAN, Alan   Division of Engineering, Brown University, Providence,
                            RI 02912.

NEMAT-NASSER, Siavouche   Department of Civil Engineering, Northwestern University,
                            Evanston, IL 60201.

ODEN, J. Tinsley   Department of Aerospace Engineering and Engineering
                            Mechanics, The University of Texas at Austin, Austin, TX
                            78712.

PATCH, Dan F.   Pacifica Technology, 11696 Sorrento Valley Road,
                            San Diego, CA 92109.

PHILLIPS, Aris	Department of Mechanical Engineering, Yale University, 15 Prospect Street, New Haven, CT 06520.
PISTER, Karl S.	College of Engineering, University of California, Berkeley, Berkeley, CA 94720.
PREVOST, Jean H.	Department of Civil Engineering, Princeton University, Princeton, NJ 08544.
READ, Harold E.	S-CUBED, 3390 Carmel Mountain Road, San Diego, CA 92121.
ROESSET, Jose M.	Department of Civil Engineering, The University of Texas at Austin, Austin, TX 78712.
RUDNICKI, John W.	Department of Civil Engineering, Northwestern University, Evanston, IL 60201.
SANDLER, Ivan S.	Weidlinger Associates, 333 Seventh Avenue, New York, NY 10001.
SCHUSTER, Sheldon H.	California Research and Technology, Inc., 20943 Devonshire St., Chatsworth, CA 91311.
SCOTT, Ronald F.	Division of Engineering and Applied Science, California Institute of Technology, Pasadena, CA 91109.
SNOWDEN, William E.	DARPA/Defense Sciences Office, 1400 Wilson Blvd., Arlington, VA 22209.
STICKLEY, C. Martin	The BDM Corporation, 7915 Jones Branch Drive, McLean, VA 22102.
TAGGART, G. Bruce	The BDM Corporation, 7915 Jones Branch Drive, McLean, VA 22102.
THIRUMALAI, K.	National Science Foundation/Division of Civil and Environmental Engineering, Washington, DC 20550.
TING, Thomas C. T.	Department of Civil Engineering, Mechanics and Metallurgy, University of Illinois at Chicago, Box 4348, Chicago, IL 60680.
TRULIO, John G.	Applied Theory, Inc., 930 South La Brea Ave., Los Angeles, CA 90036.
VALANIS, Kirk C.	College of Engineering, University of Cincinnati, Cincinnati, OH 45221.
VARDOULAKIS, Ioannis G.	Department of Civil and Mineral Engineering, University of Minnesota, Minneapolis, MN 55455.
WILCOX, Ben A.	DARPA/Defense Sciences Office, 1400 Wilson Blvd., Arlington, VA 22209.
WILKINS, Mark L.	Lawrence Livermore National Laboratory, Mail Stop L-387, P.O. Box 808, Livermore, CA 94550.